GENETIC ANALYSIS: INTEGRATING GENES AND GENOMES

Genetic Analysis: Integrating Genes and Genomes

Editor:

Dr. Prerna Pandey, Ph.D.

Delve Publishing

www.delvepublishing.com

Genetic Analysis: Integrating Genes and Genomes

Editor: **Dr. Prerna Pandey, Ph.D.**

708 3rd Avenue, 6th Floor

New York, NY 10017

United States of America

www.delvepublishing.com

ISBN: 978-1-68095-741-9

Library of Congress Control Number: 2016951404

About the Editor

Dr. Prerna Pandey, Ph.D.

Dr Prerna Pandey, a biotechnologist with several years of wet lab research experience, worked as a senior research fellow at International Center for Genetic Engineering and Biotechnology, New Delhi. Her field of research in PhD included isolation and molecular characterization of Geminiviruses, genome sequencing and gene annotation, and gene silencing using the RNA interference technology. She has published papers in peer-reviewed journals in the field, and has a submitted a number of annotated Geminiviral genome sequences in the GenBank, including two novel ones. Prerna has also completed her editing and proof-reading courses from Society for the promotion of Editors and Proof-readers, UK and now works as a freelance scientific writer and editor.

Contents

List of Contributors

Lucía F. Franchini
Instituto de Investigaciones en Ingeniería Genética y Biología Molecular (INGEBI), Consejo Nacional de Investigaciones Científicas y Técnicas (CONICET), Buenos Aires C1428, Argentina

Katherine S. Pollard
Gladstone Institutes, San Francisco, CA 94158, USA
Institute for Human Genetics, Department of Epidemiology & Biostatistics,University of California, San Francisco, CA 94158, USA

Tom R. Gaunt
MRC Integrative Epidemiology Unit (IEU) & School of Social and Community Medicine, University of Bristol, Oakfield House, Oakfield Grove, Bristol BS8 2BN, UK.

Hashem A. Shihab
MRC Integrative Epidemiology Unit (IEU) & School of Social and Community Medicine, University of Bristol, Oakfield House, Oakfield Grove, Bristol BS8 2BN, UK.

Gibran Hemani
MRC Integrative Epidemiology Unit (IEU) & School of Social and Community Medicine, University of Bristol, Oakfield House, Oakfield Grove, Bristol BS8 2BN, UK.

Josine L. Min
MRC Integrative Epidemiology Unit (IEU) & School of Social and Community Medicine, University of Bristol, Oakfield House, Oakfield Grove, Bristol BS8 2BN, UK.

Geoff Woodward
MRC Integrative Epidemiology Unit (IEU) & School of Social and Community Medicine, University of Bristol, Oakfield House, Oakfield Grove, Bristol BS8 2BN, UK.

Oliver Lyttleton
Avon Longitudinal Study of Parents and Children (ALSPAC) & School of Social and Community Medicine, University of Bristol, Oakfield House, Oakfield Grove, Bristol BS8 2BN, UK.

Jie Zheng
MRC Integrative Epidemiology Unit (IEU) & School of Social and Community Medicine, University of Bristol, Oakfield House, Oakfield Grove, Bristol BS8 2BN, UK.

Aparna Duggirala
Avon Longitudinal Study of Parents and Children (ALSPAC) & School of Social and Community Medicine, University of Bristol, Oakfield House, Oakfield Grove, Bristol BS8 2BN, UK.

Wendy L. McArdle
Avon Longitudinal Study of Parents and Children (ALSPAC) & School of Social and Community Medicine, University of Bristol, Oakfield House, Oakfield Grove, Bristol BS8 2BN, UK.

Karen Ho
Avon Longitudinal Study of Parents and Children (ALSPAC) & School of Social and Community Medicine, University of Bristol, Oakfield House, Oakfield Grove, Bristol BS8 2BN, UK.

Susan M. Ring
MRC Integrative Epidemiology Unit (IEU) & School of Social and Community Medicine, University of Bristol, Oakfield House, Oakfield Grove, Bristol BS8 2BN, UK.
Avon Longitudinal Study of Parents and Children (ALSPAC) & School of Social and Community Medicine, University of Bristol, Oakfield House, Oakfield Grove, Bristol BS8 2BN, UK.

David M. Evans
MRC Integrative Epidemiology Unit (IEU) & School of Social and Community Medicine, University of Bristol, Oakfield House, Oakfield Grove, Bristol BS8 2BN, UK.
University of Queensland Diamantina Institute, Translational Research Institute, Brisbane 4102, Australia.

George Davey Smith
MRC Integrative Epidemiology Unit (IEU) & School of Social and Community Medicine, University of Bristol, Oakfield House, Oakfield Grove, Bristol BS8 2BN, UK.

Caroline L. Relton
MRC Integrative Epidemiology Unit (IEU) & School of Social and Community Medicine, University of Bristol, Oakfield House, Oakfield Grove, Bristol BS8 2BN, UK.
Institute of Genetic Medicine, Newcastle University, Newcastle upon Tyne NE1 7RU, UK.

Erik Corona
Division of Systems Medicine, Department of Pediatrics, Stanford University School of Medicine, Stanford, California, United States of America
Program in Biomedical Informatics, Stanford University School of Medicine, Stanford, California, United States of America
Lucile Packard Children's Hospital, Palo Alto, California, United States of America

Rong Chen
Division of Systems Medicine, Department of Pediatrics, Stanford University School of Medicine, Stanford, California, United States of America
Lucile Packard Children's Hospital, Palo Alto, California, United States of America

Martin Sikora
Department of Genetics, Stanford University School of Medicine, Stanford, California, United States of America

Alexander A. Morgan
Division of Systems Medicine, Department of Pediatrics, Stanford University School of Medicine, Stanford, California, United States of America
Program in Biomedical Informatics, Stanford University School of Medicine, Stanford, California, United States of America

Chirag J. Patel
Division of Systems Medicine, Department of Pediatrics, Stanford University School of Medicine, Stanford, California, United States of America
Program in Biomedical Informatics, Stanford University School of Medicine, Stanford, California, United States of America
Lucile Packard Children's Hospital, Palo Alto, California, United States of America

Aditya Ramesh
Division of Systems Medicine, Department of Pediatrics, Stanford University School of Medicine, Stanford, California, United States of America
Lucile Packard Children's Hospital, Palo Alto, California, United States of America

Carlos D. Bustamante
Department of Genetics, Stanford University School of Medicine, Stanford, California, United States of America

Atul J. Butte
Division of Systems Medicine, Department of Pediatrics, Stanford University School of Medicine, Stanford, California, United States of America
Program in Biomedical Informatics, Stanford University School of Medicine, Stanford, California, United States of America
Lucile Packard Children's Hospital, Palo Alto, California, United States of America
Department of Genetics, Stanford University School of Medicine, Stanford, California, United States of America

Jintana Sirivarasai
Graduate Program in Nutrition, Faculty of Medicine Ramathibodi Hospital, Mahidol University, Bangkok 10400, Thailand

Sukhumpun Kaojarern
Cardiovascular and Metabolic Center, Faculty of Medicine Ramathibodi Hospital, Mahidol University, Bangkok 10400, Thailand

Suwannee Chanprasertyothin
Office of Research Academic and Innovation, Faculty of Medicine Ramathibodi Hospital, Mahidol University, Bangkok 10400, Thailand

Pachara Panpunuan
Department of Medicine, Faculty of Medicine Ramathibodi Hospital, Mahidol University, Bangkok 10400, Thailand

Krittaya Petchpoung
Research and Development Institute, Kasetsart University, Bangkok 10900, Thailand

Aninthita Tatsaneeyapant
Health Office, Electricity Generating Authority of Thailand, Nonthaburi 11130, Thailand

Krongtong Yoovathaworn
Department of Pharmacology, Faculty of Science, Mahidol University, Bangkok 10400, Thailand

Thunyachai Sura
Department of Medicine, Faculty of Medicine Ramathibodi Hospital, Mahidol University, Bangkok 10400, Thailand

Sming Kaojarern
Occupational and Environmental Toxicology Center, Faculty of Medicine Ramathibodi Hospital, Mahidol University, Bangkok 10400, Thailand

Piyamit Sritara
Department of Medicine, Faculty of Medicine Ramathibodi Hospital, Mahidol University, Bangkok 10400, Thailand

Fatimah L. C. Jackson
Department of Biology and W. Montague Cobb Research Laboratory, Howard University, Washington, DC, USA

Jens Kringelum
Department of Systems Biology, Center for Biological Sequence Analysis, buildin 208, kemitorvet, Technical University of Denmark, DK-2800 Lyngby, Denmark

Sonny Kim Kjaerulff
Department of Systems Biology, Center for Biological Sequence Analysis, buildin 208, kemitorvet, Technical University of Denmark, DK-2800 Lyngby, Denmark

Søren Brunak
Department of Disease Systems Biology, Faculty of Health and Medical Sciences, Novo Nordisk Foundation, Center for Protein Research, University of Copenhagen, Blegdamsvej 3A, DK-2200, Copenhagen, Denmark

Ole Lund
Department of Systems Biology, Center for Biological Sequence Analysis, buildin 208, kemitorvet, Technical University of Denmark, DK-2800 Lyngby, Denmark

Tudor I. Oprea
Department of Systems Biology, Center for Biological Sequence Analysis, buildin 208, kemitorvet, Technical University of Denmark, DK-2800 Lyngby, Denmark
Translational Informatics Division, Department of Internal Medicine, University of New Mexico Health Sciences Center, MSC09 5025, Albuquerque 87181, New Mexico, United States of America

Olivier Taboureau
Department of Systems Biology, Center for Biological Sequence Analysis, buildin 208, kemitorvet, Technical University of Denmark, DK-2800 Lyngby, Denmark
INSERM, UMRS-973, MTi, Universite Paris Diderot, 35 rue Helene Brion, 75205 Paris Cedex 13, Sorbonne Paris Cite, France

Giuseppe Passarino
Department of Biology, Ecology and Earth Science, University of Calabria, 87036 Rende, Italy

Francesco De Rango
Department of Biology, Ecology and Earth Science, University of Calabria, 87036 Rende, Italy

Alberto Montesanto
Department of Biology, Ecology and Earth Science, University of Calabria, 87036 Rende, Italy

Itziar Lamiquiz-Moneo
Unidad Clínica y de Investigación en Lípidos y Arteriosclerosis, Hospital Universitario Miguel Servet, Instituto de Investigación Sanitaria Aragón (IIS Aragón), Avenida Isabel La Católica 1-3, 50009 Zaragoza, Spain

Cristian Blanco-Torrecilla
Unidad Clínica y de Investigación en Lípidos y Arteriosclerosis, Hospital Universitario Miguel Servet, Instituto de Investigación Sanitaria Aragón (IIS Aragón), Avenida Isabel La Católica 1-3, 50009 Zaragoza, Spain

Ana M. Bea
Unidad Clínica y de Investigación en Lípidos y Arteriosclerosis, Hospital Universitario Miguel Servet, Instituto de Investigación Sanitaria Aragón (IIS Aragón), Avenida Isabel La Católica 1-3, 50009 Zaragoza, Spain

Rocío Mateo-Gallego
Unidad Clínica y de Investigación en Lípidos y Arteriosclerosis, Hospital Universitario Miguel Servet, Instituto de Investigación Sanitaria Aragón (IIS Aragón), Avenida Isabel La Católica 1-3, 50009 Zaragoza, Spain

Sofía Pérez-**Calahorra**
Unidad Clínica y de Investigación en Lípidos y Arteriosclerosis, Hospital Universitario Miguel Servet, Instituto de Investigación Sanitaria Aragón (IIS Aragón), Avenida Isabel La Católica 1-3, 50009 Zaragoza, Spain

Lucía Baila-Rueda
Unidad Clínica y de Investigación en Lípidos y Arteriosclerosis, Hospital Universitario Miguel Servet, Instituto de Investigación Sanitaria Aragón (IIS Aragón), Avenida Isabel La Católica 1-3, 50009 Zaragoza, Spain

Ana Cenarro
Unidad Clínica y de Investigación en Lípidos y Arteriosclerosis, Hospital Universitario Miguel Servet, Instituto de Investigación Sanitaria Aragón (IIS Aragón), Avenida Isabel La Católica 1-3, 50009 Zaragoza, Spain

Fernando Civeira
Unidad Clínica y de Investigación en Lípidos y Arteriosclerosis, Hospital Universitario Miguel Servet, Instituto de Investigación Sanitaria Aragón (IIS Aragón), Avenida Isabel La Católica 1-3, 50009 Zaragoza, Spain

Isabel de Castro-Orós
Unidad Clínica y de Investigación en Lípidos y Arteriosclerosis, Hospital Universitario Miguel Servet, Instituto de Investigación Sanitaria Aragón (IIS Aragón), Avenida Isabel La Católica 1-3, 50009 Zaragoza, Spain
Universidad de Zaragoza, Departamento de Bioquímica, Biología Molecular y Celular, 50009 Zaragoza, Spain

Maximilian Kothe
Department of Historical Anthropology and Human Ecology, Georg-August-University, 37073 Göttingen, Germany

Verena Seidenberg
Department of Historical Anthropology and Human Ecology, Georg-August-University, 37073 Göttingen, Germany

Susanne Hummel
Department of Historical Anthropology and Human Ecology, Georg-August-University, 37073 Göttingen, Germany

Oliver Piskurek
Department of Historical Anthropology and Human Ecology, Georg-August-University, 37073 Göttingen, Germany

Amanda Warr
The Roslin Institute and Royal (Dick) School of Veterinary Studies, University of Edinburgh, Easter Bush, Edinburgh UK

Christelle Robert
The Roslin Institute and Royal (Dick) School of Veterinary Studies, University of Edinburgh, Easter Bush, Edinburgh UK

David Hume
The Roslin Institute and Royal (Dick) School of Veterinary Studies, University of Edinburgh, Easter Bush, Edinburgh UK

Alan Archibald
The Roslin Institute and Royal (Dick) School of Veterinary Studies, University of Edinburgh, Easter Bush, Edinburgh UK

Nader Deeb
Genus plc., Hendersonville, Tennessee 37075

Mick Watson
The Roslin Institute and Royal (Dick) School of Veterinary Studies, University of Edinburgh, Easter Bush, Edinburgh UK

Ettie M. Lipner
Center for Genes, Environment, and Health, National Jewish Health, Denver, Colorado, United States of America
Computational Bioscience Program, University of Colorado Denver, School of Medicine, Aurora, Colorado, United States of America

Benjamin J. Garcia
Center for Genes, Environment, and Health, National Jewish Health, Denver, Colorado, United States of America
Computational Bioscience Program, University of Colorado Denver, School of Medicine, Aurora, Colorado, United States of America

Michael Strong
Center for Genes, Environment, and Health, National Jewish Health, Denver, Colorado, United States of America
Computational Bioscience Program, University of Colorado Denver, School of Medicine, Aurora, Colorado, United States of America

Albino Bacolla
Division of Pharmacology and Toxicology, College of Pharmacy, Dell Pediatric Research Institute, The University of Texas at Austin, 1400 Barbara Jordan Boulevard, Austin, TX 78723, USA

Xiao Zhu
Texas Advanced Computing Center, Austin, TX 78758-4497, USA

Hanning Chen
Department of Chemistry, George Washington University, 725 21st Street, NW, Washington, DC 20052, USA

Katy Howells
Institute of Medical Genetics, School of Medicine, Cardiff University, Cardiff CF14 4XN, UK

David N. Cooper
Institute of Medical Genetics, School of Medicine, Cardiff University, Cardiff CF14 4XN, UK

Karen M. Vasquez
Division of Pharmacology and Toxicology, College of Pharmacy, Dell Pediatric Research Institute, The University of Texas at Austin, 1400 Barbara Jordan Boulevard, Austin, TX 78723, USA

Riyue Bao
Center for Research Informatics, The University of Chicago, Chicago, IL, USA

Lei Huang
Center for Research Informatics, The University of Chicago, Chicago, IL, USA

Jorge Andrade
Center for Research Informatics, The University of Chicago, Chicago, IL, USA

Wei Tan
IBM Thomas J. Watson Research Center, Yorktown Heights, New York, USA

Warren A Kibbe
Biomedical Informatics Center (NUBIC), Clinical and Translational Sciences Institute (NUCATS), Northwestern University, Chicago, IL, USA

Hongmei Jiang
Department of Statistics, Northwestern University, Evanston, IL, USA

Gang Feng
Biomedical Informatics Center (NUBIC), Clinical and Translational Sciences Institute (NUCATS), Northwestern University, Chicago, IL, USA

David R. Deyle
Department of Medicine, University of Washington, Seattle, WA, 98195, USA

Li B. Li
Department of Medicine, University of Washington, Seattle, WA, 98195, USA

Gaoying Ren
Department of Medicine, University of Washington, Seattle, WA, 98195, USA

David W. Russell
Department of Medicine, University of Washington, Seattle, WA, 98195, USA
Department of Biochemistry, University of Washington, Seattle, WA, 98195, USA

Gregory M. Ku
Diabetes Center, University of California San Francisco, San Francisco, California, United States of America
Division of Endocrinology, Metabolism, and Diabetes, Department of Medicine, University of California San Francisco, San Francisco, California, United States of America

Zachary Pappalardo
Diabetes Center, University of California San Francisco, San Francisco, California, United States of America

Chun Chieh Luo
Diabetes Center, University of California San Francisco, San Francisco, California, United States of America

Michael S. German
Diabetes Center, University of California San Francisco, San Francisco, California, United States of America
Division of Endocrinology, Metabolism, and Diabetes, Department of Medicine, University of California San Francisco, San Francisco, California, United States of America

Michael T. McManus
Diabetes Center, University of California San Francisco, San Francisco, California, United States of America
Department of Microbiology and Immunology, University of California San Francisco, San Francisco, California, United States of America

Isabelle Güller
Centre for Physical Activity and Nutrition, School of Exercise and Nutrition Sciences, Deakin University, 221 Burwood highway, Burwood, VIC 3125, Australia.

Sarah McNaughton
Centre for Physical Activity and Nutrition, School of Exercise and Nutrition Sciences, Deakin University, 221 Burwood highway, Burwood, VIC 3125, Australia.

Tamsyn Crowley
School of Medicine, Deakin University, Waurn Ponds, Australia.
Australian Animal Health Laboratory, CSIRO Animal, Food and Health Sciences, Geelong, Victoria, Australia.

Vicente Gilsanz
Department of Radiology, Childrens Hospital Los Angeles, University of Southern California, Los Angeles, California, USA.

Shingo Kajimura
UCSF Diabetes Center, Department of Cell and Tissue Biology, University of California, San Francisco, USA.

Matthew Watt
Department of Physiology, Monash University, Clayton, Victoria 3800, Australia.

Aaron P. Russell
Centre for Physical Activity and Nutrition, School of Exercise and Nutrition Sciences, Deakin University, 221 Burwood highway, Burwood, VIC 3125, Australia.

Jessica Tyrrell
Genetics of Complex Traits, Institute of Biomedical and Clinical Science, University of Exeter Medical School, Royal Devon and Exeter Hospital, Exeter EX2 5DW, UK
European Centre for Environment and Human Health, University of Exeter Medical School, The Knowledge Spa, Truro TR1 3HD, UK

Samuel E Jones
Genetics of Complex Traits, Institute of Biomedical and Clinical Science, University of Exeter Medical School, Royal Devon and Exeter Hospital, Exeter EX2 5DW, UK

Robin Beaumont
Genetics of Complex Traits, Institute of Biomedical and Clinical Science, University of Exeter Medical School, Royal Devon and Exeter Hospital, Exeter EX2 5DW, UK

Christina M Astley
Program in Medical and Population Genetics, Broad Institute of MIT and Harvard, Cambridge, MA 02142, USA
Center for Basic and Translational Obesity Research and Division of Endocrinology, Boston Children's Hospital, Boston, MA 02115, USA

Rebecca Lovell
European Centre for Environment and Human Health, University of Exeter Medical School, The Knowledge Spa, Truro TR1 3HD, UK

Hanieh Yaghootkar
Genetics of Complex Traits, Institute of Biomedical and Clinical Science, University of Exeter Medical School, Royal Devon and Exeter Hospital, Exeter EX2 5DW, UK

Marcus Tuke
Genetics of Complex Traits, Institute of Biomedical and Clinical Science, University of Exeter Medical School, Royal Devon and Exeter Hospital, Exeter EX2 5DW, UK

Katherine S Ruth
Genetics of Complex Traits, Institute of Biomedical and Clinical Science, University of Exeter Medical School, Royal Devon and Exeter Hospital, Exeter EX2 5DW, UK

Rachel M Freathy
Genetics of Complex Traits, Institute of Biomedical and Clinical Science, University of Exeter Medical School, Royal Devon and Exeter Hospital, Exeter EX2 5DW, UK

Joel N Hirschhorn
European Centre for Environment and Human Health, University of Exeter Medical School, The Knowledge Spa, Truro TR1 3HD, UK
Program in Medical and Population Genetics, Broad Institute of MIT and Harvard, Cambridge, MA 02142, USA
Department of Genetics, Harvard Medical School, Boston, MA 02115, USA

Andrew R Wood
Genetics of Complex Traits, Institute of Biomedical and Clinical Science, University of Exeter Medical School, Royal Devon and Exeter Hospital, Exeter EX2 5DW, UK

Anna Murray
Genetics of Complex Traits, Institute of Biomedical and Clinical Science, University of Exeter Medical School, Royal Devon and Exeter Hospital, Exeter EX2 5DW, UK

Michael N Weedon
Genetics of Complex Traits, Institute of Biomedical and Clinical Science, University of Exeter Medical School, Royal Devon and Exeter Hospital, Exeter EX2 5DW, UK

Timothy M Frayling
Genetics of Complex Traits, Institute of Biomedical and Clinical Science, University of Exeter Medical School, Royal Devon and Exeter Hospital, Exeter EX2 5DW, UK

Changbing Shen
Institute and Department of Dermatology, The First Affiliated Hospital, Anhui Medical University, Hefei, China

Jing Gao
Department of Dermatology, The Second Affiliated Hospital, Anhui Medical University, Hefei, China

Yujun Sheng
Institute and Department of Dermatology, The First Affiliated Hospital, Anhui Medical University, Hefei, China

Jinfa Dou
Institute and Department of Dermatology, The First Affiliated Hospital, Anhui Medical University, Hefei, China

Fusheng Zhou
Institute and Department of Dermatology, The First Affiliated Hospital, Anhui Medical University, Hefei, China

Xiaodong Zheng
Institute and Department of Dermatology, The First Affiliated Hospital, Anhui Medical University, Hefei, China

Randy Ko
Department of Biochemistry, University of New Mexico, Albuquerque, NM, USA

Xianfa Tang
Institute and Department of Dermatology, The First Affiliated Hospital, Anhui Medical University, Hefei, China

Caihong Zhu
Institute and Department of Dermatology, The First Affiliated Hospital, Anhui Medical University, Hefei, China

Xianyong Yin
Department of Genetics and Renaissance Computing Institute, University of North Carolina at Chapel Hill, Chapel Hill, NC, USA

Liangdan Sun
Institute and Department of Dermatology, The First Affiliated Hospital, Anhui Medical University, Hefei, China

Yong Cui
Department of Dermatology, China-Japan Friendship Hospital, Beijing, China

Xuejun Zhang
Institute and Department of Dermatology, The First Affiliated Hospital, Anhui Medical University, Hefei, China
Department of Dermatology, The Second Affiliated Hospital, Anhui Medical University, Hefei, China

Yong-Wu Li
Department of Emergency Surgery, First Affiliated Hospital of Jiamusi University, Jiamusi, Heilongjiang 154002, China

Lin Bai
Experiment Teaching Center of Clinical Medicine, The First Affiliated Hospital of Chengdu Medical College, Chengdu, Sichuan 610500, China

Lyu-Xia Dai
Experiment Teaching Center of Clinical Medicine, The First Affiliated Hospital of Chengdu Medical College, Chengdu, Sichuan 610500, China

Xu He
Experiment Teaching Center of Clinical Medicine, The First Affiliated Hospital of Chengdu Medical College, Chengdu, Sichuan 610500, China

Xian-Ping Zhou
Experiment Teaching Center of Clinical Medicine, The First Affiliated Hospital of Chengdu Medical College, Chengdu, Sichuan 610500, China

René Santer
Department of Pediatrics, University Medical Center Hamburg-Eppendorf, Martinistraße 52, D - 20246 Hamburg, Germany

Marcel du Moulin
Department of Pediatrics, University Medical Center Hamburg-Eppendorf, Martinistraße 52, D - 20246 Hamburg, Germany

Tatevik Shahinyan
Arabkir Institute of Child and Adolescent Health, Yerevan, Armenia

Inga Vater
Institute of Human Genetics, University of Kiel, Kiel, Germany

Esther Maier
Department of Pediatrics, University Children's Hospital, Munich, Germany

Ania C. Muntau
Department of Pediatrics, University Medical Center Hamburg-Eppendorf, Martinistraße 52, D - 20246 Hamburg, Germany

Beat Steinmann
Department of Pediatrics, University of Zurich, Zurich, Switzerland

Preface

The new advances in the field of genetics have led to numerous medical breakthroughs in the past few years. Genetic Analysis: Integrating Genes and Genomes provides an insight on the advances made in human genetics in the last few years. The editor has comprehensively assembled recent developments on the use of modern genetic analytical techniques and tools for answering biological questions related to genomic screening, mutations, gene and environment interactions, comparative and molecular genomic approaches for detecting genetic components of diseases.

A better understanding of the genetic basis of disease can be achieved by studying genetic diversity across human populations. The interaction of gene and environment and their effects on human-specific developmental traits, genetic influences on methylation across the human life course, interplay of genes and lifestyle, genetic basis of disease in the context of migration, exposure to toxic metals like Lead, and calculation of genetic risk of diseases in unrelated individuals have been discussed. In addition, the synergistic effects of genes involved in cancer development, fructose-1,6-bisphosphatase deficiency, vitiligo, psoriasis, severe hypertriglyceridemia, and susceptibility to mycobacterial infection are explored in some chapters. Use of small interfering RNAs (siRNA), microRNAs and SINE sequences in screening systems have also been discussed to identify genes.

In one of the articles, the general framework of whole exome sequence data analysis is outlined. Finally, we also present an article on ChemProt, a publicly available compilation of chemical-protein-disease annotation resources that enables the study of a molecule from cellular to clinical levels.

Editor
Dr. Prerna Pandey, Ph.D.

CHAPTER 1

GENOMIC APPROACHES TO STUDYING HUMAN-SPECIFIC DEVELOPMENTAL TRAITS

Lucía F. Franchini[1] and Katherine S. Pollard[2,3]

[1]Instituto de Investigaciones en Ingeniería Genética y Biología Molecular (INGEBI), Consejo Nacional de Investigaciones Científicas y Técnicas (CONICET), Buenos Aires C1428, Argentina

[2]Gladstone Institutes, San Francisco, CA 94158, USA

[3]Institute for Human Genetics, Department of Epidemiology & Biostatistics, University of California, San Francisco, CA 94158, USA

ABSTRACT

Changes in developmental regulatory programs drive both disease and phenotypic differences among species. Linking human-specific traits to alterations in development is challenging, because we have lacked the tools to assay and manipulate regulatory networks in human and primate embryonic cells. This field was transformed by the sequencing of hundreds of genomes – human and non-human – that can be compared to discover the regulatory machinery of genes involved in human development. This approach has identified thousands of human-specific genome alterations in developmental genes and their regulatory regions. With recent advances in stem cell techniques, genome engineering, and genomics, we can now test these sequences for effects on developmental gene regulation and downstream phenotypes in human cells and tissues.

INTRODUCTION

Humans differ from chimpanzees, our closest living relatives, and other mammals in a variety of traits, including disease susceptibilities. Many of these differences have their origins in development. The fossil record shows that some human traits, such as the pelvic morphology associated with upright walking (Harcourt-Smith et al., 2004), evolved around the time of divergence from our common ancestor with chimpanzees about six million years ago. Other traits, such as loss of a prominent brow ridge (Lieberman, 2000), emerged only after modern humans split from Neanderthals and other extinct hominins. Some phenotypes are very recently evolved, as evidenced by variation between modern human populations. These include pigmentation (Hancock et al., 2011), keratinization (Gautam et al., 2015), hair texture (Jablonski and Chaplin, 2014; Kamberov et al., 2013) and high altitude adaptation (Huerta-Sanchez et al., 2014; Simonson et al., 2010; Yi et al., 2010). For many distinctive traits, such as social behaviors, symbolic thought and spoken language (Sterelny, 2011), we have no evidence of the time period of evolution or only indirect evidence from changes in the material culture left behind in the archaeological record.

Collectively, human-specific traits have allowed our species to dominate all climates and modify the landscape in a dramatic way never seen in the history of life on earth. At the same time,

Citation: Lucía F. Franchini and Katherine S. Pollard. Genomic approaches to studying human-specific developmental traits, Development (2015) vol. 142 no. 18, 3100-3112, DOI: http://dx.doi.org/10.1242/dev.120048

our species has acquired a unique profile of susceptibility to different diseases compared with our close relatives. Some examples are AIDS (Varki and Altheide, 2005), cardiovascular disease (Varki et al., 2009), neurodegenerative disease (Finch, 2010) and psychiatric disorders (Crow, 2000, 2007). Our high burden of neurological disease may be an 'Achilles heel' associated with cognitive adaptations (Crow, 2000, 2007). To understand the evolutionary forces that shaped human-specific traits and the molecular mechanisms through which changes occurred, we must first track down the genetic alterations underlying phenotypic differences between humans and our close relatives. As in other species, the evolution of human-specific traits must have emerged through genetic modification of developmental, physiological or behavioral programs.

Development is a highly constrained and tightly regulated process, orchestrated through complex gene regulatory networks (Davidson and Erwin, 2006). There is a remarkable conservation of development between organisms as evolutionarily distant as cnidarians, insects and mammals. The same molecular pathways control this conserved developmental program across animal lineages. For example, humans, flatworms and cnidarians use the same basic elements of paracrine signaling cascades, such as the Wnt and TGFβ pathways (Finnerty et al., 2004; Carroll, 2005). Transcription factors that regulate development are also highly conserved, and they tend to regulate the same general processes in diverse species, including regulation of body axis patterning by Hox genes (Lemons and McGinnis, 2006), light-sensing organs by *Pax6* (Gehring, 2011), head formation by Otx homologs (Yasuoka et al., 2014) and heart morphology by *Tinman/Nkx2-5* (Erwin, 1999). This high level of conservation has allowed us to study human development, despite limited access to human tissues for research, using model organisms as proxies.

Conserved developmental programs, and hence morphology and other traits, are modified through two primary mechanisms: direct alterations to genes and thereby the functions of the encoded products (proteins, RNAs), and changes in gene regulation (transcription, splicing, translation, post-translational modifications). As most developmental genes are pleiotropic and participate in multiple independent developmental processes, their evolution is highly constrained. By contrast, gene regulatory elements, including distal enhancers (Pennacchio et al., 2006), tend to function in a more limited number of cell types and stages, combining additively to control the complex expression patterns of developmental genes (Noonan and McCallion, 2010). This modularity makes gene regulation an ideal template for the evolution of morphology (Carroll, 2008). Supporting this idea, there is mounting evidence that human traits evolved largely through genetic changes in regulatory regions (Haygood et al., 2010; Horvath et al., 2014; McLean et al., 2011; Pollard et al., 2006a,b), as initially proposed by King and Wilson (1975), although the relative contributions of coding versus non-coding sequences continue to be debated.

In this Review, we discuss how genomics is transforming the study of human developmental biology to enable direct analysis of genetic variants that arose during human evolution in their native context. The transformation started with genome sequencing and the development of comparative genomic techniques for pinpointing sequences, including both genes and regulatory elements, that are unique to humans compared with chimpanzees and other mammals. Although a few developmental genes have been associated with the evolution of particular human-specific traits (discussed further below), it has been more challenging to characterize the role of regulatory sequences in human evolution and disease. Functional genomics has helped to address this challenge by generating rich information about the cell types and developmental stages in which uniquely human sequences, including non-coding sequences, function. By introducing human variants of proteins and gene regulatory sequences into model organisms and assaying their effects during embryogenesis, a handful of human-specific sequences have been partially characterized (reviewed by Devoy et al., 2012; Enard, 2014). However, significant further work is needed to test the hypothesis that these sequences altered traits during human evolution. Fortunately, we are entering an era in which it is possible to identify human-specific regulatory elements and to characterize them functionally in human and non-human primate cells using emerging techniques from stem cell biology and genome engineering. Directly studying human development will accelerate understanding of what makes our species unique.

THE GENOME SEQUENCING ERA

The availability of genome sequences for humans (International Human Genome Sequencing Consortium, 2004), primates (The Chimpanzee Sequencing and Analysis Consortium, 2005; Rhesus Macaque Genome Sequencing and Analysis Consortium et al., 2007), extinct hominins (Meyer et al., 2012; Prufer et al., 2014) and many other vertebrates has fueled the discovery of human-specific DNA sequences. Comparative genomic studies have cataloged three general classes of genomic differences between humans and other primates: large chromosomal alterations, smaller insertions and deletions (indels), and single nucleotide substitutions (Fig. 1). Developmental loci have accumulated polymorphic and fixed differences of all three types during human evolution. Although the catalog of human-specific differences is continually growing, functional proof of how genetic changes translate into phenotypic changes is scarce. In Table 1, we summarize select examples of functional studies that have been undertaken in order to assess the impact of human evolutionary changes. In a few cases, these studies have illuminated the role that human-specific changes could have had in the evolution of human traits, although a definitive causal role in altering human phenotypes has yet to be established for any of these human genomic regions.

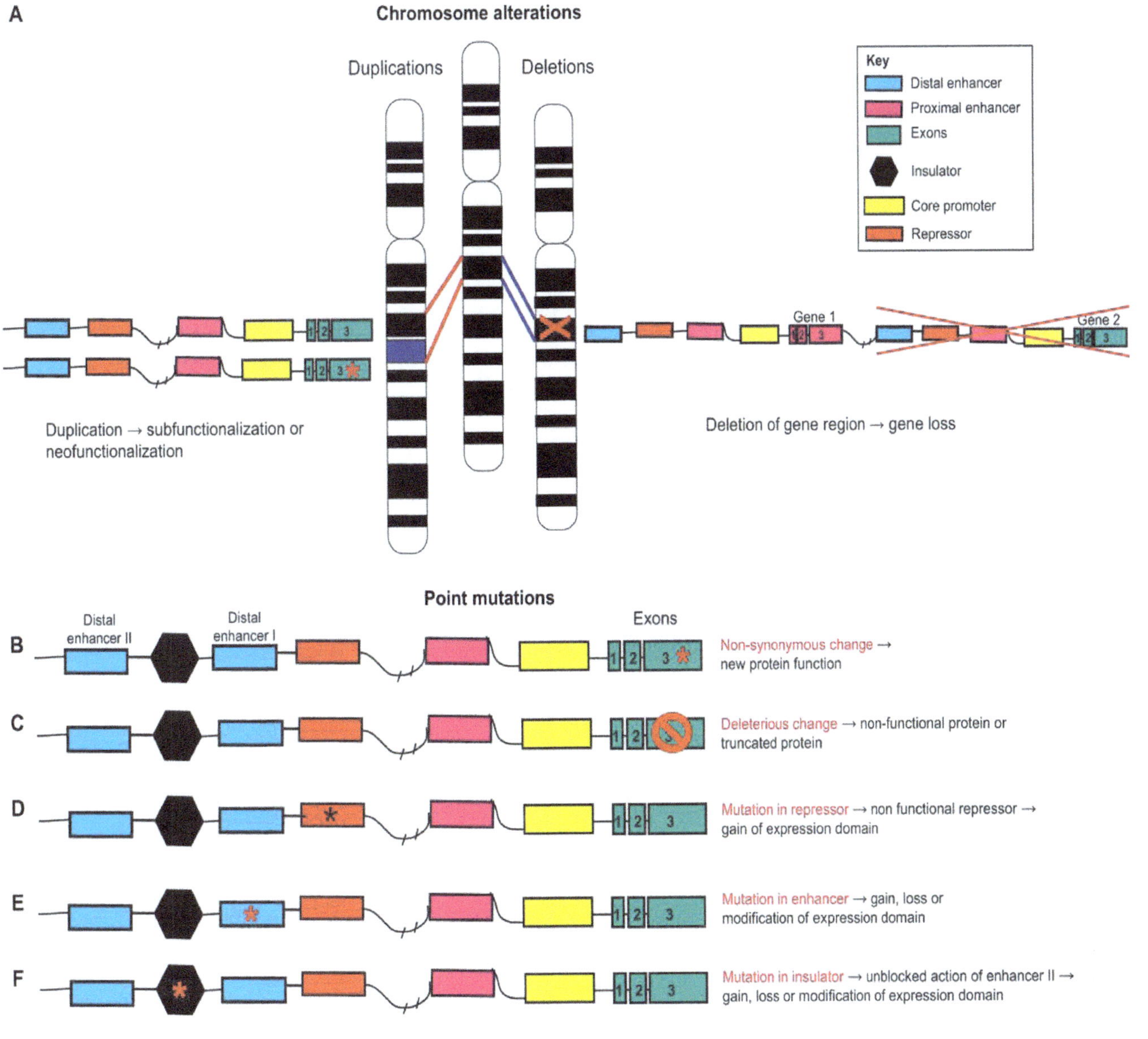

Figure 1. Type of genetic and genomic changes and possible functional consequences. (A) Schematic of large chromosome rearrangements such as duplications and deletions. Duplication of gene regions can lead to the appearance of new gene paralogs that can gain a new function after amino acid changes (neofunctionalization or subfunctionalization), or be lost after deleterious mutations, leading to pseudogenization. By contrast, deletions of gene regions can lead to loss of genes and regulatory elements. (B-F) Schematic of possible point mutation changes. (B) A non-synonymous change in the coding region of a gene leading to change in the protein function. (C) A deleterious change such as a premature stop codon that can produce a truncated protein or pseudogene. (D,E) Changes in repressor elements can lead to gain in expression domains for a gene (D), whereas mutations in an enhancer can produce gain, loss or modification of gene expression patterns (E). (F) Mutations in insulators can lead to gain, loss or modification of expression domains as a result of unblocked action of a distal enhancer on a target gene.

Table 1. Candidate genes and regulatory regions underlying human phenotypic evolution

Table 1. Candidate genes and regulatory regions underlying human phenotypic evolution

Gene or element	Mechanism of evolution	Function	Associated human phenotypic trait	Caveats	References
AMY1	Copy number variation; increased copy number in human	Encodes amylase, the enzyme that degrades starch in the mouth; in humans the amount of amylase is dependent on the number of copies of *AMY1*	Higher capability to degrade starch in humans; also differences at the human population level	Evolutionary advantages of having more *AMY1* copies is not clear, but may be related to accessing novel high-energy food sources (e.g. tubers)	(Perry et al., 2007)
ANC516/ HARE5	Accelerated evolution in non-coding region	Non-coding element upstream of the Wnt receptor gene *Fzd8*; human and chimp HARE5 behave as nervous system enhancers during mouse development: chimp enhancer displays weaker and later activity; in mice, human-HARE5::Fzd8 induced faster progenitor cell cycle in the developing brain and increased neocortical size compared with chimp-HARE5::Fzd8	Faster neural progenitor cell cycle in humans compared with chimpanzee?	Insertion site and copy number of human-HARE5:: Fzd8 and chimp-HARE5:: Fzd8 unknown – this could influence phenotype; not clear if *FZD8* is in fact differentially expressed in human and chimpanzee brains	(Bird et al., 2007; Boyd et al., 2015)
Androgen receptor (AR) enhancer	Deletion of regulatory DNA	Mouse and chimp sequences function as a developmental enhancer in transgenic mice driving expression of *lacZ* to facial vibrissae and genital tubercle	Loss of sensory vibrissae and penile spines	Regulation of AR expression by enhancer not demonstrated; not known if loss of AR expression causes loss of sensory vibrissae and penile spines	(McLean et al., 2011)
ARHGAP11A/B	Human-specific duplication	ARHGAP11A is a Rho GTPase activating protein (GAP); ARHGAP11B is a truncated version that shows no GAP activity; ARHGAP11B is expressed in the developing brain of humans; electroporation in mouse neocortex induces basal progenitor mitosis	Expansion of the human neocortex?	Function of ARHGAP11B in brain development is still poorly understood	(Sudmant et al., 2010; Florio et al., 2015)
CMAH	Pseudogene: *CMAH* gene inactivated by deletion of 92-bp exon 6	CMAH is required for biosynthesis of the sialic acid Neu5Gc	Altered sialic acid composition on all cells	Physiological impact of the absence of Neu5Gc in human cells has not been demonstrated	(Varki, 2010)
FOXP2	Two human-chimp single nucleotide substitutions subject to positive selection	FoxP2 is a transcription factor; R553H loss-of-function mutation in human apparently leads to apraxia/dyspraxia; knock-in mouse with human-specific *FOXP2* mutations displays developmental defects (see Box 2)	Evolution of speech capability?	Mutation in humans leading to speech problems not same as human-chimp differences; no clear connection between human-specific substitutions in FOXP2 and the evolution of speech or language	(Hurst et al., 1990; Lai et al., 2001; Enard et al., 2002; Enard et al., 2009)
Olfactory receptors	Gene loss and pseudogenization	Odor detection	Loss of capability to detect some odors?	Limited functional evidence of odor detection gain or loss with loss of genes	(Gilad et al., 2005; Keller and Vosshall, 2008)
SRGAP2	Copy number increase	SRGAP2 is a RhoGAP important in cortical development; mutants show increased neuronal branching; SRGAP2B and C (truncated human parologs) are expressed in the developing cortex of humans; SRGAP2C heterodimerizes with SRGAP2A and inhibits its effect on neuronal spine maturation	Human-specific changes in spine morphology and density – could have implications for cognition, learning and memory?	*Srgap2* knockout mice have reduced viability; differences in spine density have not been found in the cortex of human and chimpanzees	(Dennis et al., 2012; Charrier et al., 2012; Bianchi et al., 2013)

Chromosomal Alterations

Large genomic duplications, deletions and rearrangements (**Fig. 1**) are relatively rare, but they encompass many developmental loci owing to their size, which is usually thousands of base pairs (Coe et al., 2014; Girirajan et al., 2011, 2013; Ma et al., 2006). For example, segmental duplications are typically defined as regions greater than one kilobase (kb) with >97% sequence identity (Marques-Bonet et al., 2009). Human-specific structural variation can be challenging to identify, because these loci are difficult to assemble and align. They are therefore poorly represented in genome assemblies, and their discovery often requires targeted sequencing (Chaisson et al., 2015) and cytogenetic techniques. The first structural differences between the human and chimp genomes were discovered using chromatin-stained banding techniques and include the fusion of two ancestral ape chromosomes to form human chromosome 2, human-specific constitutive heterochromatin C bands on chromosomes 1, 9, 16 and Y, and human-specific pericentric inversions on chromosomes 1 and 18 (Yunis and Prakash, 1982). Fluorescent *in situ* hybridization (FISH) and comparative genomic hybridization (CGH) arrays identified >60 human-specific segmental duplications (Goidts et al., 2006; Jauch et al., 1992; Wilson et al., 2006) and 152 genes displaying copy number variation (Armengol et al., 2010; Fortna et al., 2004).

Many of these structural variants have altered gene expression or downstream phenotypes in humans. The pericentric inversion of chromosome 1, for example, is associated with human developmental and neurogenetic diseases and contains copy number increases of the developmental genes SLIT-ROBO Rho GTPase activating protein (*SRGAP2*) (Box 1) (Dennis et al., 2012), *HYDIN*(Doggett et al., 2006), and several DUF1220 domain-containing gene families [e.g. the neuroblastoma breakpoint family (NBPF)] (Fortna et al., 2004). This region demonstrates the complex functional consequences of structural variants. The human-specific duplication of the locus created two duplicate genes (*SRGAP2B* and *SRGAP2C*). SRGAP2C dimerizes with the ancestral SRGAP2A and phenocopies SRGAP2 inhibition, which appears to have led to changes in radial neuron migration and cellular phenotypes (discussed further inBox 1) (Charrier et al., 2012; Dennis et al., 2012). Another interesting human-specific structural variant occurred at chromosome 15q13-q14. This region contains duplications of several genes, including *ARHGAP11B*, which is a partial copy of the gene *ARHGAP11A*, truncated by the boundary of the duplication.*ARHGAP11B* is expressed in the developing brain of humans and appears to regulate brain development (Florio et al., 2015). A polymorphic human-specific duplication also created the salivary amylase gene *AMY1*, which probably enabled humans to eat a high-starch diet (Perry et al., 2007) and is associated with obesity (Falchi et al., 2014; but see Usher et al., 2015). Supporting the adaptive role of duplications in human evolution (Iskow et al., 2012), both coding (Hahn et al., 2007) and non-coding (Kostka et al., 2010) elements in duplicated loci show signatures of positive selection.

Box 1. SRGAP2 – A Highly Duplicated Gene

The gene *SRGAP2* duplicated in a series of genomic events in the lineage leading to *Homo*, between 3.4 and 1 million years ago (Dennis et al., 2012). In humans, three additional, although partial, copies of this gene are present: SRGAP2B, C and D (Charrier et al., 2012; Dennis et al., 2012). The ancestral copy, present in all mammals, is named *SRGAP2A* in humans and has been recently demonstrated to be important in brain cortical development in mice (Guerrier et al., 2009). The *SRGAP2* duplicates encode a truncated F-BAR domain that binds to *SRGAP2A* and antagonizes its function during neuronal migration and morphogenesis. The introduction of human-specific *SRGAP2C in utero* in mouse pyramidal neurons induces a reduction in dendritic spine heads, longer spine necks and higher spine density compared with control neurons. In addition, it was found that mutant mice lacking SRGAP2 display reduced width of dendritic spine heads, longer spine necks and a higher density of dendritic spines (Charrier et al., 2012). So, the expression of SRGAP2C mimics SRGAP2 deficiency during neuronal migration, leading to a deficit in branching in the leading process of migrating neurons and allowing neurons to reach their final position in the cortical plate faster than control neurons. As dendritic spines are known to enhance synaptic connectivity, enable linear integration of synaptic inputs, and implement synapse specific plasticity (Yuste, 2011), Charrier and colleagues (Charrier et al., 2012) speculate that expression of SRGAP2C might allow human cortical pyramidal neurons to receive and integrate a significantly higher number of synaptic inputs without saturation, which could have important implications for cognition, learning and memory. However, spine density in various human and chimpanzee cortical regions is similar (Bianchi et al., 2013). Charrier et al. (2012) also reported that *Srgap2* knockout mice display reduced viability, suggesting that loss of SRGAP2 activity might also have detrimental consequences. Supporting this idea, a large genomic alteration affecting the human ancestral *SRGAP2A* gene may be responsible for the early-infantile encephalopathy and associated epilepsy displayed in a patient (Saitsu et al., 2012). At this time, it is not clear what human traits, if any, were directly affected by the duplication of *SRGAP2*.

Indels

Human-specific duplications and deletions of DNA shorter than one kilobase are numerous and comprise ~3.5% of the human genome, most of which is non-coding (Britten, 2002; The Chimpanzee Sequencing and Analysis Consortium, 2005; Varki and Altheide, 2005). They contribute more base pairs to human-chimp differences than do individual DNA substitutions (see below), albeit fewer than larger chromosomal alterations. Indels can have large functional effects. Non-coding indels can alter human phenotypes by modifying or completely deleting conserved developmental enhancers (**Table 1**). A genome-wide survey revealed 510 highly conserved sequences that were lost in humans, most of which were non-coding, that included a forebrain subventricular zone enhancer near the tumor suppressor gene *GADD45G* and a sensory vibrissae and penile spine enhancer for the androgen receptor gene (McLean et al., 2011) (Box 2). Additionally, retroelements resembling transcription factor binding sites have expanded regulatory networks in a variety of species (reviewed by de Souza et al., 2013). Indels also affect genes, including changing the splicing, reading frame, start or end of genes (Varki and Altheide, 2005). These changes have generated many human-specific pseudogenes (Karro et al., 2007) and transcripts targeted to nonsense-mediated decay (Lareau et al., 2007), including the loss of hundreds of olfactory receptors during human evolution (Gilad et al., 2005; Malnic et al., 2004). Quantifying rates and patterns of human-specific chromosomal alterations and shorter indels is difficult, despite the fact that their evolution can be modeled, because estimation and inference with indel models is computationally challenging (Chindelevitch et al., 2006; Diallo et al., 2007). Additionally, indels are difficult to detect in whole-genome and exome data (Fang et al., 2014). Thus, developing bioinformatics methods for detecting indels and statistical tests for selection on indels are important areas for future research.

Box 2. Deletion of a Conserved Enhancer in the Androgen Receptor Locus

McLean and colleagues (McLean et al., 2011) discovered several non-coding functional regions that may have played a role in human evolution. They conducted a genome-wide search for conserved regions that were lost in the human genome after divergence from chimpanzees. These deleted regions are enriched near genes involved in steroid hormone receptor signaling and neural function (McLean et al., 2011). They decided to examine in detail a 60-kb human deletion downstream of the androgen receptor (*AR*) locus. Within this human-specific deletion lies an approximately 5-kb region that contains non-coding sequences that are highly conserved in other mammals. The authors cloned the corresponding chimpanzee and mouse regions and tested their capacity to drive expression of an hsp68-*lacZ*reporter gene during mouse development. Chimpanzee and mouse constructs both drove consistent *lacZ* expression in the facial vibrissae and genital tubercle of five or more independent transgenic embryos and the mouse sequence also drove expression in hair follicles. Authors found that*lacZ* expression was located in the mesoderm surrounding vibrissae follicles, and in the superficial mesoderm within the presumptive glans of the developing genital tubercle. Sixty-day-old transgenic mice showed expression in the superficial tissue underlying epidermal spines of the penis (McLean et al., 2011). Although previous studies have shown that AR is expressed in mesenchyme surrounding developing epithelial structures (Crocoll et al., 1998), colocalization studies of *lacZ* expression and AR would confirm if the enhancer drives *lacZ* to a subset of *AR* expression domains. It has been shown that humans lack micro- and macro-sensory vibrissae, whereas chimpanzees have micro-sensory vibrissae and mice have both micro- and macro-vibrissae (Muchlinski, 2010). However, the presence or absence of penile spines in humans and chimpanzees is controversial (Reno et al., 2013). In addition, it has been shown that AR is required for normal development of vibrissae, as castration shortens vibrissae in mice, and excess testosterone increases growth (Ibrahim and Wright, 1983). It has not been demonstrated that lack of vibrissae or penile spines is caused by lack of AR expression in the vibrissae follicles or developing penile spines. These results indicate that the human deletion removes a conserved enhancer sequence that directs expression in a subset of the AR expression pattern. However, functional experiments are still required to test whether deletion of this single enhancer can disrupt the formation of penile spines and vibrissae, a molecular event that may help explain the phenotypic loss of these structures in the human lineage.

Single Nucleotide Substitutions

The human and chimpanzee genomes differ by >30 million single nucleotide substitutions (1.2% of the human genome), and slightly less than half of these occurred on the human lineage, mostly in non-coding DNA (The Chimpanzee Sequencing and Analysis Consortium, 2005). Evolutionary theory posits that most substitutions are nearly neutral and are therefore unlikely to have produced uniquely human traits. To identify functional differences, research performed before whole-genome sequencing focused on non-synonymous changes to individual protein-coding sequences (Dorus et al., 2004). The first genome-wide comparative genomic analyses of humans and chimpanzees also focused on protein-coding differences and revealed that genes involved in immunity, sensory perception, and reproduction are enriched for positive selection in humans (Brown et al., 2013; Clark et al., 2003; Nielsen et al., 2005; Voight et al., 2006). Similar approaches, including ones that incorporate population genetic data (Racimo et al., 2014), have been used to identify genes that underwent selection after modern humans diverged from Neanderthals and Denisovans (Meyer et al., 2012; Prufer et al., 2014). Several developmental genes that acquired human-specific coding changes have been hypothesized to be responsible for traits that changed during human evolution (reviewed by Sikela, 2006; O'Bleness et al., 2012) (**Table 1**). These include the forkhead transcription factor *FOXP2*, which is associated with speech (Enard et al., 2002) and may have undergone positive selection, although it is not clear if selection targeted amino acid changes in the protein or nearby non-coding substitutions (Ptak et al., 2009) (Box 3). Another trait that changed significantly during human evolution is pigmentation. Several regulators of pigmentation, including the ligand for the c-KIT receptor (KITLG) (Sturm and Duffy, 2012), contain human-specific protein-coding changes that may have evolved through positive selection. Such examples are compelling, but additional work is required to show that these genes are indeed responsible for modification of the associated traits in humans.

Box 3. The 'Speech Gene' FOXP2

The gene *FOXP2* is one of the most extensively studied examples of a uniquely human genome sequence. First, a family with severe speech disabilities and an arginine-to-histidine substitution at position 553 (R553H) was identified (Hurst et al., 1990; Lai et al., 2001). To understand the function of FOXP2 in humans, mouse models that express the mutant form of FOXP2 present in affected members of the family were generated (Groszer et al., 2008). The homozygotes were severely developmentally delayed and died 3-4 weeks after birth. The cerebellum was abnormally small, with decreased foliation, and, behaviorally, the pups emitted fewer ultrasonic distress calls than did heterozygotes or wild-type mice. Further analyses of the gene identified two human-specific amino acid substitutions in comparison with chimpanzee, gorilla and macaque [a threonine-to-asparagine substitution at position 303 (T303N) and an asparagine-to-serine substitution at position 325 (N325S)] with evidence of positive selection (Enard et al., 2002). To investigate the phenotypic change resulting from the two amino acid differences, mice with humanized FOXP2 were generated and intensively studied (Enard et al., 2009). These 'humanized mice' were fully viable and fertile, in contrast to the R553H mouse model. They had no gross behavioral or anatomical abnormalities but showed increased neuronal dendritic length, increased synaptic plasticity and changes in ultrasonic vocalization in comparison with wild-type mice (Enard et al., 2009). Despite the gigantic effort taken to generate and study the consequences of the evolutionary changes in human FOXP2, there is still no clear or direct connection between the human-specific amino acid substitutions in FOXP2 and speech or language. It is clear that mutations of FOXP2 in humans result in speech impairments, indicating that FOXP2 plays a role in speech development, but the nature of this role and whether the gene participated in the evolution of language remain unknown.

As more vertebrates were sequenced, it became possible to use models of DNA evolution to scan the whole human genome for sequences that changed significantly more than expected by chance since divergence from chimpanzees (Bird et al., 2007; Bush and Lahn, 2008; Pollard et al., 2006b; Prabhakar et al., 2006a). To focus on those changes outside coding portions of genes that have a high probability to be functional, these studies analyzed regions that are highly conserved in non-human species (mammals or vertebrates) but significantly changed in humans. In the absence of functionally annotated non-coding sequences, using this signature of negative selection in other species helps to enrich for regulatory elements with constrained function (Ovcharenko et al., 2004; Prabhakar et al., 2006b; Schwartz et al., 2000; Siepel et al., 2005). These studies collectively identified >2500 non-coding regions defined as 'human accelerated regions' (HARs) (Hubisz and Pollard, 2014), most of which show signatures of positive selection but some of which were probably shaped by non-selective mechanisms, such as GC-biased gene conversion or loss of constraint (Katzman et al., 2010; Kostka et al., 2012; Pollard et al., 2006a; Ratnakumar et al., 2010; Sumiyama and Saitou, 2011). Similar techniques have also been used to analyze regions of the human genome that changed significantly since divergence from extinct hominins (Green et al., 2010). Interestingly, HARs are enriched for substitutions that pre-date the divergence from Neanderthals and Denisovans, suggesting that our genome did not evolve particularly rapidly during the emergence of modern humans (Burbano et al., 2012; Hubisz and Pollard, 2014). From a developmental perspective, HARs have a particularly interesting genomic distribution: they cluster nearby transcriptional factors and other regulatory genes expressed in embryos (Capra et al., 2013; Kamm et al., 2013b), suggesting that HAR mutations could be responsible for the evolution of human traits through modification of developmental gene regulatory networks.

THE NEXT-GENERATION SEQUENCING ERA

Comparative genomics identifies uniquely human genome sequences, but it does not tell us the cell types and developmental stages in which a human-specific sequence functions, making it challenging to link these genetic differences to human traits. Additional data are needed to develop and test hypotheses about the molecular and organismal phenotypes affected by human-specific mutations. This gap is being rapidly filled by functional genomics experiments that assay gene expression, epigenetic marks, protein-DNA binding events, and three-dimensional interactions of regulatory elements with gene promoters in many cell types and species (Box 4). For genes with human-specific changes, functional genomics has helped to shed light on tissue specificity, developmental timing, and, in the case of regulatory genes, downstream targets. For example, the gene network regulated by FOXP2 (Box 3) is becoming clearer through identification of its binding locations and DNA-binding motif (Nelson et al., 2013), as well as studies of how *FOXP2* mutations alter neuronal gene expression (Konopka et al., 2009). These datasets lay the groundwork for a deeper understanding of how FOXP2 might be involved in language acquisition and, more broadly, human evolution.

Box 4. Assaying Genome Function Across Cell Types

Functional genomics, broadly defined as sequencing experiments that probe genome activity, helps to annotate and interpret uniquely human DNA sequences in light of human development. To date, functional genomics has mostly been applied to embryonic stem cells and other homogenous cell cultures, although some embryonic tissues (Nord et al., 2013; Visel et al., 2009) and developmental cell lines (Roadmap Epigenomics Consortium et al., 2015; Romanoski et al., 2015) are being studied. Several of the primary techniques are briefly defined here.

ChIP-seq. Chromatin immunoprecipitation coupled to sequencing (Furey, 2012; Zentner and Henikoff, 2014) is used to generate genome-wide maps of protein binding for transcription factors, structural proteins, polymerases, and modified histones (The ENCODE Project Consortium, 2012; Roadmap Epigenomics Consortium et al., 2015; Vierstra et al., 2014a). These binding events are integrated by computational methods to predict regulatory elements and their activity. For example, enhancers are associated with the acetyltransferase and transcriptional co-activator p300 (Visel et al., 2009; Blow et al., 2010; Ghisletti et al., 2010), histone H3 lysine 4 monomethylation (H3K4Me1) in the absence of significant trimethylation (H3K4Me3) (Heintzman et al., 2007; Xi et al., 2007; Koch et al., 2007), and combinations of transcription factor binding (Zinzen et al., 2009). Additional chromatin modifications help to distinguish active, inactive and poised enhancers (Rada-Iglesias et al., 2011). Regulatory elements are frequently distal to the promoters they target (Bulger and Groudine, 2011; Ong and Corces, 2011), making it hard to predict the genes, pathways and phenotypes affected when they are mutated.

Chromatin capture. Three-dimensional regulatory interactions can be revealed by chromatin conformation capture (3C) and extensions thereof (4C; Cullen et al., 1993; Dekker et al., 2002; Miele and Dekker, 2009) (5C, Hi-C; Dostie et al., 2006; Lieberman-Aiden et al., 2009; Rao et al., 2014). The National Institutes of Health 4D Nucleome program promises to produce a reliable three-dimensional map of many cell types from humans and model organisms (Pennisi, 2015), and it should also be possible to make maps for chimpanzees or other primates.

DNA methylation. Chemical modification of regulatory DNA that can be assayed by a variety of techniques.

DNase-seq, FAIRE (formaldehyde-assisted isolation of regulatory elements). Measurements of open chromatin associated with regulatory elements (Buenrostro et al., 2013; He et al., 2014; Vierstra et al., 2014b).

RNA sequencing. Quantifies gene expression; can also be used to predict enhancers (Arner et al., 2015).

Functional genomics has been crucially important for the study of non-coding changes in the human genome (Sholtis and Noonan, 2010). First, it provides an alternative to simply relying on evolutionary conservation to identify regulatory regions and pinpoint changes that might affect them, which is important given the dynamic turnover of regulatory DNA (Kunarso et al., 2010; Rands et al., 2014). Functional genomics profiles can be computationally integrated to identify different types of regulatory sequences (promoters, enhancers, silencers and insulators; Riethoven, 2010) that function in particular cell types and developmental stages (Kellis et al., 2014). This approach provided strong support for the hypothesis that non-coding HARs function as gene regulatory elements (Hubisz and Pollard, 2014). A second use of functional genomics data is to generate information about the tissue- and stage-specificity of human-specific non-coding sequences. Comparing chromatin states, transcription factor binding profiles and gene expression across multiple cell types predicts that one-third or more of HARs are gene regulatory enhancers active in various embryonic tissues (Capra et al., 2013). Such predictions can help identify the phenotypes in which a human-specific non-coding mutation is involved (Trynka et al., 2013).

Functional genomics also provides a means to probe human and non-human cells directly for differences in regulatory regions and gene expression (Nowick et al., 2009; Somel et al., 2011), which can then be traced back to genetic determinants. For example, combined analysis of human and chimpanzee gene expression data and sequences suggested that pain perception and nociception may have changed in humans through differential regulation of opioid signaling (Cruz-Gordillo et al., 2010). Comparative epigenetic profiling of human, rhesus macaque and mouse corticogenesis revealed promoters and enhancers that have gained activity during human evolution (Reilly et al., 2015), although the sequence differences driving these changes are yet to be identified. Similar investigations have compared patterns of open chromatin (Vierstra et al., 2014a), associated with gene regulation, across tissues in various humans and other mammals. However, most of these studies do not include chimpanzee tissues, making it impossible to pinpoint which of the identified differences occurred after the human-chimpanzee divergence. Moreover, although these studies provide a valuable catalog of differences between humans and other mammals, more work is still required to understand how these differences actually impact on human-specific phenotypes.

THE POWER AND LIMITATIONS OF TRANSGENIC MODEL ORGANISMS

Efforts to characterize human-specific genes and non-coding sequences functionally have primarily utilized mouse or zebrafish models and low-throughput transgenic approaches. For example, humanization of the *Foxp2*gene in mice showed that the human genotype leads to changes in learning (Schreiweis et al., 2014), behavior, vocalizations and brain dopamine concentrations, suggesting alterations to cortico-basal ganglia circuits (Enard et al., 2009) (Box 2). Mouse models also play a large role in studies of the functional effects of human-specific gene duplications, including stimulation of mitosis in neocortical progenitors by ARHGAP11B (Florio et al., 2015) (**Table 1**) and antagonism of SRGAP2A function in neuronal spine maturation by SRGAP2C (Charrier et al., 2012) (Box 1).

Transgenic animals can also be used to test non-coding sequences hypothesized to play a role in human evolution. Reporter gene assays allow candidate enhancers to be validated in mouse and zebrafish embryos with transient transfections or stable lines (for mouse examples, see **Fig. 2**). Enhancer activity has been demonstrated for two conserved non-coding sequences deleted in humans – an androgen receptor enhancer (see Box 2) and a forebrain enhancer upstream of *GADD45G* (McLean et al., 2011). Multiple HARs have also been tested in this manner: HAR2/HANCS1 (which drives expression in the limb, pharyngeal arches, ear and eye; Prabhakar et al., 2008); a cluster of 14 HARs near the gene encoding the transcription factor *NPAS3* (11 of which drive expression in the nervous system; Kamm et al., 2013a); 23 HARs tested by the VISTA Enhancer Browser project (17 in the nervous system, three in limb, two in heart, eight in other tissues; Visel et al., 2007); 29 HARs with epigenetic signatures of active developmental enhancers (20 in the nervous system, eight in limb, four in heart, eight in other tissues; Capra et al., 2013); and HARE5/ANC516 (Bird et al., 2007), a non-coding region located upstream of the Wnt receptor frizzled 8 gene (*FZD8*) that is apparently neocortex specific (Boyd et al., 2015). Supporting the hypothesis that human-specific mutations in HARs may have altered their developmental regulatory functions, several show expression differences between reporter constructs carrying the chimpanzee and human HAR sequences (Boyd et al., 2015; Capra et al., 2013; Kamm et al., 2013a,b; Prabhakar et al., 2008). These include human gains of enhancer activity for *NPAS3*-associated 2xHAR.142 in forebrain at embryonic day (E) 12.5 (Kamm et al., 2013a) and for HAR2/HANCS1 at the base of the limb bud at E11.5 (Prabhakar et al., 2008), which may have resulted from the destruction of a repressor-binding site (Sumiyama and Saitou, 2011). These studies aiming to analyze the function of human-specific non-coding sequences and then to study them comparatively in model organism enhancer assays are adding valuable information about the possible functional impact of human-specific DNA changes.

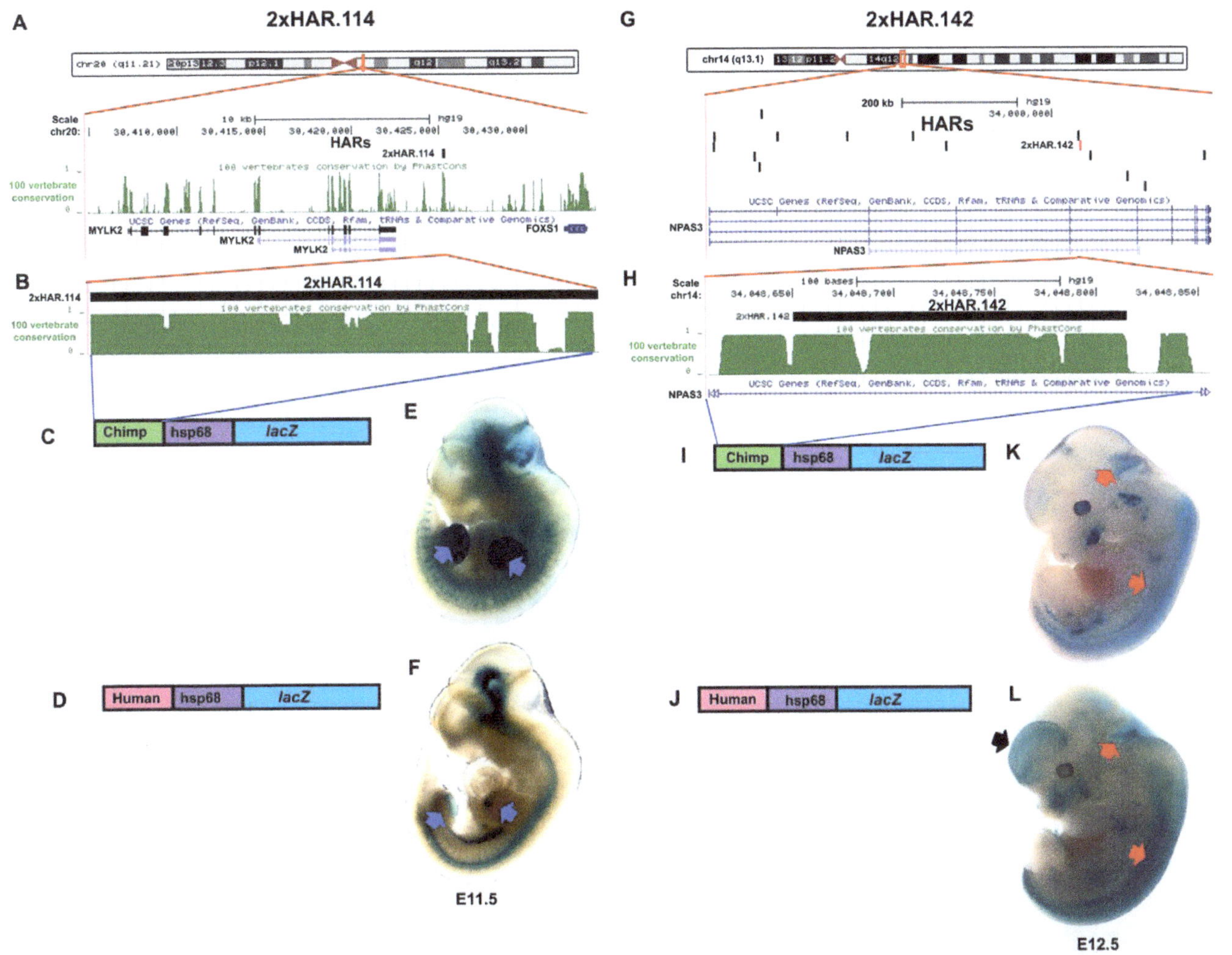

Figure 2. Human accelerated regions (HARs) from human and chimp drive different expression patterns.Examples of HARs enhancer activity in transgenic mouse reporter assays. (A) 2xHAR.114 is located on human chromosome 20 flanked by the genes *FOXS1* and *MYLK2*. (B) 2xHAR.114 shows very high conservation in mammals. (C,D) Schematics of the transgenes used to generate transgenic mice: sequences were cloned upstream of a minimal hsp68 promoter and the reporter gene *lacZ*. (E,F) *lacZ* staining shows that both the chimpanzee (E) and the human (F) sequences produce consistent staining in the limbs (blue arrows) as well as other regions, including spinal cord and developing brain of transgenic mice at E11.5, but the chimpanzee sequence drives expression to a more extensive region in the limb (E; figure modified from Capra et al., 2013). (G) 2xHAR.142 is located on chromosome 14 in the fifth intron of the *NPAS3* gene where 13 other HARs have also been found. (H) 2xHAR.142 shows very high conservation in placental mammals. (I-L) Both chimpanzee (I) and human (J) transgenes produce consistent staining in the hindbrain and spinal cord (red arrows) of transgenic mice at E12.5 (K and L, respectively), but only the human sequence drives expression in the developing cortex (L, black arrow) (figure modified from Kamm et al., 2013a).

However, to demonstrate further the role of HARs in human evolution will require additional studies of molecular and organismal phenotypes. Boyd et al. (2015) took one step in this direction in the functional study HARE5/ANC516 (**Table 1**). They showed that the human enhancer is active earlier in forebrain development than its chimpanzee ortholog, and then they generated transgenic mice expressing the mouse Fzd8-myc tagged coding region under the control of the human and chimpanzee HARE5 enhancer and analyzed them comparatively. They found that human-HARE5-Fzd8 mice display faster progenitor cell cycles in the developing brain and increased neocortical size compared with mice in which*Fzd8* is controlled by chimpanzee HARE5 or with wild-type mice (Boyd et al., 2015). However, this study also highlights several of the challenges of model organism transgenic approaches. First, as the authors used random-insertion transgenic mice and not locus-directed insertions of transgenes (Tasic et al., 2011) or knock-in strategies, we do not know if the observed phenotype is the result of differences between human and chimpanzee HARE5 sequences or the number of insertions of transgenes controlling *Fzd8*. In addition, analyzing overexpression phenotypes is always challenging. Finally, studying human and chimpanzee regulatory sequences or genes in mice is not guaranteed to recapitulate their function in primates. There are important differences between embryonic development in humans compared with model organisms, including mice (Rossant, 2015), which ultimately limit these studies to investigations of small pieces of the human genome out of context.

THE ERA OF HUMAN DEVELOPMENT IN A DISH

Until recently, it was impossible to do genetics or apply functional genomics techniques in developmentally relevant contexts in humans, chimpanzees and other primates. Comparative studies at the molecular and cellular level, therefore, have been mainly based on analysis of preserved tissues from different adult organs (frequently postmortem) or immortalized cell lines (Romero et al., 2012; Schmidt et al., 2010; Zhou et al., 2014). Human embryonic tissues and cells have also occasionally been utilized, for example in the characterization of HAR1 (Pollard et al., 2006b) and SRGAP2 (Charrier et al., 2012; Dennis et al., 2012), and for comparative epigenetic profiling (Reilly et al., 2015). However, the chimpanzee samples necessary for determining whether observed differences are human specific are not available.

Induced pluripotent stem cells (iPSCs) (Takahashi et al., 2007) and techniques allowing their *in vitro* differentiation into various cell lines and tissues (reviewed by Karagiannis and Yamanaka, 2014; Yu et al., 2014) make it feasible to study and manipulate developmental pathways in human and non-human primate cells (Marchetto et al., 2013; Wunderlich et al., 2014). This is particularly important for chimpanzees and other apes, for which we do not have the same embryonic stem cell resources as we have for humans. Various human and non-human primate cell types derived from pluripotent lines have the potential to provide an unprecedented view on how primate development unfolds at the molecular, cellular and tissue level. Initial efforts in this direction compared gene expression (Marchetto et al., 2013) and DNA methylation (Romero et al., 2015) in iPSCs from multiple individuals of different species, revealing rather limited differences between humans and non-human primates in pluripotent cells. This approach will become more relevant to human evolution and development as iPSCs are differentiated into many cell types and additional functional genomics assays are applied across developmental time courses in multiple species (**Fig. 3**).

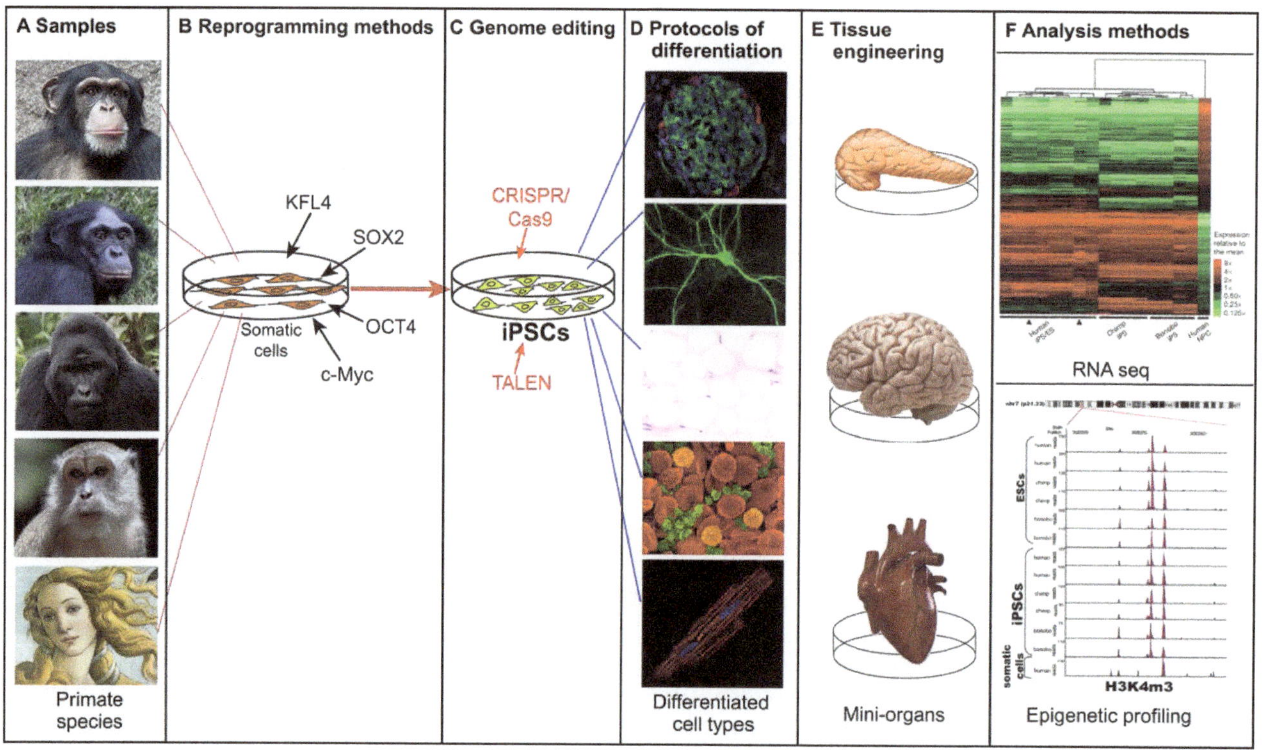

Figure 3. The present and future of comparative studies involving iPSC-derived cells and tissues from human and non-human primates. (A,B) iPSCs can be derived from humans and other primates, including chimpanzees and bonobos, our closest living relatives. (C) This will allow

the introduction of human-specific variants into a non-human primate genomic context (or chimpanzee variants into a human genomic context), through genome editing. These variants could then be screened for phenotypic effects at the molecular and cellular level. (D) In the near future, it will be also be possible to generate many differentiated cell types from primate iPSCs as is already possible with human iPSCs, allowing comparative developmental studies. (E) In many cases, it should also be possible to produce human and non-human primate organoids through tissue engineering techniques that will extend these studies to three-dimensions to recapitulate embryonic development more accurately. (F) Different analysis methods, such as gene expression and epigenetic profiling, could be used to analyze comparatively human and non-human differentiated cell types and organoids. RNA-sequencing image from Marchetto et al. (2013). All other pictures obtained from image repositories that distribute them under creative commons licenses. Image credits: Bonobo head from Kabir; common chimpanzee in the Leipzig Zoo from Thomas Lersch; human brain on white background from DJ.

In recent years, there have been dramatic advances in our ability to recapitulate organ growth *in vitro*, with culture techniques moving away from monolayers of cells towards 3D cultures and organoids, including of neural tissue, where many of human-specific genetic variants appear to be most relevant (Lancaster and Knoblich, 2014; Lancaster et al., 2013; Huch and Koo, 2015). Human cell and organoid cultures are rapidly becoming central tools for understanding human development and disease, and we predict that this will soon also transform human evolutionary studies by enabling direct comparisons of regulatory networks underlying the control of tissue-specific developmental programs across species or between genotypes engineered onto an isogenic human or chimpanzee background (see below). These systems will probably allow researchers to compare human and chimpanzee tissues directly as they develop.

FUTURE PROSPECTS

Although it is still early days for 'human development in a dish', the emerging techniques discussed above, both in the cell culture and the genomics fields, have enormous potential. As comparative developmental studies using pluripotent cell line-derived cells and tissues become routine, it will be exciting to see these approaches applied to human and non-human primates in an effort to reconstruct the genetic history of our species and connect uniquely human genotypes to phenotypes. Other promising approaches include proteomics, metabolomics and lipidomics, which have already been leveraged to compare adult tissues and cell lines across primates (Blekhman et al., 2014; Bozek et al., 2014, 2015; Khan et al., 2013). Another new method that is particularly powerful for decoding regulatory networks is the massively parallel reporter assay (MPRA), which is a high-throughput version of the transgenic enhancer assay described above that is enabled by high-fidelity DNA synthesis to make libraries of thousands of reporter constructs and low-cost RNA sequencing to assay enhancer activity using unique transcribed sequences associated with each candidate enhancer (Melnikov et al., 2012; Patwardhan et al., 2012). MPRA libraries currently can only be introduced into cell lines or by tail-vein injection into an adult mouse, and hence this technique has not yet been applied to developing embryos. Nonetheless, MPRAs promise to make screening human-specific non-coding sequences for regulatory function (e.g. in organoids or developmental cell types) much less laborious.

Potentially the most transformative breakthrough in human evolutionary genomics will be genome editing [i.e. transcription activator-like effector nucleases (TALENs), clustered regularly interspaced short palindromic repeats (CRISPR)-Cas9] (Gilbert et al., 2014; Zheng et al., 2014). Manipulating the genome to introduce individual human mutations into non-human cells or to knock out single human-specific elements in their native context, and documenting resulting changes in phenotype, is the key to demonstrating causality (**Fig. 3**). Moreover, such analyses can in principle be carried out in the context of different specified cell types using pluripotent stem cell differentiation protocols, at different stages of development, and in a comparative manner across multiple species. Coupled with other techniques, genome editing can, in principle, help researchers to understand the importance of every nucleotide in the human genome by generating and testing different genotypes on an isogenic human or non-human primate background. Editing is also accelerating the generation of model organisms (e.g. mouse, zebrafish) carrying humanized genome sequences to study *in vivo* the effects of genetic changes on development and behavior.

CONCLUSIONS

To understand fully the differences between humans and chimpanzees, and other mammals, we need to understand how genetic differences impact on molecular and cellular mechanisms of development leading to the morphological differences that separate us. Comparative and population genomics enabled the identification of thousands of genes and non-coding sequences that are uniquely human. The availability of data about the cell types in which these human-specific genotypes might be relevant allows us to develop testable hypotheses about their roles in human evolution. However, testing these hypotheses has proven to be quite challenging, primarily because of the obvious limitations that we cannot perform genetic manipulations to demonstrate causality or assay phenotypes in humans or great apes, and appropriate human and primate tissues for other studies are scarce. A handful of genes and regulatory enhancers that were lost, duplicated or changed in humans have been assayed for expression differences or other molecular and organismal phenotypes, primarily in transgenic model organisms. The majority of these are active in the developing brain, although some function in limb, eye, heart and other tissues. Emerging techniques for expanding such studies to human and non-human primate organoids and developmental cell lines, coupled with genomic techniques to increase their throughput, may be the key to bridging the gap between uniquely human DNA and the developmental events that produce the traits that are unique to our species. We conclude that the field is now poised to decipher the developmental pathways that translate human genetic changes into human traits.

REFERENCES

1. Armengol, G., Knuutila, S., Lozano, J.-J., Madrigal, I. and Caballín, M.-R. (2010).Identification of human specific gene duplications relative to other primates by array CGH and quantitative PCR. Genomics 95, 203-209. doi:10.1016/j.ygeno.2010.02.003

2. Arner, E., Daub, C. O., Vitting-Seerup, K., Andersson, R., Lilje, B., Drablos, F., Lennartsson,A., Ronnerblad, M., Hrydziuszko, O., Vitezic, M. et al. (2015). Gene regulation. Transcribed enhancers lead waves of coordinated transcription in transitioning mammalian cells.Science 347, 1010-1014. doi:10.1126/science.1259418

3. Bae, B.-I. and Walsh, C. A. (2013). Neuroscience. What are mini-brains? Science 342, 200-201. doi:10.1126/science.1245812

4. Banerji, J., Rusconi, S. and Schaffner, W. (1981). Expression of a beta-globin gene is enhanced by remote SV40 DNA sequences. Cell 27, 299-308. doi:10.1016/0092-8674(81)90413-X

5. Bershteyn, M. and Kriegstein, A. R. (2013). Cerebral organoids in a dish: progress and prospects. Cell 155, 19-20. doi:10.1016/j.cell.2013.09.010

6. Bianchi, S., Stimpson, C. D., Duka, T., Larsen, M. D., Janssen, W. G. M., Collins, Z.,Bauernfeind, A. L., Schapiro, S. J., Baze, W. B., McArthur, M. J. et al. (2013). Synaptogenesis and development of pyramidal neuron dendritic morphology in the chimpanzee neocortex resembles humans. Proc. Natl. Acad. Sci. USA 110 Suppl. 2, 10395-10401.doi:10.1073/pnas.1301224110

7. Bird, C. P., Stranger, B. E., Liu, M., Thomas, D. J., Ingle, C. E., Beazley, C., Miller, W., Hurles, M. E. and Dermitzakis, E. T. (2007). Fast-evolving noncoding sequences in the human genome. Genome Biol. 8, R118. doi:10.1186/gb-2007-8-6-r118

8. Blekhman, R., Perry, G. H., Shahbaz, S., Fiehn, O., Clark, A. G. and Gilad, Y. (2014).Comparative metabolomics in primates reveals the effects of diet and gene regulatory variation on metabolic divergence. Sci. Rep. 4, 5809. doi:10.1038/srep05809

9. Blow, M. J., McCulley, D. J., Li, Z., Zhang, T., Akiyama, J. A., Holt, A., Plajzer-Frick, I., Shoukry,M., Wright, C., Chen, F. et al. (2010). ChIP-Seq identification of weakly conserved heart enhancers. Nat. Genet. 42, 806-810. doi:10.1038/ng.650

10. Boyd, J. L., Skove, S. L., Rouanet, J. P., Pilaz, L.-J., Bepler, T., Gordân, R., Wray, G. A. andSilver, D. L. (2015). Human-chimpanzee differences in a FZD8 enhancer alter cell-cycle dynamics in the developing neocortex. Curr. Biol. 25, 772-779.doi:10.1016/j.cub.2015.01.041

11. Bozek, K., Wei, Y., Yan, Z., Liu, X., Xiong, J., Sugimoto, M., Tomita, M., Pääbo, S., Pieszek, R.,Sherwood, C. C. et al. (2014). Exceptional evolutionary divergence of human muscle and brain metabolomes parallels human cognitive and physical uniqueness. PLoS Biol. 12,e1001871. doi:10.1371/journal.pbio.1001871

12. Bozek, K., Wei, Y., Yan, Z., Liu, X., Xiong, J., Sugimoto, M., Tomita, M., Pääbo, S., Sherwood,C. C., Hof, P. R. et al. (2015). Organization and evolution of brain lipidome revealed by large-scale analysis of human, chimpanzee, macaque, and mouse tissues. Neuron 85,695-702. doi:10.1016/j.neuron.2015.01.003

13. Britten, R. J. (2002). Divergence between samples of chimpanzee and human DNA sequences is 5%, counting indels. Proc. Natl. Acad. Sci. USA 99, 13633-13635. doi:10.1073/pnas.172510699

14. Brown, E. A., Ruvolo, M. and Sabeti, P. C. (2013). Many ways to die, one way to arrive: how selection acts through pregnancy. Trends Genet. 29, 585-592.doi:10.1016/j.tig.2013.03.001

15. Buenrostro, J. D., Giresi, P. G., Zaba, L. C., Chang, H. Y. and Greenleaf, W. J. (2013).Transposition of native chromatin for fast and sensitive epigenomic profiling of open chromatin, DNA-binding proteins and nucleosome position. Nat. Methods 10, 1213-1218. doi:10.1038/nmeth.2688

16. Bulger, M. and Groudine, M. (2011). Functional and mechanistic diversity of distal transcription enhancers. Cell 144, 327-339. doi:10.1016/j.cell.2011.01.024

17. Burbano, H. A., Green, R. E., Maricic, T., Lalueza-Fox, C., de la Rasilla, M., Rosas, A., Kelso, J.,Pollard, K. S., Lachmann, M. and Pääbo, S. (2012). Analysis of human accelerated DNA regions using archaic hominin genomes. PLoS ONE 7, e32877.doi:10.1371/journal.pone.0032877

18. Bush, E. C. and Lahn, B. T. (2008). A genome-wide screen for noncoding elements important in primate evolution. BMC Evol. Biol. 8, 17. doi:10.1186/1471-2148-8-17

19. Capra, J. A., Erwin, G. D., McKinsey, G., Rubenstein, J. L. R. and Pollard, K. S. (2013). Many human accelerated regions are developmental enhancers. Philos. Trans. R. Soc. Lond. B Biol. Sci. 368, 20130025. doi:10.1098/rstb.2013.0025

20. Carroll, S. B. (2005). Evolution at two levels: on genes and form. PLoS Biol. 3, e245.doi:10.1371/journal.pbio.0030245

21. Carroll, S. B. (2008). Evo-devo and an expanding evolutionary synthesis: a genetic theory of morphological evolution. Cell 134, 25-36. doi:10.1016/j.cell.2008.06.030

22. Chaisson, M. J. P., Huddleston, J., Dennis, M. Y., Sudmant, P. H., Malig, M., Hormozdiari, F.,Antonacci, F., Surti, U., Sandstrom, R., Boitano, M. et al. (2015). Resolving the complexity of the human genome using single-molecule sequencing. Nature 517, 608-611. doi:10.1038/nature13907

23. Charrier, C., Joshi, K., Coutinho-Budd, J., Kim, J.-E., Lambert, N., de Marchena, J., Jin, W.-L., Vanderhaeghen, P., Ghosh, A., Sassa, T. et al. (2012). Inhibition of SRGAP2 function by its human-specific paralogs induces neoteny during spine maturation. Cell 149, 923-935.doi:10.1016/j.cell.2012.03.034

24. Chindelevitch, L., Li, Z., Blais, E. and Blanchette, M. (2006). On the inference of parsimonious indel evolutionary scenarios. J. Bioinform. Comput. Biol. 4, 721-744.doi:10.1142/S0219720006002168

25. Clark, A. G., Glanowski, S., Nielsen, R., Thomas, P. D., Kejariwal, A., Todd, M. A., Tanenbaum,D. M., Civello, D., Lu, F., Murphy, B. et al. (2003). Inferring nonneutral evolution from human-chimp-mouse orthologous gene trios. Science 302, 1960-1963.doi:10.1126/science.1088821

26. Coe, B. P., Witherspoon, K., Rosenfeld, J. A., van Bon, B. W. M., Vulto-van Silfhout, A. T.,Bosco, P., Friend, K. L., Baker, C., Buono, S., Vissers, L. E. L. M. et al. (2014). Refining analyses of copy number variation identifies specific genes associated with developmental delay. Nat. Genet. 46, 1063-1071. doi:10.1038/ng.3092

27. Crocoll, A., Zhu, C. C., Cato, A. C. B. and Blum, M. (1998). Expression of androgen receptor mRNA during mouse embryogenesis. Mech. Dev. 72, 175-178. doi:10.1016/S0925-4773(98)00007-0

28. Crow, T. J. (2000). Schizophrenia as the price that homo sapiens pays for language: a resolution of the central paradox in the origin of the species. Brain Res. Brain Res. Rev. 31,118-129. doi:10.1016/S0165-0173(99)00029-6

29. J. KaasCrow, T. J. (2007). Nuclear schizophrenic symptoms as the key to the evolution of the human brain. In *Evolution of Nervous Systems* (ed. J. Kaas), pp. 549-567. Oxford, UK:Elsevier.

30. Cruz-Gordillo, P., Fedrigo, O., Wray, G. A. and Babbitt, C. C. (2010). Extensive changes in the expression of the opioid genes between humans and chimpanzees. Brain Behav. Evol. 76,154-162. doi:10.1159/000320968

31. Cullen, K. E., Kladde, M. P. and Seyfred, M. A. (1993). Interaction between transcription regulatory regions of prolactin chromatin. Science 261, 203-206.doi:10.1126/science.8327891

32. Davidson, E. H. and Erwin, D. H. (2006). Gene regulatory networks and the evolution of animal body plans. Science 311, 796-800. doi:10.1126/science.1113832

33. de Souza, F. S. J., Franchini, L. F. and Rubinstein, M. (2013). Exaptation of transposable elements into novel cis-regulatory elements: is the evidence always strong? Mol. Biol. Evol. 30, 1239-1251. doi:10.1093/molbev/mst045

34. Dekker, J., Rippe, K., Dekker, M. and Kleckner, N. (2002). Capturing chromosome conformation. Science 295, 1306-1311. doi:10.1126/science.1067799

35. Dennis, M. Y., Nuttle, X., Sudmant, P. H., Antonacci, F., Graves, T. A., Nefedov, M., Rosenfeld,J. A., Sajjadian, S., Malig, M., Kotkiewicz, H. et al. (2012). Evolution of human-specific neural SRGAP2 genes by incomplete segmental duplication. Cell 149, 912-922.doi:10.1016/j.cell.2012.03.033

36. Devoy, A., Bunton-Stasyshyn, R. K. A., Tybulewicz, V. L. J., Smith, A. J. H. and Fisher, E. M. C.(2012). Genomically humanized mice: technologies and promises. Nat. Rev. Genet. 13,14-20. doi:10.1038/nrg3116

37. Diallo, A. B., Makarenkov, V. and Blanchette, M. (2007). Exact and heuristic algorithms for the Indel Maximum Likelihood Problem. J. Comput. Biol. 14, 446-461. doi:10.1089/cmb.2007.A006

38. Doggett, N. A., Xie, G., Meincke, L. J., Sutherland, R. D., Mundt, M. O., Berbari, N. S., Davy, B. E., Robinson, M. L., Rudd, M. K., Weber, J. L. et al. (2006). A 360-kb interchromosomal duplication of the human HYDIN locus. Genomics 88, 762-771.doi:10.1016/j.ygeno.2006.07.012

39. Dorus, S., Vallender, E. J., Evans, P. D., Anderson, J. R., Gilbert, S. L., Mahowald, M., Wyckoff,G. J., Malcom, C. M. and Lahn, B. T. (2004). Accelerated evolution of nervous system genes in the origin of Homo sapiens. Cell 119, 1027-1040. doi:10.1016/j.cell.2004.11.040

40. Dostie, J., Richmond, T. A., Arnaout, R. A., Selzer, R. R., Lee, W. L., Honan, T. A., Rubio, E. D.,Krumm, A., Lamb, J., Nusbaum, C. et al. (2006). Chromosome Conformation Capture Carbon Copy (5C): a massively parallel solution for mapping interactions between genomic elements. Genome Res. 16, 1299-1309. doi:10.1101/gr.5571506

41. Enard, W. (2014). Mouse models of human evolution. Curr. Opin. Genet. Dev. 29, 75-80.doi:10.1016/j.gde.2014.08.008

42. Enard, W., Gehre, S., Hammerschmidt, K., Hölter, S. M., Blass, T., Somel, M., Brückner, M. K.,Schreiweis, C., Winter, C., Sohr, R. et al. (2009). A humanized version of Foxp2 affects cortico-basal ganglia circuits in mice. Cell 137, 961-971. doi:10.1016/j.cell.2009.03.041

43. Enard, W., Przeworski, M., Fisher, S. E., Lai, C. S. L., Wiebe, V., Kitano, T., Monaco, A. P. andPääbo, S. (2002). Molecular evolution of FOXP2, a gene involved in speech and language.Nature 418, 869-872. doi:10.1038/nature01025

44. Erwin, D. H. (1999). The origin of bodyplans. Am. Zool. 39, 617-629.doi:10.1093/icb/39.3.617

45. Falchi, M., El-Sayed Moustafa, J. S., Takousis, P., Pesce, F., Bonnefond, A., Andersson-Assarsson, J. C., Sudmant, P. H., Dorajoo, R., Al-Shafai, M. N., Bottolo, L. et al. (2014). Low copy number of the salivary amylase gene predisposes to obesity. Nat. Genet. 46, 492-497. doi:10.1038/ng.2939

46. Fang, H., Wu, Y., Narzisi, G., O'Rawe, J. A., Barrón, L. T. J., Rosenbaum, J., Ronemus, M.,Iossifov, I., Schatz, M. C. and Lyon, G. J. (2014). Reducing INDEL calling errors in whole genome and exome sequencing data. Genome Med. 6, 89. doi:10.1186/s13073-014-0089-z

47. Finch, C. E. (2010). Evolution in health and medicine Sackler colloquium: evolution of the human lifespan and diseases of aging: roles of infection, inflammation, and nutrition. Proc. Natl. Acad. Sci. USA 107 Suppl. 1, 1718-1724. doi:10.1073/pnas.0909606106

48. Finnerty, J. R., Pang, K., Burton, P., Paulson, D. and Martindale, M. Q. (2004). Origins of bilateral symmetry: Hox and dpp expression in a sea anemone. Science 304, 1335-1337.doi:10.1126/science.1091946

49. Florio, M., Albert, M., Taverna, E., Namba, T., Brandl, H., Lewitus, E., Haffner, C., Sykes, A.,Wong, F. K., Peters, J. et al. (2015). Human-specific gene ARHGAP11B promotes basal progenitor amplification and neocortex expansion. Science 347, 1465-1470.doi:10.1126/science.aaa1975

50. Fortna, A., Kim, Y., MacLaren, E., Marshall, K., Hahn, G., Meltesen, L., Brenton, M., Hink, R.,Burgers, S., Hernandez-Boussard, T. et al. (2004). Lineage-specific gene duplication and loss in human and great ape evolution. PLoS Biol. 2, e207.doi:10.1371/journal.pbio.0020207

51. Fujie, Y., Fusaki, N., Katayama, T., Hamasaki, M., Soejima, Y., Soga, M., Ban, H., Hasegawa,M., Yamashita, S., Kimura, S. et al. (2014). New type of Sendai virus vector provides transgene-free iPS cells derived from chimpanzee blood. PLoS ONE 9, e113052.doi:10.1371/journal.pone.0113052

52. Furey, T. S. (2012). ChIP-seq and beyond: new and improved methodologies to detect and characterize protein-DNA interactions. Nat. Rev. Genet. 13, 840-852. doi:10.1038/nrg3306

53. Gautam, P., Chaurasia, A., Bhattacharya, A., Grover, R., Indian Genome Variation Consortium, Mukerji, M. and Natarajan, V. T. (2015). Population diversity and adaptive evolution in keratinization genes: impact of environment in shaping skin phenotypes. Mol. Biol. Evol. 32, 555-573. doi:10.1093/molbev/msu342

54. Gehring, W. J. (2011). Chance and necessity in eye evolution. Genome Biol. Evol. 3, 1053-1066. doi:10.1093/gbe/evr061

55. Ghisletti, S., Barozzi, I., Mietton, F., Polletti, S., De Santa, F., Venturini, E., Gregory, L., Lonie,L., Chew, A., Wei, C. L. et al. (2010). Identification and characterization of enhancers controlling the inflammatory gene expression program in macrophages. Immunity 32,317-328. doi:10.1016/j.immuni.2010.02.008

56. Gilad, Y., Man, O. and Glusman, G. (2005). A comparison of the human and chimpanzee olfactory receptor gene repertoires. Genome Res. 15, 224-230. doi:10.1101/gr.2846405

57. Gilbert, S. F. (2010). *Developmental Biology*. Sunderland, MA: Sinauer.

58. Gilbert, L. A., Horlbeck, M. A., Adamson, B., Villalta, J. E., Chen, Y., Whitehead, E. H.,Guimaraes, C., Panning, B., Ploegh, H. L., Bassik, M. C. et al. (2014). Genome-scale CRISPR-mediated control of gene repression and activation. Cell 159, 647-661.doi:10.1016/j.cell.2014.09.029

59. Girirajan, S., Campbell, C. D. and Eichler, E. E. (2011). Human copy number variation and complex genetic disease. Annu. Rev. Genet. 45, 203-226. doi:10.1146/annurev-genet-102209-163544

60. Girirajan, S., Johnson, R. L., Tassone, F., Balciuniene, J., Katiyar, N., Fox, K., Baker, C.,Srikanth, A., Yeoh, K. H., Khoo, S. J. et al. (2013). Global increases in both common and rare copy number load associated with autism. Hum. Mol. Genet. 22, 2870-2880.doi:10.1093/hmg/ddt136

61. Goidts, V., Armengol, L., Schempp, W., Conroy, J., Nowak, N., Müller, S., Cooper, D. N.,Estivill, X., Enard, W., Szamalek, J. M. et al. (2006). Identification of large-scale human-specific copy number differences by inter-species array comparative genomic hybridization. Hum. Genet. 119, 185-198. doi:10.1007/s00439-005-0130-9

62. Green, R. E., Krause, J., Briggs, A. W., Maricic, T., Stenzel, U., Kircher, M., Patterson, N., Li, H.,Zhai, W., Fritz, M. H. et al. (2010). A draft sequence of the Neanderthal genome. Science328, 710-722. doi:10.1126/science.1188021

63. Groszer, M., Keays, D. A., Deacon, R. M. J., de Bono, J. P., Prasad-Mulcare, S., Gaub, S.,Baum, M. G., French, C. A., Nicod, J., Coventry, J. A. et al. (2008). Impaired synaptic plasticity and motor learning in mice with a point mutation implicated in human speech deficits. Curr. Biol. 18, 354-362. doi:10.1016/j.cub.2008.01.060

64. Gudbjartsson, D. F., Helgason, H., Gudjonsson, S. A., Zink, F., Oddson, A., Gylfason, A.,Besenbacher, S., Magnusson, G., Halldorsson, B. V., Hjartarson, E. et al. (2015). Large-scale whole-genome sequencing of the Icelandic population. Nat. Genet. 47, 435-444. doi:10.1038/ng.3247

65. Guerrier, S., Coutinho-Budd, J., Sassa, T., Gresset, A., Jordan, N. V., Chen, K., Jin, W.-L.,Frost, A. and Polleux, F. (2009). The F-BAR domain of srGAP2 induces membrane protrusions required for neuronal migration and morphogenesis. Cell 138, 990-1004.doi:10.1016/j.cell.2009.06.047

66. Guruharsha, K. G., Kankel, M. W. and Artavanis-Tsakonas, S. (2012). The Notch signalling system: recent insights into the complexity of a conserved pathway. Nat. Rev. Genet. 13,654-666. doi:10.1038/nrg3272

67. Hahn, M. W., Demuth, J. P. and Han, S.-G. (2007). Accelerated rate of gene gain and loss in primates. Genetics 177, 1941-1949. doi:10.1534/genetics.107.080077

68. Hancock, A. M., Witonsky, D. B., Alkorta-Aranburu, G., Beall, C. M., Gebremedhin, A.,Sukernik, R., Utermann, G., Pritchard, J. K., Coop, G. and Di Rienzo, A. (2011). Adaptations to climate-mediated selective pressures in humans. PLoS Genet. 7, e1001375.doi:10.1371/journal.pgen.1001375

69. Harcourt-Smith, W. E. and Aiello, L. C. (2004). Fossils, feet and the evolution of human bipedal locomotion. J. Anat. 204, 403-416. doi:10.1111/j.0021-8782.2004.00296.x

70. Haygood, R., Babbitt, C. C., Fedrigo, O. and Wray, G. A. (2010). Contrasts between adaptive coding and noncoding changes during human evolution. Proc. Natl. Acad. Sci. USA 107,7853-7857. doi:10.1073/pnas.0911249107

71. He, H. H., Meyer, C. A., Hu, S. S., Chen, M.-W., Zang, C., Liu, Y., Rao, P. K., Fei, T., Xu, H., Long,H. et al. (2014). Refined DNase-seq protocol and data analysis reveals intrinsic bias in transcription factor footprint identification. Nat. Methods 11, 73-78.doi:10.1038/nmeth.2762

72. Heintzman, N. D., Stuart, R. K., Hon, G., Fu, Y., Ching, C. W., Hawkins, R. D., Barrera, L. O.,Van Calcar, S., Qu, C., Ching, K. A. et al. (2007). Distinct and predictive chromatin signatures of transcriptional promoters and enhancers in the human genome. Nat. Genet. 39, 311-318. doi:10.1038/ng1966

73. Horvath, J. E., Ramachandran, G. L., Fedrigo, O., Nielsen, W. J., Babbitt, C. C., St Clair, E. M.,Pfefferle, L. W., Jernvall, J., Wray, G. A. and Wall, C. E. (2014). Genetic comparisons yield insight into the evolution of enamel thickness during human evolution. J. Hum. Evol. 73,75-87. doi:10.1016/j.jhevol.2014.01.005

74. Hubisz, M. J. and Pollard, K. S. (2014). Exploring the genesis and functions of Human Accelerated Regions sheds light on their role in human evolution. Curr. Opin. Genet. Dev.29, 15-21. doi:10.1016/j.gde.2014.07.005

75. Huch, M. and Koo, B.-K. (2015). Modeling mouse and human development using organoid cultures. Development 142, 3113-3125. doi:10.1242/dev.118570

76. Huerta-Sanchez, E., Jin, X., Asan, X., Bianba, Z., Peter, B. M., Vinckenbosch, N., Liang, Y., Yi,X., He, M., Somel, M. et al. (2014). Altitude adaptation in Tibetans caused by introgression of Denisovan-like DNA. Nature 512, 194-197. doi:10.1038/nature13408

77. Hurst, J. A., Baraitser, M., Auger, E., Graham, F. and Norell, S. (1990). An extended family with a dominantly inherited speech disorder. Dev. Med. Child Neurol. 32, 352-355.doi:10.1111/j.1469-8749.1990.tb16948.x

78. Ibrahim, L. and Wright, E. A. (1983). Effect of castration and testosterone propionate on mouse vibrissae. Br. J. Dermatol. 108, 321-326. doi:10.1111/j.1365-2133.1983.tb03971.x

79. International Human Genome Sequencing Consortium (2004). Finishing the euchromatic sequence of the human genome. Nature 431, 931-945. doi:10.1038/nature03001

80. Iskow, R. C., Gokcumen, O. and Lee, C. (2012). Exploring the role of copy number variants in human adaptation. Trends Genet. 28, 245-257. doi:10.1016/j.tig.2012.03.002

81. Jablonski, N. G. and Chaplin, G. (2014). The evolution of skin pigmentation and hair texture in people of African ancestry. Dermatol. Clin. 32, 113-121. doi:10.1016/j.det.2013.11.003

82. Jauch, A., Wienberg, J., Stanyon, R., Arnold, N., Tofanelli, S., Ishida, T. and Cremer, T. (1992).Reconstruction of genomic rearrangements in great apes and gibbons by chromosome painting. Proc. Natl. Acad. Sci. USA 89, 8611-8615. doi:10.1073/pnas.89.18.8611

83. Kamberov, Y. G., Wang, S., Tan, J., Gerbault, P., Wark, A., Tan, L., Yang, Y., Li, S., Tang, K.,Chen, H. et al. (2013). Modeling recent human evolution in mice by expression of a selected EDAR variant. Cell 152, 691-702. doi:10.1016/j.cell.2013.01.016

84. Kamm, G. B., Lopez-Leal, R., Lorenzo, J. R. and Franchini, L. F. (2013a). A fast-evolving human NPAS3 enhancer gained reporter expression in the developing forebrain of transgenic mice. Philos. Trans. R. Soc. Lond. B Biol. Sci. 368, 20130019.doi:10.1098/rstb.2013.0019

85. Kamm, G. B., Pisciottano, F., Kliger, R. and Franchini, L. F. (2013b). The developmental brain gene NPAS3 contains the largest number of accelerated regulatory sequences in the human genome. Mol. Biol. Evol. 30, 1088-1102. doi:10.1093/molbev/mst023

86. Karagiannis, P. and Yamanaka, S. (2014). The fate of cell reprogramming. Nat. Methods 11,1006-1008. doi:10.1038/nmeth.3109

87. Karro, J. E., Yan, Y., Zheng, D., Zhang, Z., Carriero, N., Cayting, P., Harrrison, P. and Gerstein,M. (2007). Pseudogene.org: a comprehensive database and comparison platform for pseudogene annotation. Nucleic Acids Res. 35, D55-D60. doi:10.1093/nar/gkl851

88. Katzman, S., Kern, A. D., Pollard, K. S., Salama, S. R. and Haussler, D. (2010). GC-biased evolution near human accelerated regions. PLoS Genet. 6, e1000960. doi:10.1371/journal.pgen.1000960

89. Keller, A. and Vosshall, L. B. (2008). Better smelling through genetics: mammalian odor perception. Curr. Opin. Neurobiol. 18, 364-369. doi:10.1016/j.conb.2008.09.020

90. Kellis, M., Wold, B., Snyder, M. P., Bernstein, B. E., Kundaje, A., Marinov, G. K., Ward, L. D.,Birney, E., Crawford, G. E., Dekker, J. et al. (2014). Defining functional DNA elements in the human genome. Proc. Natl. Acad. Sci. USA 111, 6131-6138.doi:10.1073/pnas.1318948111

91. Khan, Z., Ford, M. J., Cusanovich, D. A., Mitrano, A., Pritchard, J. K. and Gilad, Y. (2013).Primate transcript and protein expression levels evolve under compensatory selection pressures. Science 342, 1100-1104. doi:10.1126/science.1242379

92. King, M. C. and Wilson, A. C. (1975). Evolution at two levels in humans and chimpanzees.Science 188, 107-116. doi:10.1126/science.1090005

93. Koch, C. M., Andrews, R. M., Flicek, P., Dillon, S. C., Karaoz, U., Clelland, G. K., Wilcox, S.,Beare, D. M., Fowler, J. C., Couttet, P., et al. (2007). The landscape of histone modifications across 1% of the human genome in five human cell lines. Genome Res. 17, 691-707. doi:10.1101/gr.5704207

94. Konopka, G., Bomar, J. M., Winden, K., Coppola, G., Jonsson, Z. O., Gao, F., Peng, S., Preuss,T. M., Wohlschlegel, J. A. and Geschwind, D. H. (2009). Human-specific transcriptional regulation of CNS development genes by FOXP2. Nature 462, 213-217. doi:10.1038/nature08549

95. Kostka, D., Hahn, M. W. and Pollard, K. S. (2010). Noncoding sequences near duplicated genes evolve rapidly. Genome Biol. Evol. 2, 518-533. doi:10.1093/gbe/evq037

96. Kostka, D., Hubisz, M. J., Siepel, A. and Pollard, K. S. (2012). The role of GC-biased gene conversion in shaping the fastest evolving regions of the human genome. Mol. Biol. Evol.29, 1047-1057. doi:10.1093/molbev/msr279

97. Kunarso, G., Chia, N.-Y., Jeyakani, J., Hwang, C., Lu, X., Chan, Y.-S., Ng, H.-H. and Bourque,G. (2010). Transposable elements have rewired the core regulatory network of human embryonic stem cells. Nat. Genet. 42, 631-634. doi:10.1038/ng.600

98. Lai, C. S. L., Fisher, S. E., Hurst, J. A., Vargha-Khadem, F. and Monaco, A. P. (2001). A forkhead-domain gene is mutated in a severe speech and language disorder. Nature 413,519-523. doi:10.1038/35097076

99. Lancaster, M. A. and Knoblich, J. A. (2014). Organogenesis in a dish: modeling development and disease using organoid technologies. Science 345, 1247125.doi:10.1126/science.1247125

100. Lancaster, M. A., Renner, M., Martin, C.-A., Wenzel, D., Bicknell, L. S., Hurles, M. E., Homfray,T., Penninger, J. M., Jackson, A. P. and Knoblich, J. A. (2013). Cerebral organoids model human brain development

and microcephaly. Nature 501, 373-379.doi:10.1038/nature12517

101. Lareau, L. F., Inada, M., Green, R. E., Wengrod, J. C. and Brenner, S. E. (2007). Unproductive splicing of SR genes associated with highly conserved and ultraconserved DNA elements.Nature 446, 926-929. doi:10.1038/nature05676

102. Laurell, T., VanderMeer, J. E., Wenger, A. M., Grigelioniene, G., Nordenskjöld, A., Arner, M.,Ekblom, A. G., Bejerano, G., Ahituv, N. and Nordgren, A. (2012). A novel 13 base pair insertion in the sonic hedgehog ZRS limb enhancer (ZRS/LMBR1) causes preaxial polydactyly with triphalangeal thumb. Hum. Mutat. 33, 1063-1066.doi:10.1002/humu.22097

103. Lemons, D. and McGinnis, W. (2006). Genomic evolution of Hox gene clusters. Science 313,1918-1922. doi:10.1126/science.1132040

104. O. H. Cohn Lieberman, D. E. (2000). Ontogeny, homology, and phylogeny in the hominid craniofacial skeleton: The problem of the browridge. In *Development, Growth and Evolution* (ed. O. H. Cohn), pp. 85-122. London: Academic Press.

105. Lieberman-Aiden, E., van Berkum, N. L., Williams, L., Imakaev, M., Ragoczy, T., Telling, A.,Amit, I., Lajoie, B. R., Sabo, P. J., Dorschner, M. O. et al. (2009). Comprehensive mapping of long-range interactions reveals folding principles of the human genome. Science 326,289-293. doi:10.1126/science.1181369

106. Ma, J., Zhang, L., Suh, B. B., Raney, B. J., Burhans, R. C., Kent, W. J., Blanchette, M., Haussler,D. and Miller, W. (2006). Reconstructing contiguous regions of an ancestral genome.Genome Res. 16, 1557-1565. doi:10.1101/gr.5383506

107. Malnic, B., Godfrey, P. A. and Buck, L. B. (2004). The human olfactory receptor gene family.Proc. Natl. Acad. Sci. USA 101,2584-2589. doi:10.1073/pnas.0307882100

108. Marchetto, M. C. N., Narvaiza, I., Denli, A. M., Benner, C., Lazzarini, T. A., Nathanson, J. L.,Paquola, A. C. M., Desai, K. N., Herai, R. H., Weitzman, M. D. et al. (2013). Differential L1 regulation in pluripotent stem cells of humans and apes. Nature 503, 525-529.doi:10.1038/nature12686

109. Marques-Bonet, T., Girirajan, S. and Eichler, E. E. (2009). The origins and impact of primate segmental duplications. Trends Genet. 25, 443-454. doi:10.1016/j.tig.2009.08.002

110. McLean, C. Y., Reno, P. L., Pollen, A. A., Bassan, A. I., Capellini, T. D., Guenther, C., Indjeian, V. B., Lim, X., Menke, D. B., Schaar, B. T. et al. (2011). Human-specific loss of regulatory DNA and the evolution of human-specific traits. Nature 471, 216-219.doi:10.1038/nature09774

111. Melnikov, A., Murugan, A., Zhang, X., Tesileanu, T., Wang, L., Rogov, P., Feizi, S., Gnirke, A.,Callan, C. G., Jr.., Kinney, J. B. et al. (2012). Systematic dissection and optimization of inducible enhancers in human cells using a massively parallel reporter assay. Nat. Biotechnol. 30, 271-277. doi:10.1038/nbt.2137

112. Miele, A. and Dekker, J. (2009). Mapping cis- and trans-chromatin interaction networks using chromosome conformation capture (3C). Methods Mol. Biol. 464, 105-121.doi:10.1007/978-1-60327-461-6_7

113. Meyer, M., Kircher, M., Gansauge, M.-T., Li, H., Racimo, F., Mallick, S., Schraiber, J. G., Jay, F.,Prufer, K., de Filippo, C. et al. (2012). A high-coverage genome sequence from an archaic Denisovan individual. Science 338, 222-226. doi:10.1126/science.1224344

114. Moreau, P., Hen, R., Wasylyk, B., Everett, R., Gaub, M. P. and Chambon, P. (1981). The SV40 72 base repair repeat has a striking effect on gene expression both in SV40 and other chimeric recombinants. Nucleic Acids Res. 9, 6047-6068. doi:10.1093/nar/9.22.6047

115. Muchlinski, M. N. (2010). A comparative analysis of vibrissa count and infraorbital foramen area in primates and other mammals. J. Hum. Evol. 58, 447-473. doi:10.1016/j.jhevol.2010.01.012

116. Nelson, C. S., Fuller, C. K., Fordyce, P. M., Greninger, A. L., Li, H. and DeRisi, J. L. (2013).Microfluidic affinity and ChIP-seq analyses converge on a conserved FOXP2-binding motif in chimp and human, which enables the detection of evolutionarily novel targets. Nucleic Acids Res. 41, 5991-6004. doi:10.1093/nar/gkt259

117. Nielsen, R., Bustamante, C., Clark, A. G., Glanowski, S., Sackton, T. B., Hubisz, M. J., Fledel-Alon, A., Tanenbaum, D. M., Civello, D., White, T. J. et al. (2005). A scan for positively selected genes in the genomes of humans and chimpanzees. PLoS Biol. 3, e170. doi:10.1371/journal.pbio.0030170

118. Noonan, J. P. and McCallion, A. S. (2010). Genomics of long-range regulatory elements.Annu. Rev. Genomics Hum. Genet. 11, 1-23. doi:10.1146/annurev-genom-082509-141651

119. Nord, A. S., Blow, M. J., Attanasio, C., Akiyama, J. A., Holt, A., Hosseini, R., Phouanenavong,S., Plajzer-Frick, I., Shoukry, M., Afzal, V. et al. (2013). Rapid and pervasive changes in genome-wide enhancer usage during mammalian development. Cell 155, 1521-1531. doi:10.1016/j.cell.2013.11.033

120. Nowick, K., Gernat, T., Almaas, E. and Stubbs, L. (2009). Differences in human and chimpanzee gene expression patterns define an evolving network of transcription factors in brain. Proc. Natl. Acad. Sci. USA 106, 22358-22363. doi:10.1073/pnas.0911376106

121. O'Bleness, M., Searles, V. B., Varki, A., Gagneux, P. and Sikela, J. M. (2012). Evolution of genetic and genomic features unique to the human lineage. Nat. Rev. Genet. 13, 853-866.doi:10.1038/nrg3336

122. Ong, C.-T. and Corces, V. G. (2011). Enhancer function: new insights into the regulation of tissue-specific gene expression. Nat. Rev. Genet. 12, 283-293. doi:10.1038/nrg2957

123. Ovcharenko, I., Nobrega, M. A., Loots, G. G. and Stubbs, L. (2004). ECR Browser: a tool for visualizing and accessing data from comparisons of multiple vertebrate genomes. Nucleic Acids Res. 32, W280-W286. doi:10.1093/nar/gkh355

124. Patwardhan, R. P., Hiatt, J. B., Witten, D. M., Kim, M. J., Smith, R. P., May, D., Lee, C., Andrie,J. M., Lee, S. I., Cooper, G. M. et al. (2012). Massively parallel functional dissection of mammalian enhancers in vivo. Nat. Biotechnol. 30, 265-270. doi:10.1038/nbt.2136

125. Pennacchio, L. A., Ahituv, N., Moses, A. M., Prabhakar, S., Nobrega, M. A., Shoukry, M.,Minovitsky, S., Dubchak, I., Holt, A., Lewis, K. D. et al. (2006). In vivo enhancer analysis of human conserved non-coding sequences. Nature 444, 499-502.doi:10.1038/nature05295

126. Pennisi, E. (2015). Genomics. Inching toward the 3D genome. Science 347, 10.doi:10.1126/science.347.6217.10

127. Perry, G. H., Dominy, N. J., Claw, K. G., Lee, A. S., Fiegler, H., Redon, R., Werner, J., Villanea,F. A., Mountain, J. L., Misra, R. et al. (2007). Diet and the evolution of human amylase gene copy number variation. Nat. Genet. 39, 1256-1260. doi:10.1038/ng2123

128. Pollard, K. S., Salama, S. R., King, B., Kern, A. D., Dreszer, T., Katzman, S., Siepel, A.,Pedersen, J. S., Bejerano, G., Baertsch, R. et al. (2006a). Forces shaping the fastest evolving regions in the human genome. PLoS Genet. 2, e168.doi:10.1371/journal.pgen.0020168

129. Pollard, K. S., Salama, S. R., Lambert, N., Lambot, M.-A., Coppens, S., Pedersen, J. S.,Katzman, S., King, B., Onodera, C., Siepel, A. et al. (2006b). An RNA gene expressed during cortical development evolved rapidly in humans. Nature 443, 167-172.doi:10.1038/nature05113

130. Prabhakar, S., Noonan, J. P., Paabo, S. and Rubin, E. M. (2006a). Accelerated evolution of conserved noncoding sequences in humans. Science 314, 786.doi:10.1126/science.1130738

131. Prabhakar, S., Poulin, F., Shoukry, M., Afzal, V., Rubin, E. M., Couronne, O. and Pennacchio, L. A. (2006b). Close sequence comparisons are sufficient to identify human cis-regulatory elements. Genome Res. 16, 855-863. doi:10.1101/gr.4717506

132. Prabhakar, S., Visel, A., Akiyama, J. A., Shoukry, M., Lewis, K. D., Holt, A., Plajzer-Frick, I.,Morrison, H., Fitzpatrick, D. R., Afzal, V. et al. (2008). Human-specific gain of function in a developmental enhancer. Science 321, 1346-1350. doi:10.1126/science.1159974

133. Prufer, K., Racimo, F., Patterson, N., Jay, F., Sankararaman, S., Sawyer, S., Heinze, A.,Renaud, G., Sudmant, P. H., de Filippo, C. et al. (2014). The complete genome sequence of a Neanderthal from the Altai Mountains. Nature 505, 43-49. doi:10.1038/nature12886

134. Ptak, S. E., Enard, W., Wiebe, V., Hellmann, I., Krause, J., Lachmann, M. and Paabo, S. (2009).Linkage disequilibrium extends across putative selected sites in FOXP2. Mol. Biol. Evol.26, 2181-2184. doi:10.1093/molbev/msp143

135. Racimo, F., Kuhlwilm, M. and Slatkin, M. (2014). A test for ancient selective sweeps and an application to candidate sites in modern humans. Mol. Biol. Evol. 31, 3344-3358.doi:10.1093/molbev/msu255

136. Rada-Iglesias, A., Bajpai, R., Swigut, T., Brugmann, S. A., Flynn, R. A. and Wysocka, J. (2011).A unique chromatin signature uncovers early developmental enhancers in humans.Nature 470, 279-283. doi:10.1038/nature09692

137. Rands, C. M., Meader, S., Ponting, C. P. and Lunter, G. (2014). 8.2% of the human genome is constrained: variation in rates of turnover across functional element classes in the human lineage. PLoS Genet. 10, e1004525. doi:10.1371/journal.pgen.1004525

138. Rao, S. S. P., Huntley, M. H., Durand, N. C., Stamenova, E. K., Bochkov, I. D., Robinson, J. T.,Sanborn, A. L., Machol, I., Omer, A. D., Lander, E. S. et al. (2014). A 3D map of the human genome at kilobase resolution reveals principles of chromatin looping. Cell 159, 1665-1680. doi:10.1016/j.cell.2014.11.021

139. Ratnakumar, A., Mousset, S., Glemin, S., Berglund, J., Galtier, N., Duret, L. and Webster, M. T. (2010). Detecting positive selection within genomes: the problem of biased gene conversion. Philos. Trans. R. Soc. Lond. B Biol. Sci. 365, 2571-2580.doi:10.1098/rstb.2010.0007

140. Reilly, S. K., Yin, J., Ayoub, A. E., Emera, D., Leng, J., Cotney, J., Sarro, R., Rakic, P. andNoonan, J. P. (2015). Evolutionary genomics. Evolutionary changes in promoter and enhancer activity during human corticogenesis. Science 347, 1155-1159.doi:10.1126/science.1260943

141. Reno, P. L., McLean, C. Y., Hines, J. E., Capellini, T. D., Bejerano, G. and Kingsley, D. M.(2013). A penile spine/vibrissa enhancer sequence is missing in modern and extinct humans but is retained in multiple primates with penile spines and sensory vibrissae.PLoS ONE 8, e84258. doi:10.1371/journal.pone.0084258

142. Gibbs, R. A., Rogers, J., Katze, M. G., Bumgarner, R., Weinstock, G. M., Mardis, E. R.,Remington, K. A., Strausberg, R. L., Venter, J. C. et al.Rhesus Macaque Genome Sequencing and Analysis Consortium, Gibbs, R. A., Rogers, J., Katze, M. G., Bumgarner, R.,Weinstock, G. M., Mardis, E. R., Remington, K. A., Strausberg, R. L., Venter, J. C. et al.(2007). Evolutionary and biomedical insights from the rhesus macaque genome. Science316, 222-234. doi:10.1126/science.1139247

143. Riethoven, J.-J. (2010). Regulatory regions in DNA: promoters, enhancers, silencers, and insulators. Methods Mol. Biol. 674, 33-42. doi:10.1007/978-1-60761-854-6_3

144. Kundaje, A., Meuleman, W., Ernst, J., Bilenky, M., Yen, A., Heravi-Moussavi, A., Kheradpour,P., Zhang, Z., Wang, J. et al.Roadmap Epigenomics Consortium, Kundaje, A., Meuleman,W., Ernst, J., Bilenky, M., Yen, A., Heravi-Moussavi, A., Kheradpour, P., Zhang, Z., Wang, J.et al. (2015). Integrative analysis of 111 reference human epigenomes. Nature 518, 317-330. doi:10.1038/nature14248

145. Romanoski, C. E., Glass, C. K., Stunnenberg, H. G., Wilson, L. and Almouzni, G. (2015).Epigenomics: roadmap for regulation. Nature 518, 314-316. doi:10.1038/518314a

146. Romero, I. G., Ruvinsky, I. and Gilad, Y. (2012). Comparative studies of gene expression and the

evolution of gene regulation. Nat. Rev. Genet. 13, 505-516. doi:10.1038/nrg3229

147. Romero, I. G., Pavlovic, B. J., Hernando-Herraez, I., Banovich, N. E., Kagan, C. L., Burnett, J. E., Huang, C. H., Mitrano, A., Chavarria, C. I., Ben-Nun, I. F. et al. (2015). Generation of a panel of induced pluripotent stem cells from chimpanzees: a resource for comparative functional genomics. Elife 23, 4. doi:10.7554/eLife.07103

148. Rossant, J. (2015). Mouse and human blastocyst-derived stem cells: vive les differences.Development 142, 9-12. doi:10.1242/dev.115451

149. Saitsu, H., Osaka, H., Sugiyama, S., Kurosawa, K., Mizuguchi, T., Nishiyama, K., Nishimura,A., Tsurusaki, Y., Doi, H., Miyake, N. et al. (2012). Early infantile epileptic encephalopathy associated with the disrupted gene encoding Slit-Robo Rho GTPase activating protein 2 (SRGAP2). Am. J. Med. Genet. A 158A, 199-205. doi:10.1002/ajmg.a.34363

150. Sasai, Y. (2013). Cytosystems dynamics in self-organization of tissue architecture. Nature493, 318-326. doi:10.1038/nature11859

151. Schmidt, D., Wilson, M. D., Ballester, B., Schwalie, P. C., Brown, G. D., Marshall, A., Kutter, C.,Watt, S., Martinez-Jimenez, C. P., Mackay, S. et al. (2010). Five-vertebrate ChIP-seq reveals the evolutionary dynamics of transcription factor binding. Science 328, 1036-1040. doi:10.1126/science.1186176

152. Schreiweis, C., Bornschein, U., Burguiere, E., Kerimoglu, C., Schreiter, S., Dannemann, M.,Goyal, S., Rea, E., French, C. A., Puliyadi, R. et al. (2014). Humanized Foxp2 accelerates learning by enhancing transitions from declarative to procedural performance. Proc. Natl. Acad. Sci. USA 111, 14253-14258. doi:10.1073/pnas.1414542111

153. Schwartz, S., Zhang, Z., Frazer, K. A., Smit, A., Riemer, C., Bouck, J., Gibbs, R., Hardison, R.and Miller, W. (2000). PipMaker--a web server for aligning two genomic DNA sequences.Genome Res. 10, 577-586. doi:10.1101/gr.10.4.577

154. Sholtis, S. J. and Noonan, J. P. (2010). Gene regulation and the origins of human biological uniqueness. Trends Genet. 26, 110-118. doi:10.1016/j.tig.2009.12.009

155. Siepel, A., Bejerano, G., Pedersen, J. S., Hinrichs, A. S., Hou, M., Rosenbloom, K., Clawson,H., Spieth, J., Hillier, L. W., Richards, S. et al. (2005). Evolutionarily conserved elements in vertebrate, insect, worm, and yeast genomes. Genome Res. 15, 1034-1050.doi:10.1101/gr.3715005

156. Sikela, J. M. (2006). The jewels of our genome: the search for the genomic changes underlying the evolutionarily unique capacities of the human brain. PLoS Genet. 2, e80.doi:10.1371/journal.pgen.0020080

157. Simonson, T. S., Yang, Y., Huff, C. D., Yun, H., Qin, G., Witherspoon, D. J., Bai, Z., Lorenzo, F. R., Xing, J., Jorde, L. B. et al. (2010). Genetic evidence for high-altitude adaptation in Tibet. Science 329, 72-75. doi:10.1126/science.1189406

158. Somel, M., Liu, X., Tang, L., Yan, Z., Hu, H., Guo, S., Jiang, X., Zhang, X., Xu, G., Xie, G. et al.(2011). MicroRNA-driven developmental remodeling in the brain distinguishes humans from other primates. PLoS Biol. 9, e1001214. doi:10.1371/journal.pbio.1001214

159. Sterelny, K. (2011). From hominins to humans: how sapiens became behaviourally modern.Philos. Trans. R. Soc. Lond. B Biol. Sci. 366, 809-822. doi:10.1098/rstb.2010.0301

160. Sturm, R. A. and Duffy, D. L. (2012). Human pigmentation genes under environmental selection. Genome Biol. 13, 248. doi:10.1186/gb-2012-13-9-248

161. Sudmant, P. H., Kitzman, J. O., Antonacci, F., Alkan, C., Malig, M., Tsalenko,A., Sampas, N.,Bruhn, L., Shendure, J., 1000 Genomes Project et al. (2010). Diversity of human copy number variation and multicopy genes. Science 330, 641-646.doi:10.1126/science.1197005

162. Sumiyama, K. and Saitou, N. (2011). Loss-of-function mutation in a repressor module of human-specifically activated enhancer HACNS1. Mol. Biol. Evol. 28, 3005-3007.doi:10.1093/molbev/msr231

163. Takahashi, K., Tanabe, K., Ohnuki, M., Narita, M., Ichisaka, T., Tomoda, K. and Yamanaka, S.(2007). Induction of pluripotent stem cells from adult human fibroblasts by defined factors. Cell 131, 861-872. doi:10.1016/j.cell.2007.11.019

164. Tasic, B., Hippenmeyer, S., Wang, C., Gamboa, M., Zong, H., Chen-Tsai, Y. and Luo, L. (2011).Site-specific integrase-mediated transgenesis in mice via pronuclear injection. Proc. Natl. Acad. Sci. USA 108, 7902-7907. doi:10.1073/pnas.1019507108

165. The Chimpanzee Sequencing and Analysis Consortium (2005). Initial sequence of the chimpanzee genome and comparison with the human genome. Nature 437, 69-87. doi:10.1038/nature04072

166. The ENCODE Project Consortium (2012). An integrated encyclopedia of DNA elements in the human genome. Nature 489, 57-74. doi:10.1038/nature11247

167. Abecasis, G. R., Altshuler, D., Auton, A., Brooks, L. D., Durbin, R. M., Gibbs, R. A., Hurles, M. E. and McVean, G. A.The 1000 Genomes Project Consortium, Abecasis, G. R., Altshuler,D., Auton, A., Brooks, L. D., Durbin, R. M., Gibbs, R. A., Hurles, M. E. and McVean, G. A.(2010). A map of human genome variation from population-scale sequencing. Nature467, 1061-1073. doi:10.1038/nature09534

168. Trynka, G., Sandor, C., Han, B., Xu, H., Stranger, B. E., Liu, X. S. and Raychaudhuri, S. (2013).Chromatin marks identify critical cell types for fine mapping complex trait variants. Nat. Genet. 45, 124-130. doi:10.1038/ng.2504

169. Usher, C. L., Handsaker, R. E., Esko, T., Tuke, M. A., Weedon, M. N., Hastie, A. R., Cao, H.,Moon, J. E., Kashin, S., Fuchsberger, C. et al. (2015). Structural forms of the human amylase locus and their relationship to SNPs, haplotypes, and obesity. Nat. Genet. 47,921-925. doi:10.1038/ng.3340

170. Varki, A. (2010). Colloquium paper: uniquely human evolution of sialic acid genetics and biology. Proc. Natl.

Acad. Sci. USA 107 Suppl. 2, 8939-8946.doi:10.1073/pnas.0914634107

171. Varki, A. and Altheide, T. K. (2005). Comparing the human and chimpanzee genomes: searching for needles in a haystack. Genome Res. 15, 1746-1758. doi:10.1101/gr.3737405

172. Varki, N., Anderson, D., Herndon, J. G., Pham, T., Gregg, C. J., Cheriyan, M., Murphy, J.,Strobert, E., Fritz, J., Else, J. G. et al. (2009). Heart disease is common in humans and chimpanzees, but is caused by different pathological processes. Evol. Appl. 2, 101-112.doi:10.1111/j.1752-4571.2008.00064.x

173. Vierstra, J., Rynes, E., Sandstrom, R., Zhang, M., Canfield, T., Hansen, R. S., Stehling-Sun, S.,Sabo, P. J., Byron, R., Humbert, R. et al. (2014a). Mouse regulatory DNA landscapes reveal global principles of cis-regulatory evolution. Science 346, 1007-1012. doi:10.1126/science.1246426

174. Vierstra, J., Wang, H., John, S., Sandstrom, R. and Stamatoyannopoulos, J. A. (2014b).Coupling transcription factor occupancy to nucleosome architecture with DNase-FLASH.Nat. Methods 11, 66-72. doi:10.1038/nmeth.2713

175. Visel, A., Blow, M. J., Li, Z., Zhang, T., Akiyama, J. A., Holt, A., Plajzer-Frick, I., Shoukry, M.,Wright, C., Chen, F. et al. (2009). ChIP-seq accurately predicts tissue-specific activity of enhancers. Nature 457, 854-858. doi:10.1038/nature07730

176. Visel, A., Minovitsky, S., Dubchak, I. and Pennacchio, L. A. (2007). VISTA Enhancer Browser--a database of tissue-specific human enhancers. Nucleic Acids Res. 35, D88-D92.doi:10.1093/nar/gkl822

177. Voight, B. F., Kudaravalli, S., Wen, X. and Pritchard, J. K. (2006). A map of recent positive selection in the human genome. PLoS Biol. 4, e72. doi:10.1371/journal.pbio.0040072

178. Williamson, I., Berlivet, S., Eskeland, R., Boyle, S., Illingworth, R. S., Paquette, D., Dostie, J.and Bickmore, W. A. (2014). Spatial genome organization: contrasting views from chromosome conformation capture and fluorescence in situ hybridization. Genes Dev.28, 2778-2791. doi:10.1101/gad.251694.114

179. Wilson, G. M., Flibotte, S., Missirlis, P. I., Marra, M. A., Jones, S., Thornton, K., Clark, A. G. andHolt, R. A. (2006). Identification by full-coverage array CGH of human DNA copy number increases relative to chimpanzee and gorilla. Genome Res. 16, 173-181. doi:10.1101/gr.4456006

180. Wong, L.-P., Ong, R. T.-H., Poh, W.-T., Liu, X., Chen, P., Li, R., Lam, K. K.-Y., Pillai, N. E., Sim,K.-S., Xu, H. et al. (2013). Deep whole-genome sequencing of 100 southeast Asian Malays. Am. J. Hum. Genet. 92, 52-66. doi:10.1016/j.ajhg.2012.12.005

181. Wunderlich, S., Kircher, M., Vieth, B., Haase, A., Merkert, S., Beier, J., Göhring, G., Glage, S.,Schambach, A., Curnow, E. C. et al. (2014). Primate iPS cells as tools for evolutionary analyses. Stem Cell Res. 12, 622-629. doi:10.1016/j.scr.2014.02.001

182. Xi, H., Shulha, H. P., Lin, J. M., Vales, T. R., Fu, Y., Bodine, D. M., McKay, R. D., Chenoweth, J. G., Tesar, P. J., Furey, T. S. et al. (2007). Identification and characterization of cell type-specific and ubiquitous chromatin regulatory structures in the human genome. PLoS Genet. 3, e136. doi:10.1371/journal.pgen.0030136

183. Yamanaka, S. (2012). Induced pluripotent stem cells: past, present, and future. Cell Stem Cell 10, 678-684. doi:10.1016/j.stem.2012.05.005

184. Yasuoka, Y., Suzuki, Y., Takahashi, S., Someya, H., Sudou, N., Haramoto, Y., Cho, K. W.,Asashima, M., Sugano, S. and Taira, M. (2014). Occupancy of tissue- specific cis-regulatory modules by Otx2 and TLE/Groucho for embryonic head specification. Nat. Commun. 5, 4322. doi:10.1038/ncomms5322

185. Yi, X., Liang, Y., Huerta-Sanchez, E., Jin, X., Cuo, Z. X. P., Pool, J. E., Xu, X., Jiang, H.,Vinckenbosch, N., Korneliussen, T. S. et al. (2010). Sequencing of 50 human exomes reveals adaptation to high altitude. Science 329, 75-78. doi:10.1126/science.1190371

186. Yu, C., Liu, K., Tang, S. and Ding, S. (2014). Chemical approaches to cell reprogramming.Curr. Opin. Genet. Dev. 28, 50-56. doi:10.1016/j.gde.2014.09.006

187. Yunis, J. J. and Prakash, O. (1982). The origin of man: a chromosomal pictorial legacy.Science 215, 1525-1530. doi:10.1126/science.7063861

188. Yuste, R. (2011). Dendritic spines and distributed circuits. Neuron 71, 772-781.doi:10.1016/j.neuron.2011.07.024

189. Zentner, G. E. and Henikoff, S. (2014). High-resolution digital profiling of the epigenome.Nat. Rev. Genet. 15, 814-827. doi:10.1038/nrg3798

190. Zheng, Q., Cai, X., Tan, M. H., Schaffert, S., Arnold, C. P., Gong, X., Chen, C. Z. and Huang, S.(2014). Precise gene deletion and replacement using the CRISPR/Cas9 system in human cells. BioTechniques 57, 115-124. doi:10.2144/000114196

191. Zhou, X., Cain, C. E., Myrthil, M., Lewellen, N., Michelini, K., Davenport, E. R., Stephens, M.,Pritchard, J. K. and Gilad, Y. (2014). Epigenetic modifications are associated with inter-species gene expression variation in primates. Genome Biol. 15, 547.doi:10.1186/s13059-014-0547-3

192. Zinzen, R. P., Girardot, C., Gagneur, J., Braun, M. and Furlong, E. E. (2009). Combinatorial binding predicts spatio-temporal cis-regulatory activity. Nature 462, 65-70.doi:10.1038/nature08531

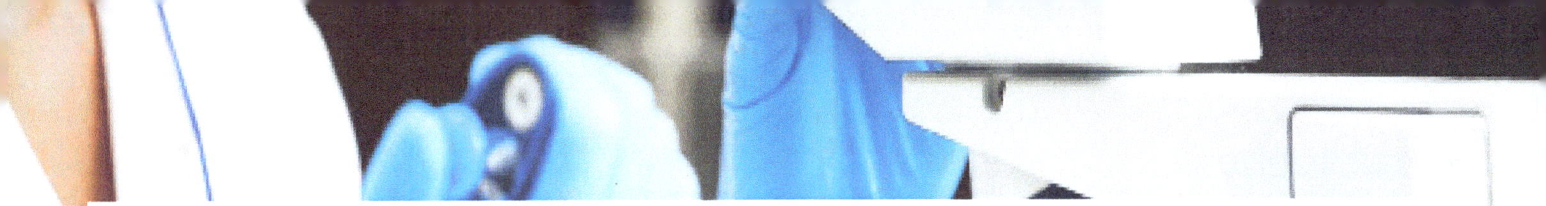

CHAPTER 2

SYSTEMATIC IDENTIFICATION OF GENETIC INFLUENCES ON METHYLATION ACROSS THE HUMAN LIFE COURSE

Tom R. Gaunt[1], Hashem A. Shihab[1], Gibran Hemani[1], Josine L. Min[1], Geoff Woodward[1], Oliver Lyttleton[2], Jie Zheng[1], Aparna Duggirala[2], Wendy L. McArdle[2], Karen Ho[2], Susan M. Ring[1,2], David M. Evans[1,3], George Davey Smith[1], and Caroline L. Relton[1,4]

[1]MRC Integrative Epidemiology Unit (IEU) & School of Social and Community Medicine, University of Bristol, Oakfield House, Oakfield Grove, Bristol BS8 2BN, UK.

[2]Avon Longitudinal Study of Parents and Children (ALSPAC) & School of Social and Community Medicine, University of Bristol, Oakfield House, Oakfield Grove, Bristol BS8 2BN, UK.

[3]University of Queensland Diamantina Institute, Translational Research Institute, Brisbane 4102, Australia.

[4]Institute of Genetic Medicine, Newcastle University, Newcastle upon Tyne NE1 7RU, UK.

ABSTRACT

Background

The influence of genetic variation on complex diseases is potentially mediated through a range of highly dynamic epigenetic processes exhibiting temporal variation during development and later life. Here we present a catalogue of the genetic influences on DNA methylation (methylation quantitative trait loci (mQTL)) at five different life stages in human blood: children at birth, childhood, adolescence and their mothers during pregnancy and middle age.

Results

We show that genetic effects on methylation are highly stable across the life course and that developmental change in the genetic contribution to variation in methylation occurs primarily through increases in environmental or stochastic effects. Though we map a large proportion of the cis-acting genetic variation, a much larger component of genetic effects influencing methylation are acting in trans. However, only 7 % of discovered mQTL are trans-effects, suggesting that the trans component is highly polygenic. Finally, we estimate the contribution of mQTL to variation in complex traits and infer that methylation may have a causal role consistent with an infinitesimal model in which many methylation sites each have a small influence, amounting to a large overall contribution.

Conclusions

DNA methylation contains a significant heritable component that remains consistent across the lifespan. Our results suggest that the genetic component of methylation may have a causal role in complex traits. The database of mQTL presented here provide a rich resource for those interested in investigating the role of methylation in disease.

Citation: Tom R. Gaunt, Hashem A. Shihab, Gibran Hemani, Josine L. Min, Geoff Woodward, Oliver Lyttleton, Jie Zheng, Aparna Duggirala, Wendy L. McArdle, Karen Ho, Susan M. Ring, David M. Evans, George Davey Smith, and Caroline L. Relton. Systematic identification of genetic influences on methylation across the human life course, Genome Biology (2016) 17:61 DOI 10.1186/s13059-016-0926-z

BACKGROUND

Epigenetic mechanisms play a central role in the regulation of cellular processes by influencing genomic activity [1]. DNA methylation, defined as the covalent bonding of a methyl group to a cytosine in the context of a CpG dinucleotide, is an important component of these mechanisms in mammals. Canonically, DNA methylation typically represses transcription, which can occur by inhibiting the binding of transcription factors or by recruiting DNA binding proteins that remodel chromatin structure. Consequently, the establishment and maintenance of DNA methylation patterns are crucial for normal cellular function and developmental processes and, indeed, these patterns are highly heterogeneous at different life stages [2] and between different tissue types [3].

Genome-wide DNA methylation can be considered to be a large set of measurable traits (one per CpG site) in which variation can arise from environmental [4], stochastic [5] or genetic [6] perturbations, and there is growing evidence that DNA methylation could mediate the relationship between these processes in influencing complex diseases [7]. An important step in understanding the processes underpinning DNA methylation is mapping the genetic factors that influence its variation. A recent study estimated that almost 20 % of reliably assayed variation in blood DNA methylation is heritable and that 50 % of CpG sites showed evidence of a significant genetic component [8]. A different perspective is provided by a study looking at the most variably methylated regions between neonates, which found 25 % were best explained by genotype alone and 75 % by a combination of genotype and environment [9]. Another large study showed that DNA methylation variation in adipose tissue was highly heritable ($h^2_{median}=0.34$) and that shared environmental effects correlated with metabolic phenotype-associated CpGs [10]. Heritability of CpG methylation levels in whole blood in young people has also been shown to correlate highly with stability of methylation at the same sites in later life [11]. However, those studies that have attempted to map genetic effects influencing DNA methylation (methylation quantitative trait loci (mQTL)) have so far only explained a small proportion of the genetic variance that is estimated to exist and it is evident that much larger sample sizes will be required to map the majority of the predicted genetic effects.

One can address the question of how genetic variation influences DNA methylation in a number of ways and here we focus on three specific research avenues. First, the genetic architecture of methylation variation can provide information about the level of complexity that underlies population-level differences. This can be examined by (a) estimating the proportion of explanatory common genetic variation (mQTL) that occurs close to the methylation site (in *cis*) versus the proportion that occurs elsewhere in the genome (in *trans*), and (b) the level of polygenicity of the genetic component for *cis* and *trans* regions.

Second, in characterising and mapping mQTL it is of interest to know the extent to which genetic effects are stable over time. Because epigenetic change is a cornerstone of mammalian development, elucidating whether genetic effects have a consistent influence across the life course or are specific to certain developmental windows is important for gauging the extent to which mQTL could be involved in epigenetic restructuring and perturbation of developmental trajectories.

Third, a comprehensive catalogue of mQTL can be used to investigate (a) whether those regions of methylation that are influenced by genetic variation are likely to be inert or are involved in cellular function and (b) if these elements are functional, then to what extent do mQTL influence complex disease as a consequence of their influence on DNA methylation.

Here we present a comprehensive genome-wide *cis* and *trans* mQTL longitudinal analysis in blood DNA at three time points in the life course of a large number of participants in the Avon Longitudinal Study of Parents and Children (ALSPAC) [12] and two time points in the life course of their mothers [13], in the form of an online searchable database (http://www.mqtldb.org/). We assess the stability of mQTL across the life course and identify the biological pathways in which they function. We evaluate the relationship of mQTL with other downstream phenotypes, including gene expression, traits and diseases, and quantify the contribution made by mQTL to genetic variance in several common complex diseases that have previously been the subject of genome-wide association studies (GWAS).

RESULTS

Cis and *Trans* mQTL Mapping

The ARIES dataset [14] represents DNA methylation levels collected at five different time points across the life course from individuals in ALSPAC: in young people we collected samples at birth (cord blood, n=771), childhood (n=834) and adolescence (n=837); in their mothers we sampled during pregnancy (n=764) and in middle age (n=742). We performed an exhaustive whole-genome mQTL analysis by testing approximately 8.3 million common single-nucleotide polymorphisms (SNPs) against each reliable CpG probe (395,625 out of 485,577) in each time point. After conservative multiple testing correction ($p<1\times10^{-14}$) we identified between 24,262 and 31,729 sentinel associations at each time point. Approximately 93 % of the mQTL were acting in *cis* (defined as within ±1 Mb of the CpG probe on the basis of previous a previous report [15] and our own observation of the distribution of SNP/CpG distances, although definitions of *cis* in the literature vary widely from a few hundred base pairs [16] to 1 Mb [17, 18]).

Table 1. Number of mQTL and associated CpGs reaching the significance threshold for each time point

Counts	Birth	Childhood	Adolescence	Pregnancy	Middle age
Sentinel mQTL					
Cis [a]	24,262	31,729	30,294	29,038	27,043
Trans [b]	1979	2658	2442	2394	2144
Conditionally independent[c]	2705	5446	5040	4454	3463
Total mQTL	28,946	39,833	37,776	35,886	32,650
Total unique CpGs	27,387	36,705	34,886	33,344	30,676

[a]Number of CpG sites with a *cis* SNP

[b]Number of independent (±1 Mb) *trans* effects

[c]Number of mQTL further detected after performing conditional analysis

We also performed conditional analysis which identified between 2705 and 5446 further mQTL at each time point that showed secondary, tertiary and quaternary effects also acting in *cis*, giving 28,946–39,833 mQTL discovered at each time point influencing a total of 43,897 CpG sites across the genome (Table 1, Fig. 1a). The effect sizes as difference in median proportion methylated between homozygote groups.

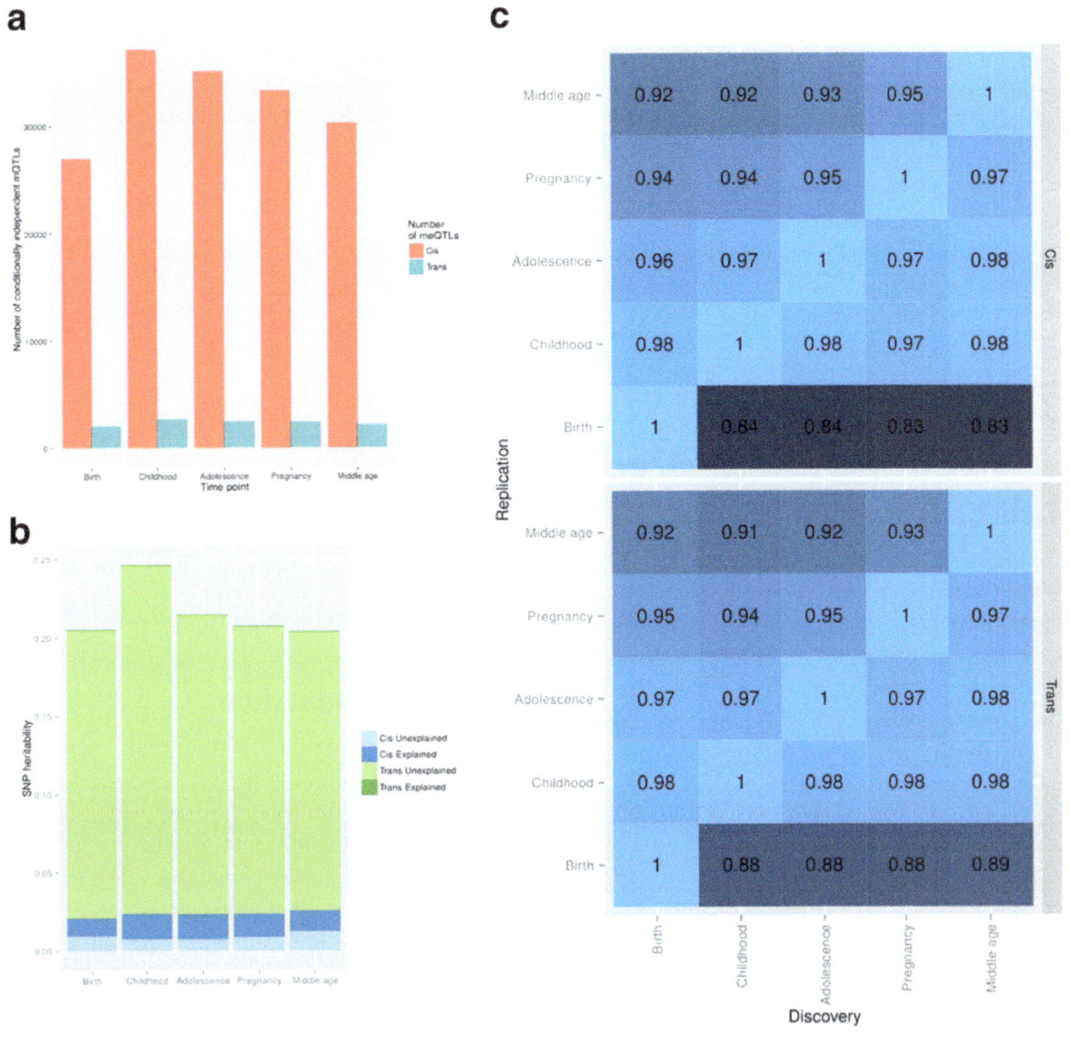

Figure. 1. Temporal pattern of mQTL. a The total number of *cis* and *trans* mQTL discovered at each time point. b Total bars represent the SNP heritability at each time point. Each bar is split into genetic variation due to common SNPs acting in *cis* (*blue*) and *trans* (*green*). *Cis* and *trans* variation is further divided into the proportion that is explained by mapped SNPs ($p<1\times10^{-14}$). c The proportion of discovered mQTL at a specific time point that replicate at $p<1\times10^{-7}$ in each of the other time points. *Darker colours* correspond to lower replication rates.

Genetic Architecture of Methylation Variation

DNA methylation may be influenced by both genetic and environmental factors. To address the question of the relative contribution of genetic variation we used genomic restricted maximum likelihood (GREML) [19]. Here we estimated how much of the total variation in each methylation probe was captured by all 1.1 million common HapMap3 [20] tagging SNPs (minor allele frequency (MAF) > 0.01) to estimate what is known as the SNP heritability. Although the standard error is high for any one probe, when performed on all reliable probes at five time points this analysis enabled us to estimate the distribution of genetic contribution to methylation variation (SNP heritability) and how it varies over time. In addition, for each probe we partitioned the genetic variance into two components, the first using *cis* SNPs only and the second using *trans* SNPs only.

The results demonstrate that although the majority of mQTL act in *cis*, the majority of estimated genetic variation that influences methylation levels is acting in *trans* (Fig. 1b). This applies even if we extend the definition of *cis* to include the entire chromosome (data not shown).

Variance Explained by Detected mQTL

Having partitioned the phenotypic variance of each CpG probe into *cis*-acting genetic variance, *trans*-acting genetic variance and environmental variance (which includes any genetic variance not captured by 1.1 million HapMap3 SNPs), we then went on to estimate how much of the genetic variation was 'explained' by the mQTL that we had detected in our association analyses. We estimated that the discovery mQTL detected in our association analysis explained over half of the proportion of total *cis* variation, whereas the *trans* mQTL explained less than 1 % of total *trans* variation. This result is consistent with the hypothesis that genetic perturbation close to the CpG site tends to have large effects whereas the complex network of regulatory interactions and a large mutational target size makes the *trans* component, on average, highly polygenic.

Stability of Genetic Effects Over Time

DNA methylation changes in response to environmental exposures. We addressed the question of how stable the mQTL effects were over time by estimating the rate of replication of discovery mQTL from one time point in all other time points. Fig. 1c show that the proportion that replicated at a threshold of $p < 1 \times 10^{-7}$ was typically in excess of 95 %, although the rate of replication of post-birth discovery mQTL in the birth time point was consistently lower (84–86 %).

The distributions of estimated SNP heritabilities illustrate that average SNP heritability gradually falls from 0.24 in childhood to 0.21 at middle age (Fig. 1a, b). A regression of SNP heritability on age indicated a reduction of heritability of −0.0009 per year from childhood to adulthood (–0.0009, se=1.6e–5). There are two simple explanations for this observation: first that the influence of genetic variation is reducing over time; or second, that the influence of environmental or stochastic perturbations is increasing over time. In the former case we would expect that the average coefficient of variation of methylation to decrease over time due to there being fewer genetic factors, whereas in the latter case we would expect it to increase due to more environmental factors or higher stochasticity. The latter explanation is supported by a clear increase (3.0 %, standard error 0.015 %) in the average coefficient of variation of methylation from childhood to middle age (data not shown).

In addition, while replication of mQTL from middle age to childhood is high, replication of mQTL from childhood is lower in later time points (Fig. 1c). Together this suggests that genetic effects are largely stable and that environmental or stochastic perturbation is gradually increasing over the life course, leading to lower SNP heritability estimates and lower power to detect mQTL as age increases. Two caveats to these observations are that the later two time points are different individuals from the earlier ones and they are comprised exclusively of women.

The stability of genetic effects is surprisingly high given the observational correlations of DNA methylation between different time points. However, we observe that the mean correlation of methylation probes that have at least one significant mQTL is substantially higher than the average value.

Long Range Influences of Methylation Levels

In contrast to what has been seen in expression quantitative trait loci (eQTL) studies, there is little evidence of individual mQTL influencing many CpG sites across the genome, with the vast majority of *trans* mQTL in our data just representing a single or small number of associations . To gauge the extent to which methylation levels were influenced by CpG sites elsewhere in the genome, we performed a mediation analysis testing for mediation of *trans* mQTL effects by *cis* methylation. Mediation analyses are particularly susceptible to measurement error in the mediator (which will attenuate estimates of mediation). However, we provide some evidence that, amongst mQTL with both *cis* and *trans* effects, a proportion demonstrate some degree of mediation of the *trans* association by *cis* CpG sites. Whilst a large number of sites follow the y=x diagonal (providing little evidence of mediation), a proportion deviate from this line, showing that if the effects of *cis* methylation are taken into account, the*trans* association is attenuated. This non-independence between *cis* and *trans* effects at these loci does not prove a causal effect of *cis* methylation on *trans* methylation.

Functional Annotation

If mQTL are functionally important, they are likely to be distributed differentially across genomic features. To address this question, we analysed both mQTL and mQTL-associated CpG sites to determine their distribution across genomic regions. This illustrates consistency in distributions between time points, but with notable differences between locations (consistent with observations by Shi et al. [21]).

The distribution of *cis* and *trans* mQTL across genic features at each time point (adjusted for overall representation within each feature) is presented in Fig. 2a–c. More than 65 % of *trans* associations were trans-chromosomal. Higher densities of *trans*-associated SNPs appear to occur on chromosomes 16, 17 and 19, illustrated in the circos bar charts in Fig. 2d showing density of associated CpGs and SNPs, but these do correspond closely with regions of higher gene density. The SNP heritability estimated by common *trans* SNPs was, on average, consistent for methylation levels at different genomic features, but there was a substantial relative increase of *cis* genetic influence on methylation levels at regulatory regions compared with coding regions (Fig. 2e).

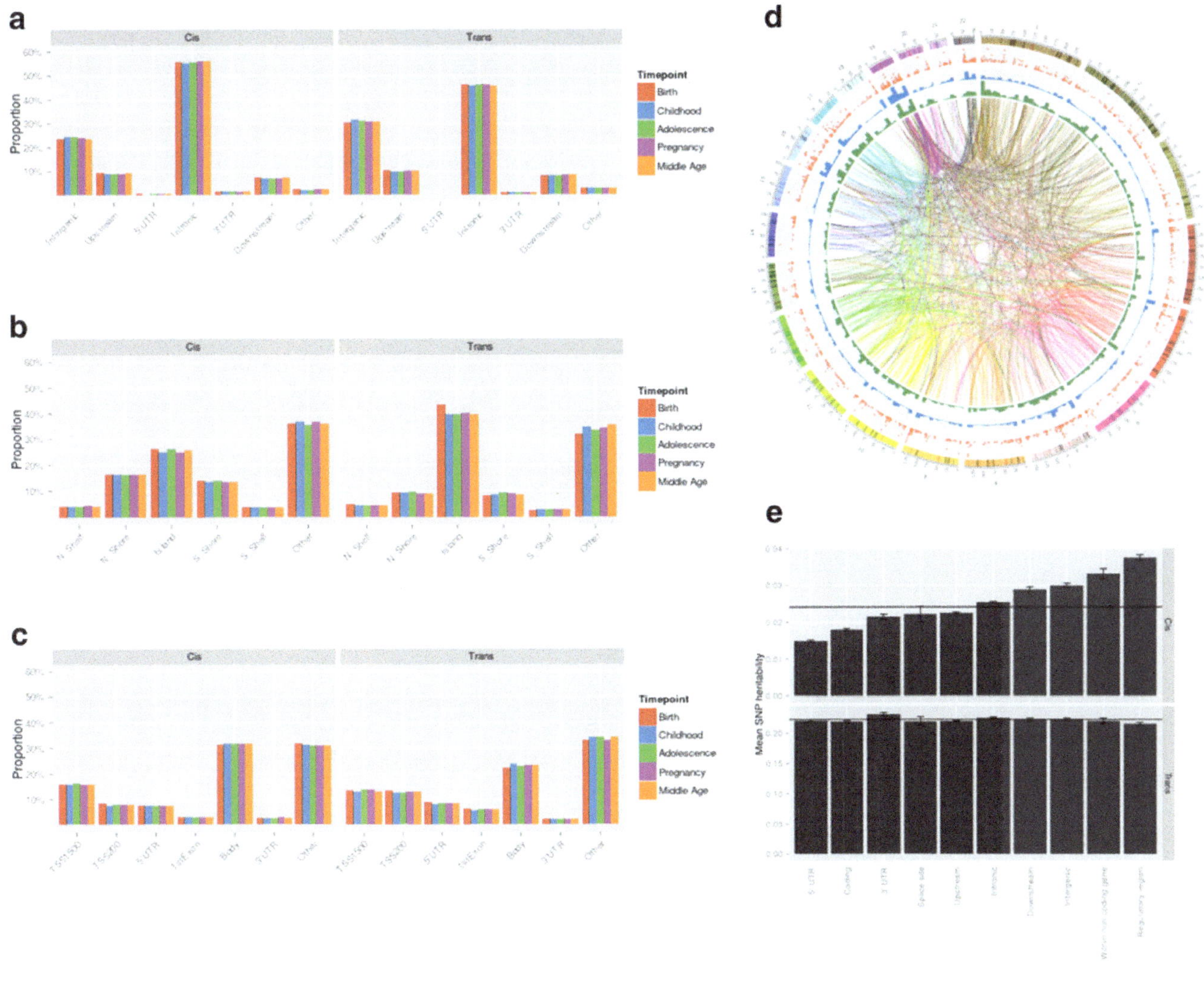

Figure 2. Genomic distribution of mQTL. a Distribution of mQTL across genomic features; b distribution of mQTL-associated CpG sites across CpG islands; c distribution of mQTL-associated CpG sites across genic features. d Circos plot illustrating *trans* mQTL at birth. From the outside: chromosomes, $-\log_{10}(p$ value) for association (*red points*), density of mQTL (*blue bars*), density of associated CpGs (*green bars*), density of genes (*gray bars*), *trans* associations between SNP and CpG (*lines*). e Average estimated *cis*(*top*) and *trans* (*bottom*) SNP heritability for methylation levels at different genomic features. Bar heights show mean heritability for each genomic feature. Error bars show standard error of the mean heritability. Horizontal lines indicate the mean heritability across all features. *UTR* untranslated region.

Regional association plots for the top 25 *cis* associations at each time point are presented, demonstrating a range of different patterns of methylation depending on linkage disequilibrium and correlation of methylation across the surrounding region. There is little evidence of individual mQTL influencing many CpG sites across the genome, with the vast majority of *trans* mQTL in our data just representing a single or small number of associations. However, this is very likely to reflect a lack of power to detect effects in the current sample size.

Overlap between our mQTL and eQTL reported previously by the GTEx consortium [22] is illustrated. Amongst *cis* QTL more loci are both eQTL and mQTL than mQTL only. In a gene ontology comparison of our mQTL with the eQTL reported by Westra et al. [23], enrichment for genes from all mQTL results showed little evidence of informative enrichment, but we found that *cis* mQTL also reported as eQTL are enriched for tissue development terms, whilst *trans* mQTL reported as eQTL are enriched for terms related to plasma membrane and cell periphery.

Influence of mQTL on Disease

Though it is clear that the genetic component of methylation variation is substantial, the broad influence of variation in methylation on distal traits and disease outcomes is yet to be established. Observational associations between methylation and outcome are insufficient to ascertain causality. An alternative approach is to test if the *cis* mQTL that are likely to have direct effects on methylation levels also influence complex traits. To test this we partitioned the SNP heritability of seven complex traits in the Wellcome Trust Case Control Consortium (WTCCC) [24] data into a component for *cis* mQTL and a component for all HapMap3 [20] tagging SNPs. For comparison we compared our results against two different null hypotheses: is the variance explained by the mQTLs more than would be expected by chance (a) from the same number of SNPs sampled from genic regions and (b) if we sample the same number of SNPs to have the same distribution of genomic annotations as the mQTLs (Fig. 3a)? In both cases we match our null SNPs to our mQTLs on allele frequency and linkage disequilibrium (LD) structure also.

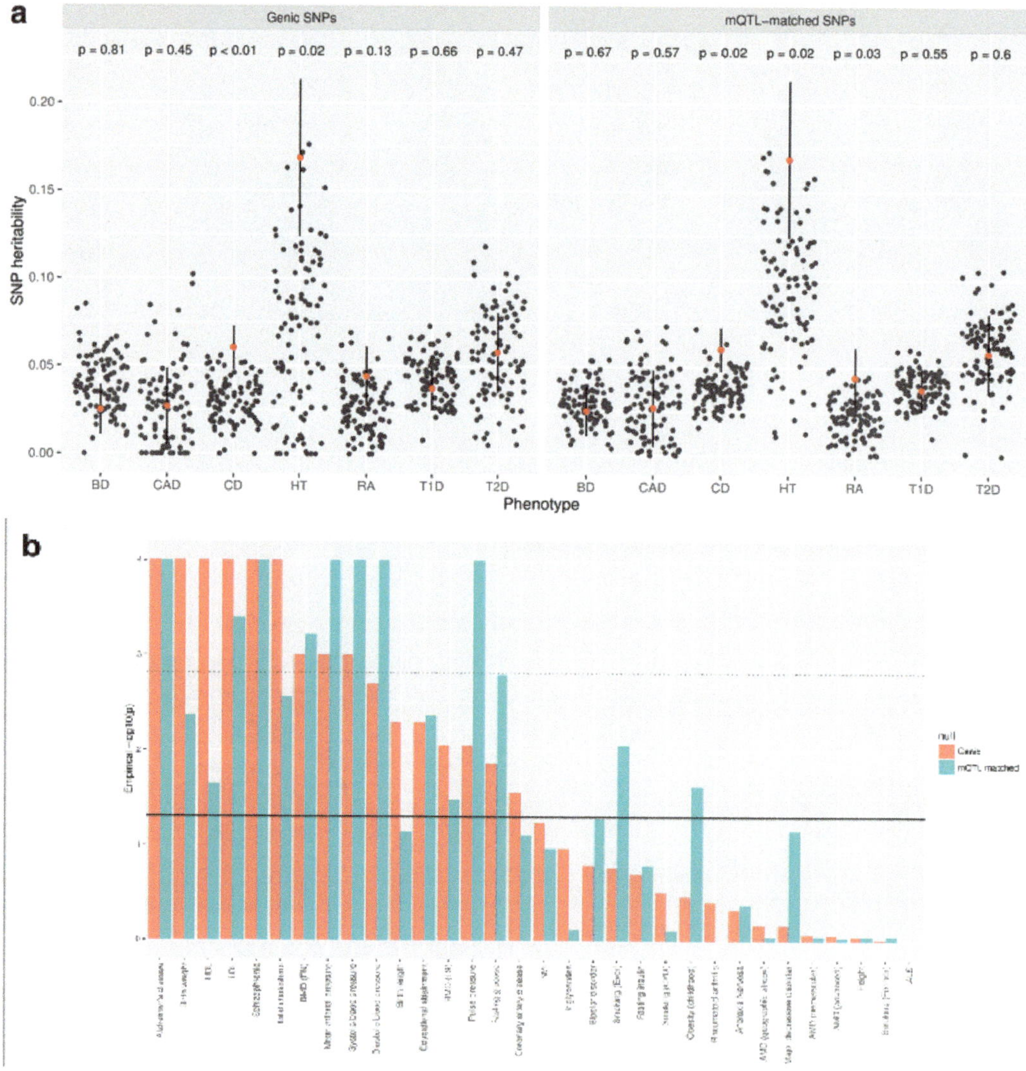

Figure 3. mQTL enrichment in diseases and traits. a Contribution of mQTL identified at each time point to variance of WTCCC common diseases bipolar disorder (*BD*), coronary artery disease (*CAD*), Crohns disease (*CD*), hypertension (*HT*), rheumatoid arthritis (*RA*), type 1 diabetes (*T1D*) and type 2 diabetes (*T2D*). *Red dots* represent the component of a trait's genetic variance attributable to *cis*-acting mQTL SNPs with significance levels of $p < 1 \times 10^{-14}$, on the liability scale, excluding chromosome 6. *Black points* depict the point estimates of SNP heritability estimates under the null hypotheses of SNPs coming from genic regions (*left plot*) or SNPs with the same proportion of genic features as the mQTLs (*right plot*). *P* values relate to the proportion of the null estimates that surpass the mQTL estimates. b Enrichment analysis of *cis*-acting mQTL SNPs with significance levels of $p < 1 \times 10^{-14}$ in large-scale GWAS summary statistics for 33 complex traits. The *solid horizontal line* denotes empirical *p* value of 0.05 and the *dotted line* shows the threshold after correcting for multiple testing. *Red bars* are based on a null of genic SNPs, *blue bars* on a null of mQTL-matched SNPs. HDL = high-density lipoprotein cholesterol, LDL = low-density lipoprotein cholesterol, BMD = bone mineral density, FN = femoral neck, LS = lumbar spine, BMI = body mass index, AMD = age-related macular degeneration and ALS = amyotrophic lateral sclerosis.

Our results suggest that, for Crohn's disease, hypertension and rheumatoid arthritis, the proportion of SNP heritability attributed to mQTL was greater than expected under the null hypothesis that their contribution was solely due to residing within a particular set of genomic annotations (null hypothesis 'b'). However, note that the evidence for these results is weak ($p > 0.05$) if multiple testing is taken into account using a Bonferroni correction. An additional limitation of this analysis is that the WTCCC controls (used for all seven diseases) were normal population samples and, therefore, may have included some cases. We next used our discovery mQTL to test for enrichment of low *p* values in publicly available GWAS results from large meta-analyses (Fig. 3b). We observed evidence that mQTL were more enriched for low *p* values in GWAS results than (a) randomly sampled genic SNPs and (b) mQTL annotation-matched SNPs for 9 and 8 out of 33 complex traits (Bonferroni corrected $p < 0.05/33$), respectively. Those complex traits showing enrichment for *cis* mQTL in GWAS results against both null hypotheses were Alzheimer's disease, schizophrenia, low-density lipoprotein cholesterol, bone mineral density and blood pressure. These results suggest that there may be a genetic influence via methylation on downstream complex traits and that it is likely that very many CpG sites contribute to complex traits, each contributing only a small effect.

DISCUSSION

We present a large-scale analysis of blood mQTL in children and their mothers at multiple time points through the life course. We found that, whilst methylation itself is not highly stable over time, the effects of mQTL are surprisingly consistent, both within the same individuals and across a generation. This stability is consistent with another study that found significant overlap in mQTL between developmental stages (in addition to between ancestral groups and tissue types) [9]. Average SNP heritability of DNA methylation exceeded 0.2 at all time points, with *trans* effects estimated to account for the majority of this heritability. This is consistent with previously reported total heritability on the Illumina Infinium HM450 array of 0.187 [8]. However, the proportion of *trans* heritability explained by our *trans* mQTL is much smaller than the proportion of *cis*heritability explained by our *cis* mQTL. This is most likely due to polygenic *trans* effects, with many *trans* mQTL explaining too little variance for us to detect in a sample of this size, whilst *cis* effects have relatively larger mono- or oligogenic effects. The effect sizes of our *trans* mQTL also tend to be smaller, although a minority still have effects exceeding a 10 % difference in proportion methylated. Overall our effect sizes exceeded 2 % difference in proportion methylated for more than 50 % of mQTL and greater than 5 % methylated for more than 25 % of mQTL. Whilst this means that some of our mQTL have a potentially small biological effect, these results will still be of value in Mendelian randomization analysis [25] and analysis of the genetic architecture of DNA methylation.

We observed two interesting patterns when we examined the similarities in genetic effects on methylation at different time points. Firstly, fewer mQTL were discovered at birth and estimates of heritability were lower at birth than in childhood. Secondly, heritability of methylation and the total number of mQTL decrease with increasing age (with the exception of birth). The potential explanations for these observations are that either genetic effects change over time or environmental effects change over time. An increase in overall variation in methylation over time suggests that the latter is more likely, with the environment having a relatively greater influence around birth and then later in life. This is consistent with the postulated period of methylation plasticity in utero and the epigenetic drift observed through the life course. The largest proportion of time point-specific effects were at birth (cord blood), with adolescence and middle age having the lowest proportion. This suggests that a small proportion of mQTL may play a more important role during early development, with their relative importance diminishing later in life.

Regional association patterns for *cis* mQTL varied widely between loci, with some showing distinct blocks of mQTL SNPs and associated CpGs (i.e. SNPs associating with many CpGs within the same LD block but very few outside). This apparent alignment of SNP LD blocks with methylation blocks may simply represent the short-range functional effect of SNPs on CpG methylation, with a number of SNPs in LD affecting a number of nearby CpG sites. Other loci show a less distinct pattern of association, with different CpGs associating with different, but overlapping, sets of SNPs.

We observed some evidence that mQTL are enriched in 5′ and 3′ untranslated regions compared with intronic and intergenic regions, whilst the mQTL-associated CpGs tend to be more frequent upstream of the transcription start site. With respect to CpG islands these mQTL-associated sites are enriched in north and south CpG island shores. Whilst we don't have other QTL data from the same samples for direct comparison, interestingly Banovich et al. [26] report notable overlaps of mQTL with QTL for other regulatory features such as histone modifications, DNase I hypersensitivity, chromatin access and expression levels. Although we don't have direct measurements of expression to perform eQTL analyses, we did perform a comparison of our mQTL with the GTEx eQTL data [22]. The approximately 5 % of GTEx eQTL that we report as mQTL is not inconsistent with the observation from a study in fibroblasts that found 12 % of genes with an eQTL have a nearby mQTL [27], although Banovich et al. [26] found that, amongst eQTL within 3 kb of a methylation probe, approximately 25 % were also mQTL.

In an additional analysis we found evidence that some *trans* SNP–CpG associations may be mediated by *cis*methylation. A recent mQTL study with a greater sample size than ours found a limited number of mQTL affecting multiple (up to 19) sites [17]. Although we observed a number of such cases, many *trans* effects do not show evidence of *cis* mediation. However, our limited power to detect such mediation mechanisms and the possibility that measurement error reduces the apparent extent of mediation should be taken into account when interpreting these findings. These observations are consistent with a previous report based on lung tissue [21] and underline the complexity of mQTL effects.

We found no evidence of SNPs affecting methylation at a large number of sites, with the majority of our *trans*mQTL representing just one SNP–CpG association. This contrasts with observations in eQTL analysis which suggest some SNPs impact on the regulation of a large number of transcripts as 'master regulators' [28]. The difference between mQTL and eQTL in this context aligns with our observation that there is only limited overlap between published blood eQTL [23] and our blood mQTL (discussed in more detail above). A previously reported study of mQTL and eQTL amongst loci associated with bipolar disorder also found orthogonality between the two types of QTL [29]. Grundberg et al. [10], profiling adipose tissue from 648 twins, revealed that 28 % of CpGs were associated with nearby SNPs and, when overlapping them with adipose eQTL from the same individuals, they found that 6 % of the loci played a role in regulating both gene expression and DNA methylation. In contrast, a systems analysis of ~418 K SNPs with ~23 K CpG sites and ~16 K expression probes using blood from 148 individuals found evidence using *cis* eQTL and mQTL to support a causal model in which genotype affects methylation, which in turn affects expression [30]. However, the same study did not investigate *trans* mQTL. There is, therefore, little evidence at this stage to suggest either that there are mQTL master regulators or that eQTL master regulators act on expression via methylation. Thus, whilst methylation is known to influence gene expression, we find limited overlap between mQTL and published eQTL and no evidence of 'master regulators' in mQTL, suggesting a more complex relationship than the simple SNP–methylation–expression model, consistent with the recently reported findings of the Epigenomics Roadmap Consortium [31].

Our results support a stable effect of genetic variation on methylation across the life course, with common SNPs accounting for approximately 20 % of variance in methylation (with the caveat that cord blood is quite different in cellular composition from peripheral blood). One important question is whether mQTL play a functional role in disease. One previous study has indicated potential roles for population-differential mQTL in age of menarche, hepatitis B infection and HIV control [18], whilst another has reported potential contributions of mQTL to autoimmune disease (although they didn't formally analyse enrichment) [18]. Our analyses suggest that mQTL are relevant to common diseases and traits, accounting for a significant proportion of the heritability of Crohn's disease, hypertension, rheumatoid arthritis and type 1 diabetes. Our observations of a potential role in autoimmune disease seem consistent with those of Lemire et al. [17], although it is crucial to consider the potential that autoimmune genotype may drive measured DNA methylation through cellular heterogeneity. The enrichment of mQTL in autoimmune diseases is unlikely to be driven by large effects and long-range LD in the major histocompatibility complex (MHC) region because reanalysis excluding chromosome 6 returned the same conclusion. Whilst this observation does not prove causality, it suggests two important phenomena. First, it provides a plausible role of methylation in mediating some of the effects of common genetic variation on disease and is consistent with one of the first mQTL analyses to use the Illumina Infinium HM450 array, which demonstrated enrichment of mQTL among disease-associated loci [21]. Second, it suggests that if methylation influences complex traits, then this is likely through variation at many genomic loci, with each locus having a small effect.

Limitations

Methylation data were only available in peripheral blood from ALSPAC participants, so we were unable to compare results between tissues (interestingly, mQTL have been reported to be enriched for CpG sites that are variable between tissues [32]). In addition, cord blood is not directly equivalent to peripheral blood. Cellular heterogeneity can account for differences in methylation between samples, affecting different CpG sites to different extents; we had no directly measured cell count data available but applied an established method [33] to adjust for the potential confounding influence of these effects. However, this method was optimised on adult samples (not children). It may also not be an appropriate approach for use in cord blood, which may explain some of the differences we observe between birth and other time points.

DNA was extracted from different blood sample types, which could potentially affect methylation measurements, although in seeking to test this possibility we illustrate that sample type does not appear to stratify the top principal components of DNA methylation.

The adults in our sample are all female, so one may wish to consider the generalizability of the reported findings to adult males when interpreting data.

Some probes on the Illumina Infinium HM450 array are known to be affected by a range of factors, including underlying SNPs [34]; we filtered out the most severely affected but recommend that this is taken into account when interpreting and subsequently utilising the data we have generated.

Whilst the number of mQTL we report in Table 1 differs from those in other publications (varying from 49 [35] to 52,708 [17] or more *cis* mQTL), it is important to note that these numbers rely on arbitrary thresholds, are susceptible to power and heavily influenced by definitions of *cis* and *trans*.

We note that even though there was generally good agreement in overall results between our two null models (null SNPs sampled from genic regions and null SNPs sampled from the same distribution of genic annotations as the true mQTLs), the enrichment p values were quite sensitive to the null model used. We recommend caution in performing and interpreting such enrichment analyses.

CONCLUSIONS

We demonstrate that genetic effects on methylation levels are highly stable over time and that methylation variation increases over time, most likely due to increased environmental or stochastic influences. We also provide evidence that methylation may play a mediating role in the influence of genetic variation on complex traits. We are releasing the results from our mQTL analysis in the form of an open database (http://www.mqtldb.org/) to encourage consistent publication and integration of results from other studies.

METHODS

Study Sample

Samples were drawn from the Avon Longitudinal Study of Parents and Children [12, 13]. Blood from 1018 mother–child pairs (children at three time points and their mothers at two time points) were selected for analysis as part of the Accessible Resource for Integrative Epigenomic Studies (ARIES,http://www.ariesepigenomics.org.uk/) [14]. Numbers of samples surviving the imposed quality control (QC) thresholds at each time point. Sample selection was on the basis of sample availability at all of the chosen time points in mother–child pairs. Cord blood and peripheral blood samples (whole blood, buffy coats, white blood cells or blood spots) were collected according to standard procedures.

Methylation Assays

Following DNA extraction, samples were bisulphite converted using the Zymo EZ DNA Methylation™ kit (Zymo, Irvine, CA, USA). Following conversion, genome-wide methylation was measured using the Illumina Infinium HumanMethylation450 (HM450) BeadChip. The arrays were scanned using an Illumina iScan, with initial quality review using GenomeStudio. During the data generation process a wide range of batch variables were recorded in a purpose-built laboratory information management system (LIMS). The LIMS also reported QC metrics from the standard control probes on the HM450 BeadChip for each sample. Samples failing QC were excluded from further analysis and the assay repeated. Data points with a low signal:noise ratio (detection $p>0.01$) or with methylated or unmethylated read counts of 0 were also excluded from analysis. As an additional QC step genotype probes on the Ilumina Infinium HM450 array were compared between samples from the same individual at different time points and against SNP-chip data (HM450 probes clustered using k-means) to identify and remove any sample mismatches. Methylation data were normalised in R with the wateRmelon package [36] using the Touleimat and Tost [37] algorithm to reduce the non-biological differences between probes. Data were then rank-normalised to remove outliers and regressed on all covariates plus bisulphite-converted DNA (BCD) plate batch to remove potential batch effects (with missing values set to probe mean).

Genotyping Assays

The ARIES participants were previously genotyped as part of the larger ALSPAC study, with QC, cleaning and imputation performed at the cohort level before extraction of the subset comprising ARIES. ARIES participants are of European white ancestry and homogeneous compared with HapMap reference populations.

Children were genotyped using the Illumina HumanHap550 quad genome-wide SNP genotyping platform (Illumina Inc., San Diego, CA, USA) by the Wellcome Trust Sanger Institute (WTSI; Cambridge, UK) and the Laboratory Corporation of America (LCA, Burlington, NC, USA). Individuals were excluded on the basis of incorrect gender assignment, abnormal heterozygosity (<0.320 or >0.345 for WTSI data; <0.310 or >0.330 for LCA data), high missingness (>3 %), cryptic relatedness (>10 % identity by descent) and non-European ancestry (detected by multidimensional scaling analysis). Following QC, the final directly genotyped dataset contained 500,527 SNP loci.

Mothers were genotyped using the Illumina human660W-quad genome-wide SNP genotyping platform (Illumina Inc., San Diego, CA, USA) at the Centre National de Génotypage (CNG; Paris, France). Individuals were excluded based on non-European ancestry, missingness, relatedness, gender mismatches and heterozygosity. PLINK (v1.07) [38] was used to carry out quality control measures on an initial set of 10,015 subjects (including non-ARIES ALSPAC participants) and 557,124 directly genotyped SNPs. Following QC, the final directly genotyped dataset contained 526,688 SNP loci.

Imputation was performed to increase the SNP density for all genotyped mothers and children combined. Genotypes were phased together using Shapelt (version 2, revision 727) and then imputed against the 1000 Genomes reference panel (phase 1, version 3, phased using Shapelt version 2, December 2013, using all populations) using Impute (v2.2.2). Genotypes were filtered to have Hardy–Weinberg equilibrium $p > 5 \times 10^{-7}$, MAF >1 % and imputation info score >0.8. Best guess genotypes were used for subsequent analysis. The final imputed dataset used for the analyses presented here contained 8,074,398 loci.

Genotype/Methylation Association Tests

Summary details of samples, SNPs, CpGs and covariates included in association tests are presented. Each SNP in the imputed datasets was analysed against all CpG sites in the Illumina Infinium HM450 array with the exception of those failing QC and those reported to map to more than one location (N = 19,834) or to contain a genetic variant at the CpG site (N = 74,182) [34]. We performed a post hoc test for analysis of non-specific probes which determined that there was no enrichment of cross-hybridising probes in our results. We opted for post hoc annotation of potential probe effects within our results database for the remaining potentially problematic CpG sites (N = 229,983) [34]. The final number of probes analysed was 395,625. Preliminary association analysis of SNPs with CpG sites was performed using an additive model (rank-normalised CpG methylation on SNP allele count) using Matrix eQTL [39] in order to perform the computationally demanding task of estimating 16 trillion associations. Analyses were batched by SNP chromosome and sample time point for parallel analysis on the University of Bristol High Performance Computing (HPC) cluster. SNP effects from this analysis that were $p < 1 \times 10^{-7}$ were then taken forward for re-analysis in PLINK1.07 to perform exact linear regression including covariates. Covariates included in all analyses were age (excluding birth), sex (children only), the top ten ancestry principal components, bisulphite conversion batch and estimated white blood cell counts (using an algorithm based on differential methylation between cell types [33]). Methylation at each CpG site was regressed on these covariates and residuals taken forward for regression on SNP genotype. To report the number of mQTL, we used a conservative threshold of 1×10^{-14}. All associations below 1×10^{-7} were stored and are available in our online mQTL database (http://www.mqtldb.org/). Analysis on rank-normalised data results in effect sizes that are not directly interpretable on the original scale.

Without access to an appropriate replication sample, we performed all subsequent enrichment and downstream analysis using only *cis* mQTL with $p < 1 \times 10^{-14}$ to reduce the possibility of including false positives (unless otherwise stated). We consider results below $p < 1 \times 10^{-14}$ in a single time point to provide informative evidence of association on the basis that this corresponds to a 0.2 % false positive rate after a Bonferonni correction for the number of tests based on directly genotyped SNPs and directly assayed CpG sites. We define replication in additional time points to be associations at $p < 1 \times 10^{-7}$ because typically we are testing in the order of 30,000 associations for replication and on the basis that these are supported by their combination with the evidence at other time points through the life course.

Code Availability

Code to run comparable mQTL analyses with ARIES data is available under the GNU GPL (v3) license athttps://github.com/MRCIEU/ariesmqtl.

Conditional mQTL Analysis

Linkage disequilibrium is expected to cause many mQTL to be represented by multiple tag SNPs. We used the conditional analysis implementation in GCTA [19, 40] to determine the most representative independent loci associated with each CpG site. We used the summary statistics from the results of our PLINK analysis (i.e. any SNP–CpG pair with a $p \leq 1 \times 10^{-14}$ at any time point) and entered them into a stepwise model using individual level genotype data to determine whether they independently account for association with the CpG site. Independent signals with $p < 1 \times 10^{-14}$ are considered significant.

Patterns of Methylation across Genomic Features and Time

The distribution of DNA methylation across CpG sites in different genic and CpG island regions was analysed using annotations from the Illumina Infinium HM450 manifest file (v1.2). Each probe was assigned to an annotation category (e.g. 5′ UTR, 3′ UTR, gene body, CpG island, etc.), the median methylation beta (proportion methylation) for that probe was calculated and then the distribution of these medians across mQTL-associated CpG sites in a category was plotted in a histogram.

To characterise the temporal pattern of methylation, the correlation of methylation between each pair of time points was calculated for each probe. The distribution of correlation coefficients (r^2) was plotted in a histogram for each pair of time points. A null distribution of correlation for each pair was derived by randomising the sample order for one of the pair of time points, and similarly plotted as a histogram of correlation coefficients.

Patterns of mQTL across Genomic Features and Time

The distribution of mQTL-associated CpG sites was analysed using annotations from the Illumina Infinium HM450 manifest file (v1.2). The proportion of CpG probes falling into each annotation category was plotted in a bar chart to evaluate the distribution across different genomic features. A similar approach was used for mQTL SNPs, which were annotated with features using the Variant Effect Predictor [41] (all associated SNPs were included).

SNP Heritability Estimation of Methylation Probes

In order to estimate the proportion of the variation in CpG probes that is due to genetic variation, we used SNP heritability analysis for each probe (after filtering) at each time point. This was performed using genomic restricted maximum likelihood (GREML) as implemented in GCTA (v1.24). The same covariates were fitted here as they were for the mQTL mapping analysis. Only unrelated individuals were used in these analyses (genetic relatedness <0.05) in order to obtain a population-based 'SNP' heritability estimate, which is typically an underestimate of true heritability in complex traits because the magnitude of the estimate is limited to the proportion of all genomic variation that is captured by the SNPs on the genotyping array. In this case we used HapMap3 [20] SNPs with MAF >0.01 and imputation quality score >0.8 (1,171,463 SNPs after filtering). The SNP heritability was estimated for each probe using two variance components, the first generated using only SNPs within ±1 Mb of the CpG site (the *cis* component) and the second generated using all remaining SNPs (the*trans* component), such that var(y) = var(g_cis) + var(g_trans) + var(e), where the genetic variance due to all SNPs var(g) = var(gi_cis) + var(g_trans), and var(g)/var(p) is the total SNP heritability for a probe at a particular time point. The same analysis was repeated for each probe at each time point.

We then make a distinction between 'explained' and 'unexplained' genetic variation (Fig. 2b). Some proportion of the total estimated SNP heritability has been 'explained' by the detected mQTLs, estimated by summing the R^2 values for each conditionally significant association at $p < 1 \times 10^{-14}$ for a particular probe at a particular time point. Thus, we term the remaining genetic variation 'unexplained'. The 'explained' genetic variation was estimated for both the *cis* and *trans* genetic components.

Proportion of trait variance explained by mQTL

In order to ascertain if mQTL contribute to variance in disease liability, a mixed model analysis was performed to calculate the proportion of the trait variance that could be explained by mQTL and non-mQTL in publicly available data from the first WTCCC publication [24]. The variance partitioning model:

$$y_i = \mu + g_{i,1} + g_{i,2} + e_i$$

was fitted using REML as implemented in GCTA (v1.24) software [19]. Here, y_i is the phenotype of individual i, $g_{i,1}$ is their genetic value due to mQTL, $g_{i,2}$ is their genetic value due to all HapMap3 [20] SNPs that were not identified as mQTL and e_i is their residual value. To avoid results being influenced by MHC, both $g_{i,1}$ and $g_{i,2}$ were estimated excluding all SNPs on chromosome 6. This analysis was performed for seven traits (bipolar disorder, Crohn's disease, type 1 diabetes, type 2 diabetes, coronary heart disease, hypertension, rheumatoid arthritis) using the WTCCC data [24]. The WTCCC data underwent the same quality control procedure described in [42]. Specifically, the data were split into seven case-control datasets where the controls were common across all traits; each dataset was filtered for MAF >0.01, genotype missingness <0.002, Hardy–Weinberg equilibrium $p > 0.01$ and any SNPs with differential missingness between cases and controls ($p < 0.05$) were removed. This left approximately 230,000 SNPs for each dataset. Pairwise genomic similarity was then estimated using scaled SNP similarity and any individual pairs with genomic relatedness >0.025 were removed. Principal components were estimated from the SNP data after removing regions of long-range LD and pruning to low LD and five rounds of outlier removal were used whereby if any individual had a value of mean ± 5 standard deviations in any of the first 20 components they were removed. Imputation was performed to 1000 Genomes reference panel (phase 1, version 3, release December 2013) in two stages: first, haplotypes of the WTCCC samples were estimated jointly using ShapeIt2 and then imputation was performed using Impute2. Then, $g_{i,2}$ was estimated using only the imputed SNPs that were present in the HapMap3 reference. To test if the mQTLs detected in ARIES made disproportionately large contributions to SNP heritability in the WTCCC datasets, we used only *cis* mQTL with $p < 1 \times 10^{-14}$ at any time point that were then LD pruned, leaving 29,805 SNPs in total. Generating an appropriate null distribution against which to compare the point estimate of the mQTL variance component in this analysis is important because the mQTL variance component may have been contaminated by variance due to genic SNPs rather than directly due to mQTL alone. To test for this possibility, we generated two different 'null' experiments. First, we sampled 29,805 SNPs that were within ±500 kb of a gene ('genic SNPs'), matching the SNPs to have the same LD and allele frequency distributions as the mQTLs and performing the entire analysis again, but instead of using mQTLs, $g_{i,1}$ was constructed using these sampled SNPs. This was performed 100 times to get a distribution of estimates under the null hypothesis that proximity to genic regions is sufficient to explain the influence of mQTLs on the trait. Second, the same was done except instead of sampling from genic regions specifically, the null SNP set was generated to have the same number of SNPs in 5′ UTR, 3′ UTR, intronic, intergenic, upstream, downstream and unannotated regions of the genome ('mQTL annotation-matched SNPs'). All SNP heritability estimates were transformed from the observed to the liability scale using the trait prevalence estimates used in [43]. Sex and the first 20 genetic principal components were included as covariates in all analyses.

Mediation of *Trans* mQTL Effects by *cis* Methylation

Potential mediation of *trans* mQTL effects by *cis* methylation has been previously reported [21]. This type of analysis is particularly susceptible to measurement error in the *cis* (mediating) CpG methylation variable so requires cautious interpretation. For all *trans* associations we took the estimate of association from a linear regression of B~G, and then performed a second linear regression B~G+Kx, where B is methylation at the*trans* CpG site, G is the genotype (or allele score if more than one independent *cis* association) and K is methylation at the *cis* CpG site. Reduction in regression coefficient for genotype when *cis* CpG methylation was added to the model was evaluated as an estimate of potential mediation (given the limitations of measurement error). We estimated the likely impact of measurement error by performing a series of analyses in which we added increasing levels of simulated measurement error (multiplication by random values sampled from a normal distribution with mean 1 and a given standard deviation).

Enrichment Analysis for mQTL in GWAS Results

To test if the mQTL were enriched for low *p* values in previously published GWAS results, the following procedure was performed. Only independent mQTL with *cis* effects of $p<1\times10^{-14}$ obtained from the conditional analysis and then filtered to remove SNPs with LD $r^2>0.1$ were used to test for enrichment. We collected data from a set of 33 complex traits from online sources. All available mQTL were extracted from each trait and Fisher's method was used to estimate a combined *p* value for each trait:

$$\chi^2_{2k} \sim -2\sum_{i=1}^{k} \ln p_i$$

In order to compare these *p* values to a realistic null distribution, the same procedure was performed but using 10,000 random draws of the same number of SNPs from (a) genic regions or (b) mQTL annotation-matched SNPs. In both cases SNPs were also matched to the mQTLs on allele frequency and number of proxies to adjust for LD structure. In order to match on LD, each mQTL was given an LD score by calculating the number of SNPs that were in LD $r^2>0.8$ and the same was done for each of the candidate genic SNPs (approximately 200,000 genic SNPs). A null distribution for each GWAS trait was generated by performing this procedure 1000 times, each time making a new random draw of genic SNPs matched for allele frequency and LD. Empirical *p* values were generated by simply finding the rank of the meQTL *p* value among the 1000 null *p*values.

Gene Ontology Enrichment

mQTL SNPs were mapped to the nearest gene (by genomic coordinate) using the Variant Effect Predictor and lists of unique gene names were analysed for enrichment of gene ontology (GO) terms relative to a human reference using the gene ontology function ('go') in the Orange bioinformatics add-on ('Orange.bio',http://orange.biolab.si/download/). CpG sites associated with mQTL were similarly (but separately) tested for GO term enrichment using the Illumina gene-name annotations for each CpG site. Analyses were repeated for time point-specific loci and those common to all time points. Potentially enriched GO terms (biological function) were plotted using REVIGO [44], where strength of evidence for enrichment ($-\log_{10}(p$ value)) is proportional to size.

Comparison of Blood mQTL with Blood eQTL

Previously reported blood eQTL [22] were downloaded from http://www.gtexportal.org/ and compared with mQTL discovered at each time point to determine overlap. *Cis* QTL were considered to be shared if the same SNP was reported to be associated at $p<1\times10^{-14}$ in mQTL data and $p<2.5\times10^{-7}$ in eQTL data (the reported GTEx threshold). For GO analyses we compared our mQTL data with blood eQTL from Westra et al. [23] as these data enabled us to evaluate enrichment of common QTLs for both *cis* and *trans*; however, a lack of data prevented us from comparing the numeric overlap between the Westra et al. eQTL and our mQTL.

Ethics Approval and Consent to Participate

Written informed consent has been obtained from all ALSPAC participants. Ethical approval for the study was obtained from the ALSPAC Ethics and Law Committee (IRB00003312) and Local Research Ethics Committees in accordance with the guidelines of The Declaration of Helsinki.

Availability of Data and Materials

A mQTL database containing all results from this study is available online at http://www.mqtldb.org, which includes both search functionality and full download of all results at $p<1\times10^{-7}$ using Matrix eQTL and GCTA. Data are also available for download at doi:10.5523/bris.r9bxayo5mmk510dczq6golkmb.

All data used in this study were provided to the authors by the ALSPAC cohort study via their standard Access Policy. Data are available to other bona fide researchers on the same basis; for details please seehttp://www.bristol.ac.uk/alspac/researchers/data-access/.

ACKNOWLEDGEMENTS

We are extremely grateful to all the families who took part in this study, the midwives for their help in recruiting them and the whole ALSPAC team, which includes interviewers, computer and laboratory technicians, clerical workers, research scientists, volunteers, managers, receptionists and nurses. GWAS data were generated by Sample Logistics and Genotyping Facilities at the Wellcome Trust Sanger Institute and LabCorp (Laboratory Corporation of America) using support from 23andMe.

This work was carried out using the computational facilities of the Advanced Computing Research Centre, University of Bristol (http://www.bris.ac.uk/acrc/).

We thank Oliver Stegle (European Bioinformatics Institute) for analytical advice.

COMPETING INTERESTS

Members of the MRC Integrative Epidemiology Unit have received contributions (funding or in-kind contribution) from the following organizations: Pfizer Inc., Astra Zeneca, Rusan Pharma Ltd, Cambridge Cognition Ltd, Janssen Pharmaceutica, GlaxoSmithKline.

AUTHORS' CONTRIBUTIONS

TRG managed methylation data processing, designed the study, coordinated the analyses and drafted the manuscript. HAS and GH participated in design of the study and performed analyses and prepared figures and tables. GH performed genotype imputation. JL advised on analyses and data QC. GW and OL performed data processing and QC. JZ advised on conditional analyses. AD, WLM and KH performed laboratory analyses of DNA methylation. SMR managed laboratory data production. DME advised on statistical genetics and complex trait comparisons. GDS and CR conceived of the study and participated in its design. All authors read and approved the final manuscript.

REFERENCES

1. Maunakea AK, Chepelev I, Cui K, Zhao K. Intragenic DNA methylation modulates alternative splicing by recruiting MeCP2 to promote exon recognition. Cell Res. 2013;23:1256–69.
2. Bjornsson HT, Sigurdsson MI, Fallin MD, Irizarry RA, Aspelund T, Cui H, et al. Intra-individual change over time in DNA methylation with familial clustering. JAMA. 2008;299:2877–83.
3. Zhang B, Zhou Y, Lin N, Lowdon RF, Hong C, Nagarajan RP, et al. Functional DNA methylation differences between tissues, cell types, and across individuals discovered using the M&M algorithm. Genome Res. 2013;23:1522–40.
4. Breitling LP, Yang R, Korn B, Burwinkel B, Brenner H. Tobacco-smoking-related differential DNA methylation: 27 K discovery and replication. Am J Hum Genet. 2011;88:450–7.
5. Fraga MF, Ballestar E, Paz MF, Ropero S, Setien F, Ballestar ML, et al. Epigenetic differences arise during the lifetime of monozygotic twins. Proc Natl Acad Sci U S A. 2005;102:10604–9.
6. Bell JT, Pai AA, Pickrell JK, Gaffney DJ, Pique-Regi R, Degner JF, et al. DNA methylation patterns associate with genetic and gene expression variation in HapMap cell lines. Genome Biol. 2011;12:R10.
7. Relton CL, Davey Smith G. Epigenetic epidemiology of common complex disease: prospects for prediction, prevention, and treatment. PLoS Med. 2010;7:e1000356.
8. McRae AF, Powell JE, Henders AK, Bowdler L, Hemani G, Shah S, et al. Contribution of genetic variation to transgenerational inheritance of DNA methylation. Genome Biol. 2014;15:R73.
9. Teh AL, Pan H, Chen L, Ong ML, Dogra S, Wong J, et al. The effect of genotype and in utero environment on interindividual variation in neonate DNA methylomes. Genome Res. 2014;24:1064–74.
10. Grundberg E, Small KS, Hedman Å, Nica AC, Buil A, Keildson S, et al. Mapping cis- and trans-regulatory effects across multiple tissues in twins. Nat Genet. 2012;44:1084–9.
11. Shah S, McRae AF, Marioni RE, Harris SE, Gibson J, Henders AK, et al. Genetic and environmental exposures constrain epigenetic drift over the human life course. Genome Res. 2014;24:1725–33.
12. Boyd A, Golding J, Macleod J, Lawlor DA, Fraser A, Henderson J, et al. Cohort profile: the 'children of the 90s'--the index offspring of the Avon Longitudinal Study of Parents and Children. Int J Epidemiol. 2013;42:111–27.
13. Fraser A, Macdonald-Wallis C, Tilling K, Boyd A, Golding J, Davey Smith G, et al. Cohort profile: the Avon Longitudinal Study of Parents and Children: ALSPAC mothers cohort. Int J Epidemiol. 2013;42:97–110.
14. Relton CL, Gaunt T, McArdle W, Ho K, Duggirala A, Shihab H, et al. Data resource profile: Accessible Resource for Integrated Epigenomic Studies (ARIES). Int J Epidemiol. 2015;44:1181–90.
15. Gibbs JR, van der Brug MP, Hernandez DG, Traynor BJ, Nalls MA, Lai SL, et al. Abundant quantitative trait loci exist for DNA methylation and gene expression in human brain. PLoS Genet. 2010;6:e1000952.
16. Kaplow IM, MacIsaac JL, Mah SM, McEwen LM, Kobor MS, Fraser HB. A pooling-based approach to mapping genetic variants associated with DNA methylation. Genome Res. 2015;25:907–17.
17. Lemire M, Zaidi SH, Ban M, Ge B, Aïssi D, Germain M, et al. Long-range epigenetic regulation is conferred by genetic variation located at thousands of independent loci. Nat Commun. 2015;6:6326.
18. Heyn H, Moran S, Hernando-Herraez I, Sayols S, Gomez A, Sandoval J, et al. DNA methylation contributes to natural human variation. Genome Res. 2013;23:1363–72.
19. Yang J, Lee SH, Goddard ME, Visscher PM. GCTA: a tool for genome-wide complex trait analysis. Am J Hum Genet. 2011;88:76–82.
20. Altshuler DM, Gibbs RA, Peltonen L, Dermitzakis E, Schaffner SF, Yu F, et al. Integrating common and rare genetic variation in diverse human populations. Nature. 2010;467:52–8.
21. Shi J, Marconett CN, Duan J, Hyland PL, Li P, Wang Z, et al. Characterizing the genetic basis of methylome diversity in histologically normal human lung tissue. Nat Commun. 2014;5:3365.
22. Consortium G. Human genomics. The Genotype-Tissue Expression (GTEx) pilot analysis: multitissue gene regulation in humans. Science. 2015;348:648–60.

23. Westra HJ, Peters MJ, Esko T, Yaghootkar H, Schurmann C, Kettunen J, et al. Systematic identification of trans eQTLs as putative drivers of known disease associations. Nat Genet. 2013;45:1238–43.

24. Consortium WTCC. Genome-wide association study of 14,000 cases of seven common diseases and 3,000 shared controls. Nature. 2007;447:661–78.

25. Davey Smith G, Hemani G. Mendelian randomization: genetic anchors for causal inference in epidemiological studies. Hum Mol Genet. 2014;23:R89–98.

26. Banovich NE, Lan X, McVicker G, van de Geijn B, Degner JF, Blischak JD, et al. Methylation QTLs are associated with coordinated changes in transcription factor binding, histone modifications, and gene expression levels. PLoS Genet. 2014;10:e1004663.

27. Wagner JR, Busche S, Ge B, Kwan T, Pastinen T, Blanchette M. The relationship between DNA methylation, genetic and expression inter-individual variation in untransformed human fibroblasts. Genome Biol. 2014;15:R37.

28. Fairfax BP, Makino S, Radhakrishnan J, Plant K, Leslie S, Dilthey A, et al. Genetics of gene expression in primary immune cells identifies cell type-specific master regulators and roles of HLA alleles. Nat Genet. 2012;44:502–10.

29. Gamazon ER, Badner JA, Cheng L, Zhang C, Zhang D, Cox NJ, et al. Enrichment of cis-regulatory gene expression SNPs and methylation quantitative trait loci among bipolar disorder susceptibility variants. Mol Psychiatry. 2013;18:340–6.

30. van Eijk KR, de Jong S, Boks MP, Langeveld T, Colas F, Veldink JH, et al. Genetic analysis of DNA methylation and gene expression levels in whole blood of healthy human subjects. BMC Genomics. 2012;13:636.

31. Kundaje A, Meuleman W, Ernst J, Bilenky M, Yen A, Heravi-Moussavi A, et al. Integrative analysis of 111 reference human epigenomes. Nature. 2015;518:317–30.

32. Smith AK, Kilaru V, Kocak M, Almli LM, Mercer KB, Ressler KJ, et al. Methylation quantitative trait loci (meQTLs) are consistently detected across ancestry, developmental stage, and tissue type. BMC Genomics. 2014;15:145.

33. Houseman EA, Accomando WP, Koestler DC, Christensen BC, Marsit CJ, Nelson HH, et al. DNA methylation arrays as surrogate measures of cell mixture distribution. BMC Bioinformatics. 2012;13:86.

34. Naeem H, Wong NC, Chatterton Z, Hong MK, Pedersen JS, Corcoran NM, et al. Reducing the risk of false discovery enabling identification of biologically significant genome-wide methylation status using the HumanMethylation450 array. BMC Genomics. 2014;15:51.

35. Fraser HB, Lam LL, Neumann SM, Kobor MS. Population-specificity of human DNA methylation. Genome Biol. 2012;13:R8.

36. Pidsley R, Wong CC Y, Volta M, Lunnon K, Mill J, Schalkwyk LC. A data-driven approach to preprocessing Illumina 450K methylation array data. BMC Genomics. 2013;14:293.

37. Touleimat N, Tost J. Complete pipeline for Infinium(®) Human Methylation 450 K BeadChip data processing using subset quantile normalization for accurate DNA methylation estimation. Epigenomics. 2012;4:325–41.

38. Purcell S, Neale B, Todd-Brown K, Thomas L, Ferreira MA, Bender D, et al. PLINK: a tool set for whole-genome association and population-based linkage analyses. Am J Hum Genet. 2007;81:559–75.

39. Shabalin AA. Matrix eQTL: ultra fast eQTL analysis via large matrix operations. Bioinformatics. 2012;28:1353–8.

40. Yang J, Ferreira T, Morris AP, Medland SE, Madden PA, Heath AC, et al. Conditional and joint multiple-SNP analysis of GWAS summary statistics identifies additional variants influencing complex traits. Nat Genet. 2012;44:369–75. S1–3.

41. McLaren W, Pritchard B, Rios D, Chen Y, Flicek P, Cunningham F. Deriving the consequences of genomic variants with the Ensembl API and SNP Effect Predictor. Bioinformatics. 2010;26:2069–70.

42. Lee SH, Wray NR, Goddard ME, Visscher PM. Estimating missing heritability for disease from genome-wide association studies. Am J Hum Genet. 2011;88:294–305.

43. Gusev A, Bhatia G, Zaitlen N, Vilhjalmsson BJ, Diogo D, Stahl EA, et al. Quantifying missing heritability at known GWAS loci. PLoS Genet. 2013;9:e1003993.

44. Supek F, Bošnjak M, Škunca N, Šmuc T. REVIGO summarizes and visualizes long lists of gene ontology terms. PLoS One. 2011;6:e21800.

CHAPTER 3

ANALYSIS OF THE GENETIC BASIS OF DISEASE IN THE CONTEXT OF WORLDWIDE HUMAN RELATIONSHIPS AND MIGRATION

Erik Corona,[1,2,3] Rong Chen,[1,3] Martin Sikora,[4] Alexander A. Morgan,[1,2] Chirag J. Patel,[1,2,3] Aditya Ramesh,[1,3] Carlos D. Bustamante,[4] and Atul J. Butte[1,2,3]

[1]Division of Systems Medicine, Department of Pediatrics, Stanford University School of Medicine, Stanford, California, United States of America

[2]Program in Biomedical Informatics, Stanford University School of Medicine, Stanford, California, United States of America

[3]Lucile Packard Children's Hospital, Palo Alto, California, United States of America

[4]Department of Genetics, Stanford University School of Medicine, Stanford, California, United States of America

ABSTRACT

Genetic diversity across different human populations can enhance understanding of the genetic basis of disease. We calculated the genetic risk of 102 diseases in 1,043 unrelated individuals across 51 populations of the Human Genome Diversity Panel. We found that genetic risk for type 2 diabetes and pancreatic cancer decreased as humans migrated toward East Asia. In addition, biliary liver cirrhosis, alopecia areata, bladder cancer, inflammatory bowel disease, membranous nephropathy, systemic lupus erythematosus, systemic sclerosis, ulcerative colitis, and vitiligo have undergone genetic risk differentiation. This analysis represents a large-scale attempt to characterize genetic risk differentiation in the context of migration. We anticipate that our findings will enable detailed analysis pertaining to the driving forces behind genetic risk differentiation.

AUTHOR SUMMARY

The environment humans inhabit has changed many times in the last 100,000 years. Migration and dynamic local environments can lead to genetic adaptations favoring beneficial traits. Many genes responsible for these adaptations can alter disease susceptibility. Genes can also affect disease susceptibility by varying randomly across different populations. We have studied genetic variants that are known to modify disease susceptibility in the context of worldwide migration. We found that variants associated with 11 diseases have been affected to an extent that is not explained by random variation. We also found that the genetic risk of type 2 diabetes has steadily decreased along the worldwide human migration trajectory from Africa to America.

INTRODUCTION

Analyzing the impact of human migration on genetic disease susceptibility is critical to the understanding of complex disease. Moving to new environments causes adaptation, which can also affect disease susceptibility. The availability of methods to measure genomic differences between worldwide populations has increased our understanding of human migration. For example, a worldwide human relationships phylogenetic tree was constructed after genotyping over 50 worldwide populations [1]. This process has enabled researchers to characterize worldwide genetic variation and has provided information regarding migrations that founded entire populations [2], [3]. At the same time, Genome-Wide

Citation: Erik Corona, Rong Chen, Martin Sikora, Alexander A. Morgan, Chirag J. Patel, Aditya Ramesh, Carlos D. Bustamante, and Atul J. Butte. Analysis of the Genetic Basis of Disease in the Context of Worldwide Human Relationships and Migration. PLOS Genetics. 2013; 9(5):e1003447. doi:10.1371/journal.pgen.1003447.

Association Studies (GWAS) have increased discoveries of disease-associated genetic loci. These developments have paved the way for studies investigating the effects of migration on the genetic basis of disease [4]–[6].

The human genome has been subjected to many selective pressures in recent history. They include changes brought about by the domestication of crops and animals, and the rise of urbanization [7]. These changes may increase the frequency of mutations that are beneficial in the new environment. They may also lead to disruptions of biological processes. When mutations confer a net increase in fitness, they are expected to increase in frequency in affected populations [8], [9]. A mutation increasing disease risk can accompany a beneficial mutation through linkage disequilibrium (LD).

Studies have shown that unlinked single nucleotide polymorphisms (SNPs) associated with a single phenotype may be affected as a group if the phenotype undergoes differentiation[10], [11]. While a large set of loci may increase susceptibility to complex disease, individual loci generally make modest contributions, and their effect sizes indicate that they would not be expected to decrease reproductive success [12]. This situation allows differences in the genetic basis of disease to build naturally via genetic drift. However, deviations from genetic drift are expected when environmental changes occur due to migration [13]. Such changes provide an opportunity to learn about factors elevating disease risk in multiple populations.

Adaptation to new environments may have caused genetic risk differences across many human populations. Despite the recent explosion of knowledge regarding disease-associated loci and the genetic structure of different world populations [14]–[19], few studies have examined population-based differences in the genetic risk factors for disease. Additionally, they have included only a modest number of diseases, populations, or genetic samples. For example, Myles et al. [6] genotyped 25 disease-associated SNPs in ~1,000 individuals from 53 populations in the HGDP-CEPH Human Genome Diversity Cell Line panel [15]. The study measured allele frequency differences in the SNPs and concluded that while risk allele differentiation was unusually high in some cases, overall, disease SNPs were not more differentiated between populations than random SNPs. However, 25 SNPs may not be sufficient to determine whether disease-associated SNPs as a whole have undergone risk allele differentiation in worldwide populations. These 25 SNPs are a small subset of variants that influence a small number of specific diseases. Other studies have examined interactions between migration, selection, and disease. One examined the impact of selection on hypertension variants during the migration out of Africa [20]. Others reported increased allele frequency differentiation of type 2 diabetes variants [21]–[23].

It is also plausible that allele differentiation may have occurred in unexamined combinations of diseases and populations. One study probed 2,186 disease-associated SNPs in the HapMap CEU cohort and concluded that they were enriched for low-frequency alleles [24]. This study used a large number of SNPs, but was conducted on a single population. A study incorporating a comprehensive catalog of SNPs and a large genetically diverse cohort would implicitly place genetic risk in the context of migration and reveal worldwide genetic risk variation with respect to many diseases.

In a previous study, we surveyed 8,377 SNPs representing 437 diseases in 11 subpopulations. We found differences in risk distribution and protective alleles in different diseases across many subpopulations [25]. The current study examines how the differences in these distributions could have arisen. We analyzed correlations between genetic risk and migration trajectories. Our goal was to identify populations with different genetic dispositions to disease, and to pinpoint migrations preceding these differences. Our analysis highlights the role evolution has played in changing disease susceptibility across populations.

Importantly, our approach controls for genetic substructure within and among diverse populations [26]. To detect differences in the genetic basis of disease across migratory events, we integrated a large curated database of geographically-annotated, disease-associated SNPs with human variation measurements in 51 populations from the Human Genome Diversity Panel (HGDP). We controlled for established genetic similarity in subpopulations in order to detect differences in genetic risk exceeding those expected under genetic drift.

RESULTS

Type 2 Diabetes

Figure 1A summarizes our findings for type 2 diabetes on a world map. Consistent with our previous work [25], we found that African populations have the highest genetic risk for type 2 diabetes, followed by people from the Middle East, Europe, and Asia. The Mandinka population had the greatest risk, at a mean 0.95 log likelihood ratio (LLR), while the Surui population had the lowest risk (LLR: −0.87). This high-level view revealed stark differences in genetic risk between African and Asian populations. European and Middle-Eastern populations had an intermediate genetic risk for this disease.

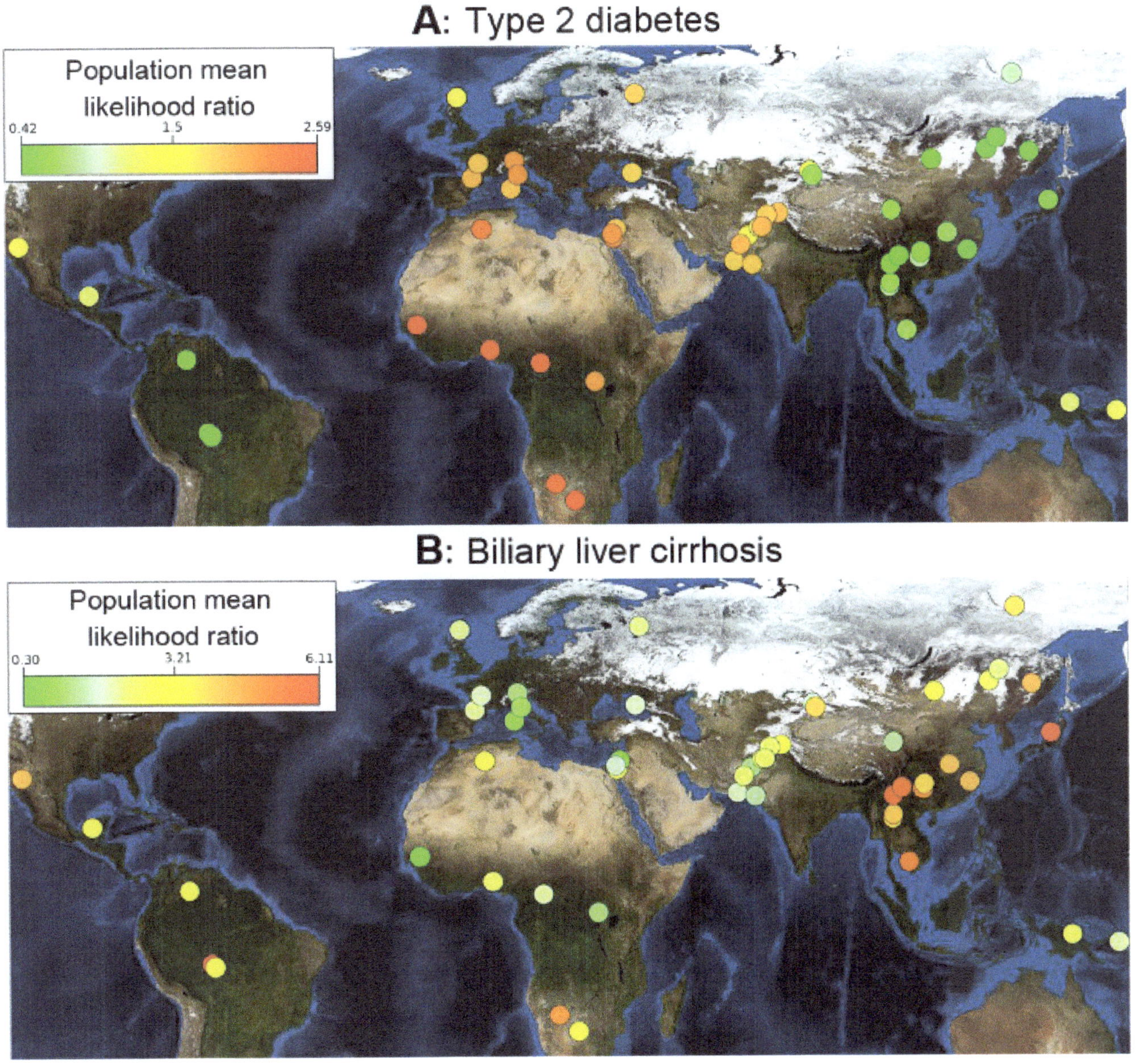

Figure 1. Differences in genetic risk among populations. Each population is ranked by risk, which is denoted by a color. Populations with the greatest risk are bright red, and those with the lowest risk are green. (A) Populations for East Asia and the Americas have lower genetic risk for type 2 diabetes than those from Africa and Europe. Genetic risk differentiation is sharply divided along major population migration events. Type 2 diabetes is represented by 16 SNPs. (B) Genetic risk for biliary liver cirrhosis is represented by 44 SNPs. Genetic risk peaks in East Asia and in the Karitiana population in South America. The background is a public domain world map from NASA Earth Observatory (http://eoimages.gsfc.nasa.gov/images/imagerecords/73000/73909/world.topo.bathy.200412.3×5400×2700.jpg); an interactive online tool is available at http://geneworld.stanford.edu using Google Maps technology.

These patterns identified worldwide trends of increases or decreases in genetic risk and show a high-level view of variation in genetic risk for different diseases. An interactive version of Figure 1A is available for over 100 diseases at geneworld.stanford.edu.

Placing genetic risk on a map does not reveal close relationships between populations.Figure 2 is a worldwide phylogenetic tree displaying the relative genetic risk for each population in the context of population relationships. Relationships were inferred in a previous study [1] by analyzing 650,000 SNPs using a maximum-likelihood approach. Results from that analysis have been incorporated here. Each branch in the figure represents the ancestral population common to all populations below it.

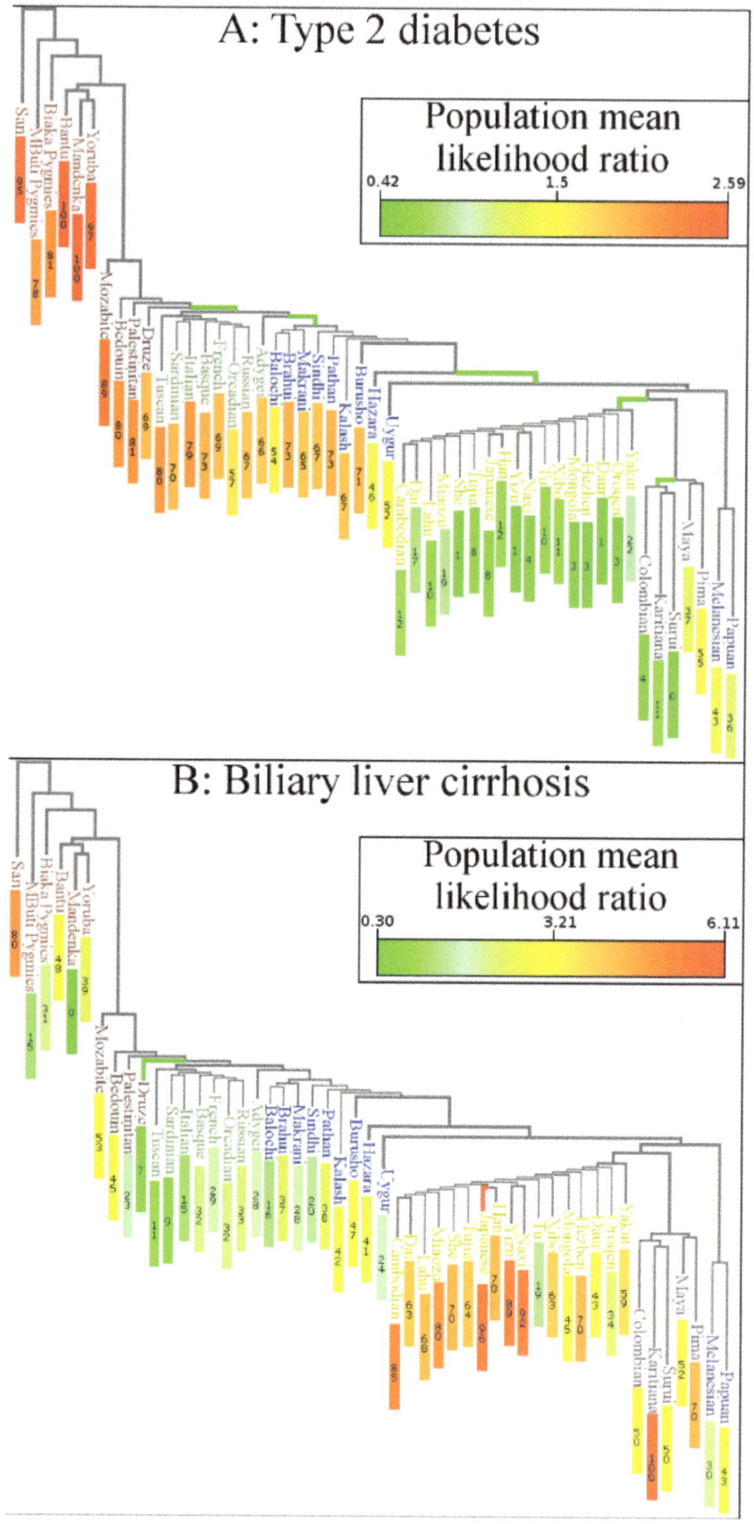

Figure 2. Genetic risk in the context of human relationships. The figure superimposes genetic risk for type 2 diabetes on a maximum likelihood phylogenetic tree based on the HGDP. Branch color indicates whether the subtree has shifted towards increased or decreased risk (q-value<0.05). Red:increased risk; green:decreased risk. Each colored branch also represents an independent genetic risk differentiation event as determined by a maximum likelihood model. Numbers in colored rectangles show percentile of genetic risk, with a zero being lowest. (A) Type 2 diabetes risk was decreased in East Asia and America. (B) Biliary Liver Cirrhosis shifted towards increased genetic risk in the Japanese population and towards decreased risk in the Druze population. The Karitiana population also showed borderline signs of genetic risk differentiation (q=0.057).

Figure 1A and Figure 2A show that genetic risk for type 2 diabetes has undergone independent differentiation multiple times as humans migrated out of Africa to East Asia, and finally to the Americas. "Independent differentiation event" was defined as genetic risk differentiation occurring de novo in a population rather than via inheritance from an ancestral population. "Dependent risk differentiation" is a genetic risk inherited from an ancestor. Table 1 shows individual populations with genetic risk differences for type 2 diabetes that are larger than expected under genetic drift; however, it does not distinguish between dependent and independent genetic risk differentiation.

Table 1. List of populations with disease risk differentiation.

Disease	Population	Genetic Risk	World Genetic Risk	Q-value	SNPs
Alopecia areata	Mozabite	−3.546	−0.121	1.000×10^{-5}	26
Biliary liver cirrhosis	Druze	−0.989	0.122	1.735×10^{-2}	30
Biliary liver cirrhosis	Japanese	1.691	0.026	1.123×10^{-2}	27
Bladder cancer	Tu	−0.535	−0.072	3.673×10^{-2}	7
Inflammatory bowel disease	Balochi	−0.679	−1.051	1.000×10^{-5}	11
Inflammatory bowel disease	Burusho	−0.886	−1.046	2.450×10^{-2}	9
Inflammatory bowel disease	Makrani	−0.792	−1.048	1.533×10^{-2}	10
Inflammatory bowel disease	Palestinian	−0.784	−1.056	3.433×10^{-4}	11
Inflammatory bowel disease	Sindhi	−0.704	−1.051	3.433×10^{-4}	10
Membranous nephropathy	French Basque	−1.876	−5.217	1.000×10^{-5}	17
Pancreatic cancer	Biaka Pygmies	0.972	−0.215	2.602×10^{-2}	9
Pancreatic cancer	Mbuti Pygmies	0.872	−0.194	2.780×10^{-2}	9
Pancreatic cancer	Mandenka	0.877	−0.204	2.244×10^{-2}	9
Pancreatic cancer	Yoruba	1.007	−0.207	1.429×10^{-2}	9
Pancreatic cancer	Sardinian	0.244	−0.191	2.998×10^{-2}	7
Pancreatic cancer	Cambodian	−0.644	−0.174	3.900×10^{-2}	7
Pancreatic cancer	Japanese	−0.679	−0.165	2.702×10^{-2}	7
Pancreatic cancer	Han	−0.601	−0.160	4.774×10^{-2}	7
Pancreatic cancer	Naxi	−0.673	−0.175	4.871×10^{-2}	7
Pancreatic cancer	Xibo	−0.722	−0.174	2.453×10^{-2}	7
Pancreatic cancer	Daur	−0.732	−0.174	2.630×10^{-2}	7
Pancreatic cancer	Maya	−0.783	−0.164	2.487×10^{-2}	7
Systemic lupus erythematosus	French Basque	−2.76	−0.975	4.012×10^{-2}	29
Systemic lupus erythematosus	Maya	1.354	−1.074	4.591×10^{-2}	29
Systemic lupus erythematosus	Pima	1.882	−1.087	3.265×10^{-2}	29
Systemic sclerosis	French Basque	−0.746	0.079	5.407×10^{-2}	15
Systemic sclerosis	Maya	1.151	0.033	3.520×10^{-2}	17
Type 2 diabetes	Bedouin	0.576	0.079	4.341×10^{-2}	13
Type 2 diabetes	Cambodian	−0.643	0.11	2.577×10^{-2}	13
Type 2 diabetes	Colombian	−0.787	0.114	3.659×10^{-2}	12
Type 2 diabetes	Daur	−0.839	0.111	2.653×10^{-2}	13
Type 2 diabetes	Han	−0.646	0.135	2.098×10^{-2}	13
Type 2 diabetes	Hezhen	−0.806	0.11	1.918×10^{-2}	13
Type 2 diabetes	Japanese	−0.723	0.126	1.734×10^{-2}	13
Type 2 diabetes	Lahu	−0.674	0.11	4.826×10^{-2}	13
Type 2 diabetes	Mongola	−0.807	0.111	3.877×10^{-2}	13
Type 2 diabetes	Mozabite	0.749	0.083	2.183×10^{-2}	15
Type 2 diabetes	Naxi	−0.78	0.11	2.051×10^{-2}	13
Type 2 diabetes	Oroqen	−0.803	0.111	1.853×10^{-2}	13
Type 2 diabetes	Palestinian	0.6	0.077	2.470×10^{-2}	13
Type 2 diabetes	She	−0.834	0.111	1.874×10^{-2}	13
Type 2 diabetes	Tu	−0.681	0.11	2.012×10^{-2}	13
Type 2 diabetes	Tujia	−0.721	0.11	2.401×10^{-2}	13
Type 2 diabetes	Xibo	−0.668	0.109	2.576×10^{-2}	13
Type 2 diabetes	Yakut	−0.466	0.116	4.458×10^{-2}	13
Type 2 diabetes	Yizu	−0.841	0.111	2.074×10^{-2}	13
Ulcerative colitis	Balochi	−0.525	−1.778	1.000×10^{-5}	27
Ulcerative colitis	Palestinian	−1.052	−1.784	9.183×10^{-2}	29
Ulcerative colitis	Sindhi	−0.749	−1.772	1.025×10^{-2}	27
Vitiligo	French Basque	0.339	−0.467	2.653×10^{-2}	8
Vitiligo	Russian	0.239	−0.466	3.673×10^{-2}	8

The populations in this table have undergone disease risk differentiation exceeding what would be expected by genetic drift for the diseases listed. Population-specific and overall genetic risks are shown. Genetic risk is the mean combined LLR for all SNPs associated with a disease. The q-value corrects for the number of populations tested. Certain diseases, such as bladder cancer, showed localized genetic risk differentiation while worldwide trends were evident for others, like pancreatic cancer and type 2 diabetes.

doi:10.1371/journal.pgen.1003447.t001

We used a maximum likelihood method to identify branches in the phylogenetic tree representing independent genetic risk differentiation events in Figure 2A. The log-likelihood of having only one event for type 2 diabetes was $l_1=-108.848$. The log-likelihoods for 2,3,4,5, and 6 events were $l_2=-66.6083$, $l_3=-40.7165$, $l_4=-33.7324$, $l_5=-26.8371$, $l_6=-23.3415$. We used the likelihood ratio test to determine the number of branches undergoing genetic risk differentiation independently that exceeded what would be expected under genetic drift. This test allows for the calculation of a p-value for *n* independent branches by converting two log-likelihood scores to a χ^2 variable as follows: $2(l_n-l_{n-1})\sim\chi^2_1$. The p-value for 2 branches versus 1 branch undergoing independent genetic risk differentiation for type 2 diabetes was less than 1.00×10^{-16}, meaning there was evidence for more than one independent genetic risk differentiation event. The p-values for 3, 4, 5, and 6 branches were 6.30×10^{-13}, 1.86×10^{-4}, 8.19×10^{-3}, and 5.64×10^{-2}, respectively. There was evidence for 5 distinct genetic risk differentiation events (highlighted in green in Figure 2A). Figure 3 shows the genetic risk of all 1043 individuals. Each type 2 diabetes-associated genotype in each individual is displayed. The x-axis shows individual genetic risk; the y-axis corresponds to West (bottom) to East (top) migration. The figure shows that genetic risk for type 2 diabetes decreases as populations move into East Asia. The figure shows that genetic risk decreases steadily, as opposed to being caused by a single genetic risk differentiation event.

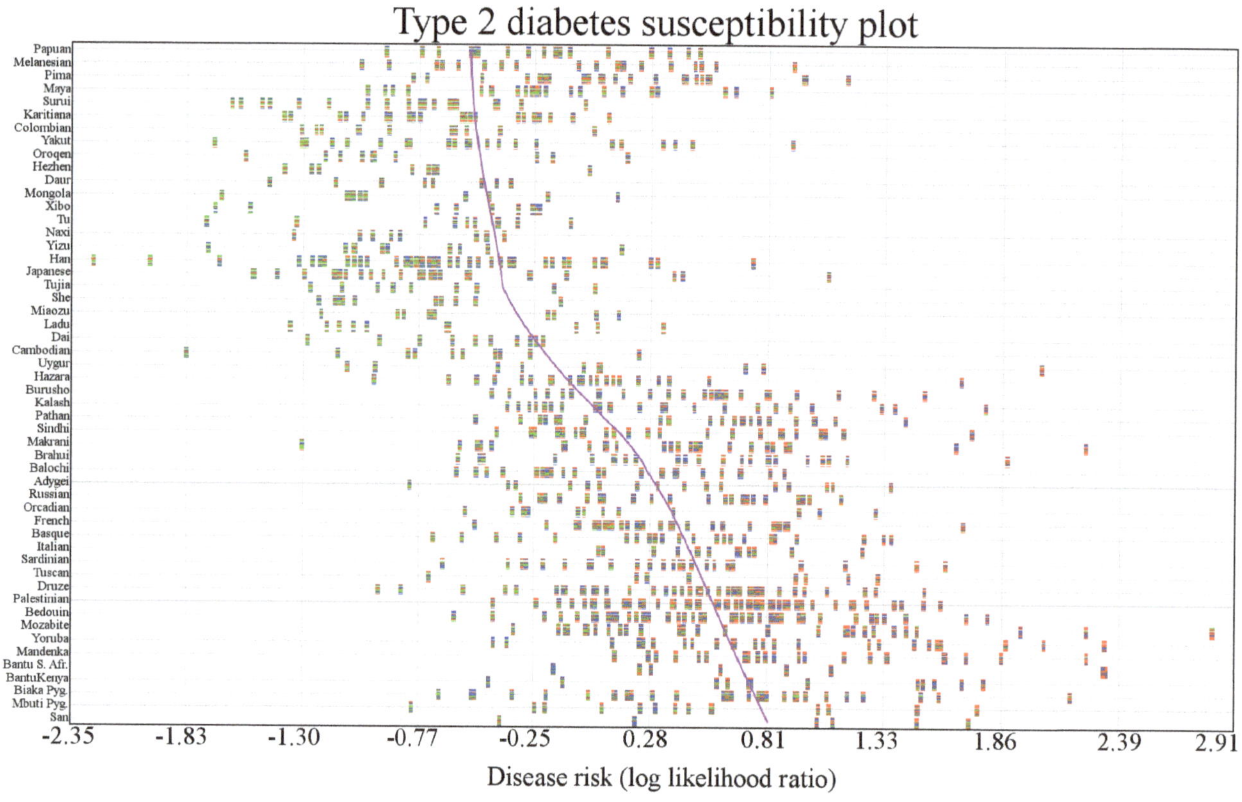

Figure 3. Variability in genetic risk for type 2 diabetes. Individuals are represented by vertical multicolored rectangles. Individual bars in each rectangle represent one of the 16 SNPs associated with type 2 diabetes. Bar colors indicate the following: red: homozygosity for a risk allele; blue:heterozygosity; green:homozygosity for a protective allele; white: missing genotype. The order of the SNPs is preserved across individuals. The x-axis shows genetic risk and the y-axis shows each population. The purple line is a locally weighted linear regression curve displaying the general direction of disease susceptibility as populations migrated from West to East. Genetic risk is lower in East Asians.

Analysis of individuals showed that a person in the Mozabite population (HGDP1255) had the highest genetic risk (LLR:2.81), and an individual in the Han population (HGDP1291) had the lowest risk (LLR: −2.21).

The effect size of each individual variant is not necessarily the same in each population[27]. While the genetic risk of disease is currently computed using all available GWASs, there is a well-known European bias, as most GWASs are based on European-derived populations [28]. Table S1 displays the populations in which SNPs associated with type 2 diabetes in this study have been replicated. In order to ensure that effect sizes unique to European populations were not solely responsible for observed levels of genetic risk differentiation, the genetic risk for type 2 diabetes was recomputed using Asian-specific effect sizes for all variants. This was accomplished by using GWASs exclusively based on Asian populations. Table S2 compares the likelihood ratio computed using GWASs in which European populations are overrepresented with the Asian-specific likelihood ratio computed using Asian based GWASs. Table S3 compares the genetic risk estimates using all available GWASs and Asian-specific GWASs. Despite major differences in overall risk estimates, significant risk differentiation was still observed in Asian populations when risks were computed with Asian-specific GWASs.

Biliary Liver Cirrhosis

Figure 1B shows the genetic risk of biliary liver cirrhosis (BLC) across worldwide populations. In contrast to type 2 diabetes, no worldwide trend is apparent. Figure 2B shows risk differences that were larger than expected under genetic drift ($q<0.05$). Branches were colored green if the deviation of a population below them shifted toward decreased risk compared to all other populations and red if the shift was toward increased risk. The maximum likelihood model was not applied to Figure 2B, as each population showing signs of genetic risk differentiation did not have descendant populations in the phylogenetic tree. The maximum likelihood test only distinguishes between inherited and independent genetic risk differentiation and is therefore only suitable for diseases displaying obvious worldwide deviation trends. Figure 2B shows that the Druze and Japanese populations have genetic risk differentiation exceeding what is expected under genetic drift. The Druze population shows significantly less risk, with only 28 out of 100,000 random draws showing less risk ($q<0.05$) than all other populations combined. The genetic risk in the Japanese population was higher, with only 18 out of 100,000 random draws showing a higher risk. Genetic risk differentiation in BLC was localized in one European and an Asian population.

We examined genetic risk in individuals to expose potential outliers (Figure 4) and to visualize variation in genetic risk estimates across all individuals. The person with the lowest risk (HGDP1201) was from the Mandinka population and had a risk score (combined LLR) of −4.372. The person with the highest risk (HGDP1279) was from the Mozabite population and had combined LLR of 9.521. This person appears to be an outlier. The individual with the second highest genetic risk (HGDP998) was from the Karitiana population (combined LLR: 5.54).

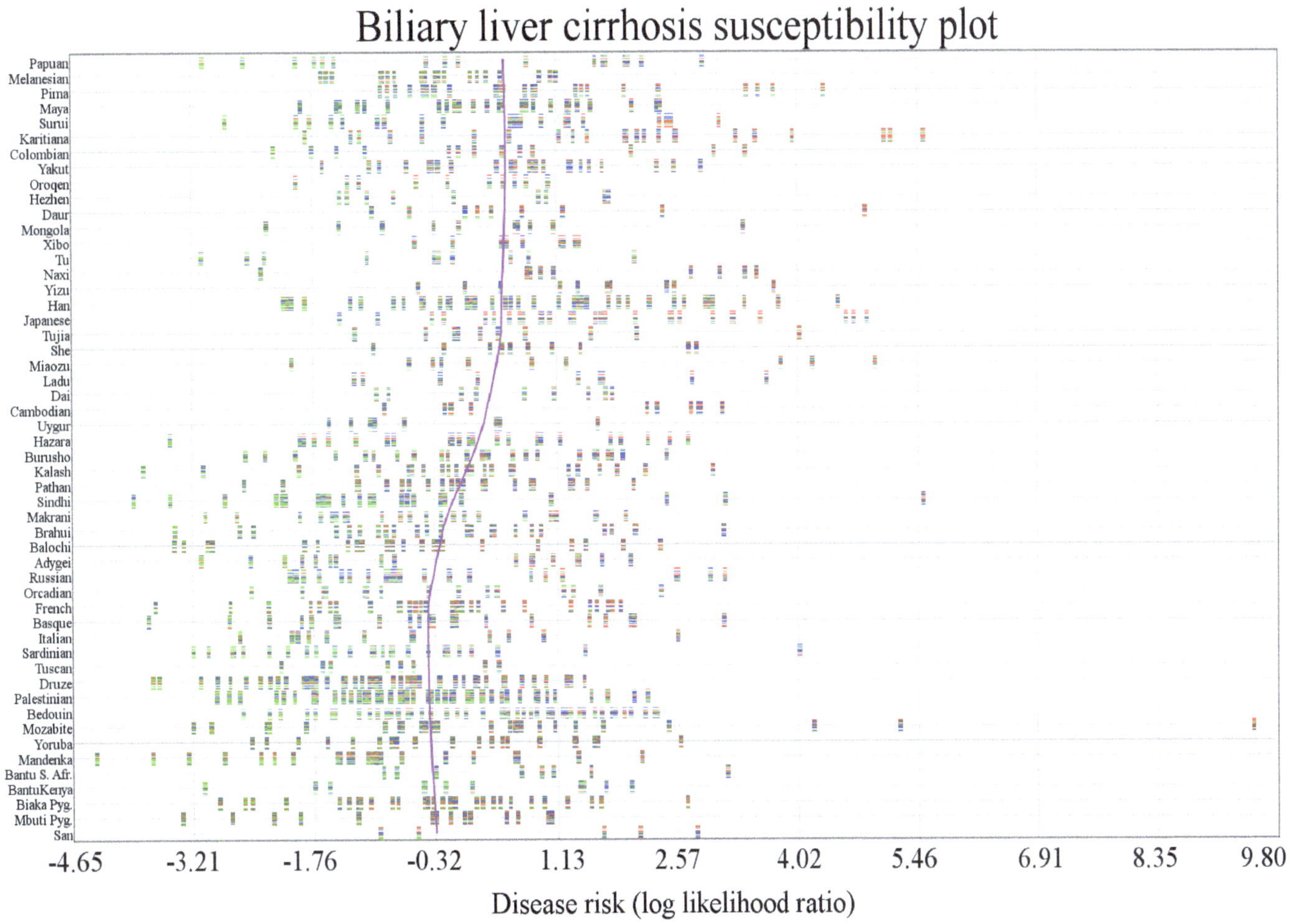

Figure 4. Variability in genetic risk in biliary liver cirrhosis. Individuals are represented by vertical multicolored rectangles. Bar colors indicate the following: red: homozygosity for a risk allele; blue: heterozygosity; green: homozygosity for a protective allele; white: missing genotype. The x-axis shows risk and the y-axis shows each population. The purple line is a locally weighted linear regression curve displaying the general direction of disease susceptibility as populations migrated from West to East. No trend is obvious, but risk appears to be higher in Cambodian, Yizu, Japanese, and San populations.

Other Diseases

We observed genetic risk trends associated with migration in other diseases, including prostate cancer, alopecia areata, melanoma, asthma, neuroblastoma, polycystic ovary syndrome, and pancreatic cancer. One of the most extreme examples of genetic risk differentiation was observed in ulcerative colitis. Figure 5 displays the distribution of the expected amount of genetic risk difference for this disease between the Sinhdi population and all others. The red vertical line represents the actual observed genetic risk difference between the two populations. Only 15 out of 100,000 randomly generated genetic risk values had a larger ulcerative colitis genetic risk difference.

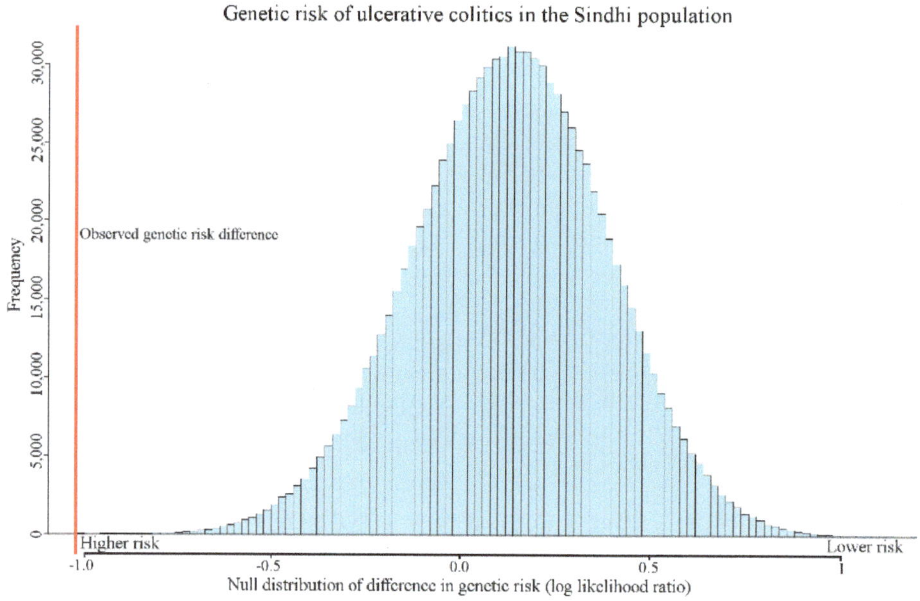

Figure 5. Expected amount of genetic risk differentiation in ulcerative colitis. Random variation caused by genetic drift may alter genetic risk between populations. The figure represents the expected amount of genetic risk difference between Sindhi and all other populations. A randomly generated genetic risk score was computed by randomly picking SNPs to represent ulcerative colitis (see Materials and Methods). The red vertical line represents the observed genetic difference between the Sindhi and all other populations combined. Only 15 out of 100,000 randomly generated genetic risk values had a larger genetic risk difference than the observed.

We found genetic risk differentiation in multiple worldwide populations for pancreatic cancer and other diseases (Table 1). Detailed information relating to genetic risk differences for many other diseases are available on (geneworld.stanford.edu).

Individual SNPs may have a large impact on overall levels of genetic risk differentiation (Text S1 and Figure S1). One notable case is rs13151961, associated with inflammatory bowel disease. Figure S1 shows examples of diseases in which a few SNPs have a disproportionate impact as well as examples in which all SNPs have a uniform impact on genetic risk differentiation.

Fst and Differentiation

Our method can detect genetic-risk differentiation that is not captured by Fst calculations.Figure S2 shows that rank normalized Fst values, after binning with SNPs having matching allele frequencies, failed to detect population structure differentiation among SNPs associated with biliary liver cirrhosis. After combining p-values for each individual SNP, the combined p-value was 0.91. Our method detects genetic-risk differentiation in the Japanese and Druze populations (Figure 2B). Figure S3 shows the rank normalized Fst values for SNPs associated with type 2 diabetes. The combined p-value for the observed type 2 diabetes Fst scores was not significant (p=0.30). Our method detects genetic-risk differentiation for multiple populations in type 2 diabetes, as can be observed in Figure 2A.Text S1 has additional details on the Fst analysis.

Individual Outliers

Individuals may be outliers with respect to genetic risk, leading to false detection of genetic risk differentiation in a population. We defined outliers as individuals deviating >1.5 times the interquartile genetic risk range from the median genetic risk and removed them from each population-disease sets showing genetic risk differentiation (Table S4). Each disease showed strong signs of genetic risk differentiation after removing outlying individuals.

Controlling for Association Ascertainment Bias

Disease-associated SNPs are likely to be biased towards genomic regions with higher LD levels where they are tagged more efficiently. We addressed this bias by resampling from SNPs reported to be associated with any phenotype (disease or not) as opposed to resampling from all SNPs. We discarded 614,563 and retained 46,192 of the genotyped SNPs in the HGDP cohort. Genetic risk differentiation was still detected after resampling exclusively from SNPs associated with a phenotype (Table S5).

DISCUSSION

Our findings place the genetic basis of disease susceptibility in the context of human migration and increase understanding of the role of population differentiation in complex disease. The HGDP-CEPH cohort represents 51 populations from around the world, with >650,000 genotyped SNPs per person. We combined this cohort with SNPs from our disease association database and were able to assess the genetic risk of disease of all populations in the HGDP-CEPH cohort. We demonstrated that differences in genetic risk for multiple diseases go well beyond what is expected by genetic drift. In addition, using a human population phylogenetic tree allowed us to incorporate a substructure of worldwide relationships.

A certain amount of variation can be expected due to random frequency variation in disease-associated genotypes. Genetic risk differentiation not caused by genetic drift is more likely to be caused by environmental adaptation, and provides insight into factors affecting disease susceptibility. Our method controls for random genetic risk variation by randomly sampling SNPs across the genome. When testing for significance in our model, the baseline was the expected amount of risk differentiation caused by genetic drift between a population of interest and all others. The low q-values in Table 1 are of particular interest, as they indicate that genetic drift is unlikely to account for the amount of observed genetic risk differentiation.

Genetic risk differentiation occurred in multiple populations in type 2 diabetes. In contrast, genetic risk differentiation in biliary liver cirrhosis was found in only two populations. Ulcerative colitis exhibits the single most extreme example of genetic risk differentiation (Figure 5). These three diseases capture the observable distinct genetic risk differentiation patterns.

Admixture

While the HGDP is an important sampling of the worldwide distribution of genotypes, it is possible that some populations are admixed. Detecting genetic risk differentiation in admixed populations may be more difficult, as the levels of genetic risk would most likely be less extreme. The genetic risk for any two populations in which admixed individuals were erroneously sampled is expected to be somewhere between the population with the least and most genetic risk. This explains why controlling for admixture is unlikely to decrease the significance of our findings.

Type 2 Diabetes

Multiple independent genetic risk differentiation events have occurred in various worldwide populations. Table 1 shows individual populations reported to have undergone or inherited an increase/decrease in genetic risk. We applied a maximum likelihood method to identify branches in a phylogenetic tree representing independent genetic risk differentiation events for type 2 diabetes (Figure 2A). Individuals migrate toward decreased genetic risk of type 2 diabetes as populations migrated East. Figure 3 shows that while there was great variability in risk within populations, there was still a steady decrease in susceptibility to type 2 diabetes.

Biliary Liver Cirrhosis

There were 44 SNPs associated with biliary liver cirrhosis used in this study. We found genetic risk differentiation in the Japanese and Druze populations (Figure 2B). Risk was increased in the Japanese population. The genetic risk score (combined LLR) was 1.691, compared to 0.026 for all other populations combined. The q-value for such a large risk difference was 0.0112 (Table 1). Consistent with results in this study, our previous work shows significantly higher biliary liver cirrhosis risk in the Japanese population (p: 0.013) in the Hap Map III cohort [29].

Ulcerative Colitis

Like all the diseases discussed in this study, ulcerative colitis has genetic and environmental components [30]. Our results suggest that genetic risk differentiation for this condition is increased in South Asian Sindhis (Figure 5). Previous studies have reported that prevalence rates are not necessarily correlated to significant increases in genetic risk[31], [32]. However, our results imply that some event affected pathophysiology of this disease in a particular population. The environment, as well as currently unidentified loci, may also affect the absolute risk. In addition to the Sindhi population, genetic risk differentiation was detected in the Palestinian and Balochi populations (Table 1).

Cause(s) of Type 2 Diabetes Genetic Risk Differentiation

Specific environmental differences inducing genetic risk differentiation in type 2 diabetes and other diseases have not been found. However, there is evidence that climate, diet, and living conditions have led to them [33]. For example, exposure to viruses may have increased risk for type 1 diabetes [34]. Autoimmune diseases show disproportionate positive selection in a trajectory toward increased versus decreased risk [35]. This finding has given rise to speculation that viral epidemics are likely to have increased the risk of these diseases by selecting for an overactive immune system. It is also established that modern cultural changes can cause drastic differences in disease prevalence in related populations [36]. However, little is known about how these changes modify risk profiles and disease prevalence over time. This study provides critical clues for the foundation for future analyses.

Detection of Disease Inter-Relationships

The concepts discussed here could be used to link diseases that may share pathophysiology and environmental triggers. It is possible that modulation of environmental features in different global regions changed the genetic risk of certain diseases. If an environmental feature affects multiple diseases, risk estimates for the diseases should correlate across the same populations, even if no common genetic basis is apparent. If the environment increases risk for one disease and decreases it for another, a negative correlation is expected. Finding diseases with genetic risk estimates correlated across worldwide populations would represent a novel and potentially highly informative approach to uncover shared pathophysiologies. This type of analysis would benefit from the largest possible catalog of genetic variants. For example, in the HGDP, East Asia is biased to detect more genetic risk differentiation than the Americas, due to the increased sensitivity that comes from 17 East Asian versus 5 American populations. In addition, full genome sequence analysis enables the inclusion of copy number polymorphisms and rare variants that may be found to contribute to complex disease susceptibility. As sequencing costs decrease, analyses expanding the scope of this study will occur.

MATERIALS AND METHODS

Data and Cohorts

We investigated >650,000 SNPs from the HGDP-CEPH, which has DNA from 1043 individuals in 51 populations on 8 continents [15]. We also used the HapMap Phase 3 cohort to analyze 1.6 million SNPs from 11 populations [29]. Finally, we used VARIMED, a database of disease-associated SNPs [37]. VARIMED was built by curating 5,478 published studies with 4,573 disease associations. At the time of this study, it contained 67,678 unique phenotype-associated SNPs, of which 51,404 were associated with a disease. Of these, after filtering by p-value and other methods, 723 unique SNPs were on the Illumina genotyping array we used to represent disease phenotypes in the HGDP [15].

For this study, we used only GWAS SNPs that had been detected across ≥2 populations with p-values<10^{-6}. SNPs were excluded if information about them was insufficient to compute likelihood ratios for the genotypes of associated SNPs. VARIMED was used to compute genetic risk estimates of the resulting diseases across all HGDP populations.

The HapMap Phase 3 cohort includes 11 populations with 1.6 million SNPs genotyped per person [29]. We used this cohort, in combination with results from previous work [25], to check our results in individual populations with elevated levels of genetic risk differentiation. All the methods with the exception of multiple hypothesis testing are applicable to this cohort.

The Combined Likelihood Ratio As a Measure of Genetic Risk

In this paper, the sample genetic risk in a population is referred to as genetic risk. Any computation of a population's genetic risk is inaccurate, due to the inability to genotype all individuals in a population and the existence of many undiscovered disease-associated variants.

The likelihood ratio (LR) represents the effect size of a particular genotype on genetic disease risk. SNPs in linkage disequilibrium (R^2≥0.2) in a population were excluded.

Computing Genetic Risk of a Disease in an Individual

For a given bi-allelic SNP, there are three possible genotypes: homozygous for the major allele, homozygous for the minor allele, or heterozygous. The function **L(*g*)** maps the genotype ***g*** to the estimated likelihood ratio. The LR used in our calculations represents the weighted mean LR reported across all studies [37]. Each LR was weighted by the square root of the sample size in each association study.

The following is our method for combining multiple LRs from multiple GWASs. ***A*** is the vector of all sample sizes in each study. P(***g*** | disease is present)$_i$ is the estimated probability that a genotype ***g*** is in the disease population of the ***i***th GWAS. P(***g*** | disease is not present)$_i$ is the estimated probability that a genotype ***g*** is present in the non-disease population of the ***i***th GWAS.

$$A=[A_1,A_2,A_3,\ldots,A_k]$$

$$L(g)=\frac{\sum_{i=1}^{k}\sqrt{A_i}\times\dfrac{P(g|\text{disease is present})_i}{P(g|\text{disease is not present})_i}}{\sum_{i=1}^{k}\sqrt{A_i}}$$

Once computed, the combined LR was used to compute the genetic disease risk for each individual, as follows. ***G*** is the vector of all genotypes in disease-associated SNPs in person***m***.

$$G_m=[G_{1m},G_{2m},G_{3m},\ldots,G_{nm}]$$

The predicted genetic risk ***r*** for person ***m*** is the log of the combined likelihood ratios for all disease-associated variants present in that person.

$$r_m=\log\left(\prod_{i=1}^{n}L(G_{im})\right)=\sum_{i=1}^{n}\log(L(G_{im}))$$

Computing the Genetic Risk of Disease in a Population

The genetic risk for a population was the mean risk of all its members. Derived populations created by combining multiple populations in the HGDP were treated as single populations, and the following computations applied equally. ***S*** is the vector of a population.

$$S=[s_1,\ s_2,\ s_3,\ \ldots\ s_z]$$

The predicted genetic risk ***R*** for an entire population ***P*** is the mean predicted risk for everyone in the population:

$$R_p=\frac{1}{|S|}\sum_{j=1}^{|S|}r_{s_j}$$

Computing the Difference in Genetic Risk across Two Populations

In order to compare genetic risk (D) between two populations A and B, we subtracted the genetic risk of one population from the other's:

$$D_{AB}=R_A-R_B$$

Population A's risk for a disease is higher than B's if D_{AB} is positive and lower if D_{AB} is negative.

Modeling the Expected Difference in Genetic Risk

We constructed a distribution of the expected difference of genetic risk between two populations in order to see if the observed difference was larger than expected by random chance:

The vector H represents all SNPs associated with a particular disease.

$$H=[h_1,h_2,h_3,\ldots,h_w]$$

Our goal was to assess the significance of a difference in risk D_{AB} across two populations Aand B. Each element in H was replaced by a SNP randomly drawn from the entire set of SNPs in the two populations. The global major allele frequency of the randomly drawn SNP was drawn to match original SNP's global major allele frequency. In every case, the risk allele's major or minor allele status in the randomly drawn SNP matched that of the SNP it replaced. In addition, each SNP was placed in one of eight functional categories (frameshift, nonsense, missense, untranslated, near-gene, intron, coding-synonymous, or unknown). Each randomly drawn SNP also matched the functional category of the SNP it replaced in vector H.

Once all elements of H had been replaced, the genetic risk of all populations was recomputed, effectively assigning a randomly generated genetic risk score to each population. Since each population was assigned a genetic risk score from the same randomly drawn set of SNPs, the expected amount of correlation between genetic risk values among all populations was preserved. We created phylogenetic trees of our results with each branch representing a migration event. We computed the genetic risk difference of each migration event by subtracting the genetic risk of all descendant populations from the risk of all ancestral populations (those above the branch). Branches on the human phylogenetic tree created from the HGDP populations were tested for genetic risk differences. We computed the difference in risk between all ancestral and descendant populations. A phylogenetic tree of all the HGDP populations was used as described previously [1]. Each branch in the tree partitions an ancestral and descendant population. The ancestral population is made up of populations above a branch; the descendant population is below it. The expected difference in genetic risk between all possible ancestral and descendant comparisons was computed by randomly replacing all disease-associated SNPs by performing a random draw of H 100,000 times.

We computed a matrix representing 100,000 randomly generated phylogenetic trees and compared it with the observed phylogenetic tree in the context of genetic risk. Let $r_{i,k}$ represent the genetic risk difference between ancestral and descendant populations in branch i computed from the kth randomly generated H.

$$R=\begin{pmatrix} r_{11} & \cdots & r_{100,1} \\ \vdots & \ddots & \vdots \\ r_{1,100000} & \cdots & r_{100,100000} \end{pmatrix}$$

This matrix of 100,000 randomly generated phylogenetic trees was then compared to the observed tree. O_i is the observed genetic risk difference between ancestral and descendant populations in the ith branch of the phylogenetic tree.

$$\boldsymbol{O}=[O_1,\ O_2,\ O_3,\ldots O_{100}]$$

The probability of observing a genetic risk difference between ancestral and descendant populations in the ith branch is as follows:

$$\Pr(O_i)=$$

$$\frac{2}{100{,}000+1}\min\left(\sum_{k=1}^{100{,}000} I(r_{i,k}\le O_i)+1,\ \sum_{k=1}^{100{,}000} I(r_{i,k}>O_i)+1\right)$$

This equation gave a probability for a specific genetic risk difference observed on branch i. However, we tested multiple strongly correlated populations simultaneously. In addition, the genetic risk differences of nearby branches in the phylogenetic tree were strongly correlated due to genetic similarity between populations. One row in matrix ***R*** represented the genetic risk difference of all branches in the tree between ancestral and descendant populations in a single random draw. The genetic risk difference was calculated with the same set of SNPs in every branch in the phylogenetic tree in each row. This preserved the correlation of genetic risk among closely related populations and enabled us to detect independent genetic risk deviations.

To assess the significance of a p-value α in the context of the entire tree, we first computed the number of branches having equal or greater significance for genetic risk differentiation.

Significant genetic risk differentiation events were calculated follows. $N(\alpha)$ is the observed number of branches with a p-value for genetic risk differentiation at or below α.

$$N(\alpha)=\sum_{k=1}^{100} I(\Pr(O_i)\le\alpha)$$

We next computed the expected number of branches with genetic risk differentiation at significance α from the matrix representing 100,000 randomly generated phylogenetic trees.

$$E(\alpha)=\frac{1+\sum_{k=1}^{100{,}000}\sum_{i=1}^{100} I(\mathrm{P}(\mathrm{r}_{i,k})\le\alpha)}{1+100{,}000}$$

$Q(\alpha)$ (the q-value) for a branch with p-value α for genetic risk differentiation was:

$$Q(\alpha)=\frac{E(\alpha)}{N(\alpha)}$$

While $Q(\alpha)$ can inform on the significance for a particular p-value, it does not estimate the total number of distinct genetic risk differentiation events in a phylogenetic tree. We estimated the most likely number of migration events independently contributing to observed genetic risk deviation in a phylogenetic tree using a maximum likelihood approach.

P is the vector containing all probabilities of the observed genetic risk difference between ancestral and descendant populations of each branch of tree t.

$$P=[\Pr(O_1),\Pr(O_2),\ldots,\Pr(O_{100})]$$

We defined a function that returns true if branch ***k*** is not a descendant of branch *z*:

$$D(k,z)=\begin{cases}\text{true,} & \text{branch k is not a descendant c} \\ \text{false} & \text{branch k is a descendant of}\end{cases}$$

Branch z represents the branch in which an independent genetic risk differentiation event occurred. Applying the principle of maximum likelihood, the branch z that maximizes the likelihood function in tree t is shown below. The number of events refers to the number of independent genetic risk differentiation events.

$$\ln(\mathrm{L}(\mathrm{t}|1\ \mathrm{Event})) = \ln(L(P|1\ \mathrm{Event})) = \sum_{i=1}^{|P|} \ln\left(\frac{\sum_{k=1}^{100,000} I(P(r_{i,\mathrm{k}}) \le \mathrm{p}_i) I(P(r_{z,k}) \le p_z \vee D(i,z))}{\sum_{k=1}^{100,000} I(P(r_{z,k}) \le p_z \vee D(i,z))}\right)$$

The log likelihood function produces the branch most likely to have undergone genetic risk differentiation, given that a single genetic risk differentiation event occurred. The 'v' symbol represents a logical disjunction that returns true if one or more of the two operands is true. The branch that maximized likelihood was the one most likely to have caused the risk differentiation. However, there may have been multiple genetic risk differentiation events. This maximum likelihood method can be generalized to detect an arbitrary number of branches that underwent genetic risk differentiation.

Find the indices $z_1, z_2, \ldots, z_y$ for branches that maximize the following likelihood function:

$$\sum_{i=1}^{|P|} \ln \left(\frac{\sum_{k=1}^{100,000} I(P(r_{i,k}) \le p_i) I\left(P(r_{z_1,k}) \le \mathrm{p}_{z_1} \vee D(\mathrm{i},z_1)\right) I\left(P(r_{z_2,k}) \le \mathrm{p}_{z_2} \vee D(\mathrm{i},z_2)\right), \ldots, I\left(P(r_{z_y,k}) \le \mathrm{p}_{z_y} \vee D(\mathrm{i},z_y)\right)}{\sum_{k=1}^{100,000} I\left(P(r_{z_1,k}) \le \mathrm{p}_{z_1} \vee D(\mathrm{i},z_1)\right) I\left(P(r_{z_2,k}) \le \mathrm{p}_{z_2} \vee D(\mathrm{i},z_2)\right), \ldots, I\left(P(r_{z_y,k}) \le \mathrm{p}_{z_y} \vee D(\mathrm{i},z_y)\right)}\right)$$

Due to the constraints this equation imposed, a very large number of random draws was required to produce an accurate estimate. All indicator functions in the numerator must return a non-zero value for the count to be incremented by one; this was computationally infeasible. In order to make the computation tractable, we made a simplifying assumption that only a single ancestral branch which has undergone de novo genetic risk differentiation (as opposed to an arbitrary combination of branches that have undergone de novo genetic risk differentiation) can pass a significantly modified genetic risk for a disease to a descendant branch. This led us to the method used for this study. The following calculation was used to find all branches with sufficient evidence of having undergone independent genetic risk differentiation.

Find the indices $z_1, z_2, \ldots, z_y$ that maximize the following likelihood function:

$$\sum_{i=1}^{|P|} \ln \left(\max \left\{ \begin{array}{c} \dfrac{\sum_{k=1}^{100,000} I\left(P(r_{i,k}) \le \mathrm{p}_i\right) I\left(P(r_{z_1,k}) \le p_{z_1} \vee D(i,z_1)\right)}{\sum_{k=1}^{100,000} I\left(P(r_{z_1,k}) \le p_{z_1} \vee D(i,z_1)\right)} \\ \dfrac{\sum_{k=1}^{100,000} I\left(P(r_{i,k}) \le \mathrm{p}_i\right) I\left(P(r_{z_2,k}) \le p_{z_2} \vee D(i,z_2)\right)}{\sum_{k=1}^{100,000} I\left(P(r_{z_2,k}) \le p_{z_2} \vee D(i,z_2)\right)} \\ \bullet \\ \bullet \\ \bullet \\ \dfrac{\sum_{k=1}^{100,000} I\left(P(r_{i,k}) \le \mathrm{p}_i\right) I\left(P(r_{z_y,k}) \le p_{z_y} \vee D(i,z_y)\right)}{\sum_{k=1}^{100,000} I\left(P(r_{z_y,k}) \le p_{z_y} \vee D(i,z_y)\right)} \end{array} \right. \right)$$

SUPPORTING INFORMATION

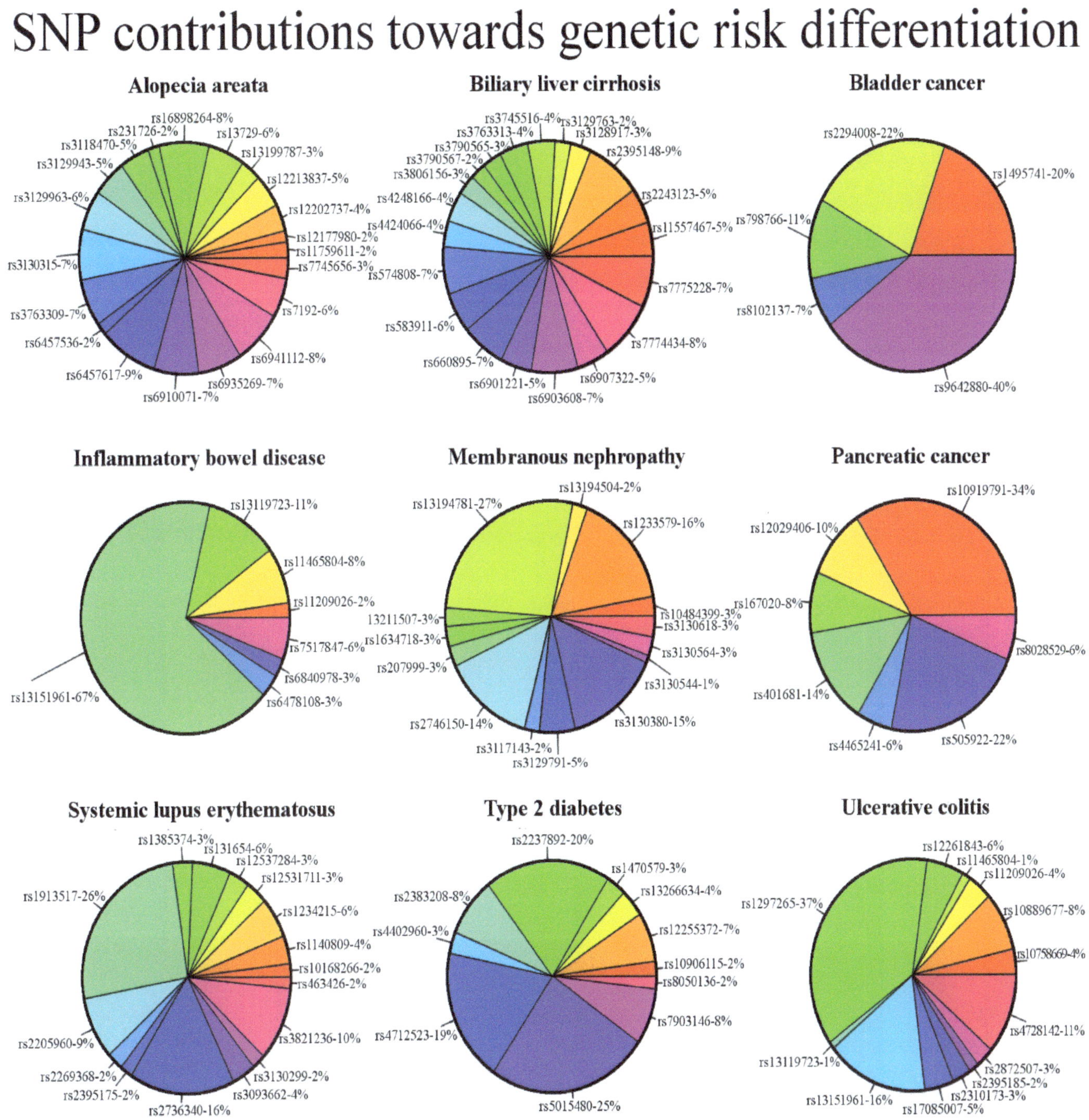

Figure S1. Impact of individual SNPs on genetic risk differentiation. The relative deviation from expected genetic risk attributable to each SNP is shown in each pie chart. The branch in the human phylogeny tree with the most significance for genetic risk differentiation was used to assess the impact of individual SNPs for each disease. The branch selected for alopecia areata includes European, Central South Asian, East Asian, Oceaniac, American, Palestinian, and Druze populations. The branch selected for inflammatory bowel disease includes the Brahui and Makrani populations. The branch selected for pancreatic cancer includes Central South Asian, East Asian, Oceaniac, and American populations. The branch selected for systemic lupus erythematosus includes the Mayan and the Pima populations. The branch selected for type 2 diabetes includes East Asian, Oceaniac, and American populations. The branch selected for ulcerative colitis includes the Brahui and Makrani populations. The branches selected for biliary liver cirrhosis, bladder cancer, and membranous nephropathy are the Druze, Tu, and French Basque populations, respectively. Each branch is compared to all other worldwide populations. Some SNPs have a disproportionate impact on genetic risk. An example is rs13151961, associated with inflammatory bowel disease.

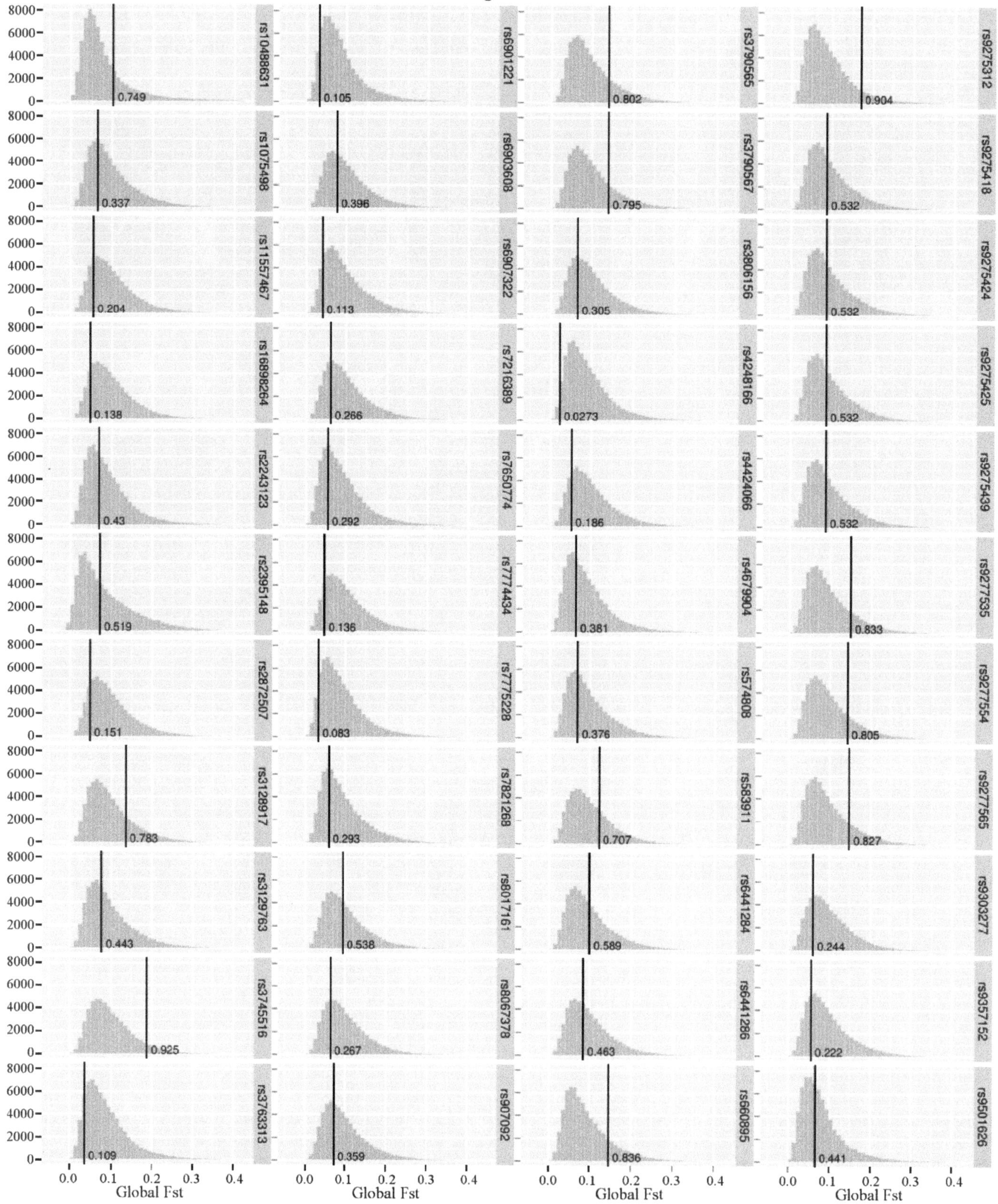

Figure S2. Fst analysis for biliary liver cirrhosis. Global Fst values using the HGDP cohort were calculated for the 44 high confidence biliary liver cirrhosis SNPs used in this study. The Fst score for each SNP was compared to all other SNPs with the same minor allele frequency. The p-value represents the fraction of SNPs with a lower Fst value. No individual lung cancer SNP appeared to be significantly differentiated. Fst analysis failed to capture the localized genetic-risk differentiation that has occurred in the Japanese and Druze populations. Combining the p-values revealed no signs that these SNPs have collectively undergone differentiation (p-value=0.91).

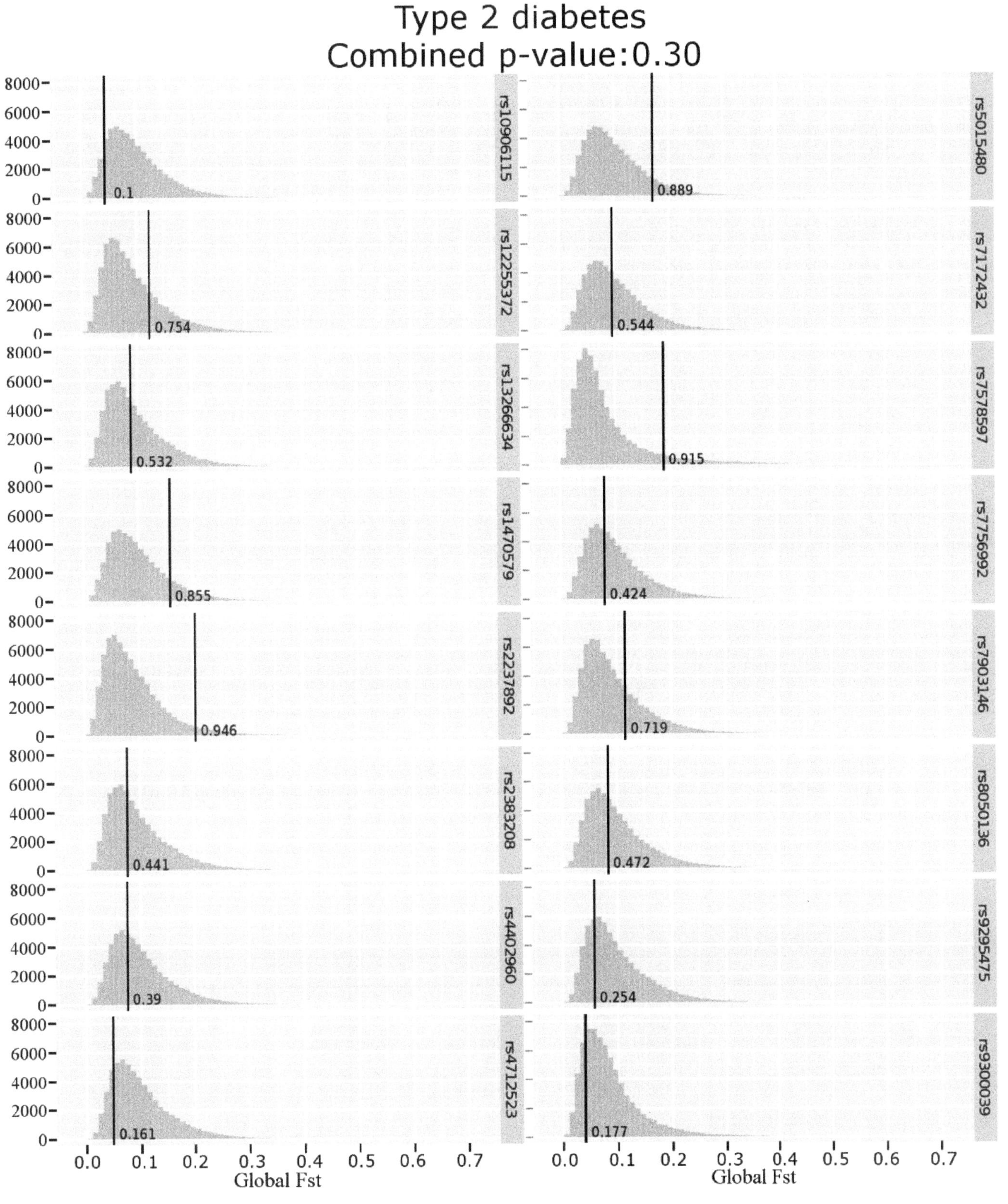

Figure S3. Fst analysis for type 2 diabetes. Global Fst values were calculated for the 16 type 2 diabetes-associated SNPs used in this study. The distribution of p-values did not reveal elevated type 2 diabetes genetic risk differentiation across worldwide populations compared to non-disease associated SNPs. The Fst score for each SNP was compared to all other SNPs with the same minor allele frequency. The combined p-value for these SNPs failed to capture the extreme extent to which genetic-risk differentiation has occurred at a global scale (p-value=0.30).

Table S1. Type 2 diabetes-associated SNPs replication demographics. The number of times any publication has found each type 2 diabetes SNP used in this study to be associated with the disease is shown. European populations have the most replications across the majority of SNPs associated with type 2 diabetes. Arabic and American Indian populations have the lowest number of replications.

dbSNP	European	Asian	Afri-can	Indian	His-panic	Arab	American Indian	Locus
1470579	5	5	0	0	0	0	0	IGF2BP2
2237892	2	12	0	0	0	0	0	KCNQ1
2383208	0	2	0	0	0	0	0	-
4402960	19	10	2	1	0	0	0	IGF2BP2
4712523	1	2	0	0	0	0	0	CDKAL1
5015480	2	5	0	0	0	0	0	-
7172432	2	6	0	0	0	0	0	-
7578597	3	1	0	0	0	0	0	THADA
7756992	6	12	0	0	0	0	0	CDKAL1
7903146	75	15	7	5	2	2	2	TCF7L2
8050136	12	1	1	0	0	0	1	FTO
9295475	0	2	0	0	0	0	0	CDKAL1
9300039	2	0	1	0	1	0	0	-
10906115	0	1	0	0	0	0	0	-
12255372	35	8	4	5	3	1	1	TCF7L2
13266634	28	14	1	0	0	0	0	SL-C30A8

Table S2. Asian-specific type 2 diabetes effect size. The table compares the overall likelihood ratio for each SNP associated with type 2 diabetes against the Asian-specific likelihood ratio. The Asian-specific likelihood ratio was computed by including only GWASs based in Asian populations. While many of the observed effect sizes did not differ, some were significantly different and may lead to a modified risk estimate for Asian populations.

DBSNP	Genotype	Overall Likelihood Ratio	Asian Likelihood Ratio
rs1470579	AA	0.89	0.90
	AC	1.07	1.06
	CC	1.28	1.31
rs2237892	TT	0.61	0.63
	CT	0.92	0.92
	CC	1.21	1.21
rs2383208	GG	0.71	0.71*
	AG	0.95	0.95*
	AA	1.27	1.27*
rs4402960	GG	0.90	0.89
	TG	1.06	1.08
	TT	1.27	1.34
rs4712523	AA	0.75	0.75*
	AG	1.03	1.03*
	GG	1.43	1.43*
rs5015480	TT	0.89	0.89
	TC	1.22	1.23
	CC	1.63	1.65
rs7172432	GG	0.89	0.90
	AG	0.98	0.98
	AA	1.09	1.07

rs7578597	CC	0.67	0.31
	TC	0.83	0.56
	TT	1.02	1.01
rs7756992	AA	0.83	0.79
	AG	1.04	0.98
	GG	1.35	1.32
rs7903146	CC	0.70	0.96
	TC	1.17	1.16
	TT	1.97	1.66
rs8050136	CC	0.88	0.89
	AC	1.05	0.95
	AA	1.26	1.00
rs9295475	AA	0.89	0.89*
	AG	0.94	0.94*
	GG	1.39	1.39*
rs9300039	AA	0.61	**NA**
	AC	0.80	**NA**
	CC	1.05	**NA**
rs10906115	GG	0.85	0.85*
	AG	0.97	0.97*
	AA	1.10	1.10*
rs12255372	GG	0.81	0.97
	TG	1.12	1.75
	TT	1.66	1.01
rs13266634	TT	0.83	0.87
	TC	0.95	0.98
	CC	1.11	1.11

NA: The SNP was never verified in an Asian based GWAS and was excluded

*no likelihood ratio difference is possible since only Asian-based studies exist

Table S3. Genetic risk of type 2 diabetes using Asian GWASs. Most GWASs are based on European cohorts. The effect size of an associated variant may differ across distinct populations. This table compares the combined likelihood ratio for type 2 diabetes (computed using all available GWASs) with the combined likelihood ratio using only GWASs in Asian populations. This approach allowed us to base the effect size of each variant on Asian populations as opposed to mainly European derived populations. Genetic risk for type 2 diabetes is still significantly lower in Asian populations when the effect size for each associated variant is taken from GWASs exclusively based in Asian populations (q-value<0.05).

	All GWASs			**Asian based GWASs**		
Population	**Likelihood Ratio**	**P-value**		**Likelihood Ratio**	**P-value**	
Bedouin	1.78	1.63×10^{-2}		1.46	2.84×10^{-2}	
Cambodian	0.53	7.91×10^{-3}		0.66	4.97×10^{-2}	
Colombian	0.46	1.13×10^{-2}		0.42	2.95×10^{-2}	
Daur	0.43	1.03×10^{-3}		0.47	1.91×10^{-3}	
Han	0.52	3.39×10^{-3}		0.64	1.82×10^{-2}	
Hezhen	0.45	1.95×10^{-3}		0.54	1.16×10^{-2}	
Japanese	0.49	1.33×10^{-3}		0.61	1.34×10^{-2}	

Lahu	0.51	1.70 x 10^{-2}		0.58	3.42 x 10^{-2}	
Mongola	0.45	6.90 x 10^{-4}		0.50	2.41 x 10^{-3}	
Mozabite	2.11	4.35 x 10^{-3}		2.27	4.91 x 10^{-3}	
A*	0.89	1.15 x 10^{-3}		0.84	5.99 x 10^{-3}	
B*	0.77	3.10 x 10^{-3}		0.76	3.44 x 10^{-3}	
C*	0.57	1.50 x 10^{-4}		0.61	1.56 x 10^{-3}	
D*	0.55	1.90 x 10^{-4}		0.60	1.61 x 10^{-3}	
East Asian	**0.51**	**2.70 x 10^{-4}**		**0.60**	**2.71 x 10^{-3}**	
E*	0.47	9.29 x 10^{-3}		0.44	2.35 x 10^{-2}	
Naxi	0.46	3.53 x 10^{-3}		0.60	3.47 x 10^{-2}	
Oroqen	0.45	2.41 x 10^{-3}		0.53	1.00 x 10^{-2}	
Palestinian	1.82	6.75 x 10^{-3}		1.62	4.96 x 10^{-3}	
She	0.43	2.65 x 10^{-3}		0.48	5.24 x 10^{-3}	
Surui	0.42	2.23 x 10^{-2}		0.39	4.63 x 10^{-2}	
Tu	0.51	4.51 x 10^{-3}		0.61	2.17 x 10^{-2}	
Tujia	0.49	6.11 x 10^{-3}		0.58	2.27 x 10^{-2}	
Xibo	0.51	4.97 x 10^{-3}		0.67	4.83 x 10^{-2}	
Yakut	0.63	1.38 x 10^{-2}		0.60	5.46 x 10^{-2}	
Yizu	0.43	1.47 x 10^{-3}		0.54	9.89 x 10^{-3}	

A*Europe, Central South Asia, East Asia, Oceania, and America

B*Central South Asia, East Asia, Oceania, and America

C*East Asia, Oceania, and America

D*East Asia and America

H*Colombia, Karitiana, and Surui

Table S4. Replication after removal of outliers. Outliers carry a disproportionate number of risk alleles compared to the rest of their population. For each population, we calculated the interquartile range (IQR) of genetic risk for each individual and removed persons with a genetic risk deviating more than 1.5 IQRs from the population median. The table compares original p-values to the p-value after removal of outliers. A total of 21 out of the original 24 genetic risk differentiation events had p-values<0.006 after removing outliers.

Disease	**Population**	**Original P-Value**	**P-value: No Outliers**	**Outliers Removed**
Alopecia areata	Mozabite	1.00 x 10^{-5}	1.00 x 10^{-5}	52
Alopecia areata	G*	5.70 x 10^{-4}	5.00 x 10^{-4}	52
Biliary liver cirrhosis	Druze	5.70 x 10^{-4}	5.69 x 10^{-3}	45
Biliary liver cirrhosis	Japanese	3.70 x 10^{-4}	7.45 x 10^{-3}	45
Bladder cancer	Tu	4.90 x 10^{-4}	4.70 x 10^{-3}	54
Inflammatory bowel disease	Balochi	1.00 x 10^{-5}	1.00 x 10^{-5}	91

Inflammatory bowel disease	Burusho	4.30×10^{-4}	6.05×10^{-1}	91
Inflammatory bowel disease	Makrani	1.50×10^{-4}	3.00×10^{-5}	91
Inflammatory bowel disease	E*	1.50×10^{-4}	1.00×10^{-5}	91
Inflammatory bowel disease	Palestinian	5.00×10^{-5}	4.74×10^{-1}	91
Inflammatory bowel disease	Sindhi	5.00×10^{-5}	4.25×10^{-1}	91
Membranous nephropathy	French Basque	1.00×10^{-5}	2.90×10^{-3}	67
Pancreatic cancer	A*	1.00×10^{-5}	1.00×10^{-5}	34
Pancreatic cancer	B*	1.00×10^{-5}	1.00×10^{-5}	34
Pancreatic cancer	F*	3.50×10^{-4}	3.30×10^{-4}	34
Pancreatic cancer	Yoruba	3.10×10^{-4}	3.90×10^{-4}	34
Systemic lupus erythematosus	Maya + Pima*	3.30×10^{-4}	2.50×10^{-4}	45
Type 2 diabetes	B*	3.10×10^{-4}	1.10×10^{-4}	48
Type 2 diabetes	C*	1.50×10^{-4}	1.90×10^{-4}	48
Type 2 diabetes	D*	1.90×10^{-4}	3.10×10^{-4}	48
Type 2 diabetes	East Asia*	2.70×10^{-4}	1.49×10^{-3}	48
Ulcerative colitis	Balochi	1.00×10^{-5}	1.00×10^{-5}	55
Ulcerative colitis	E*	1.00×10^{-5}	3.00×10^{-5}	55
Ulcerative colitis	Sindhi	1.00×10^{-5}	3.00×10^{-5}	55

A*Europe, Central South Asia, East Asia, Oceania, and America

B*Central South Asia, East Asia, Oceania, and America

C*East Asia, Oceania, and America

D*East Asia and America

E*Brahui, Makrani

F*Yoruba, Mandenka

G*Europe, Central South Asia, East Asia, Oceania, and America, Palestinian, Druze

Table S5. Replication after resampling from known associations. A resampling procedure is used to model the expected genetic risk difference between a population and the rest of the worldwide set of populations combined. However, a GWAS is more likely to detect associations in regions with higher linkage disequilibrium, which is a potential source for bias in detecting genetic risk differentiation. We address this by reproducing the results after randomly drawing only from SNPs previously reported to show association to any phenotype during the resampling step. Genetic risk differentiation is still detected after randomly drawing from other known associations during the resampling step.

Disease	Population	Original P-Value	Known Associa-tions P-value	SNPs
Alopecia areata	Mozabite	1.00×10^{-5}	1.00×10^{-5}	26
Alopecia areata	G*	5.70×10^{-4}	7.50×10^{-4}	38
Biliary liver cirrhosis	Druze	5.70×10^{-4}	1.69×10^{-3}	30
Biliary liver cirrhosis	Japanese	3.70×10^{-4}	2.77×10^{-3}	27
Bladder cancer	Tu	4.90×10^{-4}	5.30×10^{-4}	7
Inflammatory bowel disease	Balochi	1.00×10^{-5}	5.00×10^{-5}	11

Inflammatory bowel disease	Burusho	4.30×10^{-4}	2.49×10^{-3}	9
Inflammatory bowel disease	Makrani	1.50×10^{-4}	1.29×10^{-3}	10
Inflammatory bowel disease	E*	1.50×10^{-4}	3.00×10^{-5}	11
Inflammatory bowel disease	Palestinian	5.00×10^{-5}	3.90×10^{-4}	11
Inflammatory bowel disease	Sindhi	5.00×10^{-5}	3.10×10^{-4}	10
Membranous nephropathy	French Basque	1.00×10^{-5}	7.10×10^{-4}	17
Pancreatic cancer	A*	1.00×10^{-5}	1.00×10^{-5}	8
Pancreatic cancer	B*	1.00×10^{-5}	1.00×10^{-5}	8
Pancreatic cancer	F*	3.50×10^{-4}	2.30×10^{-4}	9
Pancreatic cancer	Yoruba	3.10×10^{-4}	1.50×10^{-4}	9
Systemic lupus erythematosus	Maya + Pima*	3.30×10^{-4}	3.50×10^{-4}	29
Type 2 diabetes	B*	3.10×10^{-4}	2.50×10^{-4}	14
Type 2 diabetes	C*	1.50×10^{-4}	1.30×10^{-4}	14
Type 2 diabetes	D*	1.90×10^{-4}	1.30×10^{-4}	14
Type 2 diabetes	East Asia*	2.70×10^{-4}	7.00×10^{-5}	14
Ulcerative colitis	Balochi	1.00×10^{-5}	1.00×10^{-5}	27
Ulcerative colitis	E*	1.00×10^{-5}	1.00×10^{-5}	27
Ulcerative colitis	Sindhi	1.00×10^{-5}	3.00×10^{-5}	27

A*Europe, Central South Asia, East Asia, Oceania, and America

B*Central South Asia, East Asia, Oceania, and America

C*East Asia, Oceania, and America

D*East Asia and America

E*Brahui, Makrani

F*Yoruba, Mandenka

G*Europe, Central South Asia, East Asia, Oceania, and America, Palestinian, Druze

Text S1. Supporting Online Material: Analysis of the Genetic Basis of Disease in the Context of Worldwide Human Relationships and Migration

Comparison with Fst

Although Fst is a traditional approach for detecting population differentiation [1], its use for detecting genetic risk differentiation is inappropriate for two reasons. First, it does not account for the effect size of each susceptibility allele. Fst treats all variants equally instead of weighing variants proportionately on how much they increase/decreases genetic risk of disease. In addition, there is a well-known correlational structure within the populations studied in HGDP [2]. Our Genetic Risk Variation (GRV) method found significant genetic-risk differentiation for biliary liver cancer (BLC) in 2 populations (Figure 2B), yet Fst analysis failed to detect significant deviations from the background (Figure S2). This finding is not surprising in BLC, as global Fst values are not likely to detect the localized risk differentiation we observed. However, Figure S3 shows that Fst analysis also failed to detect genetic risk differentiation in type 2 diabetes despite high levels of worldwide differentiation found by our method (Figure 2A).

GRV accounts for the effect size of each genotype on disease and controls for genetic drift. It estimates the geographic location where disease risk is most likely to have changed compared to other populations, as well as its position in the phylogenetic tree. It relies on q-value computations (the expected proportion of false positives incurred when labeling a particular p-value significant) [3] to control for multiple hypotheses.

Coop et al. also developed a method to identify allele frequency correlations with environmental features [4]. However, their method correlates allele frequency to environment, while genetic risk estimates are a function of genotype frequencies. The goal of our study was to detect genetic risk differentiation across worldwide populations. We have emphasized cases where risk differentiation exceeds what would be expected under genetic drift.

Assessing the Impact of Individual SNPs

Figure S1 shows the relative impact of each SNP on overall levels of genetic risk differentiation in nine diseases. We chose one branch in the human phylogeny tree per disease in order to assess each SNP's impact. Each branch was compared to all other worldwide populations. The branch selected to assess the impact of each individual SNP was chosen as the branch with the strongest significance (lowest q-value) with respect to each disease. The expected risk difference between two populations was ascertained by the method used for multiple SNPs, but using one SNP at a time. As before, the selected branch for a disease is compared to all other worldwide populations by measuring deviation of observed genetic risk difference (between the two populations) with the expected genetic risk difference. SNPs contributing to decreased significance for risk differentiation were excluded. Each deviation measurement was normalized by dividing by the sum of all deviation measurements and multiplying by 100 to produce percent values. Figure S1 represents the percent of deviation from the expected genetic risk that is attributable to each SNP.

Phenotypes Transcend Population Boundaries

Populations are typically examined independently without combining closely related groups of people. In reality, phenotypes are seldom strictly divided across populations. It is evident that the degree of phenotype similarity observed across different populations varies. Some characteristic phenotypes may be observed in almost all populations on a continent, such as periocular anatomy in East Asia. Traits such as disease susceptibility are not as easily observed and it is difficult to know in advance how to group different populations when searching for genetic risk deviations. This is why iteratively grouping the two most phenotypically closely related populations to search for genetic risk differentiation not only increases sample size, but represents abandoning the common practice of studying traits within established population boundaries. This can also lead to revelations regarding points on the human phylogenetic tree at which genetic risk differentiation occurred. For example, studies have indicated that East Asians have a lower genetic risk for type 2 diabetes [5]. We found that American populations also have a lower genetic risk for this disease. However, the ancestral population common to East Asians and Americans is not the most likely to have actually undergone genetic risk differentiation (Figure 2A). One of the insights gained from using a phylogenetic tree in this analysis is that it is more likely that an ancestor of East Asians and a different ancestor of Native American populations independently experienced a genetic risk differentiation event in a trajectory toward lower risk for type 2 diabetes. This possibility is counter to the idea that Native American and East Asian populations merely inherited lower genetic risk prior to the branching of the two populations.

Limitations

Prevalence information was not included in this study. Genetic risk does not necessarily correlate with disease prevalence in practice [6,7] or theory. Intuitively, one might expect prevalence to be higher in populations with higher predicted genetic risk, but we hypothesize that environmental risk and genetic risk can just as easily be negatively correlated. Higher genetic risk may accumulate in an environment strongly disfavoring disease as there is no selection pressure to act against the spread of increased genetic risk for disease. Likewise, genetic resistance to disease may develop as a natural result of living in an environment strongly favoring its occurrence. In addition, if the environment has recently undergone change or a population is physically displaced, risk/prevalence ratios would change. Such an analysis is more appropriate for a study exploring gene-environment interactions on disease in the context of differences in genetic risk. It has been shown that the environment has interacted with many disease-associated loci in different ways across distinct human populations [8].

This analysis does not incorporate information about the environment, which may further explain relationships between disease incidence and genetic risk. There is already work underway to comprehensively associate environmental factors with disease [9]. Repeating such work in the context of genetic risk estimates would likely lead to an increased ability to predict overall risk of disease as it would include a variety of environmental factors as well as all known disease associated SNPs.

A large fraction of mutations differentiated across populations are compensatory in long-term evolution [10]. The role of compensatory mutations in recent human evolution is unlikely to be as significant as in long-term evolution due to the relatively small effect size of common disease-associated SNPs. Yet the role of compensatory mutations in recent human evolution is not well established. In our study, genetic risk was measured with alleles increasing and decreasing genetic risk collectively. If variants increasing genetic risk in a particular population are present, they will not be detected if their effect is mitigated by other variants decreasing risk, potentially limiting our findings.

Because the majority of disease-associated SNPs used in this study were tag SNPs, it is also possible that varied LD patterns across populations would stop the tag SNP from tagging the correct disease-associated SNP in a different population. The best way to combat the possibility of false positives due to this problem is to increase the minimum number of SNPs required to represent a disease. In addition, the majority of disease-associated SNPs contain other reported associations in LD. We attempted to pick the associated SNP closest to the true causative SNP, thereby further decreasing the likelihood of an LD gap across distinct populations. We accomplished this task by examining effect size information for each of the associated SNPs in LD and picking the SNP with the largest effect size from the LD block while discarding the others. We considered only SNPs with p-values of association below 10^{-6}. In addition, the SNPs in this study were replicated across two or more populations (see Table S1 for type 2 diabetes), thereby making it more likely that they were in strong linkage disequilibrium to the causal SNPs and rendering them appropriate proxies.

A more obvious problem is the missing heritability problem. Known variants explain very little of the observed risk for complex disease. While there are many potential explanations for this problem (copy-number polymorphisms, multiple loci with very small individual effects, pervasive epistatic effects, etc.), it will take considerable further research to explore them all [11]. Further research will likely identify rare variants associated with many diseases. The variants in this study have a specific allele frequency and effect size combination that makes them detectable in the populations in which the GWASs were conducted. Our results reflect this particular category of disease-associated SNPs.

Despite the low predictive power caused by the missing heritability problem, genetic risk differentiation was detected across multiple diseases. In addition, the fact that most GWASs are conducted in European populations makes it difficult to properly assess risk in non-Europeans. For example, if increased risk is detected in Europeans, it is possible that Africans have undiscovered variants that offset the observed difference in risk reported in this study. However, it would still be important to note that common variants increasing the genetic risk of disease primarily in European populations show strong signs of genetic risk differentiation on a global scale, as is the case for type 2 diabetes.

ACKNOWLEDGMENTS

We thank Alex Skrenchuk for computer support throughout the work conducted on this manuscript.

AUTHOR CONTRIBUTIONS

Conceived and designed the experiments: EC AJB. Performed the experiments: EC. Analyzed the data: EC RC MS AAM CJP AR CDB AJB. Contributed reagents/materials/analysis tools: EC RC MS. Wrote the paper: EC AJB.

REFERENCES

1. Li JZ, Absher DM, Tang H, Southwick AM, Casto AM, et al. (2008) Worldwide human relationships inferred from genome-wide patterns of variation. Science 319: 1100–1104
2. Fu Q, Rudan P, Paabo S, Krause J (2012) Complete mitochondrial genomes reveal neolithic expansion into europe. PLoS ONE 7: e32473.doi:10.1371/journal.pone.0032473
3. Deguilloux MF, Leahy R, Pemonge MH, Rottier S (2012) European neolithization and ancient DNA: an assessment. Evol Anthropol 21: 24–37
4. Risch N, Burchard E, Ziv E, Tang H (2002) Categorization of humans in biomedical research: genes, race and disease. Genome Biol 3: comment2007.
5. Lohmueller KE, Mauney MM, Reich D, Braverman JM (2006) Variants associated with common disease are not unusually differentiated in frequency across populations. Am J Hum Genet 78: 130–136
6. Myles S, Davison D, Barrett J, Stoneking M, Timpson N (2008) Worldwide population differentiation at disease-associated SNPs. BMC Med Genomics 1: 22.
7. Sabeti PC, Reich DE, Higgins JM, Levine HZ, Richter DJ, et al. (2002) Detecting recent positive selection in the human genome from haplotype structure. Nature 419: 832–837
8. Currat M, Trabuchet G, Rees D, Perrin P, Harding RM, et al. (2002) Molecular analysis of the beta-globin gene cluster in the Niokholo Mandenka population reveals a recent origin of the beta(S) Senegal mutation. Am J Hum Genet 70: 207–223
9. Chevin LM, Billiard S (2008) Hospital F (2008) Hitchhiking both ways: effect of two interfering selective sweeps on linked neutral variation. Genetics 180: 301–316
10. Corona E, Dudley JT, Butte AJ (2010) Extreme evolutionary disparities seen in positive selection across seven complex diseases. PLoS ONE 5: e12236. doi:10.1371/journal.pone.0012236
11. Casto AM, Feldman MW (2011) Genome-wide association study SNPs in the human genome diversity project populations: does selection affect unlinked SNPs with shared trait associations? PLoS Genet 7: e1001266. doi:10.1371/journal.pgen.1001266
12. Hindorff LA, Sethupathy P, Junkins HA, Ramos EM, Mehta JP, et al. (2009)Potential etiologic and functional implications of genome-wide association loci for human diseases and traits. Proc Natl Acad Sci U S A 106: 9362–9367
13. Tennessen JA, Akey JM (2011) Parallel adaptive divergence among geographically diverse human populations. PLoS Genet 7: e1002127.doi:10.1371/journal.pgen.1002127
14. Sanishvili R, Nagarajan V, Yoder D, Becker M, Xu S, et al. (2008) A 7mum mini-beam improves diffraction data from small or imperfect crystals of macromolecules. Acta Crystallogr D Biol Crystallogr 64: 425–435
15. Cann HM, de Toma C, Cazes L, Legrand MF, Morel V, et al. (2002) A human genome diversity cell line panel. Science 296: 261–262
16. Pandey JP (2010) Genomewide association studies and assessment of risk of disease.N Engl J Med 363: 2076–2077; author reply 2077
17. Nelson MR, Bryc K, King KS, Indap A, Boyko AR, et al. (2008) The Population Reference Sample, POPRES: a resource for population, disease, and pharmacological genetics research. Am J Hum Genet 83: 347–358
18. Frazer KA, Ballinger DG, Cox DR, Hinds DA, Stuve LL, et al. (2007) A second generation human haplotype map of over 3.1 million SNPs. Nature 449: 851–861
19. Ward RH, Neel JV (1976) The genetic structure of a tribal population, the Yanomama Indians. XIV. Clines and their interpretation. Genetics 82: 103–121
20. Young JH, Chang YP, Kim JD, Chretien JP, Klag MJ, et al. (2005) Differential susceptibility to hypertension is due to selection during the out-of-Africa expansion. PLoS Genet 1: e82.doi:10.1371/journal.pgen.0010082
21. Pemberton TJ, Sandefur CI, Jakobsson M, Rosenberg NA (2009) Sequence determinants of human microsatellite variability. BMC Genomics 10: 612.

22. Hancock AM, Witonsky DB, Gordon AS, Eshel G, Pritchard JK, et al. (2008)Adaptations to climate in candidate genes for common metabolic disorders. PLoS Genet4: e32.doi:10.1371/journal.pgen.0040032
23. Hancock AM, Witonsky DB, Alkorta-Aranburu G, Beall CM, Gebremedhin A, et al. (2011) Adaptations to climate-mediated selective pressures in humans. PLoS Genet 7: e1001375.doi:10.1371/journal.pgen.1001375
24. Lachance J (2010) Disease-associated alleles in genome-wide association studies are enriched for derived low frequency alleles relative to HapMap and neutral expectations.BMC Med Genomics 3: 57.
25. Chen R, Corona E, Sikora M, Dudley JT, Morgan AA, et al. (2012) Type 2 Diabetes Risk Alleles Demonstrate Extreme Directional Differentiation among Human Populations, Compared to Other Diseases. PLoS Genet 8: e1002621.doi:10.1371/journal.pgen.1002621
26. Novembre J, Di Rienzo A (2009) Spatial patterns of variation due to natural selection in humans. Nat Rev Genet 10: 745–755
27. Waters KM, Stram DO, Hassanein MT, Le Marchand L, Wilkens LR, et al. (2010)Consistent association of type 2 diabetes risk variants found in europeans in diverse racial and ethnic groups. PLoS Genet 6: e1001078. doi:10.1371/journal.pgen.1001078
28. Bustamante CD, Burchard EG, De la Vega FM (2011) Genomics for the world.Nature 475: 163–165
29. International HapMap C, Altshuler DM, Gibbs RA, Peltonen L, Altshuler DM, et al. (2010) Integrating common and rare genetic variation in diverse human populations.Nature 467: 52–58
30. Tysk C, Lindberg E, Jarnerot G, Floderus-Myrhed B (1988) Ulcerative colitis and Crohn's disease in an unselected population of monozygotic and dizygotic twins. A study of heritability and the influence of smoking. Gut 29: 990–996
31. Kathiresan S, Melander O, Anevski D, Guiducci C, Burtt NP, et al. (2008)Polymorphisms associated with cholesterol and risk of cardiovascular events. N Engl J Med 358: 1240–1249
32. Wacholder S, Hartge P, Prentice R, Garcia-Closas M, Feigelson HS, et al. (2010)Performance of common genetic variants in breast-cancer risk models. N Engl J Med362: 986–993
33. Hancock AM, Witonsky DB, Ehler E, Alkorta-Aranburu G, Beall C, et al. (2010)Colloquium paper: human adaptations to diet, subsistence, and ecoregion are due to subtle shifts in allele frequency. Proc Natl Acad Sci U S A 107Suppl 2: 8924–8930
34. Nejentsev S, Walker N, Riches D, Egholm M, Todd JA (2009) Rare variants of IFIH1, a gene implicated in antiviral responses, protect against type 1 diabetes. Science324: 387–389
35. Barreiro LB, Quintana-Murci L (2010) From evolutionary genetics to human immunology: how selection shapes host defence genes. Nat Rev Genet 11: 17–30
36. Schulz LO, Bennett PH, Ravussin E, Kidd JR, Kidd KK, et al. (2006) Effects of traditional and western environments on prevalence of type 2 diabetes in Pima Indians in Mexico and the U.S. Diabetes Care 29: 1866–1871
37. Ashley EA, Butte AJ, Wheeler MT, Chen R, Klein TE, et al. (2010) Clinical assessment incorporating a personal genome. Lancet 375: 1525–1535
38. Reynolds J, Weir BS, Cockerham CC (1983) Estimation of the coancestry coefficient: basis for a short-term genetic distance. Genetics 105: 767-779.
39. Li JZ, Absher DM, Tang H, Southwick AM, Casto AM, et al. (2008) Worldwide human relationships inferred from genome-wide patterns of variation. Science 319: 1100-1104.
40. Storey JD, Tibshirani R (2003) Statistical significance for genomewide studies. Proc Natl Acad Sci U S A 100: 9440-9445.
41. Coop G, Witonsky D, Di Rienzo A, Pritchard JK Using environmental correlations to identify loci underlying local adaptation. Genetics 185: 1411-1423.
42. Klimentidis YC, Abrams M, Wang J, Fernandez JR, Allison DB Natural selection at genomic regions associated with obesity and type-2 diabetes: East Asians and sub-Saharan Africans exhibit high levels of differentiation at type-2 diabetes regions. Hum Genet 129: 407-418.
43. Kathiresan S, Melander O, Anevski D, Guiducci C, Burtt NP, et al. (2008) Polymorphisms associated with cholesterol and risk of cardiovascular events. N Engl J Med 358: 1240-1249.
44. Wacholder S, Hartge P, Prentice R, Garcia-Closas M, Feigelson HS, et al. Performance of common genetic variants in breast-cancer risk models. N Engl J Med 362: 986-993.
45. Pedersen NL Reaching the limits of genome-wide significance in Alzheimer disease: back to the environment. JAMA 303: 1864-1865.
46. Patel CJ, Bhattacharya J, Butte AJ An Environment-Wide Association Study (EWAS) on type 2 diabetes mellitus. PLoS One 5: e10746.
47. Camps M, Herman A, Loh E, Loeb LA (2007) Genetic constraints on protein evolution. Crit Rev Biochem Mol Biol 42: 313-326.
48. Eichler EE, Flint J, Gibson G, Kong A, Leal SM, et al. Missing heritability and strategies for finding the underlying causes of complex disease. Nat Rev Genet 11: 446-450.

CHAPTER 4

ENVIRONMENTAL LEAD EXPOSURE, CATALASE GENE, AND MARKERS OF ANTIOXIDANT AND OXIDATIVE STRESS RELATION TO HYPERTENSION: AN ANALYSIS BASED ON THE EGAT STUDY

Jintana Sirivarasai,[1] Sukhumpun Kaojarern,[2] Suwannee Chanprasertyothin,[3] Pachara Panpunuan,[4] Krittaya Petchpoung,[5] Aninthita Tatsaneeyapant,[6] Krongtong Yoovathaworn,[7] Thunyachai Sura,[4] Sming Kaojarern,[8] and Piyamit Sritara[4]

[1]Graduate Program in Nutrition, Faculty of Medicine Ramathibodi Hospital, Mahidol University, Bangkok 10400, Thailand

[2]Cardiovascular and Metabolic Center, Faculty of Medicine Ramathibodi Hospital, Mahidol University, Bangkok 10400, Thailand

[3]Office of Research Academic and Innovation, Faculty of Medicine Ramathibodi Hospital, Mahidol University, Bangkok 10400, Thailand

[4]Department of Medicine, Faculty of Medicine Ramathibodi Hospital, Mahidol University, Bangkok 10400, Thailand

[5]Research and Development Institute, Kasetsart University, Bangkok 10900, Thailand

[6]Health Office, Electricity Generating Authority of Thailand, Nonthaburi 11130, Thailand

[7]Department of Pharmacology, Faculty of Science, Mahidol University, Bangkok 10400, Thailand

[8]Occupational and Environmental Toxicology Center, Faculty of Medicine Ramathibodi Hospital, Mahidol University, Bangkok 10400, Thailand

ABSTRACT

Lead has been linked to the development of hypertension via oxidative stress. Catalase plays an important role in the disposal of hydrogen peroxide in erythrocyte and its activity was determined by CAT gene. The aims of this study were to investigate (1) the association between blood levels of antioxidant markers such as catalase, superoxide dismutase, glutathione, glutathione peroxidase, oxidative stress-marker (malondialdehyde), and blood lead level and (2) the influence of genetic polymorphism of CAT gene (rs769217) on change in blood pressure in general population of EGAT study project. This is a cross-sectional study of 332 normotensive, 432 prehypertensive, and 222 hypertensive male subjects. Hypertensive subjects had significantly higher blood lead level (5.28 µg/dL) compared to normotensive (4.41 µg/dL) and prehypertensive (4.55 µg/dL) subjects ($P < 0.05$). These significant findings are also found in MDA levels. Moreover, individuals with TT genotype in hypertensive group had significantly higher blood lead and MDA levels (6.06 µg/dL and 9.67 µmol/L) than those with CC genotype (5.32 µg/dL and 8.31 µmol/L, $P < 0.05$). Our findings suggested that decreased blood catalase activity in this polymorphism together with low level lead exposure induced lipid peroxidation may be responsible for hypertension.

Citation: Jintana Sirivarasai, Sukhumpun Kaojarern, Suwannee Chanprasertyothin, et al., "Environmental Lead Exposure, Catalase Gene, and Markers of Antioxidant and Oxidative Stress Relation to Hypertension: An Analysis Based on the EGAT Study," BioMed Research International, vol. 2015, Article ID 856319, 9 pages, 2015. doi:10.1155/2015/856319

INTRODUCTION

General population may be exposed to lead through various sources such as dietary contamination (via food chain and lead releasing from food containers or ceramic glaze), public water supplies contamination, herbal remedies, and manufacturing byproducts such as E-waste recycling, manufacture of batteries, sheet lead, solder, brass, and bronze plumbing, radiation shields, circuit boards, and military equipment [1]. Lead exposure occurs mainly through the respiratory and gastrointestinal tracts. Approximately 30–40 percent of inhaled lead is absorbed into the bloodstream. Gastrointestinal absorption varies depending on nutritional status (i.e., iron or calcium deficiency) and age. Once absorbed, 99 percent of circulating lead is bound to erythrocytes for approximately 30–35 days (estimating about 1% absorbed lead is found in plasma and serum) and is dispersed into the soft tissues, including renal cortex, liver, lung, brain, teeth, and bones [1].

Since bone accounts for more than 94% of the adult body burden of lead, bone lead level by K-X-ray fluorescence represents lead content in the cortex of tibia and the patella trabecular [2]. This measurement is an indicator of cumulative lead exposure and is particularly relevant to the elderly in whom elevated bone lead concentrations may represent chronic toxicity [3]. Measuring blood lead is the most commonly accepted and verifiable biomarker for lead exposure. This assessment, by industrial hygienist, was used both in current and past environmental lead exposures to quantify the intensity of the exposure [4]. In the blood stream, lead circulating is mobile whereas lead in bone is stored. Mobile lead exerts adverse effects on human body. Under conditions of more or less constant and prolonged exposure, an individual's blood lead level reflects the quantity of biological active forms of lead in their body [5]. A large number of reports revealed positive correlations between blood lead and detrimental effects on the central nervous, hematopoietic, renal, immune, and cardiovascular systems [6].

Hypertension is a multifactorial condition associated with both environmental and genetic factors. For environmental risk factors include dietary, lifestyle, obesity, and some toxicants, lead is one of the candidate metals which can be linked to the development of hypertension [7, 8]. Numerous human and animal studies found a causal relationship between low-level lead exposure and hypertension. Some evidences indicated that oxidative stress played a significant role in the etiology of lead-induced hypertension [8]. Oxidative stress is described as a physiological stage in which antioxidant defense is inadequate to detoxify the reactive oxygen species (ROS). This oxidative process results in the damaging of essential biomolecules such as protein, lipid, and DNA. Overproduction of ROS is demonstrated in lead-induced oxidative stress. Previous experimental studies revealed that lead could promote ROS production in kidney and cardiovascular tissues [9, 10]. In addition, lead influenced cell membrane alterations, such as lipid component, membrane integrity, permeability, and function, finally leading to lipid peroxidation [11, 12].

The most common group of indices used to assess oxidative stress is that of peroxidation products of lipids, usually polyunsaturated fatty acids, which are susceptible to attack by free radicals. All these products of degradation and decomposition are used in assessing oxidative stress, including hydroperoxides, F2-isoprostanes, and malondialdehyde (MDA) [13]. MDA is the principal and most studied product of polyunsaturated fatty acid peroxidation. This aldehyde is a highly toxic molecule and should be considered as more than a marker of lipid peroxidation [14]. Derivatization of MDA with thiobarbituric acid (TBA), as MDA-TBA adduct, is a wildly used method to monitor the level of lipid peroxidation in biological sample. The HPLC with fluorescence detection significantly improved the specificity and overcame overestimation of the MDA-TBA adduct, as indicated by much more homogenous results obtained in various publications [15]. By measurement of F2-isoprostanes, TBA-MDA adduct, or lipid hydroperoxides, there were some reports that showed correlations between TBA-MDA adduct and F2-isoprostanes or lipid hydroperoxides [12].

Another mechanism of lead-induced oxidative stress is the effect on antioxidant defense systems of cells. Lead exhibits a high affinity for sulfhydryl (SH) groups and can interfere with antioxidant activities by inhibiting functional SH groups in several enzymes such as superoxide dismutase (SOD), catalase (CAT), glutathione peroxidase (GPx), glucose-6-phosphate dehydrogenase (G6PD), and ALAD [16]. A large number of researches were conducted to further understand the imbalance between antioxidant and oxidant stages with risks of chronic diseases, especially in the field of genetic variations of antioxidant enzymes [17–19]. Catalase is a well-known antioxidant enzyme that plays a role in the conversion of H_2O_2 to H_2O and O_2 [20]. The CAT gene is located in chromosome 11p13 and consists of 13 exons. The C111T polymorphism of this gene, in exon 9 (rs769217), is responsible for alteration in its activity [21]. To our knowledge, data related to the influence of CAT C111T polymorphism on antioxidant system/oxidative stress and hypertension in environmental lead exposure are very limited. Moreover, this is the first study in Thai population that emphasizes on biomarkers of lead exposure and of susceptibility with CAT gene. The aims of the present study were to investigate (1) the association between blood levels of antioxidant markers such as CAT, SOD, GPx, glutathione (GSH), MDA, and blood lead level and (2) the possible influence of CAT polymorphism on change in blood pressure among general population.

MATERIALS AND METHODS

Study Population

The Electric Generating Authority of Thailand (EGAT) study was the first cohort study of chronic disease in Thailand, originally designed in 1985 (known as EGAT 1), and mainly covered multidisciplinary researches related to cardiovascular disease (CVD) risks such as nutrition and toxicology. The 986 male subjects were participants in the third survey of EGAT 2 in 2009 (the first survey started in 1998 and the second survey in 2003). This study was approved by the Committee on Human Rights Related to Researches Involving Human Subjects, Faculty of Medicine Ramathibodi Hospital, Mahidol University, Thailand. All participants completed a self-administered questionnaire and underwent a physical examination and performed laboratory analysis, including tests for diabetes and liver and kidney diseases [22]. Toxicological profile of heavy metals and genetic analysis were determined.

Ten milliliters of blood was collected by venipuncture into EDTA and heparinized tubes from each subject and immediately centrifuged at 2000 g. Buffy coat, erythrocytes, and plasma were separated and stored at −20°C until genotyping analysis and biochemical measurements were performed.

According to the Joint National Committee 7 (JNC 7), hypertension was defined as systolic blood pressure (SBP) ≥140 mmHg or diastolic blood pressure (DBP) ≥90 mmHg, prehypertension was defined as SBP 120–139 mmHg or DBP 80–89 mmHg, and normal blood pressure was defined as SBP <120 mmHg and DBP < 80 mmHg [23]. Based on this criterion, participants were classified into 3 groups: 332 normotensive, 432 prehypertensive, and 222 hypertensive subjects.

Determination of Blood Lead

Blood lead concentration was measured by graphite furnace atomic absorption spectrometry (GFAAS) with Zeeman background correction. The analytical procedure was based on the method described by Subramanian and Meranger [24]. The measurement was calculated as micrograms per deciliter (μg/dL) and expressed by means of total blood lead. The intra-assay coefficients of variation (CV) ranged from 2.8 to 5.9% and interassay CV ranged from 3.2 to 6.4%.

Determination of Glutathione by the DTNB Method

Whole blood (0.1 mL) was added to distilled water (1.9 mL) together with 3 mL of precipitating solution (1.67 g glacial metaphosphoric acid, 0.2 g disodium ethylenediaminetetraacetic acid, EDTA, and 30 g sodium chloride). Then, the filtrate (0.5 mL) was added to 0.3 M phosphate buffer, pH 6.4 (2 mL). Finally, 1 mM DTNB (0.25 mL) was added, mixed well, and the absorbance was read at 412 nm within 4 min [25]. The intra-assay and interassay CV were 4.3% and 6.0%, respectively.

Determination of Catalase Activity

Catalase activity was measured by the decrease in absorbance at 240 nm due to H_2O_2 consumption, according to Aebi method [26]. Hemolysate (0.1 mL) was added to cuvette containing 1.9 mL of 50 mM phosphate buffer (pH 7.0). Enzymatic reaction was started by the addition of 1.0 mL of freshly prepared 30 mM H_2O_2. The rate of decomposition of H_2O_2 was measured by spectrophotometer from changes in absorbance at 240 nm. Activity of catalase was expressed as U/gHb. The intra-assay and interassay CV were 3.2% and 4.7%, respectively.

Determination of SOD Activity

Superoxide dismutase (SOD) activity was measured by the method of Winterbourn et al. [27]. This method used nitroblue tetrazolium (NBT) as indicator and riboflavin as superoxide-generating system. Result was expressed as unit of SOD per gram of Hb. One unit is defined as the amount of enzyme causing half the maximum inhibition of NBT reduction. The intra-assay and interassay CV were 3.6% and 5.5%, respectively.

Determination of GPx Activity

GPx activity was assayed by method of Beutler [28], using t-butyl hydroperoxide (t-BuOOH) and glutathione (GSH) as substrates, and this was followed by measuring the oxidation of NADPH at 340 nm at 37°C with a spectrophotometer in the presence of glutathione reductase. The reaction mixture consisted of 0.5 mM EDTA, 0.1 M Tris-HCl (pH 8.0), 2.0 mM NADPH, 2 mM GSH, 1 U glutathione reductase, and hemolysate in a total volume of 0.99 mL. The enzymatic reaction was initiated by the addition of 10 μL t-BuOOH. One unit of GPx activity is defined as the amount of enzyme that oxidizes 1 μmol NADPH/min. GPx activity was expressed as U/gHb. The intra-assay and interassay CV were 4.6% and 6.8%, respectively.

Determination of MDA

MDA was determined using an HPLC method with fluorescence detector, as described by Khoschsorur et al. [29]. Detection limit was 0.25 μmol/L and this method exhibited a linear response of MDA in a range of concentration from 1.50 to 15.0 μmol/L and calibration curve presented high correlation coefficient (($r^2 > 0.90$, $P = 0.001$; $n = 10$). The intra-assay and interassay CV were 3.9% and 4.7%, respectively.

Genotyping Assay

The genomic DNA was extracted from lymphocytes by a modified salting-out procedure [30] and frozen at −20°C until analysis. The genetic polymorphism of CAT (rs769217) was performed by real-time polymerase chain reaction (real-time PCR) according to the method of TaqMan SNP Genotyping Assays on an ABI 7500 instrument (Applied Biosystems, Foster City, CA, USA), in 96-well format. The TaqMan Assay included the forward target-specific polymerase chain reaction (PCR) primer, the reverse primer, and the TaqMan MGB probes labeled with 2 special dyes: FAM and VIC. The concentrations of probes were 0.04 μM. Amplification of 20 ng of DNA was performed during 40 cycles in a reaction volume of 10 μL. TaqMan Universal PCR Master Mix was used for analysis. Thermocycling conditions were 95°C for 15 seconds, followed by 60°C for 1 minute. Information of specific probe and primers is available on the National Cancer Institute's SNP500 database web page at http://snp500cancer.nci.nih.gov/ [31]. Quality control procedure included repeat genotyping of at least 10% of DNA samples. On every single 96-well microtiter plate, we include negative control in the form of water. The percentage of successful rate for 95 cases and negative control was 100 and this assay was good allelic discrimination among individuals. For overall SNP genotyping assay, the reported error rate for 983 cases was 0.33% which involved DNA extracts of poor quantity and quality.

Statistical Analysis

Statistical analyses were carried out using the SPSS 16.0 for window software (SPSS, Inc., Chicago, IL). Most of the study parameters were presented as mean ± SE. Because of skewed distribution, lead level was transformed to normal distribution and expressed as geometric mean ± SE. The comparisons between variables were examined by Student's *t*-test and analysis of variance (ANOVA). Genotype distribution was analyzed with χ^2. Pearson's correlation was performed to determine the strength of the association between blood lead and other significantly correlated parameters. A P value of 0.05 was used as the criterion for statistical significance.

RESULTS

The clinical characteristics and biochemical profile in normotensive, prehypertensive, and hypertensive groups are presented in Table 1. Mean age for subjects with hypertension was comparable to those with prehypertension and normotension. More than 50% of the three groups were in the range of 40–55 years. The control group (23.58 kg/m^2) has significantly lower BMI than prehypertensive (25.09 kg/m^2) and hypertensive subjects (26.26 kg/m^2, $P < 0.05$). SBP and DBP in hypertensive group (144.9 and 93.7 mmHg) also showed statistically higher than those in the prehypertensive (126.1 and 80.94 mmHg) and control groups (110.2 and 70.3 mmHg, $P < 0.05$). No significant differences were found with respect to distributions of alcohol consumption and smoking status among three groups. Furthermore, subjects with prehypertension and hypertension had statistically higher blood levels of triglyceride, fasting glucose, creatinine, and uric acid than the normotensive subjects.

Table 1: General characteristics and biochemical parameters of study population according to hypertension.

Variables	Normotension ($N = 332$)	Prehypertension ($N = 432$)	Hypertension ($N = 222$)
Age, yrs (AM ± SE)	51.58 ± 0.23	52.01 ± 0.20	52.96 ± 0.33
Age group, N (%)			
40–55 yrs	267 (80.4)	328 (75.9)	152 (68.5)
>55 yrs	65 (19.6)	104 (24.1)	70 (31.5)
BMI, kg/m^2 (AM ± SE)	23.58 ± 0.16	25.09 ± 0.15^a	26.26 ± 0.25^a
Drinking status, N (%)			
Nondrinkers	125 (37.7)	165 (38.2)	67 (30.2)
Drinkers	207 (62.3)	267 (61.8)	155 (69.8)
Smoking status, N (%)			
Nonsmokers	247 (74.4)	342 (79.2)	156 (70.3)
Smokers	85 (25.6)	90 (20.8)	66 (20.8)
Biochemical tests (AM ± SE)			
Triglyceride, mg/dL	132.81 ± 4.04	165.78 ± 5.35^a	191.47 ± 8.81^{a,b}
HDL cholesterol, mg/dL	49.57 ± 0.57	48.74 ± 0.49	49.32 ± 0.071
LDL cholesterol, mg/dL	152.31 ± 2.14	148.49 ± 1.91	146.61 ± 2.72
Total cholesterol, mg/dL	228.55 ± 2.31	230.33 ± 2.09	233.02 ± 2.79
Fasting glucose, mg/dL	100.58 ± 1.35	105.91 ± 1.24^a	112.65 ± 3.09^{a,b}
Creatinine, mg/dL	1.10 ± 0.07	1.14 ± 0.01^a	1.14 ± 0.01^a
Uric acid, mg/dL	5.97 ± 0.06	6.40 ± 0.06^a	6.61 ± 0.08^{a,b}

^{a,b}Significantly different from normotension and prehypertension, respectively, $P < 0.05$.

As shown in Table 2, there was a significant increase in blood lead and MDA levels among the three groups, but statistical changes in antioxidant parameters were not found. Hypertensive subjects had significantly higher blood lead level (5.28 μg/dL) compared to normotensive (4.41 μg/dL) and prehypertensive (4.55 μg/dL) subjects ($P < 0.05$). In addition, blood MDA level in hypertensive group (9.64 μmol/L) was statistically higher than those in prehypertensive (8.02 μmol/L) and normotensive groups (8.23 μmol/L, $P < 0.01$).There were no significant differences in the means of CAT, SOD, GPx, and GSH among the three groups. Our results illustrated that SBP ($r = 0.218$, $P < 0.01$), DBP ($r = 0.195$, $P < 0.05$), and MDA concentrations ($r = 0.147$, $P < 0.01$) were positively correlated with blood lead level (Figure 1). There was no significant association between blood MDA and SBP or DBP ($r = 0.131$ and $P = 0.08$ and $r = 0.111$ and $P = 0.10$, resp.) (Figure 2).

Table 2: Blood lead level, antioxidants, and oxidative stress determinants of study population, according to hypertension.

Variables	Normotension (N = 332)	**Prehypertension (N = 432)**	**Hypertension (N = 222)**
Blood lead, μg/dL	4.41 ± 0.10	4.55 ± 0.09	5.28 ± 0.21[a,b]
Antioxidant biomarkers			
CAT, U/gHb	29966 ± 549	28543 ± 422	28084 ± 600
SOD, U/gHb	2417 ± 80	2365 ± 71	2363 ± 96
GPx, U/gHb	34.56 ± 0.85	34.92 ± 0.78	36.11 ± 1.14
GSH, mg/dL	31.42 ± 0.34	31.41 ± 0.32	31.34 ± 0.46
Oxidative stress biomarker			
MDA, μmol/L	8.23 ± 0.34	8.02 ± 0.26	9.64 ± 0.45[a,b]

[a,b]Significantly different from normotension and prehypertension, respectively, $P < 0.05$.

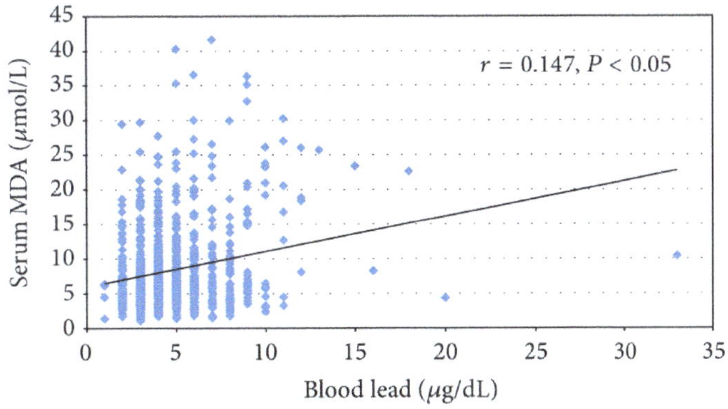

Figure 1: Association between blood lead and serum MDA levels in the study population.

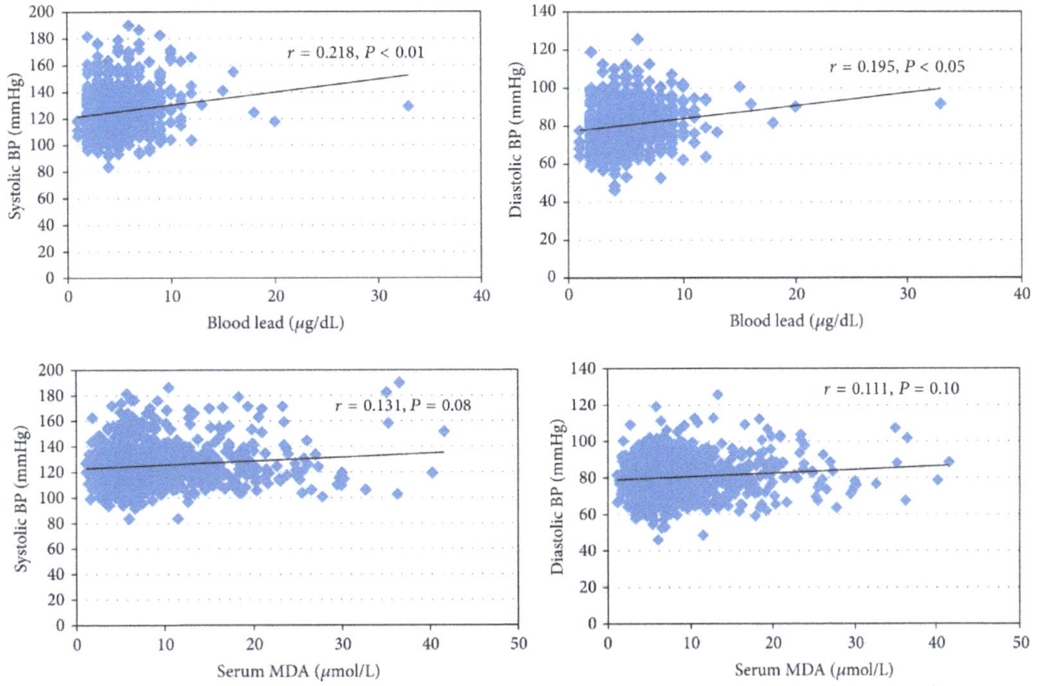

Figure 2: Association between blood lead, serum MDA, and SBP and DBP in the study population.

Genotype frequencies of CAT C111T polymorphism in this study population were 32.0% for CC, 46.2% for CT, and 21.7% for TT. The genotype distribution among controls was according to the Hardy-Weinberg equilibrium ($P = 0.28$). Catalase activity was significantly lower in individuals with TT genotype (27044 U/gHb) compared to those with CT genotype (29103 U/gHb) or CC genotype (30625 U/gHb) ($P < 0.01$). However, the means of blood lead, MDA, SBP, and DBP of individuals with CC, CT, or TT genotypes did not reveal differences (Figure 3).

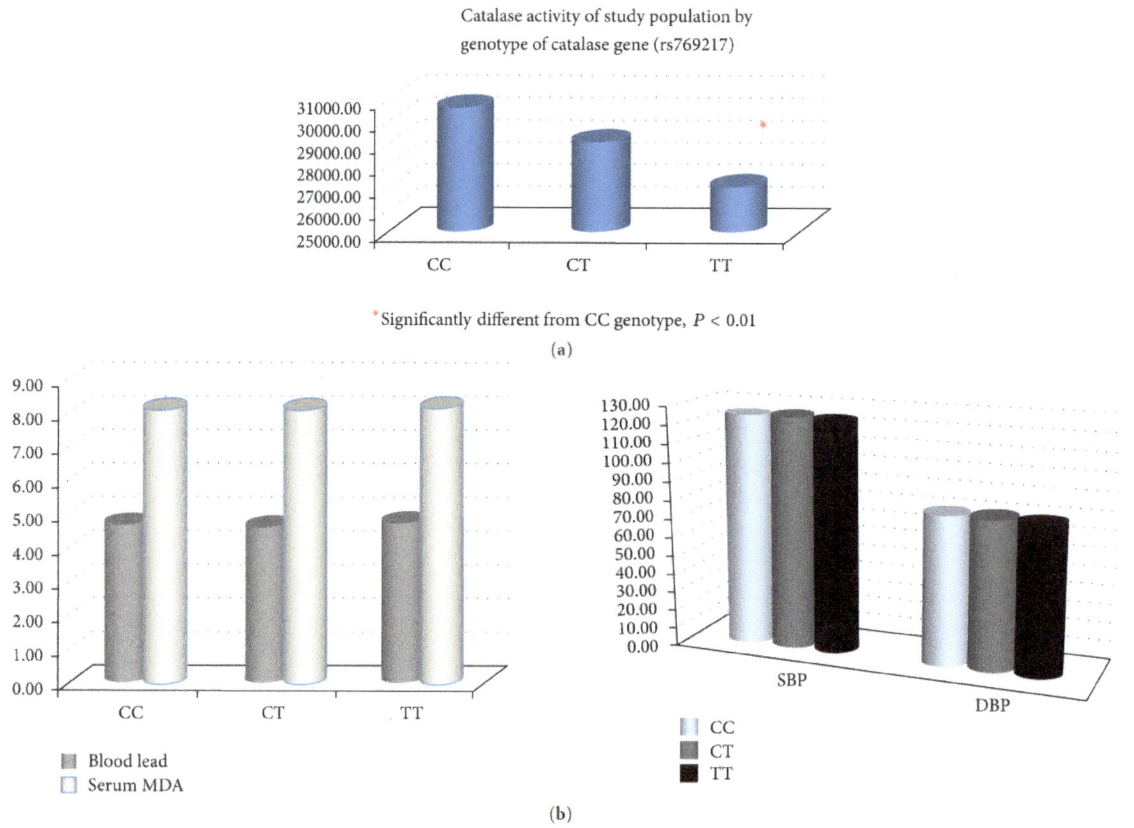

Figure 3: Genetic polymorphism of CAT gene (rs769217) and its activity (upper) and the comparison of blood lead, MDA, and BP of all study populations classified by genotype (lower).

Table 3 illustrates the effect of CAT genotypes on biochemical profile, blood lead, and antioxidant and oxidative stress determinants in the three groups classified by CAT genotypes. All groups with CC, CT, and TT genotypes had no significant ($P > 0.05$) differences for lipid profiles and other biochemical parameters (FBS, Cr, and UA). Further analysis in the same fashion with oxidative stress and antioxidant defense found that there was significant change of catalase activity with higher activity in wild-type allele (32341 U/gHb) and lower activity in those with mutant allele (27423 U/gHb, $P < 0.05$) in controls. Similar ($P < 0.05$) blood catalase activities were found for CC, CT, and TT genotypes in prehypertensive and hypertensive subjects. In contrast, SOD activity in the hypertensive group showed different trend (2055 U/gHb for CC versus 2616 U/gHb for TT, $P < 0.05$). The CAT C111T significantly modified the effect of lead on MDA, as seen only in hypertensive group, in which individuals with TT genotype had significantly higher blood lead and MDA levels (6.06 μg/dL and 9.67 μmol/L) than those with CC genotype (5.32 μg/dL and 8.31 μmol/L, $P < 0.05$).

Table 3: Biochemical tests, blood lead, and determinants of antioxidant and oxidative stress, classified by CAT genotypes among three groups.

		Normotension		Prehypertension				**Hypertension**	
	CC (*N* = 104)	**CT (*N* = 145)**	**TT (*N* = 83)**	**CC (*N* = 139)**	**CT (*N* = 207)**	**TT (*N* = 86)**	**CC (*N* = 73)**	**CT (*N* = 104)**	**TT (*N* = 45)**
Biochemical tests*									
Triglyceride, mg/dL	135.0 ± 7.7	129.6 ± 6.2	135.5 ± 7.1	161.0 ± 8.9	166.7 ± 8.1	171.0 ± 11.7	166.8 ± 9.8	206.9 ± 14.9	195.9 ± 20.7
HDL cholesterol, mg/dL	49.36 ± 1.04	49.91 ± 0.87	49.25 ± 1.14	50.37 ± 0.93	49.56 ± 0.66	48.91 ± 1.16	50.83 ± 1.32	48.70 ± 1.01	48.32 ± 1.48
LDL cholesterol, mg/dL	149.4 ± 4.2	156.1 ± 3.2	149.21 ± 3.6	149.3 ± 2.9	148.1 ± 2.9	147.1 ± 4.4	151.1 ± 4.9	145.5 ± 4.1	141.9 ± 5.3
Total cholesterol, mg/dL	225.8 ± 4.6	232.0 ± 3.4	225.7 ± 4.0	232.4 ± 3.3	229.2 ± 3.2	229.8 ± 4.7	235.1 ± 5.1	234.2 ± 4.3	226.9 ± 4.6
Fasting glucose, mg/dL	102.2 ± 2.9	99.94 ± 2.06	99.91 ± 1.76	105.4 ± 2.1	106.1 ± 1.9	106.4 ± 2.6	104.6 ± 2.1	119. ± 5.9	109.1 ± 5.5

Creatinine, mg/dL	1.10 ± 0.01	1.11 ± 0.01	1.08 ± 0.01	1.11 ± 0.01	1.15 ± 0.01	1.15 ± 0.02	1.15 ± 0.02	1.13 ± 0.02	1.14 ± 0.02
Uric acid, mg/dL	5.95 ± 0.12	5.94 ± 0.09	6.03 ± 0.13	6.51 ± 0.09	6.31 ± 0.09	6.46 ± 0.14	6.57 ± 0.13	6.60 ± 0.13	6.72 ± 0.19
Blood lead, μg/dL**	4.32± 0.17	4.51 ± 0.16	4.36 ± 0.21	4.71 ± 0.16	4.42 ± 0.14	4.59 ± 0.21	5.32 ± 0.31	5.38 ± 0.22	6.06 ± 0.41[a]
Antioxidant determinants**									
CAT, U/gHb	32341 ± 1017	29717 ± 766[a]	27423 ± 1129[a,b]	29621 ± 563	28014 ± 425	26741 ± 498[a]	28745 ± 765	28012 ± 568	26104 ± 925[a]
SOD, U/gHb	2301 ± 133	2571 ± 136	2295 ± 135	2286 ± 114	2436 ± 107	2321 ± 161	2055 ± 151	2469 ± 131	2616 ± 265[a]
GPx, U/gHb	34.31 ± 1.63	36.19 ± 1.33	32.03 ± 1.42	36.05 ± 1.41	34.58 ± 1.14	33.91 ± 1.74	37.65 ± 1.87	34.51 ± 1.59	37.32 ± 2.97
GSH, mg/dL	31.16 ± 0.62	32.05 ± 0.49	30.64 ± 0.73	31.46 ± 0.55	31.76 ± 0.48	30.50 ± 0.64	30.75 ± 0.69	30.97 ± 0.70	33.17 ± 1.12
Oxidative stress determinant**									
MDA, μmol/L	7.95 ± 0.54	8.71 ± 0.59	7.76 ± 0.64	8.32 ± 0.49	7.88 ± 0.35	7.86 ± 0.60	8.31 ± 0.65	9.75 ± 0.73[a]	9.67 ± 0.88[a]

*AM ± SE, **GM ± SE.

[a,b]Significantly different from individuals with CC and CT, respectively, $P < 0.05$.

DISCUSSION

Exposure to low level of lead in environmental manner has been reported to cause dysfunction in many target organs. The overt toxicity of this metal may result from its potentially induced oxidative stress as seen by an increased prevalence of chronic kidney disease, cardiovascular disease, peripheral arterial disease, diabetes, hyperuricemia, or hypertension [7–9]. The positive correlation between blood lead level and SBP or DBP has been well documented. These associations have been reported in both occupational workers and population with low exposure to environmental lead. In this study, hypertensive group has higher blood lead level (5.28 μg/dL) than controls (4.41 μg/dL) (Table 2). In addition, the positive correlations between blood lead level, SBP, and DBP are shown in Figure 2. Skoczynska [32] suggested various cardiovascular mechanisms of the hypertensive effect by low doses of lead, including changes in metabolisms of catecholamines, increased activity of the central adrenergic system, decreased activity of ATP-ase, changes in transmembranal transport of ions, inhibition of Na+ secretion and increase in blood volume, increased plasma rennin activity, and enhanced free radicals generation. Moreover, our previous study revealed the relationship between the effect of lead exposure on inflammatory marker (high-sensitivity C-reactive protein, hs-CRP) and the adverse change in SBP [33].

Lipid peroxidation is a well-established mechanism of cellular injury and is used as an indicator of oxidative stress in cells and tissues. In addition, measurement of MDA is widely used as an indicator of lipid peroxidation. The findings of the present study also supported the evidence of lead-induced ROS generation which represented oxidative stress (Table 2 and Figure 1). There was an increase in the MDA levels in the hypertensive group (9.46 μmol/L) in comparison to those in the control group (8.23 μmol/L) (Table 2). In addition, MDA level showed a significant increase with blood lead level (Figure 2). However, there were no significant differences in GPx activity and GSH level between controls and prehypertensive or hypertensive group. Catalase and SOD activities showed significant changes in hypertensive subjects. The results indicated adaptive mechanism in case oxidative stress occurred in cell. Impairing these enzyme activities may lead to the excess condition of free radicals which can escape antioxidant defense system, contribute to alteration of nitric oxide bioavailability, and affect structure/function of endothelium, resulting in high blood pressure [34].

Catalase has a predominant role in the disposal of hydrogen peroxide in human erythrocytes. Deficiency of erythrocyte catalase causes increased hydrogen peroxide concentration. A few polymorphisms have been revealed for the catalase-encoding gene. Genetic variations of this gene can be responsible for the change of its activity and expression. The analyzed polymorphism in this population is the C111T polymorphism in exon 9 (rs769217). Yung et al. [20] revealed the difference in catalase activity between wild-type and mutant alleles. Allele frequencies for the C and T reported in our study (55% and 45%, resp.) were similar to those reported previously in Japanese (C allele: 48%; T allele: 52%) [35] and Korean (C allele: 60%; T allele: 40%) [36] studies. In contrast to some countries, the T alleles were less frequent in Hungry (28%) [37], USA (12%) [38], and UK (17%) [20]. Based on genetic determinant of enzyme activity, this study revealed that individuals with TT genotype of C111T polymorphism associated with low catalase activity compared to those with wild-type allele (Figure 3). These findings can be explained by the data in which this nucleotide change may cause slower transcription from the mutant allele than from the wild-type allele [37].

To our knowledge, the present study is the first research that investigates the effect of CAT polymorphism on blood lead and blood pressure among nonoccupational population. Statistical differences in mean enzyme activities by genotypes were detected. The C111T polymorphism has an effect on blood catalase activity in all study groups (Table 3) but, for the hypertensive group, SOD activity in mutant allele (2616 U/gHb) was significantly higher than the wild-type allele (2055 U/gHb, $P < 0.05$). As mentioned above, these enzyme activities were partially determined by genetic factors and further modified by other various factors, such as exercise, stress condition, dietary intake of antioxidant, and coexposures to some toxicants. Moreover, individuals with TT genotype in hypertensive group showed significantly higher blood lead and MDA level than those with CC genotype. Decreased blood catalase activity in this polymorphism together with low-level lead exposure induced lipid peroxidation may be responsible for hypertension.

This study has an important strength, because EGAT study is the first cohort study in Thailand (EGAT 2 started in 1998), focusing on chronic disease with a five-year time series of survey. Thus, it also provides the opportunity to reevaluate the relationship between lead exposure and blood pressure with the large sample size. A limitation of our study is that blood lead concentration, as a biomarker of exposure, is not an appropriate marker for cumulative exposure when compared to bone lead. However, in vivo K-X-ray fluorescence for bone lead measurement is expensive and impractical for large-scale population studies. In addition, previous studies have shown strong association between blood lead and bone lead [39, 40] and most epidemiological studies to date wildly used blood lead level for biomonitoring of environmental lead exposure.

In conclusion, the evidence in this study was sufficient to infer a causal association between lead exposure and elevated blood pressure. Our finding suggested that exposure to low lead level associated with oxidative stress and deficiency of the enzyme catalase may contribute to hypertension. Further studies related to other genes involving antioxidant enzymes or elements will elucidate the relationship between environmental lead exposure, oxidative stress, and adverse health outcomes. For public health implication, regulatory plan and health intervention program should be developed and implemented for reduction of lead exposure and health risks such as cardiovascular, kidney, and other diseases.

ACKNOWLEDGMENTS

The authors wish to thank the EGAT and their people for participating in and establishing this study. The authors would like to express their gratitude to all research staffs, especially Miss Nisakron Thongmung and Ms. Yupin Wisetpanit, Office of Research Academic and Innovation, Faculty of Medicine Ramathibodi Hospital, for providing subjective data and technical assistance in specimen collection and preparation. This work was supported by the Cooperative Research Network (CRN) scholarship, the project for Higher Education Research Promotion and National Research University Development, Office of the Higher Education Commission, Ministry of Education, Thailand, and The Thailand Research Fund.

REFERENCES

1. F. Barbosa Jr., J. E. Tanus-Santos, R. F. Gerlach, and P. J. Parsons, "A critical review of biomarkers used for monitoring human exposure to lead: advantages, limitations, and future needs," Environmental Health Perspectives, vol. 113, no. 12, pp. 1669–1674, 2005.
2. E. J. O'Flaherty, "Physiologically based models for bone-seeking elements. V. Lead absorption and disposition in childhood," Toxicological Apply Pharmacology, vol. 131, no. 7, pp. 297–308, 1993.
3. A. Rosin, "The long-term consequences of exposure to lead," Israel Medical Association Journal, vol. 11, no. 11, pp. 689–694, 2009.
4. R. W. Leggett, "An age specific kinetic model of lead metabolism in humans," Environmental Health Perspective, vol. 101, no. 7, pp. 598–616, 1995.
5. K. P. Mishra, V. K. Singh, R. Rani et al., "Effect of lead exposure on the immune response of some occupationally exposed individuals," Toxicology, vol. 188, no. 2-3, pp. 251–259, 2003.
6. L. Patrick, "Lead toxicity, a review of the literature. Part I: exposure, evaluation, and treatment,"Alternative Medicine Review, vol. 11, no. 1, pp. 2–22, 2006.
7. H. Hu, A. Aro, M. Payton et al., "The relationship of bone and blood lead to hypertension: the normative aging study," The Journal of the American Medical Association, vol. 275, no. 15, pp. 1171–1176, 1996.
8. A. A. Alghasham, M. A. Meki, and H. A. Ismail, "Association of blood lead level with elevated blood pressure in hypertensive patients," International Journal Health Science, vol. 5, no. 1, pp. 17–27, 2011.
9. N. D. Vaziri, "Mechanisms of lead-induced hypertension and cardiovascular disease," American Journal of Physiology: Heart and Circulatory Physiology, vol. 295, no. 2, pp. H454–H465, 2008.
10. H. C. Gonick, Y. Ding, A. C. Bondy, Z. Ni, and N. D. Vaziri, "Lead-induced HTN: interplay of nitric oxide and reactive oxygen species," Hypertension, vol. 30, no. 6, pp. 1487–1492, 1997.
11. C. M. Bolin, R. Basha, D. Cox et al., "Exposure to lead and the developmental origin of oxidative DNA damage in the aging brain," The FASEB Journal, vol. 20, no. 6, pp. 788–790, 2006.
12. S. Kasperczyk, J. Kasperczyk, A. Ostałowska et al., "The role of the antioxidant enzymes in erythrocytes in the development of arterial hypertension among humans exposed to lead," Biological Trace Element Research, vol. 130, no. 2, pp. 95–106, 2009.
13. Y. Dotan, D. Lichtenberg, and I. Pinchuk, "Lipid peroxidation cannot be used as a universal criterion of oxidative stress," Progress in Lipid Research, vol. 43, no. 3, pp. 200–227, 2004.
14. D. D. Rio, A. J. Stewart, and N. Pellegrini, "A review of recent studies on malondialdehyde as toxic molecule and biological marker of oxidative stress," Nutrition, Metabolism and Cardiovascular Diseases, vol. 15, no. 4, pp. 316–328, 2005.
15. J. Lykkesfeldt, "Malondialdehyde as biomarker of oxidative damage to lipids caused by smoking," Clinica Chimica Acta, vol. 380, no. 1-2, pp. 50–58, 2007.
16. R. C. Patra, A. K. Rautray, and D. Swarup, "Oxidative stress in lead and cadmium toxicity and its amelioration," Veterinary Medicine International, vol. 2011, Article ID 457327, 9 pages, 2011.

17. A. Crawford, R. G. Fassett, D. P. Geraghty et al., "Relationships between single nucleotide polymorphisms of antioxidant enzymes and disease," Gene, vol. 501, no. 2, pp. 89–103, 2012.

18. L. Forsberg, U. de Faire, and R. Morgenstern, "Oxidative stress, human genetic variation, and disease,"Archives of Biochemistry and Biophysics, vol. 389, no. 1, pp. 84–93, 2001.

19. M. L. Mansego, G. D. M. Solar, M. P. Alonso et al., "Polymorphisms of antioxidant enzymes, blood pressure and risk of hypertension," Journal of Hypertension, vol. 29, no. 3, pp. 492–500, 2011.

20. L. M. Yung, F. P. Leung, X. Yao, Z.-Y. Chen, and Y. Huang, "Reactive oxygen species in vascular wall,"Cardiovascular & Hematological Disorders: Drug Targets, vol. 6, no. 1, pp. 1–19, 2006.

21. N. G. Gavalas, S. Akhtar, D. J. Gawkrodger, P. F. Watson, A. P. Weetman, and E. H. Kemp, "Analysis of allelic variants in the catalase gene in patients with the skin depigmenting disorder vitiligo," Biochemical and Biophysical Research Communications, vol. 345, no. 4, pp. 1586–1591, 2006.

22. P. Sritara, S. Cheepudomwit, N. Chapman et al., "Twelve-year changes in vascular risk factors and their associations with mortality in a cohort of 3499 Thais: the Electricity Generating Authority of Thailand study," International Journal of Epidemiology, vol. 32, no. 3, pp. 461–468, 2003.

23. A. V. Chobanian, "The seventh report of the joint national committee on prevention, detection, evaluation, and treatment of high blood pressure," Tech. Rep. 04-5230, National High Blood Pressure Education Program, 2004.

24. K. S. Subramanian and J. C. Meranger, "A rapid electrothermal atomic absorption spectrophotometric method for cadmium and lead in human whole blood," Clinical Chemistry, vol. 27, no. 11, pp. 1866–1871, 1981.

25. E. Beutler, O. Duron, and B. M. Kelly, "Improved method for the determination of blood glutathione,"The Journal of Laboratory and Clinical Medicine, vol. 61, pp. 882–888, 1963.

26. H. Aebi, "Catalase," in Methods of Enzymatic Analysis, H. U. Bergmeyer, Ed., pp. 671–684, Academic Press, London, UK, 1974.

27. C. C. Winterbourn, R. E. Hawkins, M. Brian, and R. W. Carrell, "The estimation of red cell superoxide dismutase activity," Journal of Laboratory and Clinical Medicine, vol. 85, no. 2, pp. 337–341, 1974.

28. E. Beutler, "Gluthaione peroxidase," in Red Cell Metabolism: A Manual of Biochemical Methods, E. Beutler, Ed., Grune and Stratton, New York, NY, USA, 1975.

29. G. A. Khoschsorur, B. M. Winklhofer-Roob, H. Rabl, T. Auer, Z. Peng, and R. J. Schaur, "Evaluation of a sensitive HPLC method for the determination of malondialdehyde, and application of the method to different biological materials," Chromatographia, vol. 52, no. 3-4, pp. 181–184, 2000.

30. S. A. Miller, D. D. Dykes, and H. F. Polesky, "A simple salting out procedure for extracting DNA from human nucleated cells," Nucleic Acids Research, vol. 16, no. 3, article 1215, 1988.

31. B. R. Packer, M. Yeager, B. Staats et al., "SNP500Cancer: a public resource for sequence validation and assay development for genetic variation in candidate genes," Nucleic Acids Research, vol. 32, pp. D528–D32, 2004.

32. A. Skoczynska and M. Skoczynska, "Low-level exposure to lead as a cardiovascular risk factor," inCardiovascular Risk Factors, A. Gasparyan, Ed., InTech, 2012.

33. J. Sirivarasai, W. Wananukul, S. Kaojarern et al., "Association between inflammatory marker, environmental lead exposure, and glutathione S-transferase gene," BioMed Research International, vol. 2013, Article ID 474963, 6 pages, 2013.

34. S. Taddei, A. Virdis, L. Ghiadoni et al., "Age-related reduction of NO availability and oxidative stress in humans," Hypertension, vol. 38, no. 2, pp. 274–279, 2001.

35. Y. Kishimoto, Y. Murakami, K. Hayashi, S. Takahara, T. Sugimura, and T. Sekiya, "Detection of a common mutation of the catalase gene in Japanese acatalasemic patients," Human Genetics, vol. 88, no. 5, pp. 487–490, 1992.

36. T.-H. Kim, J. M. Hong, B. Oh et al., "Genetic association study of polymorphisms in the catalase gene with the risk of osteonecrosis of the femoral head in the Korean population," Osteoarthritis and Cartilage, vol. 16, no. 9, pp. 1060–1066, 2008.

37. L. Góth, T. Nagy, Z. Kósa et al., "Effects of rs769217 and rs1001179 polymorphisms of catalase gene on blood catalase, carbohydrate and lipid biomarkers in diabetes mellitus," Free Radical Research, vol. 46, no. 10, pp. 1249–1257, 2012.

38. C. B. Casp, J.-X. She, and W. T. McCormack, "Genetic association of the catalase gene (CAT) with vitiligo susceptibility," Pigment Cell Research, vol. 15, no. 1, pp. 62–66, 2002.

39. A. C. Todd, E. L. Moshier, B. K. Lee, G. S. Lee, K. D. Ahn, and B. S. Schwartz, "Predictors of DMSA chelatable lead, tibial lead, and blood lead in 802 Korean lead workers," Occupational and Environmental Medicine, vol. 58, no. 2, pp. 73–80, 2001.

40. F. Kamel, D. M. Umbach, T. L. Munsat, J. M. Shefner, H. Hu, and D. P. Sandler, "Lead exposure and amyotrophic lateral sclerosis," Epidemiology, vol. 13, no. 3, pp. 311–319, 2002.

CHAPTER 5

GENE–ENVIRONMENT INTERACTIONS IN HUMAN HEALTH: CASE STUDIES AND STRATEGIES FOR DEVELOPING NEW PARADIGMS AND RESEARCH METHODOLOGIES

Fatimah L. C. Jackson

Department of Biology and W. Montague Cobb Research Laboratory, Howard University, Washington, DC, USA

ABSTRACT

The synergistic effects of genes and the environment on health are explored in three case studies: adult lactase persistence, autism spectrum disorders, and the metabolic syndrome, providing examples of the interactive complexities underlying these phenotypes. Since the phenotypes are the initial targets of evolutionary processes, understanding the specific environmental contexts of the genetic, epigenetic, and environmental changes associated with these phenotypes is essential in predicting their health implications. Robust databases must be developed on the local scale to deconstruct both the population substructure and the unique components of the environment that stimulate geographically specific changes in gene expression patterns. To produce these databases and make valid predictions, new, locally focused, and information-dense models are needed that incorporate data on evolutionary ecology, environmental complexity, local geographic patterns of gene expression, and population substructure.

INTRODUCTION

The impact of gene–environment interactions in important human biological conditions affecting health are underappreciated due to a current imbalance in the knowledge needed to integrate and synthesize molecular, ecological, and sociocultural databases. In addition to the need for more precise data, there is also a great need for more specific interdisciplinary paradigms and research methodologies to identify and address these gene–environment interactions (Jackson, 2004). Environmental exposures of both high intensity and long duration are closely tied to health impacts. When these occur at the population level, such exposures can be of public health relevance. Even shorter exposures, however, can alter the signatures of epigenetic dysregulation in peripheral blood samples, within a few hours of environmental exposure (Baccarelli and Ghosh, 2012). Sustained behavioral and social processes and events can directly manipulate the epigenome and modulate the expression genome. Research on health risk assessment can benefit by developing ways to discern the differences in these effects.

A number of excellent examples exist of these kinds of interactions. We begin with the first case study of the genomic consequences among some proportion of humanity in response to generations of continued milk drinking beyond the weaning stage.

Citation: Jackson F.L.C. Gene–environment interactions in human health: case studies and strategies for developing new paradigms and research methodologies. Frontiers in Genetics. 2014; 5:271. doi:10.3389/fgene.2014.00271.

CASE STUDY: ADULT LACTASE PERSISTENCE (LP): EVOLUTIONARY ECOLOGY, SUBSISTENCE PRACTICES, AND GENETIC DIVERSITY

Among almost all mammals, the ability to digest lactose declines sharply after infancy. High lactose digestion capacity in adult humans is common only in certain North Atlantic European, Mediterranean, Central Asian, and selected African populations, and overall, it is thought to be an evolutionary adaptation to millennia of drinking milk from domestic livestock (Holden and Mace, 1997). Prior to the domestication and use of milk producing species, there was no evolutionary advantage in the LP phenotype. Whereas afterward, there was and this has driven an entirely new dynamic in genetic evolution. When variability in the ability to digest and derive nutritional benefit from lactose was discovered among Western scientists, discussions of regular successful adult consumption of milk suggested that such practices were the human norm by using such terms as "adult lactose intolerance" and "adult lactase deficiency" to, in fact, describe what is the true human (and mammalian) behavioral and genetic norm. As the scope of the research was broadened to include non-European individuals and groups in the analyses as well as investigations among other mammals, it was realized that adult lactase persistence and adult lactose tolerance were the evolutionary exception. Something quite special had to have happened at a social, behavioral, and genetic levels in a small subset of humanity to allow adults to consume and derive nutritional benefit from the ingestion of animal milk without incurring the negative health consequences of lactose intolerance.

As the health-associated symptoms of the behavioral practice of adult milk drinking began to be examined by researchers with genetic training and evolutionary interests, we began to recognize that the practice of adult milk drinking had likely emerged during the early stages of animal domestication in many locales, and yet was geographically restricted to those groups with regular access to animal milk. The ability of human adults to successfully consume the milk of their domesticated livestock necessitated a genetic change in these humans, foremost of which was persistence of gene or genes for the production of the enzyme lactase into adulthood. LP is one of the clearest examples of niche construction in humans (Gerbault et al., 2011) where human modification of the environment significantly changes the direction of genetic selection thereby increasing the biological fitness associated with having specific genetic mutations.

Once the behavioral practice of regular adult consumption of animal milk was in place in these diverse groups, concomitant changes in the gene expression patterns of the domesticated livestock were required for the enhanced production of the additional milk needed to meet the dietary demands of adult human consumers. We now know that there is substantial geographic coincidence between high diversity in cattle milk genes, locations of the European Neolithic cattle farming sites (>5,000 years ago) and present-day LP among adult Europeans (Beja-Pereira et al., 2003). The reciprocal influences of genetics and the cultural environments in these early pastoralists facilitated the subsequent social and behavioral shifts observed as a result of this changed dietary pattern. In certain populations of domesticated cattle and humans their genes had coevolved in response to changes in behavior that became imbedded in the emerging social fabric of these selected groups.

Genetics has now been able to provide researchers with the specific molecular variants involved in each of the geographical regions that this coevolutionary pattern has emerged and their calculated emergence times [discussed in Ingram et al. (2009)]. These data have, in turn, allowed us to refine and expand our interpretations of the social consequences of these behavioral shifts and their ramifications on regional development and health, reemphasizing the dynamic feedback loop between "genes" and "environment."

In European dairying groups and their descendants, a single genetic mutation explains the distribution of the LP phenotype [see Anagnostou et al. (2009)]. This single nucleotide polymorphism is located 13.9 kb upstream from *LCT* (13910T) was proposed to be the cause for LP. The -13910*T allele, which is widespread in Europe was found to be located on an extended haplotype of 500 kb or more (Anagnostou et al., 2009). In Central Asia the causal polymorphism for LP is the same as in Europe (-13.910C > T, rs4988235; Heyer et al., 2011), suggesting genetic diffusion between the two geographical regions.

However, elsewhere among milk-consuming groups, several other mutations are associated with LP. For example, in some South Africans, the -14010*C allele, is associated with LP (Jensen et al., 2011). In East Africa, the primary mutation appears to be (-13907G; Enattah et al., 2008). The European (-13910T) and the East African (-13907G) allele share the same ancestral background and most likely a similar social and behavioral history, probably related to the same cattle domestication event. In Saudi Arabia, the European allele (-13910T) is absent and instead two new mutations are found as a compound allele: (-13915T/G) within the -13910 enhancer region and a synonymous SNP in the exon 17 of *MCM6* (-3712T/C), -3712 base pairs from *LCT* (Enattah et al., 2008). This compound allele is highly prevalent among many Arab populations and shows a different, highly divergent ancestral haplotype, perhaps in response to camel domestication and milk consumption (Enattah et al., 2008).

Recent studies in Tibet (Peng et al., 2012), however, suggest that three novel SNPs (-13838G/A, -13906T/A, and -13908C/T) are responsible for LP among Tibetans, placing them at variance with European, East African, and Middle Eastern genetic patterns. This emphasizes the concept of local geographic-based responses to common environmental challenges and reinforces the need for taking population history into consideration in the search for genes responsible for health-related conditions.

The worldwide geographical variability in the genetics underlying the LP phenotype suggests that this phenotype emerged independently several times in human history. Furthermore, several independent mutations tied to regional social events coupled with local behavioral choices and existing patterns of genetic variation are involved in its current multiethnic distribution. These genetic data can both broaden and deepen the scope of environmental analyses of LP by permitting researchers to identify the origins of this phenotype in any particular individual or group and to correlate a specific genetic variant of LP with the unique historical and evolutionary circumstances that characterize a specific milk-consuming group.

CASE STUDY: AUTISM SPECTRUM DISORDERS (ASDS): ENVIRONMENTAL COMPLEXITY AND DYNAMISM IN GENE EXPRESSION

One of the most poignant examples of environment interactions influencing important clinical abnormalities are the autism spectrum disorders (ASDs). ASDs are a complex group of neurodevelopment disorders, which are still poorly understood (Centers for Disease Control and Prevention, 2012). ASDs typically appear before 3 years of age, and affect the brain's normal development of social and communication skills. ASDs appear to be rising in frequency and are refractory to current treatments. In 2008, the overall estimated prevalence of ASDs was 11.3 per 1,000 (one in 88) children aged 8 years. Overall ASD prevalence estimates varied widely geographically and ASD prevalence estimates also varied widely by sex and by biosocial/ethnic group (Centers for Disease Control and Prevention, 2012).

From studies of both monozygotic and dizygotic twins, it is evident that identical twins are much more likely than fraternal twins or siblings to both have autism. Similarly, language abnormalities are more common in relatives of autistic children. Chromosomal abnormalities and other neurological problems are also more common in families with autism. Genetic factors clearly play a role as well in ASDs. Some of the recently proposed genetic factors in ASDs include *DRD2* and *PPP1R1B* in males (Hettinger et al., 2012), chromosomal (Yang et al., 2012), copy number variations in candidate genes (Griswold et al., 2012), microRNAs (miRNAs) associated with brain development and maturation (Mellios and Sur, 2012), gene-disrupting mutations (nonsense, splice site, and frame shifts; Iossifov et al., 2012), deficient GABA neurotransmission (Mendez et al., 2012), and functional mutations in postsynaptic scaffolding proteins at excitatory synapses (Ting et al., 2012).

Extensive research has been so far unable to explain the etiology of ASDs, whereas a growing body of evidence suggests that the involvement of local environmental exposures to specific, potent bioactive factors plays a major role. Phthalates, given their extensive use and their persistence, are ubiquitous environmental contaminants implicated in the etiology of ASDs. These are endocrine-disrupting chemicals suspected to interfere with neurodevelopment. Therefore, they represent interesting candidate environmental risk factors for ASDs pathogenesis (Testa et al., 2012). Some other recently suggested environmental triggers for ASDs include prenatal ethanol exposure (Middleton et al., 2012), mercury poisoning, changes in the digestive tract, diet (Pennesi and Klein, 2012), oxidative stress (Frustaci et al., 2012), maternal fever during pregnancy (Zerbo et al., 2012), paternal age (Eriksson et al., 2012; Iossifov et al., 2012), inadequate breast feeding (Al-Farsi et al., 2012), and persistent exposure to organic polybrominated diphenyl ethers (PBDEs) used in commercial flame retardants (Woods et al., 2012).

Overall, the behavioral polytypism associated with ASDs suggests that an array of genetic and environmental factors influence the individual autism phenotype [seeBuchholz et al. (2013)]. The problem for scientists working on ASDs is that this spectrum of disorders are rarely diagnosed in children younger than 2 years, since diagnosis is based entirely on behavioral tests (Frustaci et al., 2012). Changes in the epigenome (Rangasamy et al., 2013) can serve as indicators of these gene–environmental interactions and help clarify the contributions of each set of factors (individually and interactively) to autism [see Hall and Kelley (2013)].

As mentioned, environmental factors that mediate epigenetic changes appear to play a strong role in the etiology of ASDs (Ladd-Acosta et al., 2013). Unbiased investigations of a multitude of novel candidate genes encoding nuclear factors implicated in chromatin remodeling, histone demethylation, histone variants, and chromatin alterations suggest that DNA methylation underlies many neurodevelopmental aberrations, including the ASDs (LaSalle, 2013a). Additionally, long non-coding RNA (lncRNAs) contribute to ASDs risk (Wilkinson and Campbell, 2013). Future analyses of these non-coding RNAs as well as studies of a variety of epigenetic modifications around genetic risk factors augmented with quantified measures of environmental exposures and methylome analyses are expected to be important for understanding the complex etiology of autism (Hu, 2013; LaSalle, 2013b).

CASE STUDY: METABOLIC SYNDROME (METS): REDEFINING THE PARAMETERS OF POPULATION BEHAVIORAL AND BIOLOGICAL SUBSTRUCTURE:

The most common diseases affecting the US population are complex disorders that develop as a result of defects in multiple genetically controlled systems in response to environmental challenges (Berdasco and Esteller, 2013). MetS represents a set of disorders that frequently occur together and increase the risk of further cardiovascular and metabolic disease. MetS is a global pandemic of enormous medical, economic, and social concern affecting a significant portion of the world's population (Alberti et al., 2009). A number of organizations have developed definitions for the MetS to guide clinician and researchers in identifying the syndrome. One of the earliest was the World Health Organization in 1998 (Grundy, 2007). Later, the National Cholesterol Education Program/Adult Treatment Panel III (NCEP/ATPIII) developed diagnostic criteria for the MetS (Farook et al., 2012; Forti et al., 2012). The International Diabetes Federation and Joint Interim Statement also proposed criteria for identifying MetS (Ruemmele and Garnier-Lengliné, 2012).

All of these definitions of MetS include abdominal obesity, hyperglycemia, hypertriglyceridemia, low high-density lipoprotein cholesterol, and hypertension (Ruemmele and Garnier-Lengliné, 2012) and are considered the five indicators of the MetS. If an individual possesses three or more of the five risk indicators, then that person is defined as having MetS and is said to be at increased risk for the development of cardiovascular disease and also for Type 2 diabetes mellitus. Details of these five risk indicators include: elevated triglyceride levels ($\geq$150 mg/dl), elevated systolic ($\geq$130 mmHg) and diastolic ($\geq$85 mmHg) blood pressures, elevated fasting blood glucose level (>110 mg/dl), reduced blood levels of High Density Lipoproteins (<40 mg/dl for men) and <50 mg/dl for women), and abdominal obesity [large waist circumference (WC), initially defined as >102 cm for men and >88 cm for women].

However, these definitions were developed from studies of Europeans (and their descendants; Wang et al., 2009) and the cut-off values were promulgated as universal until recently when studies of Chinese and Koreans demonstrated that some East Asians develop diabetes and several risk indicators of cardiovascular disease at lower body mass index (BMI) and WC values than do European-descended populations (Zorzi et al., 2009; Kim et al., 2012). This has given rise to several new studies that result in our now suggesting that BMI and WC cut-offs be made more population-specific (Grundy, 2007). Although the cut-off values are still subject to change with additional research, most Asians (whether East or South Asians) have lower BMI and WC cut-off values than do European Americans or other Europeans when they begin to express the abnormal clinical indicators of MetS. Alternative patterns for MetS may also exist for aboriginal Canadian AmerIndians (Nikitin et al., 2012). This additional biological information on human diversity in BMI and WC may call into question the reports, in Siberia and other East Asian regions, for example, that the overall prevalence of both diabetes and MetS among the indigenous East Asian population is lower than among Europeans (Roth and Sathyanarayana, 2012). It may simply be that the detection criteria are different and that some East Asian cases are being missed using European-based physiological definitions.

Different BMI and WC cut-offs suggest that human populations are highly variable and that diet and other environmental factors have a huge impact on body shape, body composition, and susceptibility to disease. This is not unexpected since signaling peptides produced in peripheral tissues such as gut, adipose tissue, and pancreas communicate with brain centers, such as hypothalamus and hindbrain to manage energy homeostasis and these processes are highly sensitive to the influence of environmental variables (Du et al., 2012). The regulatory mechanisms of energy intake and storage have evolved during extended periods of famine in human evolution to allow for optimal adaptation and ultimately protect the species from extinction. It is now clear that these circuitries are influenced by prenatal and postnatal environmental factors including dietary adequacy and endocrine disruptive chemicals (Du et al., 2012).

Hypothalamic appetite regulatory systems develop and mature in utero and early infancy, and involve signaling pathways that are important also for the regulation of the onset of puberty (Du et al., 2012). Metabolic pathways involved in the regulation of growth, body weight gain, and sexual maturation are largely affected by epigenetic programming that can impact both current and future generations (Du et al., 2012). Therefore, researchers studying MetS cross-culturally would be well served to integrate both ecological and evolutionary perspectives in their investigations since population substructure can dramatically influence the initiation of MetS's downstream events. In particular, intrauterine and early infantile developmental phases are highly plastic and susceptible to factors that affect metabolic programming. These, therefore, affect metabolic function throughout the lifespan [discussed in Du et al. (2012)]. Thus, the underlying mechanisms regulating human metabolic biology are likely to be evident in many different groups although the specific genes involved will vary between geographical groups. Definitions of metabolic disease indicators should not be generalized to all world populations, particularly when the standards are disproportionately based a minority of the world human biodiversity and are largely restricted to measurements on individuals from economically developed regions.

Although genetics, physiology, and environmental components play a major role in the onset of the MetS diseases due to excessive fat accumulation, little is known about how or to what extent each of these factors contributes to the actual syndrome (Alberti et al., 2009) and which specific genes may serve as triggers. Currently, the search for patterns in the genetics of MetS is in its infancy. What has emerged is that a host of different genes on different chromosomes in different human groups that appear to be associated to varying degrees with MetS. For example, in Mexican Americans, chromosome 7q21 shows strong evidence for linkage to MetS, especially between markers D7S2212 and D7S821 (Forti et al., 2012). Six chromosomal regions exhibit potential evidence for linkage with MetS in this group. 29 single-nucleotide polymorphisms (SNPs) from the fatty acid translocase [*FAT* or *CD36*, 18 SNPs] and guanine nucleotide binding protein and α transducing 3 [*GNAT3*, 11 SNPs] gene (Forti et al., 2012), are associated with MetS and its related traits in Mexican Americans. In this same population, several SNPs were also associated with MetS and its related traits with the strongest associations being linked with rs11760281 in *GNAT3* and rs1194197 near *CD36* (Forti et al., 2012). This accounted for approximately 18% of the MetS linkage on chromosome 7q21 and together these genetic changes conferred nearly a threefold increase in MetS risk for Mexican Americans (Forti et al., 2012).

Among Han Chinese (Dujic et al., 2012), however, an association study suggests that *ADIPOQ* variants are associated with the risk of MetS and include genotypes rs266729CG; rs1063539GC, GC/CC; rs16861205AA, and rs7649121AT. Functional studies of these molecular variants of *ADIPOQ* are still pending.

In a recent study of Bosnian (Eastern Europe) children with MetS, a common rs45487298 polymorphism in *HSD1181* is thought to possibly have a protective effect against insulin resistance (Min et al., 2012). In an effort to identify the gene networks involved in MetS in British Europeans, researchers conducted Min et al. (2012) andAlkharfy et al. (2012) conducted whole-genome expression and genotype profiling and found nine coexpresssed genes in abdominal tissue and six in gluteal adipose tissue, but no coexpressed genes in whole blood. The rs10282458 polymorphism, affecting expression of *RARRES2* (encoding chemerin), was associated with BMI and rs2395185, affecting inter-depot differences of HLA-DRB1 expression, was associated with high-density lipoprotein and the BMI-adjusted waist-to-hip ratio in this British population.

Among Saudi Arabians, a significant association of 894G > T, 4a/b, and -786T > C polymorphisms with MetS in *eNOS* is evident (Csaba, 2011). A genetic predisposition to develop abnormal metabolic phenotypes, consistent with an increased prevalence of metabolic phenotypes has reportedly been detected in the Saudi Arab population with these genetic polymorphisms. Researchers have also reported a high prevalence of MetS among Kuwaiti adults (Al Zenki et al., 2012), presumably with a similar genetic or environmental basis.

From this cursory overview of gene-association studies, it appears that great diversity exists in specific susceptibility genes that characterize the genetic underpinnings of MetS across geographical groups (see da Costa et al., 2014). Epigenetically, MetS seems to reflect some level of hormonal imprinting (Lacaria et al., 2012) which, when faulty can produce disturbances in methylation patterns. CNVs also appear to be involved in some cases of MetS (Wang et al., 2009; Rottiers and Näär, 2012). Dysregulation of miRNA may also contribute to the metabolic abnormalities associated with MetS and thus may provide a potential source of cardiometabolic therapeutic targets.

As the technology continues to improve, the ability to integrate these epigenetic data into the analyses of MetS should prove extremely important, permitting more precise assessments of the epigenetic consequences of particular events on MetS phenotypes [see Turcot et al. (2012) and Wang et al. (2012)].

STRATEGIES FOR DEVELOPING NEW PARADIGMS AND RESEARCH METHODOLOGIES

Epigenetic factors that control chromatin dynamism are essential for the proper functioning of tissue homeostasis, cell identity, and development. Deregulation of epigenetic profiles is intimately associated with metabolic pathologies and neurological diseases, among other complex disorders (Cox et al., 2013). From the case studies presented here, we can see that both our assumptions about the "gene" and the "environment" as well as their interactions need to be more nuanced, more localized to the specific populations under consideration, and more sensitive to the particular relevant local environmental factors of influence.

Conflict of Interest Statement

The author declares that the research was conducted in the absence of any commercial or financial relationships that could be construed as a potential conflict of interest.

REFERENCES

1. Al Zenki S., Al Omirah H., Al Hooti S., Al Hamad N., Jackson R. T., Rao A., et al. (2012). High prevalence of metabolic syndrome among kuwaiti adults – a wake-up call for public health. intervention. *Int. J. Environ. Res. Public Health* 9 1984–1996 10.3390/ijerph9051984
2. Alberti K. G., Eckel R. H., Grundy S. M., Zimmet P. Z., Cleeman J. I., Donato K. A., et al. (2009). Harmonizing the metabolic syndrome: a joint interim statement of the International Diabetes Federation Task Force on Epidemiology and Prevention; National Heart, Lung, and Blood Institute; American Heart Association; World Heart Federation; International Atherosclerosis Society; and International Association for the Study of Obesity. *Circulation* 120 1640–1645 10.1161/CIRCULATIONAHA.109.192644
3. Al-Farsi Y. M., Al-Sharbati M. M., Waly M. I., Al-Farsi O. A., Al-Shafaee M. A., Al-Khaduri M. M., et al. (2012). Effect of suboptimal breast-feeding on occurrence of autism: a case-control study. *Nutrition* 8:e27–e32 10.1016/j.nut.2012.01.007
4. Alkharfy K. M., Al-Daghri N. M., Al-Attas O. S., Alokail M. S., Mohammed A. K., Vinodson B., et al. (2012). Variants of endothelial nitric oxide synthase gene are associated with components of metabolic syndrome in an Arab population.*Endocr. J.* 59 253–263 10.1507/endocrj.EJ11-0278
5. Anagnostou P., Battaggia C., Coia V., Capelli C., Fabbri C., Pettener D., et al. (2009). Tracing the distribution and evolution of lactase persistence in Southern Europe through the study of the T(-13910) variant. *Am. J. Hum. Biol.* 21 217–219 10.1002/ajhb.20851
6. Baccarelli A., Ghosh S. (2012). Environmental exposures, epigenetics and cardiovascular disease. *Curr. Opin. Clin. Nutr. Metab. Care.* 15 323–329 10.1097/MCO.0b013e328354bf5c
7. Beja-Pereira A., Luikart G., England P. R., Bradley D. G., Jann O. C., Bertorelle G., et al. (2003). Gene-culture coevolution between cattle milk protein genes and human lactase genes. *Nat. Genet.* 35 311–313 10.1038/ng1263
8. Berdasco M., Esteller M. (2013). Genetic syndromes caused by mutations in epigenetic genes. *Hum. Genet.* 132 359–383 10.1007/s00439-013-1271-x
9. Buchholz V., Votsiari A., Bleich S., Frieling H. (2013). Nature meets nurture: the importance of epigenetics for the aetiology of psychiatric diseases. *Fortschr. Neurol. Psychiatr.* 81 368–380 10.1055/s-0033-1335739
10. Centers for Disease Control and Prevention. (2012). Prevalence of autism spectrum disorders–Autism and Developmental Disabilities Monitoring Network, 14 sites, United States, 2008. *MMWR Surveill. Summ.* 61 1–19
11. Cox L. A., Cumuzzie A. G., Havill L. M., Karere G. M., Spradling K. D., Mahaney M. C., et al. (2013). Baboons as a model to study genetics and epigenetics of human disease. *ILAR J.* 54 106–121 10.1093/ilar/ilt038
12. Csaba G. (2011). The biological basis and clinical significance of hormonal imprinting, an epigenetic process. *Clin. Epigenetics* 2 187–196 10.1007/s13148-011-0024-8
13. da Costa K. A., Corbin K. D., Niculescu M. D., Galanko J. A., Zeisel S. H. (2014).Identification of new genetic polymorphisms that alter the dietary requirement for choline and vary in their distribution across ethnic and racial groups. *FASEB J.* 282970–2978 10.1096/fj.14-249557
14. Du J., Ye XH, Li Q., Yu X., Cheng J., Ma J., Gao Y., et al. (2012). Genetic variants in the ADIPOQ gene and the risk of metabolic syndrome: a case-control study of a Chinese Han population. *Ann. Hum. Genet.* 76 101–109 10.1111/j.1469-1809.2012.00699.x
15. Dujic T., Bego T., Mlinar B., Semiz S., Malenica M., Prnjavorac B., et al. (2012).Association between 11beta-hydroxysteroid dehydrogenase type 1 gene polymorphisms and metabolic syndrome in Bosnian population. *Biochem. Med. (Zagreb).* 22 76–85 10.11613/BM.2012.008
16. Enattah N. S., Jensen T. G., Nielsen M., Lewinski R., Kuokkanen M., Rasinpera H., et al. (2008). Independent

introduction of two lactase-persistence alleles into human populations reflects different history of adaptation to milk culture. *Am. J. Hum. Genet.* 82 57–72 10.1016/j.ajhg.2007.09.012

17. Eriksson M. A., Westerlund J., Anderlid B. M., Gillberg C., Fernell E. (2012).First-degree relatives of young children with autism spectrum disorders: some gender aspects. *Res. Dev. Disabil.* 33 1642–1648 10.1016/j.ridd.2012.03.025
18. Farook V. S., Puppala S., Schneider J., Fowler S. P., Chittoor G., Dyer T. D., et al. (2012). Metabolic syndrome is linked to chromosome 7q21 and associated with genetic variants in CD36 and GNAT3 in mexican americans. *Obesity (Silver Spring)* 20 2083–2092 10.1038/oby.2012.74
19. Forti P., Pirazzoli G. L., Maltoni B., Bianchi G., Magalotti D., Muscari A., et al. (2012). Metabolic syndrome and all-cause mortality in older men and women. *Eur. J. Clin. Invest.* 42 1000–1009 10.1111/j.1365-2362.2012.02688.x
20. Frustaci A., Neri M., Cesario A., Adams J. B., Domenici E., Dalla Bernardina B., et al. (2012). Oxidative stress-related biomarkers in autism: systematic review and meta-analyses. *Free Radic. Biol. Med.* 52 2128–2141 10.1016/j.freeradbiomed.2012.03.011
21. Gerbault P., Liebert A., Itan Y., Powell A., Currat M., Burger J., et al. (2011).Evolution of lactase persistence: an example of human niche construction. *Philos. Trans. R. Soc. B Biol. Sci.* 366 863–877 10.1098/rstb.2010.0268
22. Griswold A. J., Ma D., Cukier H. N., Nations L. D., Schmidt M. A., Chung R. H., et al. (2012). Evaluation of copy number variations reveals novel candidate genes in autism spectrum disorder associated pathways. *Hum. Mol. Genet.* 21 3513–3523 10.1093/hmg/dds164
23. Grundy S. M. (2007). Metabolic syndrome: a multiplex cardiovascular risk factor.*J. Clin. Endocrinol. Metab.* 92 399–404 10.1210/jc.2006-0513
24. Hall L., Kelley E. (2013). The contribution of epigenetics to understanding genetic factors in autism. *Autism* [Epub ahead of print]10.1177/1362361313503501
25. Hettinger J. A., Liu X., Hudson M. L., Lee A., Cohen I. L., Michaelis R. C., et al. (2012). DRD2 and PPP1R1B (DARPP-32) polymorphisms independently confer increased risk for autism spectrum disorders and additively predict affected status in male-only affected sib-pair families. *Behav. Brain Funct.* 8 19 10.1186/1744-9081-8-19
26. Heyer E., Brazier L., Ségurel L., Hegay T., Austerlitz F., Quintana-Murci L., et al. (2011). Lactase persistence in central Asia: phenotype, genotype, and evolution.*Hum. Biol.* 83 379–92 10.3378/027.083.0304
27. Holden C., Mace R. (1997). Phylogenetic analysis of the evolution of lactose digestion in adults. *Hum. Biol.* 69 605–628 10.3378/027.081.0609
28. Hu V. W. (2013). The expanding genomic landscape of autism: discovering the 'forest' beyond the 'trees'. *Future Neurol.* 8 29–42 10.2217/fnl.12.83
29. Ingram C. J., Mulcare C. A., Itan Y., Thomas M. G., Swallow D. M. (2009).Lactose digestion and the evolutionary genetics of lactase persistence. *Hum. Genet.* 124 579–591 10.1007/s00439-008-0593-6
30. Iossifov I., Ronemus M., Levy D., Wang Z., Hakker I., Rosenbaum J., et al. (2012). De novo gene disruptions in children on the autistic spectrum. *Neuron* 74285–299 10.1016/j.neuron.2012.04.009
31. Jackson F. (2004). Human genetic variation and health: ethnogenetic layering as a way of detecting relevant population substructuring. *Br. Med. Bull.* 69 215–235 10.1093/bmb/ldh012
32. Jensen T. G., Liebert A., Lewinsky R., Swallow D. M., Olsen J., Troelsen J. T. (2011). The -14010*C variant associated with lactase persistence is located between an Oct-1 and HNF1α binding site and increases lactase promoter activity.*Hum. Genet.* 130 483–493 10.1007/s00439-011-0966-0
33. Kim C. H., Kim H. K., Bae S. J., Kim E. H., Park J. Y. (2012). Independent impact of body mass index and metabolic syndrome on the risk of type 2 diabetes in koreans. *Metab. Syndr. Relat. Disord.* 10 321–325 10.1089/met.2011.0143
34. Lacaria M., Saha P., Potocki L., Bi W., Yan J., Girirajan S., et al. (2012). A duplication CNV that conveys traits reciprocal to metabolic syndrome and protects against diet-induced obesity in mice and men. *PLoS Genet.* 8:e100271310.1371/journal.pgen.1002713
35. Ladd-Acosta C., Hansen K. D., Briem E., Fallin M. D., Kaufmann W. E., Feinberg A. P. (2013). Common DNA methylation alterations in multiple brain regions in autism. *Mol. Psychiatry* 19 862–871 10.1038/mp.2013.114
36. LaSalle J. M. (2013a). Autism genes keep turning up chromatin. *OA Autism* 1 1410.13172/2052-7810-1-2-610
37. LaSalle J. M. (2013b). Epigenomic strategies at the interface of genetic and environmental risk factors for autism. *J. Hum. Genet.* 58 396–401 10.1038/jhg.2013.49
38. Mellios N., Sur M. (2012). The emerging role of microRNAs in Schizophrenia and autism spectrum disorders. *Front. Psychiatry* 3:39 10.3389/fpsyt.2012.00039
39. Mendez M. A., Horder J., Myers J., Coghlan S., Stokes P., Erritzoe D., et al. (2012). The brain GABA-benzodiazepine receptor α-5 subtype in autism spectrum disorder: a pilot [(11)C]Ro15-4513 positron emission tomography study.*Neuropharmacology* 68 195–201 10.1016/j.neuropharm.2012.04.008
40. Middleton F. A., Varlinskaya E. I., Mooney S. M. (2012). Molecular substrates of social avoidance seen following prenatal ethanol exposure and Its reversal by social enrichment. *Dev. Neurosci.* 34 115–128 10.1371/journal.pgen.1002505
41. Min J. L., Nicholson G., Halgrimsdottir I., Almstrup K., Petri A., Barrett A., et al. (2012). Coexpression network analysis in abdominal and gluteal adipose tissue reveals regulatory genetic loci for metabolic syndrome and related phenotypes.*PLoS Genet.* 8:e1002505 10.1371/journal.pgen.1002505

42. Nikitin I. U. P., Voevoda M. I., Simonova G. I. (2012). Diabetes mellitus and metabolic syndrome in Siberia and in the Far East (translated from Russian). *Vestn. Ross. Akad. Med. Nauk* 1 66–74
43. Peng M. S., He J. D., Zhu C. L., Wu S. F., Jin J. Q., Zhang Y. P. (2012). Lactase persistence may have an independent origin in Tibetan populations from Tibet, China. *J. Hum. Genet.* 57 394–397 10.1038/jhg.2012.41
44. Pennesi C. M., Klein L. C. (2012). Effectiveness of the gluten-free, casein-free diet for children diagnosed with autism spectrum disorder: based on parental report.*Nutr. Neurosci.* 15 85–91 10.1179/1476830512Y.0000000003
45. Rangasamy S., D'Mello S. R., Narayanan V. (2013). Epigenetics, autism spectrum, and neurodevelopmental disorders. *Neurotherapeutics* 10 742–756 10.1007/s13311-013-0227-220
46. Roth C. L., Sathyanarayana S. (2012). Mechanisms affecting neuroendocrine and epigenetic regulation of body weight and onset of puberty: potential implications in the child born small for gestational age (SGA). *Rev. Endocr. Metab. Disord.* 13129–140 10.1007/s11154-012-9212-x
47. Rottiers V., Näär A. M. (2012). MicroRNAs in metabolism and metabolic disorders. *Nat. Rev. Mol. Cell Biol.* 13 239–250 10.1038/nrm3313
48. Ruemmele F. M., Garnier-Lengliné H. (2012). Why are genetics important for nutrition? Lessons from epigenetic research. *Ann. Nutr. Metab.* 60(Suppl. 3), 38–43 10.1159/000337363
49. Testa C., Nuti F., Hayek J., De Felice C., Chelli M., Rovero P., et al. (2012). Di(2-ethylhexyl)phthalate and autism spectrum disorders. *ASN Neuro* 4 223–229 10.1042/AN20120015
50. Ting J. T., Peça J., Feng G. (2012). Functional consequences of mutations in postsynaptic scaffolding proteins and relevance to psychiatric disorders. *Annu. Rev. Neurosci.* 35 49–71 10.1146/annurev-neuro-062111-150442
51. Turcot V., Tschernof A., Deshaies Y., Perusse L., Belisle A., Marceau S., et al. (2012). LINE-1 methylation in visceral adipose tissues of severely obese individuals in association with metabolic syndrome status and related phenotypes.*Clin. Epigenetics* 4 10 10.1186/1868-7083-4-10
52. Wang J. W., Hu D. Y., Sun Y. H., Wang J. H., Wang G. L., Xie J., et al. (2009).Obesity criteria for identifying metabolic risks. *Asia Pac. J. Clin. Nutr.* 18 105–113
53. Wang J., Wu Z., Li D., Li N., Dindot S. V., Satterfield M. C., et al. (2012).Nutrition, epigenetics, and metabolic syndrome. *Antioxid. Redox Signal.* 17 282–301 10.1089/ars.2011.4381
54. Wilkinson B., Campbell D. B. (2013). Contribution of long noncoding RNAs to autism spectrum disorder risk. *Int. Rev. Neurobiol.* 113 35–59 10.1016/B978-0-12-418700-9.00002-2
55. Woods R., Vallero R. O., Golub M. S., Suarez J. K., Ta T. A., Yasui D. H., et al. (2012). Long-lived epigenetic interactions between perinatal PBDE exposure and Mecp2308 mutation. *Hum. Mol. Genet.* 21 2399–2411 10.1093/hmg/dds046
56. Yang Y., Wang C., Wang F., Zhu L., Liu H., He X. (2012). Novel chromosomal translocation t(11; 9)(p15; p23) involving deletion and duplication of 9p in a girl associated with autism and mental retardation. *Gene* 502 154–158 10.1016/j.gene.2012.04.036
57. Zerbo O., Iosif A. M., Walker C., Ozonoff S., Hansen R. L., Hertz-Picciotto I. (2012). Is maternal influenza or fever during pregnancy associated with autism or developmental delays? Results from the CHARGE (CHildhood Autism Risks from Genetics and Environment) study. *J. Autism Dev. Disord.* 43 25–33 10.1007/s10803-012-1540-x
58. Zorzi A., Wahi G., Macnab A. J., Panagiotopoulos C. (2009). Prevalence of impaired glucose tolerance and the components of metabolic syndrome in Canadian Tsimshian Nation youth. *Can. J. Rural. Med.* 14 61–67

CHAPTER 6

CHEMPROT-3.0: A GLOBAL CHEMICAL BIOLOGY DISEASES MAPPING

Jens Kringelum[1], Sonny Kim Kjaerulff[1], Søren Brunak[2], Ole Lund[1], Tudor I. Oprea[1],[3] and Olivier Taboureau[1],[4],

[1]Department of Systems Biology, Center for Biological Sequence Analysis, buildin 208, kemitorvet, Technical University of Denmark, DK-2800 Lyngby, Denmark

[2]Department of Disease Systems Biology, Faculty of Health and Medical Sciences, Novo Nordisk Foundation, Center for Protein Research, University of Copenhagen, Blegdamsvej 3A, DK-2200, Copenhagen, Denmark

[3]Translational Informatics Division, Department of Internal Medicine, University of New Mexico Health Sciences Center, MSC09 5025, Albuquerque 87181, New Mexico, United States of America

[4]INSERM, UMRS-973, MTi, Universite Paris Diderot, 35 rue Helene Brion, 75205 Paris Cedex 13, Sorbonne Paris Cite, France

ABSTRACT

ChemProt is a publicly available compilation of chemical-protein-disease annotation resources that enables the study of systems pharmacology for a small molecule across multiple layers of complexity from molecular to clinical levels. In this third version, ChemProt has been updated to more than 1.7 million compounds with 7.8 million bioactivity measurements for 19 504 proteins. Here, we report the implementation of global pharmacological heatmap, supporting a user-friendly navigation of chemogenomics space. This facilitates the visualization and selection of chemicals that share similar structural properties. In addition, the user has the possibility to search by compound, target, pathway, disease and clinical effect. Genetic variations associated to target proteins were integrated, making it possible to plan pharmacogenetic studies and to suggest human response variability to drug. Finally, Quantitative Structure–Activity Relationship models for 850 proteins having sufficient data were implemented, enabling secondary pharmacological profiling predictions from molecular structure.

INTRODUCTION

Many chemical biology initiatives in Europe and the USA aim to screen large compound collections with dedicated bioassays i.e. EU Lead Factory (1), EU-Openscreen (2) or BARD in the USA (3). Such large initiatives generate large amounts of data that support academic and industrial research in the discovery of safer chemicals, with better efficacy. To make chemical biology information accessible to scientists, several repositories of bioactive small molecules have been developed: ChEMBL (4), PubChem (5), ChemSpider (6) and OpenPhacts (7) are the largest, more general databases available to the public. The National Institutes of Health's Molecular Libraries Program (MLP) funding developed the BioAssay Research Database (BARD), focusing on assay ontologies for PubChem bioassays (3).

Advances in chemical biology and systems biology have shown that most drugs interact with multiple targets and that the pharmacological profile of a drug is not as reductionist as once believed (8). Moreover, proteins rarely operate in isolation within and outside cells but function in interconnected pathways instead. Given the integration afforded by systems biology, it is now possible to consider a more general physiological environment for protein targets and biological processes. As massive amounts of data are generated and accumulated via new experimental technologies such as transcriptomic, proteomics and genomics (through next-generation sequencing), drug action can be explored across multiple scale of complexity, from molecular and cellular to tissue and organism levels (9–11).

Citation: Jens Kringelum, Sonny Kim Kjaerulff, Søren Brunak, Ole Lund, Tudor I. Oprea, and Olivier Taboureau. ChemProt-3.0: a global chemical biology diseases mapping. Database (2016) Vol. 2016: article ID bav123; doi:10.1093/database/bav123

Multi-target pharmacology exploration increases when information linking the relationship between chemical and target spaces is readily available. As archived data are processed and homogenized, our total knowledge on protein–ligand interactions is increasing at an amazing pace (12, 13). Scientists having access to these data, approaches such as chemogenomics, proteochemometrics and polypharmacology have started to emerge (14, 15). These help to mine evaluate and ultimately distil this vast amount of protein–ligand interactions data, enabling the predictions of single ligands against a set of heterogeneous targets (16).

This third version of ChemProt is not a simple update for disease chemical biology data. Rather, we provide a friendly platform to navigate through the various data sources, from global evaluations to a focused analysis. Several computational approaches are included: ligand-based similarity, target-based promiscuity, QSAR (Quantitative Structure–Activity Relationship) methodology and network biology-based enrichment analyses. These approaches support novel hypotheses generation for bioactivity of novel and already-annotated compounds, and the ability to identify additional genes that may play major roles in modulating chemical perturbations in man. The updates and new methods introduced in ChemProt-3.0 are presented below.

DATA SOURCES

We updated all the chemical protein interactions data from the open source databases ChEMBL (version 19) (4), BindingDB (17), PDSP Ki database (18), DrugBank (version 4) (19), PharmGKB (20), IUPHAR-DB database (21) and STITCH (version 4) (22). Clinical information from the Anatomical Therapeutic Chemical Classification System (23) developed by the World Health Organization, as well as side effect data from Sider 2 were also integrated (24).

From a biological perspective, we updated our internal human interactome platform to reach 14421 genes interacting through 507142 unique PPIs (25). OMIM (26), the human disease network (27) GeneCards (28), KEGG (29), Reactome (30), UniPathway (31) and Gene Ontology (32) databases were also downloaded, curated and included in our system. Overall, the integrated data sources were increased by over 60% compared to the earlier version.

As many different data types were aggregated in ChemProt, a 'zChemProt' value for each compound-bioactivity interactions was computed for visualization in the several heatmaps developed. Basically, for each of the 11 most prevalent data types (IC_{50}, EC_{50}, Potency, AC_{50}, pIC_{50}, Log K_i, pK_i, pEC_{50}, K_d, K_i), a zChemProt value was computed using the mean and standard deviation calculated from the distribution of the associated data types for each target in a similar way described in CARLSBAD database (33). IC_{50}, EC_{50}, Potency, AC_{50}, K_d and K_i were log transformed before computing the zChemProt values. Large values indicate strong chemical–protein interactions and are represented in orange. Low value (weak interactions) is depicted in blue.

PREDICTIONS METHODS

Daylight-like 1024 bit fingerprints was computed with RDkit (www.rdkit.org) and the chemical similarity between two compounds was quantitatively assessed using the Tanimoto coefficient (34). The Similarity Ensemble Approach (SEA) (35) has been re-compiled on the ChemProt server and updated according to the novel zChemProt data and integrated into ChemProt 3.0. Only proteins with >10 chemicals were included for SEA prediction, using the same protocol as described in the previous version (36). For sequence analyses, protein sequences were obtained from Uniprot (37). Sequences comparisons were computed using BLASTP and estimated to be similar when their E value was lower than 10^{-10} (38). All compounds were decomposed into ring scaffolds based on an internal implementation of the 'Scaffold Hunter' hierarchical classification algorithm (39–40) with the addition of decomposition of non-ring molecule based on rules 7–10, as described by Schuffenhauer *et al.* (41). This hierarchical decomposition allows the generation of scaffold trees enabling an easy and interactive navigation of the chemical biology space in large datasets and the identification of potential new compound classes with desired bioactivity.

For this release, QSAR models were trained for each protein with >20 chemicals (in total 850 proteins). A Naïve Bayes classifier was trained using 5-fold cross-validation for performance assessment. Features selection, five different computational fingerprints (Daylight and Morgan fingerprints) and three different cutoffs (−log 10 value: 4, 5 and 6) for classifying active and non-active compounds, were used to produce overall 15 classification models for each target. To predict new compound, each model was weighted by the cross-validated performance measure resulting in a prediction value between 0 and 1 (where 1 is high predicted binding). To avoid bias toward negative or positive data, each of the three datasets used for training were balanced by including as many negative compounds as positive including random compounds from the ChemProt database. The performance of the developed QSAR ensemble approach was tested on a dataset of hERG binders and showed an improved performance compared to a previous reported study (42) (Aroc = 0.827, Matthews Correlation Coefficient (MCC) = 0.488 using 5-fold cross-validation). Furthermore, the method was benchmarked against the SEA implementation on a dataset consisting of 143 proteins with associated activity values from the ChemProt-3.0 dataset. In a 5-fold cross-validation scheme on each of the 143 proteins, the QSAR models outperform SEA (P_{val} = 2.2e^{-16}, paired t-test of Spearman correlation coefficients for each protein predictions). However, models developed from limited amounts of data might not provide reliable predictions. The user can consult the 'prediction info' tab (by clicking on the protein of interest on the heatmap) to obtain the number of molecules used for training the QSAR models.

VISUAL INTERFACE

In ChemProt 3.0, the front page was modified to have all the functionality available on the page. The user has the possibility to search information about a compound, a protein and a clinical outcome, or he can choose to perform a QSAR prediction for a specific compound. A molecule can be imported as a SMILES code, or alternatively it can be drawn or uploaded from a compound structure file via the SD file format. A new function called 'Heatmap' was integrated, which allows the user to have a global view of chemical-protein interactions (Figure 1). In this graphical interface, the user has the possibility to localize the bioactivity associated to a requested compound, a set of defined compounds or a set of similar compounds based on the chemical structure. Several layers of granularity have been implemented on the heatmap. The proteins have been categorizes by families, using the protein classification tree implemented in ChEMBL and the compounds have been decomposed in scaffold and chemical groups based on the scaffold tree implementation similar to CARLSBAD. This gives the user the opportunity to visualize scaffold-protein activity relationships. A color spectrum from blue (low activity) to orange (strong activity) is used to indicate the activity. All compounds structures and protein IDs (based on Uniprot) are clickable, which gives access to more detailed information about physicochemical properties and protein function respectively.

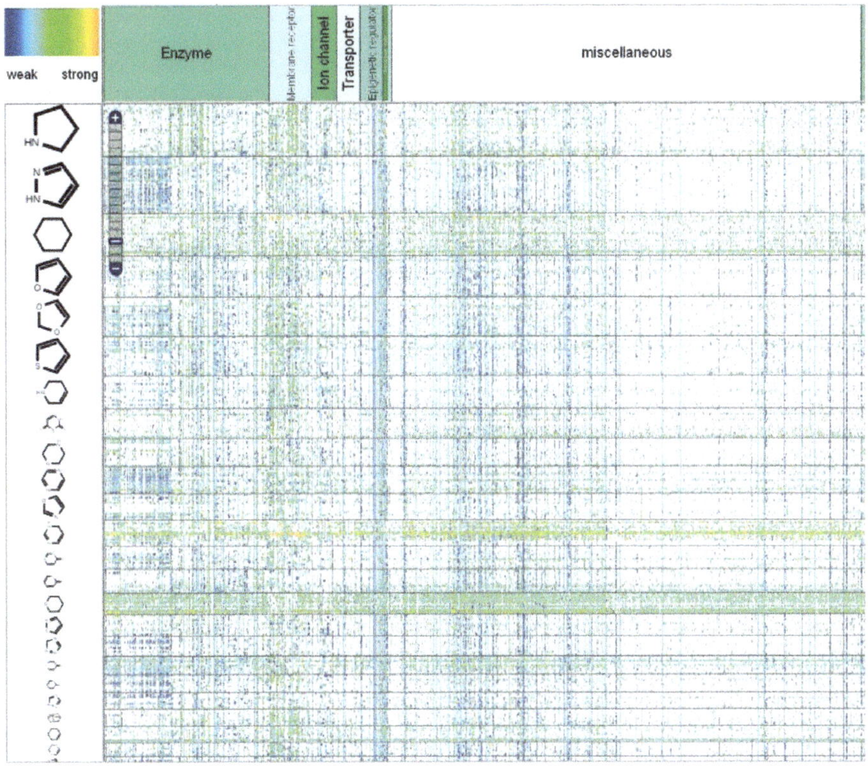

Figure 1. Global view of the chemical-protein interactions heatmap in ChemProt.

An interesting feature with this graphical interface is the possibility to match other biological to chemical data. Instead of choosing ‘for drugs’, the user can select targets, pathways, diseases or side effects and see the association between chemicals and these endpoints.

From the protein ID, the user has access to the proteins complex (represented as a protein’s network from protein–protein interaction data). The complex of proteins is then mapped to biological terms such as diseases, GO terms and pathways with a corrected *P* value to evaluate the significance of these associations.

Finally, the user has the possibility to download the results in flat-file format to perform others analyses.

Methodology: Daylight-like fingerprints, defined by 1024 fragments were computed using RDkit (www.rdkit.org). The QSAR models were trained using scikit-learn software (http://scikit-learn.org/stable/). The visual interface was implemented using HTML 5 and JavaScript. The webserver is limited for an input file of 50 molecules (in SMILES or sdf file) per query to limit the time necessary to get the output. For larger queries, the user is advised to contact us.

APPLICATIONS

Caffeine, a well-known natural product extracted from coffee beans or tea leaves, is often used as a central nervous system stimulant (43). Several outputs can be displayed like those shown in Figure 2. Typing the compound name in the ‘Compound’ field and clicking on the Submit button, the user is redirected to the global chemical-protein heatmap with the query compound showing up in the compound list as default. Any compound can be added to the list by writing a new compound name in the ‘search’ box. Clicking on the ‘flag’ next to a compound name in the compound list prompts the heatmap to zoom in or out from that specific compound, enabling a fast way to visualize the proteins signature for the queried compound, as well as for compounds sharing scaffold similarities. Clicking on the ‘fingerprint’ logo in the vicinity of the compound name, a chemical structure similarity profiling can be performed, enabling the user to visualize and to navigate within that pharmacological heatmap. For the Caffeine example, 105 similar compounds (with a Tanimoto coefficient>0.85) were found, with bioactivities associated to 449 proteins (from weak in blue to strong in orange). The user is able to zoom in the heatmap and to narrow the information from the classification proteins tree to specific proteins (defined by uniprot ID). The user has also the possibility to navigate inside the heatmap. One option is to fill out missing values by choosing ‘SEA’ or ‘QSAR’ under prediction in the top of the page.

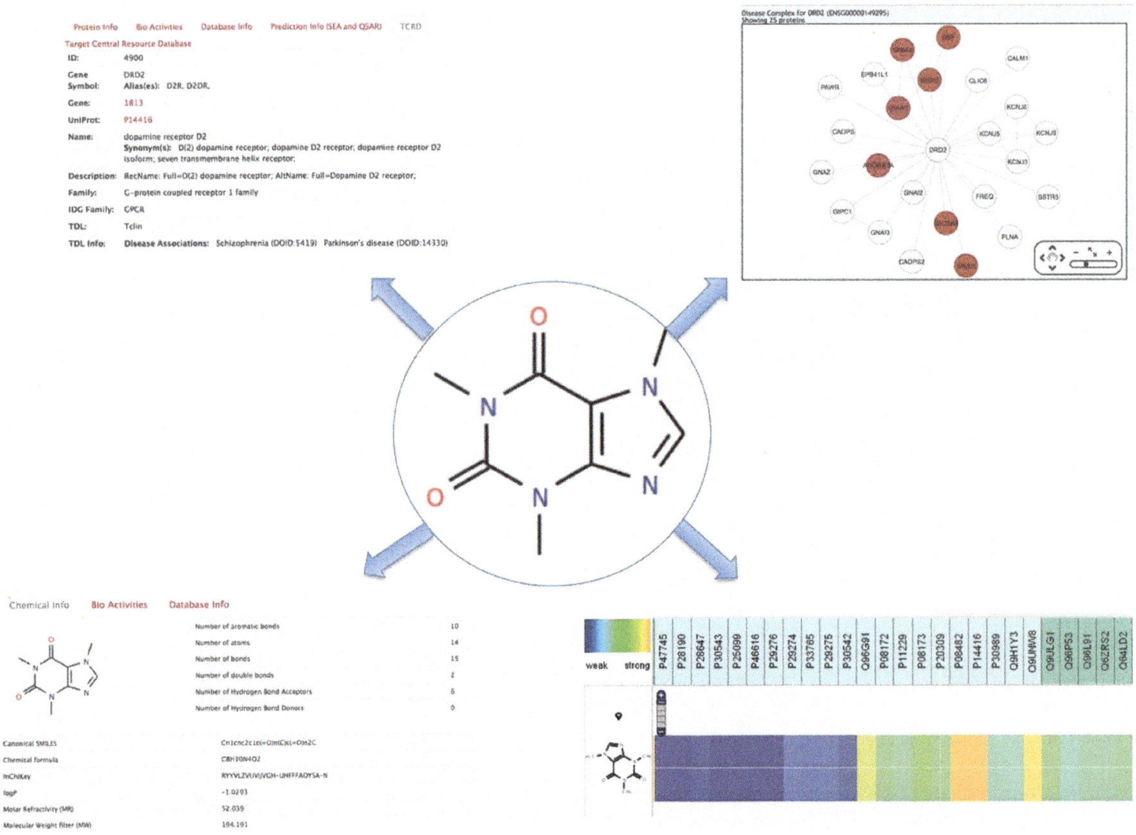

Figure 2. Information that can be collected from a search on caffeine. Top left, functional information on bioactive proteins for the query compound is depicted. Bottom left, chemical and physicochemical information is gathered. Top right, protein's complex associated to the chemical is shown and the bottom right is depicted the protein's annotation and prediction (through QSAR) for caffeine.

By clicking on the compound structure, physicochemical features [such as the Lipinski rules (44)], number of proteins with bioactivities and the databases from which the information was gathered, are shown. Similarly, the user can click on a specific protein name and get more information on the function of the protein, diseases associated to this protein and predictions based on SEA and QSAR. For example, under 'family A GPCR', caffeine is shown to be potent (35.5 nM) on the rat muscarinic M1 acetylcholine receptor (P08482). It also shows a strong association with the dopamine D2 receptor (P14416) based on the STITCH system. By clicking on this protein, the user is presented with information on this dopamine receptor. Notably, disease associations are queried through the TCRD (Target Central Ressource Database Application:http://juniper.health.unm.edu/tcrd/) database and the genetic variation through the Ensembl database (45). A complex disease network is also associated to this protein. Clicking on this link, diseases (such as schizophrenia and epilepsy) and GO terms (plasma membrane, ligand-gated ion channel activity, etc.) are shown.

Instead of looking for 'functional' protein annotation, it is further possible to select 'pathway' or 'diseases' for caffeine (also for the set of 105 similar compounds). The heatmap will be depicted according to the query. Each protein annotation (functional, pathway, disease) is presented in a tree format. Proteins have been categorized from families to proteins using the protein target tree implemented in ChEMBL. It has been done similarly for the pathway using the unipathway (31) implementation tree and disease using the human disease network (27). For example, using the disease heatmap, strong associations are found between caffeine and ventricular tachycardia, slow acetylation, glycogen storage disease, dystonia and thyroid carcinoma.

Finally, from the ChemProt-3 front page, the user can write the caffeine's SMILES in the QSAR prediction box, click on 'Submit QSAR' and then get a prediction of positive and negative bioactivities for the ensemble of proteins in ChemProt where reliable QSAR models can be produced. This option allows the user to have a direct QSAR prediction for a new compound not present in the ChemProt database.

CONCLUSION

Given that access to many chemogenomics databases is possible, linking them to biological resources and using a number of machine learning tools, scientists can now estimate the bioactivity profile of molecules across a large number of targets, pathways, diseases and other clinical outcomes using ligand-based, target-based and network-based models. Such multi-target, multi-layer strategies are becoming more and more accepted by the scientific community. Within ChemProt, it is possible to navigate the chemogenomics space and to link chemically induced target perturbations to diseases and other biological outcomes. Such tools might be of interest for drug discovery, drug safety and also chemical risk assessment. ChemProt 3.0 supports predicting bioactivities on targets and off-targets for new compounds and can assist in the associations to phenotypes and side effects relationships.

REFERENCES

1. Mullard A. (2013) European lead factory opens for business. Nat. Rev. Drug. Discov., 12, 173–175.
2. Frank R. (2014) EU-Openscreen—a European infrastructure of open screening platforms for chemical biology. ACS Chem. Biol., 9, 853–854.
3. de Souza A., Bittker J.A., Lahr D.L. et al. (2014) An overview of the challenges in designing, integrating and delivering BARD: a public chemical-biology resource and query portal for multiple organizations, locations and disciplines. J. Biomol. Screen., 19, 614–627.
4. Gaulton A., Bellis L.J., Bento A.P. et al. (2012) ChEMBL: a large-scale bioactivity database for drug discovery. Nucleic Acids Res., 40, D1100–D1107.
5. Sayers E.W., Barrett T., Benson D.A. et al. (2012) Database resources of the National Center for Biotechnology Information. Nucleic Acids Res., 40, D13–D25.
6. Williams A., Tkachenko V. (2014) The Royal Society of Chemistry and the delivery of chemistry data repositories for the community. J. Comput. Aided Mol. Des., 28, 1023–1030.
7. William A.J., Harland L., Groth P. et al. (2012) Open PHACTS: semantic interoperability for drug discovery. Drug Discov. Today, 17, 1188–1198.
8. Wild D.J., Ding Y., Sheth A.P. et al. (2012) Systems chemical biology and the semantic web: what they mean for the future of drug discovery research. Drug Discov. Today, 17, 469–474.
9. Keith C.T., Borisy A.A., Stockwell B.R. (2005) Multicomponent therapeutics for networked systems. Nat. Rev. Drug Discov., 4, 71–78.
10. Zimmerman G.R., Lehár J., Keith C.T. (2007) Multi-target therapeutics: when the whole is greater than the sum of the parts. Drug Discov. Today, 12, 34–42.
11. Fitzgerald J.B., Schoeberl B., Nielsen U.B. et al. (2006) Systems biology and combination therapy in the quest for clinical efficacy. Nat. Chem. Biol., 2,458–466.
12. Ekins S., Williams A.J. (2010) When pharmaceutical companies publish large datasets: an abundance of riches or fool's gold? Drug Discov. Today, 15,812–815.
13. Wang J., Li Z., Qiu C. et al. (2012) The relationship between rational drug design and drug side effects. Brief. Bioinform., 13, 377–382.
14. Rognan D. (2007) Chemogenomic approaches to rational drug design. Br. J. Pharmacol., 152, 38–52.
15. Keiser M.J., Setola V., Irwin J.J. et al. (2009) Predicting new molecular targets for known drugs. Nature, 462, 175–181.
16. Schneider G. (2010) Virtual screening: an endless staircase? Nat. Rev. Drug Discov., 9, 273–276.
17. Liu T., Lin Y., Wen X. et al. (2007) Binding DB: a web-accessible database of experimentally determined protein-ligand binding affinities. Nucleic Acids Res.,35, D198–D201.
18. Roth B., Lopez E., Beischel S. et al. (2004) Screening the receptorome to discover the molecular targets for plant-derived psychoactive compounds: a novel approach for CNS drug discovery. Pharmacol. Ther., 102, 99–110.
19. Law V., Knox C., Djoumbou Y. et al. (2011) DrugBank 4.0: shedding new light on drug metabolism. Nucleic Acids Res., 42, D1091–D1097.
20. McDonagh E.M., Whirl-Carrillo M., Garten Y. et al. (2011) From pharmacogenomic knowledge acquisition to clinical applications: the PharmGKB as a clinical pharmacogenomic biomarker resource. Biomark. Med., 5, 795–806.
21. Pawson A.J., Sharman J.L., Benson J.L. et al. (2014) The IUPHAR/BPS guide to PHARMACOLOGY: an expert-driven knowledgebase of drug targets and their ligands. Nucleic Acids Res., 42, D1098–D1106.
22. Kuhn M., Szklarczyk D., Pletscher-Frankild S. et al. (2014) STITCH 4: integration of protein-chemical interactions with user data. Nucleic Acids Res.,42, D401–D407.
23. De Smet P.A.G.M. (1993) New applications of the ATC/DDD methodology in the Netherlands part 1. ATC/DDD principles and computerized medication surveillance. Int. Pharm. J., 7, 196–199.
24. Kuhn M., Campillos M., Letunic I. et al. (2010) A side effect resource to capture phenotypic effects of drugs. Mol. Syst. Biol., 6, 343
25. Lage K., Karlberg E.O., Størling Z.M. et al. (2007) A human phenome-interactome network of protein complexes implicated in genetic disorders. Nat. Biotechnol., 25, 309–316.
26. Amberger J., Bocchini C., Hamosh A. (2011) A new face and new challenges for Online Mendelian Inheritance in Man (OMIM). Hum. Mutat., 32, 564–567.
27. Goh K.I., Cusick M.E., Valle D. et al. (2007) The human disease network. *Proc.*Natl. Acad. Sci. USA, 104, 8685–8690.
28. Stelzer G., Dalah I., Stein T.I. et al. (2011) In-silico human genomics with GeneCards. Hum. Genomics, 5, 709–717.
29. Kanehisa M., Goto S., Sato Y. et al. (2012) KEGG for integration and interpretation of large-scale molecular data sets. Nucleic Acids Res., 40, D109–D114.
30. Matthews L., Gopinath G., Gillespie M. et al. (2009) Reactome knowledgebase of human biological pathways and processes. Nucleic Acids Res., 37, D619–D622.
31. Morgat A., Coissac E., Coudert E. et al. (2012) Unipathway: a resource for the exploration and annotation of metabolic pathways. Nucleic Acids Res., 40, D761–D769.
32. Dimmer E.C., Huntley R.P., Alam-Faruque Y. et al. (2012) The UniProt-GO Annotation database in 2011. Nucleic Acids Res., 40, D565–D570.
33. Mathias S.L., Hines-Kay J., Yang J.J. et al. (2013) The CARLSBAD database: a confederated database of chemical bioactivities. Database, 2013, bat044.
34. Willet P. (2006) Similarity-based virtual screening using 2D fingerprints. Drug Discov. Today, 11, 1046–1053.

35. Keiser M.J., Roth B.L., Armbruster B.N. et al. (2007) Relating protein pharmacology by ligand chemistry. Nat. Biotechnol., 25, 197–206.

36. Kjærulff K.S., Wich L., Kringelum J. et al. (2013) ChemProt-2.0: visual navigation in a disease chemical biology database. Nucleic Acids Res., 41, D464–D469.

37. UniProt Consortium. (2011) Ongoing and future developments at the universal protein resource. Nucleic Acids Res., 39, D214–D219.

38. Altschul S.F., Madden T.L., Schäffer A.A. et al. (1997) Gapped BLAST and PSI-BLAST: a new generation of protein database search programs. Nucleic Acids Res., 25, 3389–3402.

39. Klein K., Koch O., Kriege N. et al. (2013) Visual analysis of biological activity data with Scaffold Hunter. Mol. Inf., 32, 964–975.

40. Wetzel S., Klein K., Renner S. et al. (2009) Interactive exploration of chemical space with Scaffold Hunter. Nat. Chem. Biol., 5, 581–583.

41. Schuffenhauer A., Ertl P., Roggo S. et al. (2007) The scaffold tree—visualization of the scaffold universe by hierarchical scaffold classification. J. Chem. Inf. Model., 47, 47–58.

42. Li Q., Jorgensen F.S., Oprea T. et al. (2008) hERG classification model based on a combination of support vector machine method and GRIND descriptors.Mol. Pharm., 5, 117–127.

43. Nehlig A., Daval J.L., Debry G. (1992) Caffeine and the central nervous system: mechanisms of action, biochemical, metabolic and psychostimulant effects. Brain Res. Brain Res. Rev., 17, 139–170.

44. Lipinski C.A., Lombardo F., Dominy B.W. et al. (2001) Experimental and computational approaches to estimate solubility and permeability in drug discovery and development settings. Adv. Drug Deliv. Rev., 46, 3–26.

45. Cunningham F., Amode M.R., Barrell D. et al. (2015) Ensembl 2015. Nucleic Acids Res., 42(Database issue):D749–D755.

CHAPTER 7

HUMAN LONGEVITY: GENETICS OR LIFESTYLE? IT TAKES TWO TO TANGO

Giuseppe Passarino, Francesco De Rango, and Alberto Montesanto

Department of Biology, Ecology and Earth Science, University of Calabria, 87036 Rende, Italy

ABSTRACT

Healthy aging and longevity in humans are modulated by a lucky combination of genetic and non-genetic factors. Family studies demonstrated that about 25 % of the variation in human longevity is due to genetic factors. The search for genetic and molecular basis of aging has led to the identification of genes correlated with the maintenance of the cell and of its basic metabolism as the main genetic factors affecting the individual variation of the aging phenotype. In addition, studies on calorie restriction and on the variability of genes associated with nutrient-sensing signaling, have shown that ipocaloric diet and/or a genetically efficient metabolism of nutrients, can modulate lifespan by promoting an efficient maintenance of the cell and of the organism. Recently, epigenetic studies have shown that epigenetic modifications, modulated by both genetic background and lifestyle, are very sensitive to the aging process and can either be a biomarker of the quality of aging or influence the rate and the quality of aging.

On the whole, current studies are showing that interventions modulating the interaction between genetic background and environment is essential to determine the individual chance to attain longevity.

BACKGROUND

The research on aging, and in particular the search for the determinants of successful aging and longevity, has been continuously growing in the last decades also due to the social and medical burden correlated to the continuous increase of lifespan in western countries and the consequent grow of the elderly population. One of the main questions in this field is the correlation between the genetic background and lifestyle in determining the individual chance of a delayed aging (possibly without age-related diseases and disabilities) and longevity. The results obtained by biogerontologists in these years, which highlighted most of the biological and biochemical mechanisms involved in the aging process, allowed to better understand such correlation. This has brought to elaborate important strategies focused on possible interventions to improve lifestyle in order to increase the chance to attain longevity by modulating the basic molecular mechanisms of aging.

The Genetics of Aging

Before the 1990ies it was largely spread the idea that aging is ineluctable and that genetics does not control it. It was important, in this view, the idea that aging occurs after reproduction, and then there is no need, but also no opportunity, for selection to act on genes that are expressed during this late period of life [1].

Citation: Giuseppe Passarino, Francesco De Rango, and Alberto Montesanto. Human longevity: Genetics or Lifestyle? It takes two to tango, Immunity & Ageing (2016) 13:12, DOI: 10.1186/s12979-016-0066-z

The researcher who pioneered the genetics of aging and longevity was Tom Johnson, who studied groups of *C. elegans* where he was able to separate long living individuals from short living subjects. The analysis of hybrids obtained from different strains of *C. elegans,* allowed to estimate that the heritability of life-span was between 20 and 50 % [2, 3]. Subsequently, he started the analysis of different mutants and, with M. Klass, found a number of mutants with longer lifespan. Subsequently, Tom Johnson found out that most of the mutants with long lifespan had mutations in the *age1* gene [4]. This gene turned out to be the catalytic subunit of class-I phosphatidylinositol 3-kinase (PI3K).

The studies of Johnson clearly demonstrated that genetic variability could indeed affect lifespan. This triggered many studies in model organisms in order to disentangle the different biochemical pathways which could affect lifespan, and to highlight the genes coding for the proteins involved in such pathways. In particular, yeast, *C. elegans*, drosophila and mice were analyzed and this highlighted numerous genes which could affect lifespan if mutated (for an updated list of these genes seehttp://genomics.senescence.info/genes/models.html). Most of these genes are related to the maintenance of the integrity of the cell (especially the integrity of DNA). In *C. elegans*, however, some of the main genes which have been found to modulate lifespan (*daf2*, *daf16*) are related to the ability to enter the *dauer* status [5, 6], that is a quiescent status (usually entered in case of nutrient deprivation) with a minimum energy expense, which causes an arrest of the reproduction process and allows the organism to live longer "expecting" for the availability of nutrients. This suggested that longevity can be attained by means of an efficient maintenance of the cell but also by diverting resources from reproduction to self maintenance, in line with previous findings that dietary restriction can extend lifespan. After the characterization of these genes in *C. elegans*, it was found that in mice the ortholog of *daf16* (FOXO) could affect lifespan. In mammals, FOXO is correlated to the Insulin/IGF1 axis which is stimulated by nutrient availability and, through FOXO, promotes protein synthesis [7,8, 9, 10, 11].

It is of note that some Authors suggested these molecular mechanisms modulating lifespan could be due to a pleiotropic effect of genes which have evolved for different purposes (such as the genes in the IGF-1 pathway which have evolved to face presence/absence of nutrients) but can, ultimately affect lifespan; others proposed that some genes may have evolved to program aging and avoid "immortality", as this would hamper the continuous substitution of old subjects with new, younger, ones [12, 13].

It was obviously inevitable that the research of the genetic basis of longevity turned to human beings and investigated whether the common genetic variability of human populations could affect inter individual differences in lifespan but also whether the genes found to prolong lifespan in model organisms, on turn, were correlated to human lifespan.

As to the first question (does common genetic variability affect lifespan, and in particular does it affect longevity?), this has been studied by two approaches. The first one was the reconstruction of the sibships of long-lived subjects [14, 15] and the comparison of their survival curves with those of the birth cohorts born in the same geographical area. This approach demonstrated that brothers and sisters of the long-lived subjects had a clear survival advantage (at any age) with respect to the general population. The second approach, with intrafamily controls, was started in order to distinguish the genetic from the "familiar" effect. Montesanto et al. [15] compared the survival function of brothers of centenarians with those estimated for their brothers in law, that is with the men who married their sisters; these men were supposed to share with the brothers of the long lived subjects the familiar environment. By using this second approach, it has been found that the survival advantage of siblings of long-lived subjects was not completely shared by their brothers in law, despite they shared the same environment for most of their life. This suggested that beyond the family environment, there are genetic factors influencing survival and, consequently, lifespan. Interestingly, in this study, the survival curve of the sisters of long-lived subjects did not differ from the one of sisters in law, suggesting that the genetic component does explain lifespan in men more than in women. The genetic component of lifespan in humans has also been analyzed by comparing the age of death of monozygotic and dizygotic twins. This has allowed to estimate that about 25 % of the variation in human longevity can be due to genetic factors and indicated that this component is higher at older ages and is more important in males than in females [16, 17, 18].

In parallel to these studies, many researches have been carried out to search the genetic variants responsible of modulating human longevity. Most of them were carried out by a case/control approach, by comparing the frequency of specific polymorphisms in long-lived subjects and in younger geographically matched controls. The rationale of this study design is that as the population ages, alleles favorable for survival will be present at higher frequency among long-living people, while unfavorable alleles will be eliminated [19, 20, 21]. The candidate genes analyzed by this approach were either genes involved in age-related diseases (such as APOE, which had been observed to be involved in the predisposition to Alzheimer Disease and other age-related cognitive impairments), or genes implicated in pathways related to longevity in studies with model organisms (IGF-1, FOXO, Sirtuins) [22, 23, 24, 25]. This study design has indeed led to find numerous polymorphic genes the variability of which affects longevity. However, each of these polymorphisms turned out to explain only a very small fraction of the longevity variability. Indeed high-throughput Genome-wide analyses, which have recently been carried out have identified many genes positively associated with longevity but only a very few ones could hold multiple test significance and successfully replicated in different studies and across different populations [26, 27, 28, 29]. Population stratification and inadequate sample sizes are among the main plausible explanations [30]. The adoption of innovative study design and the development of new statistical and computational tools for effective processing of genetic data arising from high-throughput DNA technologies will help to better understand the complex genetic architecture underlying human longevity [31,32].

A new way of looking at the genetic data has been proposed by Raule et al. [33] who analyzed the complete sequences of mitochondrial DNA from long-lived subjects coming from different areas of Europe. The availability of complete sequences allowed to evaluate for the first time the cumulative effects of specific, concomitant mitochondrial DNA (mtDNA) mutations, including those that *per se* have a low, or very low, impact. The analysis indicated that the presence of single mutations on mtDNA complex I may be beneficial for longevity, while the co-occurrence of mutations on both complexes I and III or on both I and V might lower the individual's chances for longevity. Previous analyses on single mutations falling on complex I (either specific mutations or mutations defining groups of haplotypes) had given contrasting results, showing association with longevity in some cases but not in others. It is likely that positive results were obtained in populations were mutations on complex I were not associated with mutations on complex III or V, while negative results were obtained in populations with high prevalence of mtDNA haplotypes carrying mutations on complex I in association with mutations in complex III and V. This approach confirmed that most of the genetic variants have a very limited effect on longevity, and that only their cumulative effect can give a consistent appreciable effect and suggests that a limit of previous analyses has been to search for single mutations instead of cumulative effects. On the other hand, it is very difficult to think of using such approach, which has been successful for mitochondrial DNA, on genomic DNA unless small fractions (or specific regions harboring genes involved in relevant pathways) are analyzed.

On the whole, the genetic association studies suggested that, also in humans, mutations in genes correlated with the maintenance of the cell and of its basic metabolism are essential in modulating lifespan. Indeed, genes involved in DNA repair [34], telomere conservation [35, 36, 37], heat shock response [38, 39], and the management of free radicals' levels [33, 40] were found to contribute to longevity or, in case of reduced functionality, to accelerated senescence (cellular aging) and the consequent organism aging. In addition, as suggested by the studies in mice, the pathways involved in nutrient-sensing signaling and in regulating transcription, such as IGF-1/insulin axis [41] and TOR (target of rapamycin) [42] showed to be involved in modulating human longevity. Besides these genes involved in cellular maintenance/metabolism and senescence, concurrent efforts, especially from clinical studies, also showed that genes implicated in important organismal process may have a strong impact on aging and longevity. For instance genes involved in lipoprotein metabolism (especially APOE), cardiovascular homeostasis, immunity, and inflammation have been found to play an important role in aging, age-related disorders, and organism longevity [43, 44, 45, 46].

Human Longevity and Life Style

Life expectancy at birth has been increasing for most of the last century in western societies, thanks to the continuous amelioration of medical assistance, to the improvement of the environment (in particular clean, safe water and food), and to the improvement of nutrients. For instance, in Italy life expectancy went from 29 years in 1861 to 82 in 2011 (Table 1 reports the evolution of this data in women and men). Similarly, the extreme longevity has been growing in these years. Indeed, the number of centenarians (still in Italy) remarkably increased from 165 in 1951 to more than 15000 in 2011. These results have been attained first by a dramatic reduction of infectious diseases, which, on turn, has dramatically reduced infantile mortality, but also mortality in adult age. In fact, in 2011 less than 10 % of deaths occurred in subjects under 60 years of age, while the corresponding figures were 74 % in 1872, 56 % in 1901 and 25 % in 1951. However, in the last decades, the continuous extension of lifespan was mainly due to the improvement of medical assistance with respect to age-related diseases, especially Cardiovascular Diseases and Cancer, which allowed to increase lifespan of 5 years in the last 2 decades and of 2 years in the last 10 years (data from www.mortality.org andwww.istat.it).

Table 1. Evolution of lifespan expectancy in Italy from 1861

Year	Male	Female	Total
1861	28	29	29
1871	30	31	30
1881	35	35	35
1891	38	39	38
1901	43	43	43
1911	46	46	46
1921	48	50	49
1931	53	56	55
1941	55	58	56
1951	63	67	65
1961	67	72	69
1971	69	75	72
1981	71	78	75
1991	74	80	77
2001	77	83	80
2011	79	84	82

These data clearly show that environmental factors have a very strong impact on lifespan and on longevity in humans. However, the extension of lifespan that there has been in the last decades have not been accompanied by a similar extension of healthy lifespan. Indeed, in most cases this lifespan extension is due to the chronicit of the age-related diseases. This has brought the community of biogerontologists to study interventions, possibly modulated on the knowledge emerged from the studies on the genetic and biomolecular basis of longevity, to extend not only lifespan but also healthy lifespan, or, with a new word, "healthspan". In fact, model organisms with mutations that extend lifespan have a healthy life also when they are old. This suggested that health span extension could be attained by targeting (stimulating or silencing) the genes, which had been highlighted to be involved in life extension in both model organisms and humans [47]. In support of this hypothesis, it has been reported that dietary restricted mice, which live much longer and show a very delayed aging phenotype than mice fed *at libitum*, at old age have an expression pattern very different from mice of the same age for a number of genes correlated with life extension, such as those related to DNA repair, stress response, immune response and others [48, 49]. Thus, dietary restriction can trigger a molecular-genetic response which postpones aging and age-related phenotypes. This has brought to search for drugs or interventions which may act on these mechanisms without the side effects of calorie restriction. Among the most important interventions which have been considered in this context, we may name the protein restriction, the use of drugs targeting different genes of IGF-1 axis or of the FOXO/TOR pathway [47]. In addition, these studies have allowed to reconsider previous data on some areas characterized by exceptional longevity (such as Okinawa, Sardinia and Calabria) which are characterized by traditional ipoproteic diets, such as the "Mediterranean diet" [50, 51, 52, 53]. In these cases, then, the environment, that is the traditional diet, has allowed to stimulate the molecular mechanisms which can increase life span.

Among the several changes that occur with the aging process, in the last decade Epigenomics has attracted the interest of many researchers. This was mainly due to the fact that epigenetic modifications summarizing, at least in part, the interaction between the individual genetic background and lifestyle characteristics, should be potentially able to capture part of the unexplained susceptibility observed today for complex diseases (the so-called missing heritability problem).

Starting from the pioneeristic observations that epigenetic modifications affect not only the aging process but also its quality (successful aging) [54], EpiGenome-Wide Association Studies identified hundreds of sites spread along the entire genome in which methylation levels change between oldest old and younger subjects. In particular, Horwat and co-workers, on the basis of the methylation levels of 353 CpG units, formulated a mathematical model, the so-called epigenetic clock, that showed some important properties [55]. First, it was able predict the chronological age of a subject starting from the methylation level of several cells and tissues of his body. Second, it represents one of the most accurate biomarker of age (also superior to the estimates obtained from the telomere length). Third, using methylation levels of blood and brain tissues from subjects affected by Down syndrome, it showed that an accelerated aging occur in such a syndrome [56]. Fourth, it was able to predict all-cause mortality also after adjusting for traditional risk factors [57]. Finally, when it was used to estimate the biological age of several tissues from supercentenarians, it has been demonstrated that brain and muscle represent the youngest tissues of these exceptional individuals [58].

However, even if the cause-effect relationship between methylation process and aging is still not clear, the potential applications of this discovery are very wide, ranging from detailed monitoring of changes occurring with age within individual systems or organs (muscle, brain, etc.) to forensic purposes. For this and several other reasons, future advances in this field could help the understanding of the complex physiology of aging, lifespan and age-associated diseases.

CONCLUSIONS

On the whole, although the common variability accounts for only 25 % of human lifespan variability, the knowledge of the genetic basis modulating longevity may give significant hints on modulating lifestyle in order to attain longevity and extend healthspan. That is, a few subjects can attain longevity because a lucky combination of polymorphisms which allow them to have an efficient metabolism or an efficient response to different stress. Most of the others can attain a similar result by targeting the same pathways with appropriate life style or interventions. In this context, the importance of epigenetic factors, both as biomarkers of aging and target of interventions will certainly grow in the forthcoming future.

ACKNOWLEDGEMENTS

This work was partially supported by the European Union's Seventh Framework Programme (FP7/2007–2011) [grant number 259679] and by funds from Programma Operativo Nazionale [01_00937] - MIUR"Modelli sperimentali biotecnologici integrati per lo sviluppo e la selezione di molecole di interesse per la salute dell'uomo".

AUTHORS' CONTRIBUTIONS

The authors equally contributed to the drafting and finalization of the manuscript. All authors read and approved the final manuscript.

REFERENCES

1. Johnson TE. A personal retrospective on the genetics of aging. Biogerontology. 2002; 3(1–2):7–12.
2. Johnson TE, Wood WB. Genetic analysis of life-span in Caenorhabditis elegans. Proc Natl Acad Sci USA. 1982; 79(21):6603–7.
3. Johnson TE. Aging can be genetically dissected into component processes using long-lived lines of Caenorhabditis elegans. Proc Natl Acad Sci USA. 1987; 84(11):3777–81.
4. Friedman DB, Johnson TE. Three mutants that extend both mean and maximum life span of the nematode, Caenorhabditis elegans, define the age-1 gene. J Gerontol. 1988; 43(4):B102–9.
5. Kenyon C, Chang J, Gensch E, Rudner A, Tabtiang R. A C. elegans mutant that lives twice as long as wild type. Nature. 1993; 366(6454):461–4. doi: 10.1038/366461a0.

6. Gottlieb S, Ruvkun G. daf-2, daf-16 and daf-23: genetically interacting genes controlling Dauer formation in Caenorhabditis elegans. Genetics. 1994; 137(1):107–20.

7. Taguchi A, Wartschow LM, White MF. Brain IRS2 signaling coordinates life span and nutrient homeostasis. Science. 2007; 317(5836):369–72. doi: 10.1126/science.1142179.

8. Lin K, Dorman JB, Rodan A, Kenyon C. daf-16: An HNF-3/forkhead family member that can function to double the life-span of Caenorhabditis elegans. Science. 1997; 278(5341):1319–22.

9. Ogg S, Paradis S, Gottlieb S, Patterson GI, Lee L, Tissenbaum HA, et al. The Fork head transcription factor DAF-16 transduces insulin-like metabolic and longevity signals in C. elegans. Nature. 1997; 389(6654):994–9. doi: 10.1038/40194.

10. Kenyon C. The plasticity of aging: insights from long-lived mutants. Cell. 2005; 120(4):449–60. doi: 10.1016/j.cell.2005.02.002.

11. Kenyon CJ. The genetics of ageing. Nature. 2010; 464(7288):504–12. doi: 10.1038/nature08980.

12. Rose G, Passarino G, Franceschi C, De Benedictis G. The variability of the mitochondrial genome in human aging: a key for life and death? Int J Biochem Cell Biol. 2002; 34(11):1449–60.

13. Skulachev VP. The programmed death phenomena, aging, and the Samurai law of biology. Exp Gerontol. 2001; 36(7):995–1024.

14. Schoenmaker M, de Craen AJ, de Meijer PH, Beekman M, Blauw GJ, Slagboom PE, et al. Evidence of genetic enrichment for exceptional survival using a family approach: the Leiden Longevity Study. Eur J Human Genet EJHG. 2006; 14(1):79–84. doi: 10.1038/sj.ejhg.5201508.

15. Montesanto A, Latorre V, Giordano M, Martino C, Domma F, Passarino G. The genetic component of human longevity: analysis of the survival advantage of parents and siblings of Italian nonagenarians. Eur J Human Genet EJHG. 2011; 19(8):882–6. doi: 10.1038/ejhg.2011.40.

16. vB Hjelmborg J, Iachine I, Skytthe A, Vaupel JW, McGue M, Koskenvuo M, Kaprio J, Pedersen NL, Christensen K. Genetic influence on human lifespan and longevity. Hum Genet. 2006; 119(3):312–21.

17. Herskind AM, McGue M, Holm NV, Sorensen TI, Harvald B, Vaupel JW. The heritability of human longevity: a population-based study of 2872 Danish twin pairs born 1870–1900. Hum Genet. 1996; 97(3):319–23.

18. Skytthe A, Pedersen NL, Kaprio J, Stazi MA, Hjelmborg JV, Iachine I, et al. Longevity studies in GenomEUtwin. Twin Res Off J Int Soc Twin Studies. 2003; 6(5):448–54. doi: 10.1375/136905203770326457.

19. Capri M, Salvioli S, Sevini F, Valensin S, Celani L, Monti D, et al. The genetics of human longevity. Ann N Y Acad Sci. 2006; 1067:252–63. doi: 10.1196/annals.1354.033.

20. Passarino G, Montesanto A, Dato S, Giordano S, Domma F, Mari V, et al. Sex and age specificity of susceptibility genes modulating survival at old age. Hum Hered. 2006; 62(4):213–20. doi: 10.1159/000097305.

21. Perls TT, Wilmoth J, Levenson R, Drinkwater M, Cohen M, Bogan H, et al. Life-long sustained mortality advantage of siblings of centenarians. Proc Natl Acad Sci U S A. 2002; 99(12):8442–7. doi: 10.1073/pnas.122587599.

22. Pawlikowska L, Hu D, Huntsman S, Sung A, Chu C, Chen J, et al. Association of common genetic variation in the insulin/IGF1 signaling pathway with human longevity. Aging Cell. 2009; 8(4):460–72. doi: 10.1111/j.1474-9726.2009.00493.x.

23. Ziv E, Hu D. Genetic variation in insulin/IGF-1 signaling pathways and longevity. Ageing Res Rev. 2011; 10(2):201–4. doi: 10.1016/j.arr.2010.09.002.

24. Morris BJ, Willcox DC, Donlon TA, Willcox BJ. FOXO3: A Major Gene for Human Longevity - A Mini-Review. Gerontology. 2015. doi: 10.1159/000375235.

25. Bellizzi D, Rose G, Cavalcante P, Covello G, Dato S, De Rango F, et al. A novel VNTR enhancer within the SIRT3 gene, a human homologue of SIR2, is associated with survival at oldest ages. Genomics. 2005; 85(2):258–63. doi: 10.1016/j.ygeno.2004.11.003.

26. Sebastiani P, Solovieff N, Dewan AT, Walsh KM, Puca A, Hartley SW, et al. Genetic signatures of exceptional longevity in humans. PLoS One. 2012; 7(1):e29848. doi: 10.1371/journal.pone.0029848.

27. Beekman M, Blanche H, Perola M, Hervonen A, Bezrukov V, Sikora E, et al. Genome-wide linkage analysis for human longevity: Genetics of Healthy Aging Study. Aging Cell. 2013; 12(2):184–93. doi: 10.1111/acel.12039.

28. Hindorff LA, Sethupathy P, Junkins HA, Ramos EM, Mehta JP, Collins FS, et al. Potential etiologic and functional implications of genome-wide association loci for human diseases and traits. Proc Natl Acad Sci USA. 2009; 106(23):9362–7. doi: 10.1073/pnas.0903103106.

29. Deelen J, Beekman M, Uh HW, Broer L, Ayers KL, Tan Q, et al. Genome-wide association meta-analysis of human longevity identifies a novel locus conferring survival beyond 90 years of age. Hum Mol Genet. 2014; 23(16):4420–32. doi: 10.1093/hmg/ddu139.

30. Novelli V, Viviani Anselmi C, Roncarati R, Guffanti G, Malovini A, Piluso G, et al. Lack of replication of genetic associations with human longevity. Biogerontology. 2008; 9(2):85–92. doi: 10.1007/s10522-007-9116-4.

31. Ferrario A, Villa F, Malovini A, Araniti F, Puca AA. The application of genetics approaches to the study of exceptional longevity in humans: potential and limitations. Immun Ageing. 2012; 9(1):7. doi: 10.1186/1742-4933-9-7.

32. Villa F, Carrizzo A, Spinelli CC, Ferrario A, Malovini A, Maciąg A, et al. Genetic Analysis Reveals a Longevity-Associated Protein Modulating Endothelial Function and Angiogenesis. Circ Res. 2015; 117(4):333–45. doi: 10.1161/CIRCRESAHA.117.305875.

33. Raule N, Sevini F, Li S, Barbieri A, Tallaro F, Lomartire L, et al. The co-occurrence of mtDNA mutations on

different oxidative phosphorylation subunits, not detected by haplogroup analysis, affects human longevity and is population specific. Aging Cell. 2014; 13(3):401–7. doi: 10.1111/acel.12186.

34. Debrabant B, Soerensen M, Flachsbart F, Dato S, Mengel-From J, Stevnsner T, et al. Human longevity and variation in DNA damage response and repair: study of the contribution of sub-processes using competitive gene-set analysis. Eur J Human Genet EJHG. 2014; 22(9):1131–6. doi: 10.1038/ejhg.2013.299.

35. Soerensen M, Thinggaard M, Nygaard M, Dato S, Tan Q, Hjelmborg J, et al. Genetic variation in TERT and TERC and human leukocyte telomere length and longevity: a cross-sectional and longitudinal analysis. Aging Cell. 2012; 11(2):223–7. doi: 10.1111/j.1474-9726.2011.00775.x.

36. Crocco P, Barale R, Rose G, Rizzato C, Santoro A, De Rango F, et al. Population-specific association of genes for telomere-associated proteins with longevity in an Italian population. Biogerontology. 2015; 16(3):353–64. doi: 10.1007/s10522-015-9551-6.

37. Atzmon G, Cho M, Cawthon RM, Budagov T, Katz M, Yang X, et al. Evolution in health and medicine Sackler colloquium: Genetic variation in human telomerase is associated with telomere length in Ashkenazi centenarians. Proc Natl Acad Sci U S A. 2010; 107 Suppl 1:1710–7. doi: 10.1073/pnas.0906191106.

38. Altomare K, Greco V, Bellizzi D, Berardelli M, Dato S, DeRango F, et al. The allele (A)(−110) in the promoter region of the HSP70-1 gene is unfavorable to longevity in women. Biogerontology. 2003; 4(4):215–20.

39. Ross OA, Curran MD, Crum KA, Rea IM, Barnett YA, Middleton D. Increased frequency of the 2437 T allele of the heat shock protein 70-Hom gene in an aged Irish population. Exp Gerontol. 2003; 38(5):561–5.

40. Rose G, Crocco P, De Rango F, Montesanto A, Passarino G. Further support to the uncoupling-to-survive theory: the genetic variation of human UCP genes is associated with longevity. PLoS One. 2011; 6(12):e29650. doi: 10.1371/journal.pone.0029650.

41. Junnila RK, List EO, Berryman DE, Murrey JW, Kopchick JJ. The GH/IGF-1 axis in ageing and longevity. Nat Rev Endocrinol. 2013; 9(6):366–76. doi: 10.1038/nrendo.2013.67.

42. Johnson SC, Rabinovitch PS, Kaeberlein M. mTOR is a key modulator of ageing and age-related disease. Nature. 2013; 493(7432):338–45. doi: 10.1038/nature11861.

43. Schachter F, Faure-Delanef L, Guenot F, Rouger H, Froguel P, Lesueur-Ginot L, et al. Genetic associations with human longevity at the APOE and ACE loci. Nat Genet. 1994; 6(1):29–32. doi: 10.1038/ng0194-29.

44. Garasto S, Berardelli M, DeRango F, Mari V, Feraco E, De Benedictis G. A study of the average effect of the 3'APOB-VNTR polymorphism on lipidemic parameters could explain why the short alleles (<35 repeats) are rare in centenarians. BMC Med Genet. 2004; 5:3. doi: 10.1186/1471-2350-5-3.

45. Cardelli M, Cavallone L, Marchegiani F, Oliveri F, Dato S, Montesanto A, et al. A genetic-demographic approach reveals male-specific association between survival and tumor necrosis factor (A/G)-308 polymorphism. J Gerontol A Biol Sci Med Sci. 2008; 63(5):454–60.

46. Di Bona D, Vasto S, Capurso C, Christiansen L, Deiana L, Franceschi C, et al. Effect of interleukin-6 polymorphisms on human longevity: a systematic review and meta-analysis. Ageing Res Rev. 2009; 8(1):36–42. doi: 10.1016/j.arr.2008.09.001.

47. Longo VD, Antebi A, Bartke A, Barzilai N, Brown-Borg HM, Caruso C, et al. Interventions to Slow Aging in Humans: Are We Ready? Aging Cell. 2015; 14(4):497–510. doi: 10.1111/acel.12338.

48. Prolla TA, Mattson MP. Molecular mechanisms of brain aging and neurodegenerative disorders: lessons from dietary restriction. Trends Neurosci. 2001; 24(11 Suppl):S21–31.

49. Ghosh S, Wanders D, Stone KP, Van NT, Cortez CC, Gettys TW. A systems biology analysis of the unique and overlapping transcriptional responses to caloric restriction and dietary methionine restriction in rats. FASEB J Off Publication Federation Am Soc Exp Biol. 2014; 28(6):2577–90. doi: 10.1096/fj.14-249458.

50. Willcox DC, Willcox BJ, Todoriki H, Suzuki M. The Okinawan diet: health implications of a low-calorie, nutrient-dense, antioxidant-rich dietary pattern low in glycemic load. J Am Coll Nutr. 2009; 28(Suppl):500S–16S.

51. Poulain M, Pes GM, Grasland C, Carru C, Ferrucci L, Baggio G, et al. Identification of a geographic area characterized by extreme longevity in the Sardinia island: the AKEA study. Exp Gerontol. 2004; 39(9):1423–9. doi: 10.1016/j.exger.2004.06.016.

52. Pes GM, Tolu F, Dore MP, Sechi GP, Errigo A, Canelada A, et al. Male longevity in Sardinia, a review of historical sources supporting a causal link with dietary factors. Eur J Clin Nutr. 2015; 69(4):411–8. doi: 10.1038/ejcn.2014.230.

53. Montesanto A, Passarino G, Senatore A, Carotenuto L, De Benedictis G. Spatial analysis and surname analysis: complementary tools for shedding light on human longevity patterns. Ann Hum Genet. 2008; 72(Pt 2):253–60. doi: 10.1111/j.1469-1809.2007.00405.x.

54. Bellizzi D, D'Aquila P, Montesanto A, Corsonello A, Mari V, Mazzei B, et al. Global DNA methylation in old subjects is correlated with frailty. Age. 2012; 34(1):169–79. doi: 10.1007/s11357-011-9216-6.

55. Horvath S. DNA methylation age of human tissues and cell types. Genome Biol. 2013; 14(10):R115. doi: 10.1186/gb-2013-14-10-r115.

56. Horvath S, Garagnani P, Bacalini MG, Pirazzini C, Salvioli S, Gentilini D, et al. Accelerated epigenetic aging in Down syndrome. Aging Cell. 2015; 14(3):491–5. doi: 10.1111/acel.12325.

57. Marioni RE, Shah S, McRae AF, Chen BH, Colicino E, Harris SE, et al. DNA methylation age of blood predicts all-cause mortality in later life. Genome Biol. 2015; 16:25. doi: 10.1186/s13059-015-0584-6.

58. Horvath S, Mah V, Lu AT, Woo JS, Choi OW, Jasinska AJ, et al. The cerebellum ages slowly according to the epigenetic clock. Aging. 2015; 7(5):294–306.

CHAPTER 8

FREQUENCY OF RARE MUTATIONS AND COMMON GENETIC VARIATIONS IN SEVERE HYPERTRIGLYCERIDEMIA IN THE GENERAL POPULATION OF SPAIN

Itziar Lamiquiz-Moneo[1], Cristian Blanco-Torrecilla[1], Ana M. Bea[1], Rocío Mateo-Gallego[1], Sofía Pérez-Calahorra[1], Lucía Baila-Rueda[1], Ana Cenarro[1], Fernando Civeira[1], and Isabel de Castro-Orós[1,2]

[1]Unidad Clínica y de Investigación en Lípidos y Arteriosclerosis, Hospital Universitario Miguel Servet, Instituto de Investigación Sanitaria Aragón (IIS Aragón), Avenida Isabel La Católica 1-3, 50009 Zaragoza, Spain

[2]Universidad de Zaragoza, Departamento de Bioquímica, Biología Molecular y Celular, 50009 Zaragoza, Spain

ABSTRACT

Background

Hypertriglyceridemia (HTG) is a common complex metabolic trait that results of the accumulation of relatively common genetic variants in combination with other modifier genes and environmental factors resulting in increased plasma triglyceride (TG) levels. The majority of severe primary hypertriglyceridemias is diagnosed in adulthood and their molecular bases have not been fully defined yet. The prevalence of HTG is highly variable among populations, possibly caused by differences in environmental factors and genetic background. However, the prevalence of very high TG and the frequency of rare mutations causing HTG in a whole non-selected population have not been previously studied.

Methods

The total of 23,310 subjects over 18 years from a primary care-district in a middle-class area of Zaragoza (Spain) with TG >500 mg/dL were selected to establish HTG prevalence. Those affected of primary HTG were considered for further genetic analisys. The promoters, coding regions and exon-intron boundaries of LPL, LMF1, APOC2, APOA5, APOE and GPIHBP1 genes were sequenced. The frequency of rare variants identified was studied in 90 controls.

Results

One hundred ninety-four subjects (1.04 %) had HTG and 90 subjects (46.4 %) met the inclusion criteria for primary HTG. In this subgroup, nine patients (12.3 %) were carriers of 7 rare variants in LPL, LMF1, APOA5, GPIHBP1 or

Citation: Itziar Lamiquiz-Moneo, Cristian Blanco-Torrecilla, Ana M. Bea, Rocío Mateo-Gallego, Sofía Pérez-Calahorra, Lucía Baila-Rueda, Ana Cenarro, Fernando Civeira, and Isabel de Castro-Orós. Frequency of rare mutations and common genetic variations in severe hypertriglyceridemia in the general population of Spain. Lipids in Health and Disease. 2016; 15:82. DOI: 10.1186/s12944-016-0251-2.

APOE genes. Three of these mutations are described for the first time in this work. The presence of a rare pathogenic mutation did not confer a differential phenotype or a higher family history of HTG.

Conclusion

The prevalence of rare mutations in candidate genes in subjects with primary HTG is low. The low frequency of rare mutations, the absence of a more severe phenotype or the dominant transmission of the HTG would not suggest the use of genetic analysis in the clinical practice in this population.

BACKGROUND

Hypertriglyceridemia (HTG) is a common complex metabolic trait resulting in increased plasma triglyceride (TG) levels, atherosclerosis risk, metabolic syndrome and in severe cases, high risk of acute pancreatitis [1]. HTG has been generally classified as primary, when a familial or inherited basis is suspected, whereas secondary HTG refers to cases in which coexist one or more identifiable conditions favouring HTG, including metabolic syndrome, type 2 diabetes, excess alcohol consumption, obesity, renal disease or certain drugs [2]. Selected primary very high HTG cases from lipid clinics, usually defined with TG>500 mg/dL, are the result of the accumulation of relatively common genetic variants, with small to modest effect on TG, and/or the presence of rare mutations with large effect on TG in combination with other modifier genes and environmental factors [3, 4]. Hence, genetic studies in most HTG are predicted to add little information in their clinical management. In contrast, primary severe HTG with plasma triglyceride concentrations>1000 mg/dL [5] concentrate rare mutations with major effects in genes involved in triglyceride-rich lipoprotein metabolism, including LPL, APOC2, APOA5,LMF1, APOE and GPIHBP1 genes [6–9]. Homozygosity or compound heterozygosity for severe mutations in these genes comprise the most severe cases of HTG, with TG well over 1,000 mg/dL and recurrent episodes of pancreatitis since childhood in many occasions. However, these autosomal recessive defects are extremely rare, approximately 1:1,000,000, and most cases of very high HTG do not fulfil these criteria [10].

The prevalence of very high HTG is highly variable among populations, possibly caused by differences in environmental factors like diabetes prevalence or alcohol consumption, genetic background, and cohort selection. However, the prevalence of very high TG and the frequency of rare mutations causing HTG in a whole non-selected population have not been previously studied. This information could be very useful to establish the role of genetic testing and familial cascade screening for primary HTG.

METHODS

Study Subjects

The Spanish public health system distributes the population in different health districts, approximately 20,000–30,000 patients per district. Each health district includes 100 % of citizens living in that district. They share a single primary care facility, a single laboratory with computerized data since 2005, and a centralized registry of drugs. For this study, we selected all subjects of a primary care district (Centro Salud Almozara) in a middle class area of Zaragoza city in northern Spain. This facility is responsible of the primary health care of 23,310 subjects, which includes voluntary periodic examinations with fasting blood test every 5 years. All adult subjects (≥18 years of age) of Centro Salud Almozara were included in the study to establish the frequency of HTG in general population.

HTG Definition and HTG Subjects Selection

Very high HTG was defined with fasting TG>500 mg/dL as proposed by Adult Treatment Panel III guideline recommendations [5]. HTG subject was defined as the presence in the last 10 years of very high TG, and at least two other measurements with TG >200 mg/dL. Subjects under treatment with drugs to decrease triglycerides (fibrates, omega-3 fatty acids or niacin) were considered as HTG in presence of a single TG value>500 mg/dL. Secondary very high HTG was an exclusion criterion, and was considered in presence of uncontrolled diabetes (Hb1Ac >7.5 %) , obesity (body mass index >30 Kg/m2), alcohol abuse (>30 g/day for men and >20 g/day for women), renal disease (glomerular filtration rate <30 mL/min), liver disease (ALT >3 times upper normal limit) except non-alcoholic fatty liver, hypothyroidism (TSH >6 μUI/mL), pregnancy, hemochromatosis, autoimmune diseases, and concomitant use of estrogens (except contraceptive pills), any dose of oral corticosteroids or protease inhibitors.

Normolipemic Subjects

The normolipemic group consisted of healthy, unrelated men and women volunteers from the same facility, matched by age with the HTG group. Exclusion criteria for control subjects were TG >100 mg/dL, diabetes, current or past history of lipid-lowering drug consumption and APOE genotype other than ε3/ε3.

All selected subjects, cases and controls, signed informed consent to a protocol previously approved by our local ethical committee (Comité Ético de Investigación Clínica de Aragón, Zaragoza, Spain).

Clinical and Laboratory Determinations

All participants were assessed for personal and familial history of cardiovascular disease, history of pancreatitis, medication use, cardiovascular risk factors and anthropometric measurements. Dietary intake in cases and controls was determined by interview with one single registered dietician. In this interview, a Spanish validated 137-item food frequency questionnaire (FFQ) was used [11].

EDTA plasma and serum samples were collected after 10–12 hours of fasting. Cholesterol and TG levels were determined by standard enzymatic methods. HDL cholesterol was measured directly by an enzymatic reaction using cholesterol oxidase (UniCel DxC 800; Beckman Coulter, Inc., Brea, CA, USA). Apo A1, apo B, lipoprotein(a) and C-reactive protein were determined by IMMAGE kinetic nephelometry (Beckman-Coulter, Inc.).

Genetic Analysis

DNA was extracted by standard methods. Promoters, coding regions and intron-exon boundaries of LPL (NM_000237), LMF1 (NM_022773), APOA5 (NM_052968), APOC2(NM_000483.4) and GPIHBP1 (NM_178172) were amplified by PCR and purified by ExoSap-IT (USB) using primers shown in. Amplified fragments were sequenced by Sanger method using the BigDye 3.1 sequencing kit (Applied Biosystems) in an automated ABI 3500xL sequencer (Applied Biosystems). DNA sequences were analyzed using VariantReporter™ software (Applied Biosystems).APOE genotype was determined as previously described [12].

Rare and common variants were defined by frequencies <1 % and >1 % respectively of that obtained from the 1000 Genomes project [13]. Variants that were not previously described and absent in our control population were considered as rare. Rare and common variants were considered pathogenic if previously associated with HTG or considered pathogenic by the evaluation with predictive software in a bioinformatic analysis.

To evaluate the pathogenicity of newly identified genetic variants, we used PolyPhen-2 [14] and Mutation Taster [15]. The effect of variants in potential splicing sites was predicted with NetGene2 [16] and NNSplice [17]. We refer to non-synonymous variants as sequence variations causing amino acid substitutions (missense variants) or introducing of a premature stop codon (nonsense variants). Intronic variants were considered when they were located in intron-exon boundaries.

In order to compare the frequency of identified variants with that of the general population we have compiled the allele frequencies of the identified variants from the 1000 Genomes Project [13].

Statistical Analysis

Analyses were performed using SPSS version 20.0 (IBM). The nominal level for significance was $p<0.05$. Normal distribution of variables was checked by Kolmogorov–Smirnov test. Quantitative variables with normal distribution were expressed as mean ± standard deviation and were analyzed by T-Student test. Variables with a skewed distribution were expressed as median and interquartile range and were analyzed by Mann–Whitney U test. Qualitative variables were expressed as percentage and were analyzed by Chi-square test.

RESULTS

From the total selected population of 23310 subjects, 18804 subjects were aged 18–80 years, 194 (1.04 %) of them presented very high HTG in the last 10 years, and 90 subjects (46.4 %) met the inclusion criteria for primary HTG. The selection of study subjects group was carried out as indicated in the flow chart of Fig. 1. Finally, 73 patients with primary very high HTG were studied by sequencing the LPL, LMF1, APOA5, APOC2, GPIHBP1 and APOE genes.

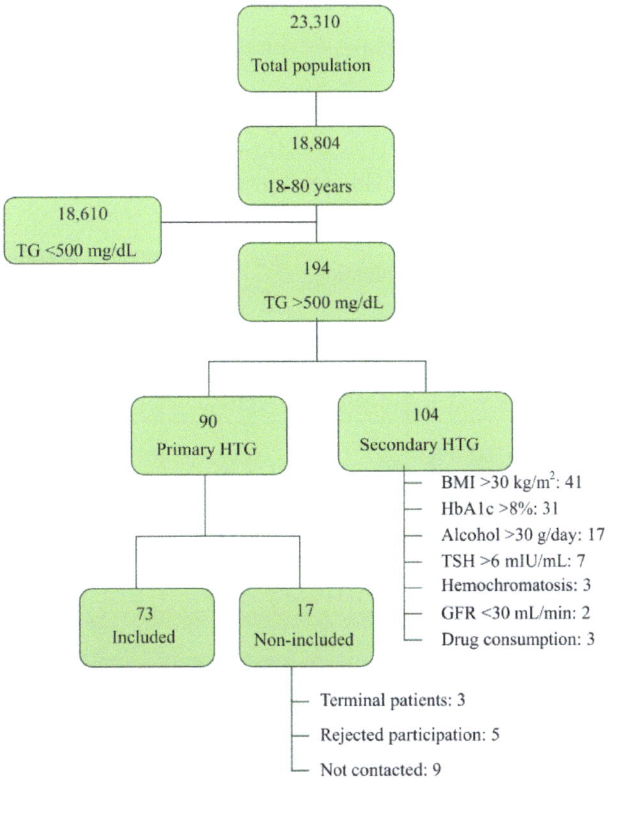

Figure 1. Flow chart of the population selected for this study.

Anthropometric and clinical characteristics of the patients with primary HTG and control subjects are shown in Table 1. In the primary HTG group, there was a clear predominance of males (79.5 %) with a mean age of 56.4 years. Cardiovascular disease, diabetes and hypertension were more frequent among cases than in controls (Table 1).

Table 1. Clinical and biochemical characteristics of primary HTG and control groups[a]

	Primary HTG	Control	*p*
	N=73	*N*=90	
Age, years	56.4±12.0	53.5±16.2	0.210
Males, n (%)	58 (79.4)	45 (50.0)	0.000
Body mass index, kg/m^2	27.6±2.35	26.6±3.25	0.044
Total cholesterol, mg/dL	264±48.0	196±34.1	0.000
Triglycerides, mg/dL	654 (550–810)	63.0 (49.7–74.0)	0.000
HDL cholesterol, mg/dL	40.1±10.5	55.2±12.6	0.000
Glucose, mg/dL	100±16.1	91.3±12.7	0.000
Hb1Ac (%)	5.6 (5.3–6.15)	5.3 (5.1–5.5)	0.000
Diabetes, n (%)	18 (24.7)	2 (2.3)	0.000
Hypertension, n (%)	31 (42.5)	13 (14.4)	0.000
Cardiovascular disease, n (%)	11 (15.1)	3 (3.3)	0.005
APOE genotype (%)			0.000
ε2- ε2	0	0	
ε2- ε3	19.7	0	
ε3- ε3	57.7	100	
ε3- ε4	16.9	0	
ε4- ε4	1.4	0	
ε4- ε2	4.2	0	

Quantitative variables are expressed as mean±standard deviation, except for variables not following normal distribution, expressed as median (interquartile range). Qualitative variables are expressed as %. The p value was calculated by Student's t test, Mann–Whitney U and Chi - square as appropriate.

[a]Lipid values refer to the most recent values without lipid-lowering treatment

Sequencing analysis of LPL, LMF1, APOA5, GPIHBP, APOC2 and APOE genes revealed that nine patients out of 73 (12.3 %), were carriers of 7 rare variants: c.1018+1G>A in LPL, p.(Leu69Leu) and p.(Pro562Arg) in LMF1, p.(Leu173Pro) and p.(Gln97*) in APOA5, p.(Arg76Cys) in GPIHBP1 and p.(Arg154Ser) in APOE . Three of them were described for the first time in this work: c.1018+1G>A in LPL, p.(Leu173Pro) in APOA5, and p.(Arg76Cys) in GPIHBP1. Only the variant p.(Pro562Arg) in LMF1 was also present in controls (Table 2).

Table 2. Rare gene variants identified in the studied HTG population

Gene	Localization	Variant	Predicted amino-acid change	HTG group N (%)	Bioinformatic Analysis	Control group
LPL	Intron 6	c.1018+1G>A		2 (2.74 %)	Damage	0
LMF1	Exon 2	c.205	p.(Leu69Leu)	1 (1.37 %)	Damage	0
	Exon 11	c.1685C>G	p.(Pro562Arg)	1 (1.37 %)	Damage	2 (2.2 %)
APOA5	Exon 4	c.289C>T	p.(Gln97*)	1 (1.37 %)	Damage	0
		c.518T>C	p.(Leu173Pro)	1 (1.37 %)	Damage	0
GPIHBP1	Exon 2	c.226C>T	p.(Arg76Cys)	1 (1.37 %)	Damage	0
APOE	Exon 4	c.460C>A	p.(Arg154Ser)	2 (2.74 %)	Damage	0

Prediction of deleterious effects was performed using SIFT (http://sift.jcvi.org/), Polyphen-2 (http://genetics.bwh.harvard.edu/pph2/) and MutationTaster (http://www.mutationtaster.org)

For all the identified variants, the in silico analysis with MutationTaster and PolyPhen-2 was performed to evaluate the pathogenicity. The following variants were predicted as potentially harmful: c.-281G>C, c.1018+1 G>A, p.(Asp36Asn) and p.(Asn318Ser) inLPL; p.(Ser19Trp), p.(Gln97*), and p.(Leu173Pro) in APOA5, p.(Ser144Phe) and p.(Arg76Cys) in GPIHBP1 and p.(Arg364Gln), p.(Pro562Arg), p.(Leu69Leu) in LMF1. This last variant in LMF1, although silent variant, was predicted as potentially harmful, since splicing score was increased (score wild type: 0.43 versus score mutant: 0.48).

Table 3 shows identified variants with allelic frequencies statistically different from those described in the 1000 Genomes project: p.(Asp36Asn) and c.-281T>G in LPL, and p.(Ser19Trp) and c.-3A>G in APOA5. Other polymorphisms previously associated with HTG [18, 19] were also present in our HTG population: p.(Ser474X) in LPL; and p.(Arg364Gln), p.(Arg354Trp) and p.(Pro562Arg) in LMF1, but their frequencies were not statistically different from those described in the 1000 Genomes project. The protector polymorphism, p.(Ser474X) in LPL, was found in our HTG population with lower frequency (11.1 %) that in the 1000 Genomes project (12.2 %), although this difference did not reach statistical significance.

Table 3. Comparison of allele frequencies of variants in our HTG population and 1000 Genomes project

Gene	Nucleotide change	Predicted aminoacid change	Frequency in HTG population	Frequency in the population 1000 Genomes	*p*	Bioinformatic Analysis
LPL	c.106G>A	p.(Asp36Asn)	0.055	0.013	0.001	Benign
	c.−281T>G		0.062	0.013	<0.001	Damage
LMF1	c.205C>T	p. (Leu69Leu)	0.007	0.0001	0.017	Damage
	c.288+298C>T		0.007	0.394	<0.001	Benign
APOA5	C.−3 A>G		0.185	0.080	<0.001	Benign
	c.56C>T	p.(Ser19Trp)	0.185	0.057	<0.001	Damage
	c.132C>A	p.(Ile44Ile)	0.178	0.056	<0.001	Benign
	c.162-43A>G		0.134	0.080	0.040	Benign
GPIHBP1	c.−509 G>A		0.164	0.442	<0.001	Unknown
	c. −208 G>C		0.240	0.443	<0.001	Unknown
	c.138G>T	p.(Val46Val)	0.281	0.443	0.0003	Benign
	c.295+19G>A		0.007	0.000	0.0226	Benign
	c.295+28G>A		0.007	0.000	0.0226	Benign
APOC2	c.−116T>A		0.50	0.376	0.004	Benign
	c.−89C>G		0.027	0.009	0.048	Benign
	c.56–30G>A		0.007	0.000	0.008	Benign
	c.216–81T>C		0.344	0.491	0.007	Benign

All variants are described in accordance with the latest recommendations of HGVS (http://www.hgvs.org/mutnomen). The p value was calculated by $\chi 2$ test

Prediction of deleterious effects was performed using SIFT (http://sift.jcvi.org/), Polyphen-2 (http://genetics.bwh.harvard.edu/pph2/) and MutationTaster (http://www.mutationtaster.org)

The clinical characteristics of the 9 patients carrying the 7 rare variants are shown in Table 4. The anthropometric and clinical data did not significantly differ between carriers and non-carriers of rare functional variants, including family history of HTG (Table 5).

Table 4. Clinical characteristics of patients carrying rare variants in LPL, LMF1, APOA5, APOE and GPIHBP1genes

Gene	Nucleotide change	Predicted aminoacid change	Age (years)	Sex	BMI (kg/m²)	TC, mg/dL	TG mg/dL	*APOE* Genotype
LPL	c.1018+1G>A		74	F	26.3	247	505	ε2/ε3
			59	M	29.7	374	1736	ε3/ε4
LMF1	c.205C>T	p. (Leu69Leu)	65	M	24.2	257	654	ε3/ε4
	c.1685C>G	p.(Pro562Arg)	28	M	26.85	286	932	ε2/ε3
APOE	c.460C>A	p.(Arg154Ser)	47	M	24.45	354	572	ε3/ε3
			64	M	29.1	287	586	ε3/ε3
APOA5	c.289C>T	p.(Gln97*)	53	M	23.9	252	1215	ε2/ε3
	c.518T>C	p.(Leu173Pro)	43	M	28.9	269	1410	ε3/ε4
GPIHBP1	c.226C>T	p.(Arg76Cys)	57	M	29.8	251	781	ε2/ε3

BMI Body Mass Index, TC Total Cholesterol, TG Triglycerides, F female, M male

Table 5. Clinical characteristics of HTG subjects according to presence/absence of rare variants

Clinical characteristics	Mutation carriers (n=9)	Non-mutation carriers (n=64)	p
Age (years)	59.8± 11	55,9± 12,2	0.017
Male n, (%)	8 (88.9 %)	50 (78.1 %)	0.058
BMI (kg/m²)	26.3 (24.0–29.4)	28,0 (25.9–29.7)	0.276
APOE genotype (%)			0.336
ε2-ε3	33.3	17.2	
ε3-ε3	33.3	62.5	
ε3-ε4	33.3	14.1	
ε4-ε4	0	1.6	
ε4-ε2	0	4.7	
Triglycerides, mg/dL	583 (529–1095)	669 (580–848)	0.416
Total cholesterol, mg/dL	281±48.9	262±48.1	0.743
HDL cholesterol, mg/dL	40.4±9.67	40.1±10.6	0.881
Diabetes, %	11.1	26.6	0.314
Hypertension, %	22.2	43.7	0.219
Cardiovascular disease, %	22.2	14	0.522
Familial history (%)	22.2	12.5	0.421

Quantitative variables are expressed as mean±standard deviation, except for variables without normal distribution (median, interquartile range). The qualitative variables are expressed as %. BMI: body mass index

P value was calculated by Student's t test, Mann - Whitney U and Chi – square, as appropriate

DISCUSSION

The prevalence of very high HTG in subjects aged 18 to 80 from the general population was 1.04 % in our study. To our knowledge, there are not similar studies including all (>99 %) subjects of a large community followed for 10 years. Due to the large variability of TG levels, we have considered those subjects with at least two separated determinations to confirm the high TG levels. The specific design of our study allows us to be confident about the consistency of our results. The characteristic of our national health system and the highly stable population, without significant immigration or emigration processes in the area of study in the last 10 years, reinforce our calculations. Furthermore, our results are concordant with other transversal studies such as the NHANES in a representative USA population, with an overall prevalence of 1.1 % of subjects with TG>500 mg/dL [18]. Moreover, the frequency of HTG was 1.12 %, when defined as TG >400 mg/dL, in stratified sampling with a transversal design in the Spanish population [19]. This prevalence seems to be higher in other populations, especially Mexican American middle-aged men, probably due to differences in genetic background, diet, and prevalence of obesity [18].

Primary HTG is defined as a lipoprotein metabolism disturbance caused by alteration of genes and proteins that regulate lipid metabolism. Of the 194 subjects who had TG>500 mg/dL, 104 met criteria for secondary HTG. This means that less than half of the HTG subjects (43.6 %) considered to have a predominantly genetic cause in our population, and is consistent with previous reports [20]. In our population, the major risk factor for secondary HTG was obesity (41.1 %). This percentage is similar to that found in the American population, in which 42.9 % of subjects with TG 200 mg/dL were classified as obese [18].

According to our results, the prevalence of HTG, both primary and secondary, was considerably lower among women than in men. In our HTG population, there was a clear predominance of males (79.5 %), as in the NHANES study, in which the 75.3 % of HTG were men [21], and also largely corroborated in other studies [22, 23]. The mechanism of this difference is not fully established. No locus related to triglycerides in sexual chromosomes has been found in GWAS, so the differences are probably due to different interaction of common gene variations with certain environmental factors. For example, a significantly greater efficiency of skeletal muscle to the postprandial clearance of both total and meal-derived triglycerides in women relative to men has been described and could explain a different impact of obesity on TG between genders [24].

Another objective our study was to determine the frequency of pathogenic mutations causing high TG levels in LPL, LMF1, APOA5, GPIHBP1, APOC2 and APOE genes in subjects with primary HTG. The analysis of the pathogenicity of the identified variants revealed that 9 out of 73 subjects were carriers of 7 disease causing variants in the candidate genes. Therefore, 12.3 % of our population had a disease causing variant. This frequency is similar to that observed by Wang et al., who studied 110 subjects with severe HTG and analyzing rare variants in LPL, APOA5 and APOC2 genes obtained a mutation frequency of 10.9 % [22].

Of the 7 variants disease causing, only one of them, p.(Pro562Arg) in LMF1, is present in controls. This mutation has been associated with HTG in previous studies [4–25] and its frequency in the general population of the 1000 Genomes project is lower than 1 % [13]. However, it would be necessary to perform further functional analysis to analyze the potencial effect of this variant.

The silent variant p.(Leu69Leu) in LMF1 was found to be pathogenic according to bioinformatic analysis. It does not generate an amino acid change, but it could generate a change in the physiological splicing according to bioinformatic analysis. Silent variants have been described as pathogenic by previous studies, showing that these variants can influence binding of regulatory factors on DNA, mRNA secondary structure and stability, ribosome traffic on mRNA and its interactions with specific ligands (as in riboswitches), including other RNAs and proteins [26]. It would be necessary to perform a functional analysis to study the potential effect of this variant.

Among the mutations identified in our population, three of them have been described for the first time in this work: p.(Leu173Pro), located in APOA5 exon 4, that generates an amino acid change predicted as harmful by bioinformatic analysis; c.1018+1 G>A, that causes a change in splicing site in LPL intron 6, and p.(Arg76Cys), located in GPIHBP1exon 2 and producing an amino acid change also predicted as harmful by in silicoanalysis. Among our cases, none homozygous or compound heterozygous subject was found. Taking into account that these mutations were not found in the normolipemic population, we considered that these variants are playing an important role in the HTG phenotype of carrier subjects, and that severe HTG could be considered a co-dominant disease with variable penetrance. There is a large heterogeneity in the variants identified in our study. Only the changes c.1018+1G>A in LPL and p.(Arg154Ser) in APOE were present in two unrelated affected subjects. This is in agreement with observations in other genetic diseases in our population and could be due to the absence of a common genetic background in Spain [27].

One important issue in our results is the lack of clinical differences between cases with and without mutation. Therefore, the presence of a rare pathogenic mutation did not confer a differential phenotype or risk with respect to those subjects without mutations, neither useful information in the management of these subjects.

CONCLUSIONS

In conclusion, we are far from knowing the full genetic basis of severe HTG, but the presence of rare mutations in candidate genes do not appear to be responsible in most cases. The low frequency of rare mutations in candidate genes and the absence of a more severe phenotype and family history of dominant transmission in those with rare mutations would not suggest the use of genetic analysis in clinical practice in this type of HTG. It would be possible that the study of known genes with minor effects on triglycerides like GPD1, GCKR, CREB3L3, MET or INSR the discovery of new HTG genes, or the use of new large scale tecniques may change this approach in the future [28, 29].

ACKNOWLEDGEMENTS

Grants from the Spanish Ministry of Health FIS PI13/02507, FIS PI15/01983, RD12/0042/0055 and Cuenca Villoro Fundation supported this work.

AUTHORS' CONTRIBUTIONS

ILM performed the experiments and contributed to the writing of the manuscript. FC and IDC contributed substantially to the conception and design; all authors participated in acquisition, analysis and interpretation of the data, in the drafting of the article or critical revision and in final approval of the version to be published.

REFERENCES

1. Wilson PW, Grundy SM. The metabolic syndrome: a practical guide to origins and treatment: Part II. Circulation. 2003; 108:1537–1540. doi: 10.1161/01.CIR.0000089506.12223.F1.
2. Hegele RA, Pollex RL. Hypertriglyceridemia: phenomics and genomics. Moll Cell Biochem. 2009; 236:35–43. doi: 10.1007/s11010-008-0005-1.
3. De Castro-Orós I, Cenarro A, Tejedor MT, Baila-Rueda L, Mateo-Gallego R, Lamiquiz- Moneo I, et al. Common genetic variants contribute to primary hypertriglyceridemia without differences between familial combined hyperlipidemia and isolated hypertriglyceridemia. Circ Cardiovasc Genet. 2014; 7:814–821. doi: 10.1161/CIRCGENETICS.114.000522.
4. Surendran RP, Visser ME, Heemelaar S, Wang J, Peter J, Defesche JC, et al. Mutations in LPL, APOC2, APOA5, GPIHBP1 and LMF1 in patients with severe hypertriglyceridemia. J Intern Med. 2012; 272:185–196. doi: 10.1111/j.1365-2796.2012.02516.x.
5. National Cholesterol Education Program (NCEP) Expert Panel on Detection, Evaluation, and Treatment of High Blood Cholesterol in Adults (Adult Treatment Panel III) Third Report of the National Cholesterol Education Program (NCEP) Expert Panel on Detection, Evaluation, and Treatment of High Blood Cholesterol in Adults (Adult Treatment Panel III) final report. Circulation. 2002; 106:3143–3421.
6. Gary FL, Changting X, Hegele RA. Hypertriglyceridemia in the Genomic Era: A New Paradigm. Endocr Rev. 2015; 36:131–147. doi: 10.1210/er.2014-1062.
7. Calandra S, Priore Oliva C, Tarugi P, Bertolini S. APOA5 and triglyceride metabolism, lesson from human APOA5 deficiency. Curr Opin Lipidol. 2006; 17:122–127. doi: 10.1097/01.mol.0000217892.00618.54.
8. Peterfy M, Ben-Zeev O, Mao HZ, Weissglas-Volkov D, Aouizerat BE, Pullinger CR, et al. Mutations in LMF1 cause combined lipase deficiency and severe hypertriglyceridemia. Nature Genet. 2007; 39:1483–1487. doi: 10.1038/ng.2007.24.
9. Davies BS, Beigneux AP, Barnes RH, Tu Y, Gin P, Weinstein MM, et al. GPIHBP1 is reponsible for the entry of lipoprotein lipase into capillares. Cell Metab. 2010; 12:42–52. doi: 10.1016/j.cmet.2010.04.016.
10. Borge GN, Varbo A. Lipids and cardiovascular disease 3. Lancet. 2014; 384:626–635. doi: 10.1016/S0140-6736(14)61177-6.
11. De la Fuente-Arrillaga C, Ruiz MV, Bes-Rastrollo M, Sampson L, Martinez-González MA. Reproducibility of an FFQ validated in Spain. Public Health Nutr.2010; 28:1–9.
12. Wenham PR, Price WH, Blundell G. Apolipoprotein E genotyping by one-stage PCR. Lancet. 1999; 337:1158–1159. doi: 10.1016/0140-6736(91)92823-K.
13. 1000 Genomes Project Consortium. Abecasis GR, Auton A, Brooks LD, DePristo MA, Durbin RM. An integrated map of genetic variation from 1,092 human genomes. Nature. 2012; 491:56–65. doi: 10.1038/nature11632.
14. Adzhubei IA, Schmidt S, Peshkin L, Ramensky VE, Gerasimova A, Bork P, et al. A method and server for predicting damaging missense mutations. Nat Methods.2010; 7:248–249. doi: 10.1038/nmeth0410-248.
15. Schwarz JM, Rödelsperger C, Schuelke M, Seelow D. MutationTaster evaluates disease-causing potential of sequence alterations. Nat Methods. 2010; 7:575–576. doi: 10.1038/nmeth0810-575.
16. Brunak S, Engelbrecht J, Knudsen S. Prediction of human mRNA donor and acceptor sites from the DNA sequence. J Mol Biol. 1991; 220:49–65. doi: 10.1016/0022-2836(91)90380-O.
17. Reese MG, Eeckman FH, Kulp D, Haussler D. Improved splice site detection inGenie. J Comput Biol. 1997; 4:311–323. doi: 10.1089/cmb.1997.4.311.
18. Miller M, Stone NJ, Ballantyne C, Bittner V, Criqui MH, Ginsberg HN, et al. Triglycerides and cardiovascular disease: a scientific statement from the American Heart Association. Circulation. 2011; 123:2292–2333. doi: 10.1161/CIR.0b013e3182160726.
19. Guallar-Castillón P, Pérez RF, López García E, León-Muñoz LM, Aguilera MT, Graciani A, et al. Magnitude and management of metabolic syndrome in Spain in 2008–2010: the ENRICA study. Rev Esp Cardiol. 2014; 67:367–373. doi: 10.1016/j.recesp.2013.08.014.
20. Valdivielso P, Pintó X, Mateo-Gallego R, Masana L, Álvarez-Sala L, Jarauta E, et al. Clinical features of patients with hypertriglyceridemia referred to lipid units: registry of hypertrigliceridemia of the Spanish Arteriosclerosis Society. Med Clin. 2011; 136:231–238. doi: 10.1016/j.medcli.2010.09.028.
21. Christian JB, Bourgeois N, Snipes R, Lowe KA. Prevalence of severe (500 to 2,000 mg/dl) hypertriglyceridemia in United States adults. Am J Cardiol. 2011; 107:891–897. doi: 10.1016/j.amjcard.2010.11.008.
22. Wang J, Cao H, Ban RM, Kennedy BA, Zhu S, Anand S, et al. Resequencing Genomic DNA of patients with severe Hypertriglyceridemia (MIM 144650) Arterioscler Thromb Vasc Biol. 2007; 27:2450–2455. doi: 10.1161/ATVBAHA.107.150680.
23. Carrol MD, Lacher DA, Sorlie PD, Cleeman JI, Gordon JD, Wolz M, et al. Trends in Serum Lipids and liporproteins of adults, 1960–2002. JAMA. 2005; 294:1773–1781. doi: 10.1001/jama.294.14.1773.

24. Horton TJ, Commerford R, Pagliassotti MJ, Bessesen DH. Postprandial leg uptake of triglyceride is greater in women than in men. Am J Physiol Endrocrinol Metab.2002; 283:1192–1202. doi: 10.1152/ajpendo.00164.2002.

25. Hegele RA, Ban RM, Hsueh N, Kennedy BA, Cao H, Yong Zou G, et al. A polygenic basis for four classical Fredrickson hyperlipoproteinemia phenotypes that are characterized by hypertriglyceridemia. Hum Mol Genet. 2009; 21:4189–4194. doi: 10.1093/hmg/ddp361.

26. Sauna ZE, Kimchi-Sarfaty C. Understanding the contribution of synonymous mutations to human disease. Nat Rev Genet. 2011; 12:683–691. doi: 10.1038/nrg3051.

27. Cenarro A, Jensen HK, Casao E, Civeira F, González-Bonillo J, Rodríguez-Rey JC, et al. Identification of recurrent and novel mutations in the LDL receptor gene in Spanish patients with familial hypercholesterolemia. Hum Mutat. 1998; 11:413–418. doi: 10.1002/(SICI)1098-1004(1998)11:5<413::AID-HUMU17>3.0.CO; 2-F.

28. Willer C, Schmidt E, Sengupta S, Peloso G, Gustafsoon S, Kanoni S, et al. Discovery and Refinement of Loci Associated with Lipid Levels. Nat Genet. 2013; 45:1–24. doi: 10.1038/ng.2797.

29. Wu JW, Yang H, Wang SP, Soni KG, Brunel-Guitton C, Mitchell GA. Inborn errors of cytoplasmic triglyceride metabolism. J Inherit Metab Dis. 2015; 38:85–98. doi: 10.1007/s10545-014-9767-7.

CHAPTER 9

ALU SINE ANALYSES OF 3,000-YEAR-OLD HUMAN SKELETAL REMAINS: A PILOT STUDY

Maximilian Kothe, Verena Seidenberg, Susanne Hummel, and Oliver Piskurek

Department of Historical Anthropology and Human Ecology, Georg-August-University, 37073 Göttingen, Germany

ABSTRACT

Background

As Short Interspersed Elements (SINEs), human-specific Alu elements can be used for population genetic studies. Very recent inserts are polymorphic within and between human populations. In a sample of 30 elements originating from three different Alusubfamilies, we investigated whether they are preserved in prehistorical skeletal human remains from the Bronze Age Lichtenstein cave in Lower Saxony, Germany. In the present study, we examined a prehistoric triad of father, mother and daughter.

Results

For 26 of the 30 Alu loci investigated, definite results were obtained. We were able to demonstrate that presence/absence analyses of Alu elements can be conducted on individuals who lived 3,000 years ago. The preservation of the ancient DNA (aDNA) is good enough in two out of three ancient individuals to routinely allow the amplification of 500 bp fragments. The third individual revealed less well-preserved DNA, which results in allelic dropout or complete amplification failures. We here present an alternative molecular approach to deal with these degradation phenomena by using internal Alu subfamily specific primers producing short fragments of approximately 150 bp.

Conclusions

Our data clearly show the possibility of presence/absence analyses of Alu elements in individuals from the Lichtenstein cave. Thus, we demonstrate that our method is reliably applicable for aDNA samples with good or moderate DNA preservation. This method will be very useful for further investigations with more Alu loci and larger datasets. Human population genetic studies and other large-scale investigations would provide insight into Alu SINE-based microevolutionary processes in humans during the last few thousand years and help us comprehend the evolutionary dynamics of our genome.

Citation: Kothe M, Seidenberg V, Hummel S, and Piskurek O. Alu SINE analyses of 3,000-year-old human skeletal remains: a pilot study. Mobile DNA. 2016; 7:7. DOI: 10.1186/s13100-016-0063-y.

BACKGROUND

After the discoveries of Barbara McClintock on *Zea Mays* [1, 2], much research has been conducted in the field of transposable elements (TEs). It is now known that TEs, long classified as junk DNA [3], have major effects on the genomes of all organisms. For example, they may affect gene functions or alternate transcription rates [4–9]. In eukaryotes, TEs are mostly inherited vertically from generation to generation and in rare cases horizontally, e.g. via a viral vector [10]. In humans, TEs make up a great part of the total genome. Estimates vary from ~45 % [11] to ~69 % [12]. Thousands of new TE loci have recently been identified in the human 1,000 Genome Project [13, 14]. The transposition mechanism of TEs can generally be divided into two classes: class I retrotransposons and class II DNA transposons. While DNA transposons move via a "cut-and-paste" mechanism, the retrotransposons move by a "copy-and-paste" mechanism. As class I elements, the non-autonomous Short Interspersed Elements (SINEs) present the largest group of TEs in eukaryotic genomes in terms of copy numbers [11]. More than 200 SINE families have been identified so far [15]. Their sequence information can be retrieved at SINEBase [15] and RepBase [16]. The retrotransposition mechanism of a SINE requires a Long Interspersed Element (LINE)-encoded protein from a LINE partner with reverse transcriptase and endonuclease activity [17].

The absence of an element at a specific locus can be described as the ancestral state, whereas presence is the derived state [18, 19]. Due to the irreversibility of an insertion and its homoplasy-free character, SINE insertions are a powerful tool for phylogenetic analyses [20, 21]. The most abundant SINEs in humans are the primate-specific *Alu*elements, reaching a copy number of about 1.1 million [11, 22]. Their partner LINEs are L1 elements that represent a family of mammalian retrotransposons that have been replicating and evolving for more than 100 Myr [23]. *Alu* elements usually have a length of approximately 300 base pairs. They began to expand with the primate radiation 65 Mya and peaked in activity 40 Mya. It is believed that only a few "Master Genes" are retropositionally competent [24]. Due to accumulations of new mutations, over evolutionary time, new *Alu* subfamilies are created. The 7SL RNA-derived *Alu* elements can be classified in three subfamilies J, S and Y, with *Alu*J being the oldest, followed by*Alu*S and *Alu*Y as the youngest and only active subfamily [22]. Within the *Alu*Y elements, the subfamilies *Alu*Ya5 and *Alu*Yb8 are the groups with the largest numbers of copies. Some of these elements retrotransposed so recently that they are absent in other primate lineages and are even polymorphic between and within human populations [25–28]. These polymorphic elements are perfectly suited for population genetic and phylogenetic studies. In cases of rapid radiation of taxa or simultaneous lineage divergence, some TEs may not show the real phylogenetic state. This phenomenon is called incomplete lineage sorting [29–32]. Nonetheless, polymorphic *Alu* elements are excellent ancestry markers for resolving relationships within and between human populations [33]. In a genome-wide study of polymorphic TEs in 2,504 individuals across 26 human populations, Rishishwar et al. [14] recently showed that the genetic diversity represented by TE polymorphisms, mainly by *Alu* elements, reflects known patterns of human evolution. *Alu* elements and TEs in general insert nearly randomly into the genome, exist in large copy numbers, and are mostly non-autonomous [34, 35]. Our genome is constantly in evolutionary change [36]. Normally, the long-term effects of gene evolution and function alternation become visible [37, 38]. The effects of short-term or microevolutionary processes can be detected by analyzing the presence/absence situations of human specific *Alu* elements.

For such analyses, human remains with well-preserved DNA are required. Usually, DNA degradation in bones is too advanced for analyses of fragments that exceed 200–300 base pairs [39, 40], but it was proven, for example, that larger fragments of 397 bp from bone samples of the Lichtenstein cave can be amplified, too [41]. The main causes of DNA loss in remains are autolysis directly after death, hydrolisis, and oxidation [42, 43]. The degree of post-mortem DNA degradation depends on environmental factors such as acid conditions, microbial activity, and high mean temperatures [42]. On the other hand, constant low temperatures and neutral or slightly alkaline pH-values provide optimal conditions for DNA preservation [40, 42]. These conditions are found in the Lichtenstein cave near Osterode in Lower Saxony, Germany. For thousands of years, the cave has had a constant temperature of 6–8°C. Additionally, the skeletal remains were coated with a gypsum layer, which causes a slightly basic environment and is thus perfect conditions for preserving bones and DNA. Previous studies on these remains revealed kinship relations between many individuals [44, 45]. These results are based on genetic fingerprinting, mtDNA, and Y-haplotypes [46, 47]. In the present work, a triad of father, mother, and daughter [44, 45] was chosen for the investigations. Besides kinship calculations, STR fingerprints are used for personal identification due to the unique pattern of STRs. In this study, a genetic STR fingerprint multiplex analysis is used to ensure the authenticity of DNA extracts by monitoring for potential contaminations from the laboratory staff.

Considering the rules of Mendelian inheritance, the known kinship relation between the chosen individuals is helpful for revealing potential false negative results. Especially in aDNA analyses, the phenomenon of allelic dropout is common. Large alleles are more frequently affected by allelic dropout than short alleles are, depending on the degree of DNA fragmentation of the remains [39].

In the present work, the presence/absence situation of 30 *Alu* loci was investigated for three members of a prehistoric family (father, left femur DO 1911; mother, left femur DO 3756; daughter, left femur DO 3750) and two modern individuals of Caucasian origin as positive controls. A presence band is defined as the *Alu* locus where the element is inserted, resulting in a long amplification fragment, an absence band as the locus where the *Alu* element is not inserted, which appears as a shorter fragment on the gel. We show that it is possible to amplify, *Alu* loci, including flanking regions with fragment lengths up to 500 bp, for the 3,000-year-old remains in the Lichtenstein cave. We also demonstrate an alternative approach for cases in which, due to DNA degradation, the classic PCR approach failed to amplify the longer presence fragments. Additionally, we give a brief statement about questions to be raised in further investigations.

RESULTS/DISCUSSION

Presence/Absence Analyses

The study illustrates the presence or absence of 30 *Alu* loci in three prehistoric and two modern individuals. The positions are based on the human genome assembly GRCh38.p5 (see online database ensembl.org) [48]. The results of the classical PCR approach and internal *Alu* amplification are presented in Table 1 (for the molecular approach, see methods). In addition to the three prehistoric samples, two modern positive controls were investigated (CAU_1 and CAU_2). CAU_1 originates from a Caucasian American person; CAU_2 is a person of Central European origin. Randomly selected loci were chosen and verified by cloning and sequencing (accession numbers KU323383-KU323387) to ensure the authenticity of the bands.

Table 1. *Alu* presence/absence results for all examined individuals

Locus	Alu subfamily	DO 1911 (F)	DO 3750 (D)			DO 3756 (M)	CAU_1	CAU_2
		FAP	FAP	IAP	CR	FAP	FAP	FAP
***Alu*_1**	Ya5	+	+			+	+/+	+/+
***Alu*_2**	Ya5	+	-	n.b.	-	+/-	+/+	+/+
***Alu*_3**	Ya5	+	n.b.	+	+	+	+/+	+/+
***Alu*_4**	Ya5	+	n.b.			n.b.	+/-	+/-
***Alu*_5**	Ya5	-	-			-	+/-	+/-
***Alu*_6**	Ya5	+	+			+	+/+	+/+
***Alu*_7**	Ya5	+	-	n.b.	-	-	+/-	+/-
***Alu*_8**	Ya5	-	-			+/-	-/-	-/-
***Alu*_9**	Ya5	-	-			-	+/+	+/+
***Alu*_10**	Ya5	-	-			-	+/-	-/-
***Alu*_11**	Ya5	-	-			-	-/-	-/-
***Alu*_12**	Ya5	-	-			-	+/-	+/-
***Alu*_13**	Ya5	-	-			-	-/-	+/-
***Alu*_14**	Yb8	+	n.b.	+	+	+	+/+	+/+
***Alu*_15**	Ya5	-	-			-	+/-	+/-
***Alu*_16**	Ya5	+	-	+	+/-	-	+/+	+/+
***Alu*_17**	Ya5	+	+			+	+/+	+/+
***Alu*_18**	Ya5	-	-			-	-/-	-/-
***Alu*_19**	Ya5	+	n.b.	+	+	+	+/+	+/+
***Alu*_20**	Ya5	+	n.b.	+	+	+	+/-	+/-
***Alu*_21**	Ya5	-	-			-	+/-	+/+
***Alu*_22**	Ya5	+/-	+			+/-	+/+	+/+
***Alu*_23**	Yb8	-	-			-	+/-	-/-
***Alu*_24**	Yc1	-	-			-	+/-	-/-
***Alu*_25**	Ya5	+/-	n.b.			+	+/+	+/+
***Alu*_26**	Yb8	+	-	+	+/-	-	+/-	+/-
***Alu*_27**	Yb8	-	-	+	+/-	+	-/-	-/-
***Alu*_28**	Yb8	-	-			-	-/-	-/-
***Alu*_29**	Ya5	-	-			-	-/-	+/-
***Alu*_30**	Ya5	-	-			-	-/-	-/-

-; -/-: genomic sequence without *Alu* insert

+; +/+: genomic sequence including *Alu* insert

n.b.: no band

FAP: result after amplification with flanking *Alu* primers

Yellow boxes: amplification with internal *Alu* primer (IAP)

CR: Combined result of both approaches for DO 3750

Bold: complete result

Red boxes: not further investigated (see Results/Discussion)

The homozygous results for the prehistoric individuals are represented only by "+" or "-", to include the possibility of allelic dropout events. In contrast, the homozygous results for the modern samples are indicated by "+/+" or "-/-", due to the reliability of analyses of modern DNA. In this study, a "definite result" is defined as successful product amplification for all three samples (father, mother and daughter) per *Alu* locus based on both molecular approaches (FAP and if necessary IAP, see Table 1) upon condition that the family situation is congruent. *Alu* loci that are marked yellow show the incongruence of the family situation in relation to Mendelian inheritance, or the amplification failed completely. This is best explained by the phenomenon of allelic dropout, which is known and common in aDNA analyses. The presence band is periodically not amplified because allelic dropout usually impacts larger alleles. A low number of intact targets is one reason why some alleles may not occur at all or may not reach the detection limits of the devices of electrophoresis [39].

Obviously incomplete and incongruent results were subjected to an alternative molecular approach. Using an internal *Alu* primer, the predicted fragment length of the amplicon was reduced to ~150 bp (Fig. 1). The internal primers were designed based on an alignment of *Alu* sequences of the respective subfamily and consequently are very specific to each *Alu*Y subfamily as described by Nelson et al. [49] or Kass and Batzer [50]. This type of amplification worked in seven cases for the sample DO 3750 (Fig. 2). The heterozygous results for *Alu*_16, *Alu*_26 and *Alu*_27 for the daughter ('CR' in Table 1) represent a combination of both amplification approaches. Further internal *Alu* primer analyses were not possible, due to a depleted DNA extract (*Alu*_4, *Alu*_25; marked red). Loci with exclusively absence bands for the prehistoric individuals, in particular, should be checked by internal *Alu* amplification. The advantage of this method is that the amplification of short fragments (usually ~150 bp) still proves the presence of an insert. In this study, this approach was applied only in those cases where the *Alu* amplification results are not in concordance with the family situation or where the amplification totally failed for DO 3750. Based on previous analyses on this prehistoric triad, it is known that the DNA is less well-preserved in DO 3750 and best-preserved in DO 1911. Consequently, the chance of allelic dropout events for DO 3750 is more likely than for DO 3756 and DO 1911. Fragments of such short lengths (~150 bp) are usually not affected by allelic dropout. However, the internal primer approach cannot be applied in isolation because it does not indicate heterozygous states.

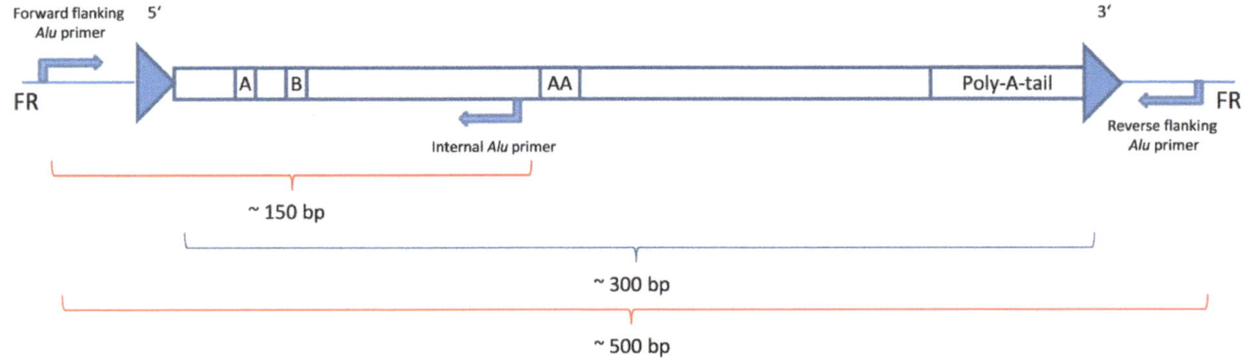

Figure 1. The amplification via an internal *Alu* primer results in amplicons of ~150 bp. The reverse flanking *Alu* primer is replaced by an internal *Alu*subfamily specific primer. The internal primer is located at the 3′ sequence of the left *Alu* monomer prior to the A-rich region in the middle of the element. The amplification via flanking *Alu* primers results in amplicons of ~500 bp. The large arrows at the 5′ and 3' ends indicate the target site duplications.

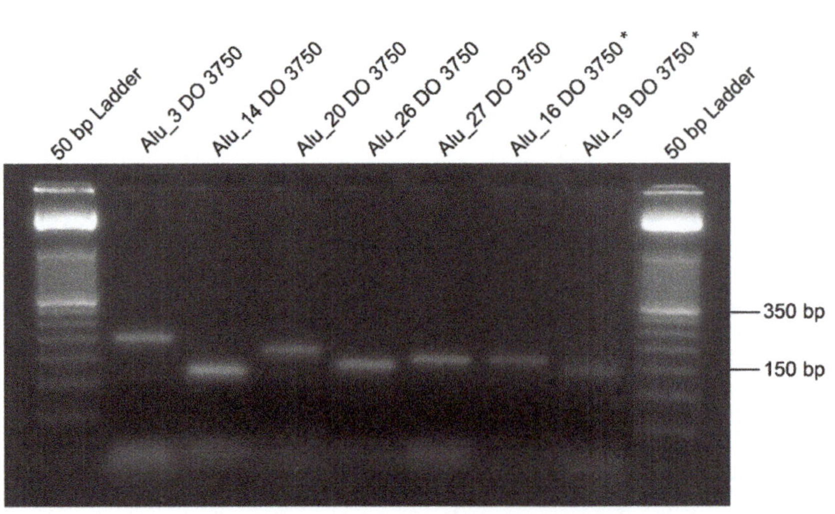

Figure 2. The photo shows seven successfully amplified amplicons of an internal *Alu* primer-based amplification. The expected fragment lengths vary from 118-194 bp. The marks on the base pair ladder are situated at 150 bp and 350 bp. For these seven *Alu* loci, the presence band

for DO 3750 was proved via internal *Alu* amplification. The asterisks indicate reverse *Alu* inserts. In these cases, the primer pairings are an internal*Alu* primer with the reverse *Alu* flanking primer, whereas the samples without asterisk were amplified with an internal *Alu* primer and the forward *Alu* flanking primer

Proof of Kinship in the Prehistoric Samples and Authenticity

The authenticity of the aDNA was ensured by amplifying STR-based genetic fingerprints. Table 2 presents the consensus results of the Heptaplex STR analyses for the DNA extracts used. Table 3 shows the consensus results for seven additional STR systems. Nearly all amplifications were conducted with DNA material taken from the left femur. The genetic fingerprint results of the STR systems D16S539, D2S1338 and D19S433 for the daughter were not achieved with DNA material from her left femur (DO 3750), but from her left humerus (DO 3994). The results of all 13 STR systems for the three prehistoric individuals were used for a kinship calculation resulting in a kinship probability of 99.999 %. All single allele frequencies were taken from the online database allstr.de [51]. Given this proven kinship, false homozygous *Alu* results for the daughter can be clearly detected by contradiction between the parental alleles and the Mendel's laws of inheritance.

Table 2. Heptaplex based fingerprint results for all investigated individuals

	Individual	Amelo	D13S317	D21S11	D18S51	TH01	D5S818	FGA
prehistoric	DO1911 (F)	X/Y	12/12	30.2/32.2	15/17	9.3/9.3	11/12	21/22
	DO3756 (M)	X/X	8/9	28/29	16/16	9/9.3	12/12	21/23
	DO3750 (D)	X/X	9/12	29/32.2	16/17	9/9.3	11/12	21/23
modern	Cau_1	X/Y	12/13	28/32.2	15/18	9/9.3	11/13	20/21
	Cau_2	X/Y	10/11	28/31.2	14/14	9.3/9.3	8/12	22/24

Table 3. Additional fingerprint results for the prehistorical individuals

Individual	D3S1358	VWA	D8S1179	D7S820	D16S639	D2S1338	D19S433
DO 1911 (F)	16/18	17/19	13/13	8/11	12/13	18/19	14/15
DO 3756 (M)	18/18	17/19	12/15	10/10	12/13	24/25	13/14
DO 3750 (D)	16/18	17/17	12/13	8/10	12/13*	19/25*	14/14*

*: The data for these alleles originate from the left humerus (DO 3994) of the same individual

CONCLUSION

The study clearly demonstrates the possibility of presence/absence analyses of TEs in 3,000-year-old human remains from the Lichtenstein cave. These and earlier results indicate and prove the high quality of DNA preservation and the applicability of molecular analyses using the remains from this cave [44, 45, 52], but could not yet show the amplification of 500 bp fragments. Of 30 loci, we initially achieved 22 definite results (FAP in Table 1). With additional amplification using internal *Alu* primers, we could add four more results (*Alu*_3, *Alu*_14, *Alu*_19 and *Alu*_20), thus, 26 definite results (FAP and IAP in Table 1). The following *Alu* loci were incongruent with the family situation: *Alu*_2, *Alu*_7, *Alu*_16, *Alu*_26 and *Alu*_27. Amplification with internal *Alu*primers could place *Alu*_16, *Alu*_26 and *Alu*_27 in congruence with the family situation. The proposed verification technique is to check for possible presence bands by amplification with an internal *Alu* primer to get short target sequences of ~150 bp. Fragment lengths of more than 200 bp tend to be affected more often by allelic dropout events; therefore, short amplicons should be used. Thus, in further analyses, results that show only absence bands should be subjected to this strategy. Even less well-preserved DNA can be analyzed by this approach. The present study constitutes the basis for further investigations with more *Alu* loci and larger samples for microevolutionary studies in Central Europe. Such large-scale investigations would provide insight into *Alu*SINE-based microevolutionary processes in humans during the last few thousand years and help us comprehend the evolutionary dynamics of our genome. Current projects, like the 1,000 Human Genome Project, investigate human genetic variation and the interrelation of genotypes and phenotypes as well as variants in annotated genes and inherited genetic disorders [13, 53]. Through computational biology, the 1,000 Genome Project recently provided a genome-wide catalog of *Alu* polymorphisms for human populations [14]. A database with these group-specific insertions of polymorphic *Alu*elements is useful for future analyses with a larger dataset of Bronze Age Lichtenstein individuals – for instance, to investigate the geographic origin of Lichtenstein family members, who belong to the longest known family tree in the world. Through large-scale*Alu* element analyses of many individuals from the Bronze Age Lichtenstein cave, we may be able to detect human variability and evolution within one geographic region on a timeline. These data would represent a great supplement to recent human population genetic studies based on TEs.

METHODS

Samples and DNA Extraction

Samples

The skeletal material used for the present thesis originates from the Bronze Age Lichtenstein cave near Osterode in Lower Saxony, Germany. All bone material from the cave is stored at -20°C in the Department of Historical Anthropology and Human Ecology of the Göttingen University, Lower Saxony, Germany. The DNA of the ancient individuals was extracted from three different members of a prehistoric family: father (left femur DO 1911), mother (left femur DO 3756) and daughter (left femur DO 3750). In all three cases, the DNA was extracted from the middle of the diaphysis. The modern DNA of the person from the United States of America was extracted from lymphocytes (CAU_1), and was provided with full written consent. This sample was ordered from "The Interstate Companies" (Memphis, Tennessee, USA) blood bank. The DNA of the modern positive control CAU_2 was extracted from cells of the buccal mucosa.

aDNA Extraction from Skeletal Material with the QIAvac-24-Plus

Fragments about 1 cm² in size are sawed out of the middle of the diaphysis of the left femora. All outer surfaces of the fragments are removed to minimize the risk of a contamination with modern human DNA from, e.g., the excavation personal. The fragments are crushed with a steel mortar before they are powdered in a swing mill for 3 min at 24 swings per second. Afterwards, 0.25 g of the powder is transferred into a 15 ml FalconTube and 3900 µl of EDTA UltraPure™ 0.5 M pH 8 (Invitrogen™) and 100 µl of Proteinase K (600mAnson-U/ml) are added. This mixture is incubated for 18 h at 37°C in a rotator. Now, an additional 50 µl of Proteinase K are added and the mixture is incubated at 56°C for 2 h in a rotator. 50 µl of SDS (10 mg/ml) are added, followed by an incubation time of 5 min at 65°C. The lysate is centrifuged at 3300rcf for 3 min to sediment surplus organic material. The lysate is transferred into a 50 ml FalconTube that contains 16 ml PB-Buffer (Qiagen) and 100 µl sodium acetate buffer (pH 5.2, 3 M, Sigma). After manually mixing the lysate, it is centrifuged at 3300rcf for 3 min. The DNA clean-up is conducted with minElute spin columns and funnels for large volumes using the QIAvac-24-plus (Qiagen). Deviating from the protocol, three wash steps with PE-buffer (Qiagen) are performed. The DNA is eluted in 60 µl RNase-free water (also cf. [54]).

Modern DNA

Blood Sample

The DNA of the blood sample from CAU_1 is extracted with the Wizard Genomic DNA Purification Kit (Promega) following the producer's protocol for extraction from whole blood samples (300 µl).

Buccal Mucosa Swab Sample

A buccal mucosa swab from CAU_2 is transferred into a 2 ml reaction tube. 400 µl of G2-buffer (Qiagen) and 10 µl of Proteinase K are added, followed by incubation for 1 h at 56°C and 350 rpm on an Eppendorf thermomixer comfort. Afterwards, 200 µl of the lysate is transferred into a clean tube and 1 ml of PB-buffer and 100 µl of sodium acetate buffer are added. After manually mixing the lysate, it is centrifuged at 3300rcf for 3 min. Now the DNA is cleaned up with minElute spin columns and large volume funnels as described above.

Alu Loci and Primer Design

Alu loci were chosen based on earlier publications with a population-genetics focus [23–26]. The site-specific *Alu* sequence was determined by using RepeatMasker [55]. An additional 500 bp flanking sequence on each site was extracted from the human reference genome (hg38) in NCBI [56]. The locus-specific primers were designed with PrimerSelect, version 10.1.2 (DNASTAR). The primer characteristics are a strong 5′ and a weak 3′ end by not exceeding a length of 30 bp; further, primer dimerization and hairpin formation were avoided to enhance the specificity and sensitivity of the reaction [38]. The total length of the target sequence (including the *Alu* insert) should be as short as possible, which usually resulted in amplicons of 450 bp to 500 bp.

The internal *Alu* primers were designed based on a highly conserved region of the *Alu*sequence that is specific to the respective subfamily. Therefore, randomly selected *Alu*inserts of the respective *Alu*Y subfamily were aligned. The amplification always includes the *Alu* head.

PCR

All PCRs are conducted under the same conditions apart from the annealing temperatures. Depending on the energy profiles and melting temperatures of the primer sets and based on preliminary primer tests, different annealing temperatures, varying from 52 to 60°C, are chosen. The amplification is conducted using the following cycling program: Initial hot start at 95°C for 5 min; 40 cycles with denaturation at 94°C for 1 min, annealing at 52 – 60°C for 1 min, elongation at 72°C for 1 min; a final soak at 10°C for 10 min. The PCR is composed of 12.5 µl of Multiplex PCR Mastermix (Qiagen), 1 µl each forward and reverse primer, (both 20 µM working solution), 5 µl DNA for aDNA samples and 0.5 µl DNA (plus 4.5 µl RNase free water) for modern DNA samples and 5.5 µl of RNase-free water to get a final volume of 25 µl per reaction.

The amplification with an internal *Alu* primer was performed with an elongation time of 20 s. All other parameters are identical to the classical PCR approach.

For proof of authenticity, every DNA extract used in the study here presented was subjected to STR-typing by a multiplex amplification as described previously [57]. Deviating from this work, the sex discriminating amelogenin gene is arranged in the blue dye panel. The reaction mix is composed of 12.5 µl Multiplex PCR Mastermix (Qiagen), 2.85 µl multiplex primer mix, 4.65 µl RNase-free water and 5 µl DNA extract.

Gel electrophoresis and fragment length estimation

Each amplification result is checked by ethidium bromide stained agarose gel electrophoresis (2.5 %). The fragment length determination is performed with a 50 bp molecular ladder (Invitrogen). For the electrophoresis, usually a voltage of 120 V and a run time of 30 min are applied.

The STR products are separated in a 50 cm capillary on an ABI 3500 Genetic Analyzer (Applied Biosystems) using POP-7™ Polymer for 3500/3500xL Genetic Analyzers and the 3500 Data collection Software (all Applied Biosystems).

Allele determination is performed with GeneMapper Software 5 (Applied Biosystems).

Cloning and Sequencing

Cloning of PCR products is conducted with the Blue/White-Selection based pGEM®-T Easy Vector System (Promega). Deviating from the manufacturer's protocol, 300 µl SOC medium (Invitrogen) is used to suspend the cells. Additionally, 50 µl – 100 µl of the cell suspension are plated. The Colony-PCR Mastermix is identical to the other PCRs except for the PCR primers. The primers pUC/M13 forward and reverse (Promega) are used in working concentrations of 20 µM. One colony replaces the DNA inset. The Colony PCR is conducted with the following program: Initial denaturation at 94°C for 3 min; 30 cycles with denaturation at 94°C for 30 s, annealing at 55°C for 1 min, elongation at 72°C for 50 s; final elongation at 72°C for 2 min and final soak at 10°C for 10 min. PCR products are purified with an isopropanol purification protocol: the PCR product is incubated with 83 µl HPLC water, 100 µl isopropanol (100 %) and 10 µl sodium acetate (3 M) for 10 min, then centrifuged at 13,200 rpm for 10 min in a conventional tabletop microcentrifuge. The supernatant is discarded and 150 µl of ethanol (70 %) is added. After another 10 min of centrifugation at 13,200 rpm, the supernatant is discarded, the pellet is dried and the desired amount of RNase-free water is added for resuspension. The sequencing reaction is composed of 4 µl Sequencing Buffer (5x), 2 µl BigDyeTerminator v1.1, 0.3 µl primer (20 µM), 6.7 µl HPLC water and 7 µl purified PCR product. Sequencing is performed in forward and reverse direction with the following program: Initial heating step at 94°C for 3 min; 33 cycles with denaturation at 94°C for 30 s, annealing at 55°C for 1 min and elongation at 72°C for 2.5 min; soak at 10°C. Sequencing products are purified with NucleoSeq® columns (Macherey-Nagel). The products are separated in a 50 cm capillary on an ABI 3500 Genetic Analyzer (Applied Biosystems) using POP-7™ Polymer for 3500/3500xL Genetic Analyzers and the 3500 Data collection Software (all Applied Biosystems). The sequences are edited in BioEdit version 7.2.5 [58] and submitted to a BLAST analysis. Finally, sequence data with the following accession numbers were deposited in GenBank: KU323383-KU323387.

Kinship Calculation

For kinship calculation (Reverse Parentage Index; RPI), the genotype probabilities are calculated: RPI=X/Y. The numerator (X) is the probability that a woman randomly selected from a population is type AB, that a man randomly selected from a population is type CD and that the child is type BC. The child gets one of the two alleles of the father and the mother, respectively. The probability that one allele of one parent is inherited by the child is 0.5. The denominator (Y) is the probability that a woman randomly selected from a population and unrelated to the child is type AB, that a man randomly selected from a population and unrelated to the child is type CD, and that a child randomly selected from a population is type BC (also cf. [59, 60]). The reverse parentage index for one STR system is calculated as follows:

$$\mathrm{RPI} = \frac{X}{Y} = \frac{2P_AP_B \times 2P_CP_D \times 0.5 \times 0.5}{2P_AP_B \times 2P_CP_D \times 2P_BP_C}$$

All single RPIs are multiplied to get a combined RPI.

ACKNOWLEDGEMENTS

The *Alu* SINE-based work was supported by the University of Göttingen with funding provided by the Lower Saxony Ministry for Science and Culture (*Ministerium für Wissenschaft und Kultur, MWK)* to OP. The research on the skeletal remains from the Lichtenstein cave was funded by the *MWK* to SH. We want to thank Stefan Flindt from the *Landkreis Osterode am Harz (Kreisarchäologie)* Lower Saxony for enabling this research by allowing the use of the invaluable sample material. His long-term cooperation in research on human remains from the Lichtenstein cave is highly appreciated.

AUTHORS' CONTRIBUTIONS

MK carried out the molecular genetic studies and generated all results that are based on presence/absence analyses and drafted most parts of the manuscript. OP designed the project and participated in molecular studies and data interpretation. Information about the samples from the Lichtenstein cave as well as STR amplifications was generated by VS. In addition VS was involved in data interpretation. SH participated in project design and data interpretation. This manuscript was critically reviewed by OP, VS and SH. All authors read and approved the final manuscript.

REFERENCES

1. McClintock B. The origin and behavior of mutable loci in maize. Proc Natl Acad Sci U S A. 1950; 36:344–55. doi: 10.1073/pnas.36.6.344.
2. McClintock B. Induction of Instability at Selected Loci in Maize. Genetics.1953; 38:579–99.
3. Doolittle WF, Sapienza C. Selfish genes, the phenotype paradigm and genome evolution. Nature. 1980; 284:601–3. doi: 10.1038/284601a0.
4. Callinan PA, Batzer MA. Retrotransposable Elements and Human Disease. In: Volff JN, editor. Genome and Disease. Genome Dyn. Basel: Karger; 2006. pp. 104–15.
5. Volff JN. Turning junk into gold: domestication of transposable elements and the creation of new genes in eukaryotes. Bioessays. 2006; 28:913–22. doi: 10.1002/bies.20452.
6. Piskurek O, Jackson DJ. Transposable Elements: From DNA Parasites to Architects of Metazoan Evolution. Genes. 2012; 3:409–22. doi: 10.3390/genes3030409.
7. Heras SR, López MC, Olivares M, Thomas MC. The L1Tc non-LTR retrotransposon of *Trypanosoma cruzi* contains an internal RNA-pol II-dependent promoter that strongly activates gene transcription and generates unspliced transcripts. Nucleic Acids Res.2007; 35(7):2199–214. doi: 10.1093/nar/gkl1137.
8. Estecio MRH, Gallegos J, Dekmezian M, Lu Y, Liang S, Issa JPJ. SINE Retrotransposons Cause Epigenetic Reprogramming of Adjacent Gene Promoters. Mol

Cancer Res. 2012; 10(10):1332–42. doi: 10.1158/1541-7786.MCR-12-0351.

9. Sasaki T, Nishihara H, Hirakawa M, Fujimura K, Tanaka M, Kokubo N, Kimura-Yoshida C, Matsuo I, Sumiyama K, Saitou N, Shimogori T, Okada N. Possible involvement of SINEs in mammalian-specific brain formation. Proc Natl Acad Sci U S A. 2008; 105(11):4220–25. doi: 10.1073/pnas.0709398105.
10. Piskurek O, Okada N. Poxviruses as possible vectors for horizontal transfer of retroposons from reptiles to mammals. Proc Natl Acad Sci U S A. 2007; 104:12046–51. doi: 10.1073/pnas.0700531104.
11. Lander ES, Linton LM, Birren B, Nusbaum C, Zody MC, Baldwin J, et al. Initial sequencing and analysis of the human genome. Nature. 2001; 409:860–921. doi: 10.1038/35057062.
12. De Koning AP, Gu W, Castoe TA, Batzer MA, Pollock DD. Repetitive Elements May Comprise Over Two-Thirds of the Human Genome. PLoS Genet. 2011; 67:183–93.
13. Abecasis GR, Altshuler D, Auton LD, Durbin RM, Gibbs RA, Hurles ME, et al. A map of human genome variation from population-scale sequencing. Nature.2010; 467(7319):1061–73. doi: 10.1038/nature09534.
14. Rishishwar L, Tellez Villa CE, Jordan IK. Transposable element polymorphisms recapitulate human evolution. Mobile DNA. 2015; 6:21. doi: 10.1186/s13100-015-0052-6.
15. Vassetzky NA, Kramerov DA. SINEBase: a database and tool for SINE analysis.Nucleic Acids Res. 2013; 41:D83–D89. doi: 10.1093/nar/gks1263.
16. Jurka J, Kapitonov VV, Pavlicek A, Klonowski P, Kohany O, Walichiewicz J. Repbase Update, a database of eukaryotic repetitive elements. Cytogenet Genome Res.2005; 110:462–7. doi: 10.1159/000084979.
17. Okada N, Hamada M, Ogiwara I, Ohshima K. SINEs and LINEs share common 3′ sequences: A review. Gene. 1997; 205:229–43. doi: 10.1016/S0378-1119(97)00409-5.
18. Okada N, Shedlock AM, Nikaido M. Retroposon mapping in molecular systematics.Methods Mol Biol. 2004; 260:189–226.
19. Shedlock AM, Kazuhiko T, Okada N. SINEs of speciation: tracking lineages with retroposons. TRENDS Ecol Evol. 2004; 19(10):545–53. doi: 10.1016/j.tree.2004.08.002.
20. Shedlock AM, Okada N. SINE insertions: powerful tools formolecularsystematics.Bioessays.2000;22(2):148–60. doi: 10.1002/(SICI)1521-1878(200002)22:2<148::AID-BIES6>3.0.CO; 2-Z.
21. Schmitz J, Ohme M, Zischler H. SINE insertions in cladistic analyses and the phylogenetic affiliations of *Tarsius bancanus* to other primates. Genetics. 2001; 157(2):777–84.
22. Deininger PL. *Alu* elements: know the SINEs. Genome Biol. 2011; 12:236. doi: 10.1186/gb-2011-12-12-236.
23. Furano AV. The biological properties and evolutionary dynamics of mammalian LINE-1 retrotransposons. Prog Nucleic Acids Res Mol Biol. 2000; 64:255–94. doi: 10.1016/S0079-6603(00)64007-2.
24. Okada N, Shedlock AM, Nikaido M. Retroposon Mapping in Molecular Systematics. In: Mobile Genetic Elements: Protocols and Genomic Applications, Methods in Molecular Biology. Totowa, New Jersey, U.S.A.: Humana Press; 2004. vol. 260. p. 189–226.
25. Carrol ML, Roy-Engel AM, Nguyen SV, Salem AH, Vogel E, Vincent B, Myers J, et al. Large-scale Analysis of the *Alu* Ya5 and Yb8 Subfamilies and their Contribution to Human Genomic Diversity. J Mol Biol. 2001; 311:17–40. doi: 10.1006/jmbi.2001.4847.
26. Carter AB, Salem AH, Hedges DJ, Nguyen Keegan C, Kimball B, Walker JA, et al. Genome wide analysis of the human *Alu* Yb8 lineage. Hum Genomics. 2004; 1:167–78.
27. Otieno AC, Carter AB, Hedges DJ, Walker JA, Ray DA, Garber RK, Anders BA, et al. Analysis of the human *Alu* Ya-lineage. J Mol Biol. 2004; 342:109–18. doi: 10.1016/j.jmb.2004.07.016.
28. Garber RK, Hedges DJ, Herke SW, Hazard NW, Batzer MA. The *Alu* Yc1 subfamily: sorting the wheat from the chaff. Cytogenet Genome Res. 2005; 110:537–42. doi: 10.1159/000084986.
29. Nikaido M, Piskurek O, Okada N. Toothed whale monophyly reassessed by SINE insertion analysis: The absence of lineage sorting effects suggest a small population of a common ancestral species. Mol Phylogenet Evol. 2007; 43:216–24. doi: 10.1016/j.ympev.2006.08.005.
30. Nishihara H, Maruyama S, Okada N. Retroposon analysis and recent geological data suggest near-simultaneous divergence of the three superorders of mammals. Proc Natl Acad Sci U S A. 2009; 106(13):5235–40. doi: 10.1073/pnas.0809297106.
31. Churakov G, Kriegs JO, Baertsch R, Zemann A, Brosius J, Schmitz J. Mosaic retroposon insertion patterns in placental mammals. Genome Res. 2009; 19(5):868–75. doi: 10.1101/gr.090647.108.
32. Matzke A, Churakov G, Berkes P, Arms EM, Kelsey D, Brosius J, et al. Retroposon Insertion Patterns of neoavian Birds: Strong Evidence for an Extensive Incomplete Lineage Sorting Era. Mol Biol Evol. 2012; 29(6):1497–501. doi: 10.1093/molbev/msr319.
33. Watkins WS, Rogers AR, Ostler CT, Wooding S, Bamshad MJ, Brassington AM, et al. Genetic variation among world populations: inferences from 100 Alu insertion polymorphisms. Genome Res. 2003; 13(7):1607–18. doi: 10.1101/gr.894603.
34. Kajikawa M, Okada N. LINEs Mobilize SINEs in the Eel through a Shared 3′ Sequence. Cell. 2002; 111(3):433–44. doi: 10.1016/S0092-8674(02)01041-3.
35. Dewannieux M, Esnault C, Heidmann T. LINE-mediated retrotransposition of marked *Alu* sequences. Nat Genet. 2003; 35:41–8. doi: 10.1038/ng1223.
36. Simonti CN, Capra JA. The evolution of the human genome. Curr Opin Genet Dev.2015; 35:9–15. doi: 10.1016/j.gde.2015.08.005.

37. Nekrutenko A, Li WH. Transposable elements are found in a large number of human protein-coding genes. Trends Genet. 2001; 17:619–21. doi: 10.1016/S0168-9525(01)02445-3.

38. Cordaux R, Batzer MA. The impact of retrotransposons on human genome evolution.Nat Rev Genet. 2009; 10(10):691–703. doi: 10.1038/nrg2640.

39. Hummel S. Ancient DNA Typing. Methods, Strategies and Applications. Berlin, Germany: Springer; 2003.

40. Höss M, Jaruga P, Zastawny TH, Dizdaroglu M, Pääbo S. DNA damage and DNA sequence retrieval from ancient tissues. Nucleic Acid Res. 1996; 24(7):1304–7. doi: 10.1093/nar/24.7.1304.

41. Haack K, Hummel S, Hummel B. Ancient DNA fragments longer than 300 bp.Anthrop Anz. 2000; 58:51–6.

42. Burger J, Hummel S, Herrmann B, Henke W. DNA preservation: A microsatellite-DNA study on ancient skeletal remains. Electrophoresis. 1999; 20:1722–8. doi: 10.1002/(SICI)1522-2683(19990101)20:8<1722::AID-ELPS1722>3.0.CO; 2-4.

43. Lindahl T. Instability and decay of the primary structure of DNA. Nature.1993; 32(422):709–11. doi: 10.1038/362709a0.

44. Schultes T. Typisierung alter DNA zur Rekonstruktion von Verwandtschaft in einem bronzezeitlichen Skelettkollektiv. 2000.

45. Schilz F. Molekulargenetische Verwandtschaftsanalysen am prähistorischen Skelettkollektiv der Lichtensteinhöhe. 2006.

46. Schultes T, Hummel S, Herrmann B. Ancient DNA-typing approaches for the determination of kinship in a disturbed collective burial site. Anthrop Anz.2000; 58(1):37–44.

47. Schilz F, Hummel S, Herrmann B. Design of a multiplex PCR for genotyping 16 short tandem repeats in degraded DNA samples. Anthrop Anz. 2004; 62(4):369–78.

48. ensembl.org. http://www.ensembl.org/index.html.

49. Nelson DL, Ledbetter SA, Corbo L, Victoria MF, Ramirez-Solis R, Webster TD, Ledbetter DH, Caskey CT. *Alu* polymerase chain reaction: A method for rapid isolation of human-specific sequences from complex DNA sources. Proc Natl Acad Sci U S A.1989; 86:6686–90. doi: 10.1073/pnas.86.17.6686.

50. Kass DH, Batzer MA. Inter-*Alu* Polymerase Chain Reaction: Advancements and Applications. Anal Biochem. 1995; 228:185–93. doi: 10.1006/abio.1995.1338.

51. allstr Autosomal Database for short tandem repeats. http://allstr.de/allstr/searchMarker.seam. Accessed 3rd September 2015.

52. Hummel S, Schmidt D, Kremeyer B, Herrmann B, Oppermann M. Detection of the CCR5- Δ32 HIV resistance gene in Bronze Age skeletons. Genes Immun. 2005; 6:371–4. doi: 10.1038/sj.gene.6364172.

53. Konkel MK, Walker JA, Hotard AB, Ranck MC, Fontenot CC, Storer J et al. Sequence Analysis and Characterization of Active Human *Alu* subfamilies Based on the 1000 Genomes Pilot Project. Genome Biology and Evolution. 2015; doi:10.1093/gbe/evv167.

54. Frischalowski M, Seidenberg V, Großkopf B, Wulf FW, Hummel S. Molekulargenetische Untersuchung des Verwandtschaftsverhältnisses von möglichen Mutter-Kind-Bestattungen aus dem frühzeitlichen Eldagsen. Nachrichten aus Niedersachsens Urgeschichte. 2015; 84:(in press).

55. Smit AFA, Hubley R, Green P. RepeatMasker Open-4.0. 2013-2015.http://www.repeatmasker.org.

56. National Center for Biotechnology Information.http://www.ncbi.nlm.nih.gov/assembly/GCF_000001405.26.

57. Seidenberg V, Schilz F, Pfister D, Georges L, Fehren-Schmitz L, Hummel S. A new miniSTR heptaplex system for genetic fingerprinting of ancient DNA from archaeological human bone. J Archaeol Sci. 2012; 39(10):3224–9. doi: 10.1016/j.jas.2012.05.019.

58. Hall TA. BioEdit: a user-friendly biological sequence alignment editor and analysis program for Windows 95/97/NT. Nucl Acids Symp. 1999; 41:95–98.

59. Brenner CH. Symbolic Kinship Program. Genetics. 1997; 145:535–42.

60. Gjertson DW, Brenner CH, Baur MP, Carracedo A, Guidet F, Luque JA, et al. ISFG: Recommendations on biostatistics in paternity testing. Forensic Sci Int Genet.2007; 1:223–31. doi: 10.1016/j.fsigen.2007.06.006.

CHAPTER 10 EXOME SEQUENCING: CURRENT AND FUTURE PERSPECTIVES

Amanda Warr[1], Christelle Robert[1], David Hume[1], Alan Archibald[1], Nader Deeb[2], and Mick Watson[1]

[1]The Roslin Institute and Royal (Dick) School of Veterinary Studies, University of Edinburgh, Easter Bush, Edinburgh UK

[2]Genus plc., Hendersonville, Tennessee 37075

The completion of a reference genome sequence for humans took more than 200 scientists more than a decade in a project that cost almost $3 billion to complete (International Human Genome Sequencing Consortium 2004). During the past decade the price of genome sequencing has plummeted, much of the work has become automated, and methods have improved. These advances mean complete genomes can be sequenced quickly and affordably. Human genome sequencing is a special case and enjoys much lower cost-per-gigabase than other species because of Illumina's HiSeq X platform (Watson 2014). Sequencing can allow for the identification of genetic variants that affect heritable phenotypes, including important disease-causing mutations and natural variation that can be exploited to improve crops and livestock. Despite the significant improvement in sequencing technology, sequencing whole genomes to a depth sufficient to find variants that affect phenotypic expression is expensive compared with targeted sequencing. This review will focus on exome sequencing, a method that targets only a subset of the genome, often the protein coding portion, significantly reducing the sequencing space and subsequently the cost. Details of the method, available platforms, uses in humans and other species, and its benefits over whole-genome sequencing (WGS) will be discussed.

Citation: Warr A, Robert C, Hume D, Archibald A, Deeb N, and Watson M. Exome Sequencing: Current and Future Perspectives. G3: Genes|Genomes|Genetics. 2015; 5(8):1543-1550. DOI: 10.1534/g3.115.018564.

WHAT IS EXOME SEQUENCING?

The exome has been defined traditionally as the sequence encompassing all exons of protein coding genes in the genome and covers between 1 and 2% of the genome, depending on species. It also may be extended to target functional nonprotein coding elements (*e.g.*, microRNA, long intergenic noncoding RNA, etc.) as well as specific candidate loci. There are two main categories of exome capture technology: solution-based and array-based.

In solution-based, whole-exome sequencing (WES), DNA samples are fragmented and biotinylated oligonucleotide probes (baits) are used to selectively hybridize to target regions in the genome. Magnetic streptavidin beads are used to bind to the biotinylated probes, the nontargeted portion of the genome is washed away, and the polymerase chain reaction (PCR) is used to amplify the sample, enriching the sample for DNA from the target region. The sample is then sequenced before proceeding to bioinformatic analysis. Array-based methods are similar except that the probes are bound to a high-density microarray. The array-based method was the first to be used in exome capture (Albert *et al.* 2007), but it has largely been supplanted by solution-based methods, which require less input DNA and are consequently potentially more efficient; however, studies byAsan *et al.* (2011) and Bodi *et al.* (2013) found that NimbleGen's Sequence Capture Array performed better than the solution-based alternatives in low GC content regions; had high sensitivity and read mapping rates; and single-nucleotide polymorphism (SNP) detection from these reads was more specific to the target region. This suggests that a niche may remain for the older technology. Array-based capture has been used successfully and accurately to identify rare and common variants and identify candidate genes for monogenic diseases in small cohorts (Ng *et al.* 2009); however, the array-based methods are less scalable owing to the limitation of the number of probes that can be accommodated on the array and additional equipment and time required to process the microarrays.

EXOME CAPTURE PLATFORMS

There are several differences between the available platforms, which are constantly being updated and improved. The major providers of exome capture platforms are NimbleGen, Agilent, and Illumina, and each have different designs and strengths (summarized in Table 1). The discussion of the characteristics of each platform will focus on the solution-based human kits; the performance of kits for other species has not been subjected to the same level of comparison.

Table 1. Summary of the differences between the solution-based, exome-sequencing platforms

	NimbleGen's SeqCap EZ Exome Library	Agilent's Sure Select Human All Exon Kit	Illumina's TruSeq Exome Enrichment Kit	Illumina's Nextera Rapid Capture Exome Kit
Probe size, bp[a]	55–105	114–126	95	95
Probe type	DNA	RNA	DNA	DNA
Coverage strategy	High-density, overlapping probes	Adjacent probes	Gaps between probes	Gaps between probes
Fragmentation method	Ultrasonication	Ultrasonication	Ultrasonication	Transposomes
Target region size (human), Mb[b]	64	50	62	62
Reads remaining after filtering[c]	66%	71.7%	54.8%	40.1%
Major strengths	(i) High sensitivity and specificity (ii) Most uniform coverage in difficult regions	(i) Better coverage of indels (ii) High alignment rate (iii) Fewer duplicate reads than other platforms	(i) Good coverage of UTRs and miRNAs	(i) Good coverage of UTRs and miRNAs
Major weaknesses	(i) More duplicate reads than Agilent (ii) Lower alignment rate than Agilent	(i) Fewer high-quality reads than NimbleGen	(i) High off-target enrichment	(i) High off-target enrichment (ii) Coverage bias for high GC content areas reducing uniformity
Nonhuman supported	Yes	Yes	No	No

UTR, untranslated region; miRNA, microRNA.

[a]As described in Clark *et al.* (2011), NimbleGen SeqCap EZ v2.0, Agilent SureSelect Human All Exon 50Mb, Illumina TruSeq Exome Enrichment.

[b]For NimbleGen SeqCap EZ v3.0, Agilent SureSelect V5, and Illumina TruSeq and Illumina Nextera original versions.

[c]Filtering for duplicates, multiple mappers, improper pairs, and off-target reads, data from Chilamakuri *et al.* (2014) .

NimbleGen's SeqCap EZ Exome Library has the greatest bait density of any of the platforms and uses short (55–105 bp), overlapping baits to cover the target region (Clark*et al.* 2011). This approach has been found to be an efficient method for enrichment with the least amount of sequencing needed to cover the target region and sensitively detect variants (Clark *et al.* 2011) and also has a high level of specificity, showing fewer off-target reads than other platforms (Clark *et al.* 2011; Sulonen *et al.* 2011). Importantly, this bait design has been found to show greater genotype sensitivity and more uniformity of coverage in difficult to sequence regions, such as areas of high GC content, than the other platforms (Asan *et al.* 2011; Sulonen *et al.* 2011; Bodi *et al.* 2013).

Agilent's SureSelect Human All Exon Kit is the only platform to use RNA probes, with the other platforms opting for DNA probes. The baits used are longer than those used in NimbleGen's platform (114–126 bp) and the corresponding target sequences are adjacent to one another rather than overlapping (Clark *et al.* 2011). This design has been found to be good at identifying insertions and deletions (indels), because longer baits can tolerate larger mismatches (Clark *et al.* 2011; Bodi *et al.* 2013; Chilamakuri *et al.* 2014); it has been suggested that this also may reduce reference allele bias at heterozygous sites compared with other bait designs; however, in practice the allele bias has been similar to other platforms (Asan *et al.* 2011; Hedges *et al.* 2011). The platform has been found to produce fewer duplicate reads than NimbleGen but also fewer high-quality reads (Sulonen *et al.* 2011). Bodi *et al.* (2013) found that Agilent had a greater alignment rate and fewer PCR duplicates than NimbleGen but also had less uniform coverage.

Illumina's TruSeq Exome Enrichment Kit uses 95-bp probes that leave small gaps in the target region, with paired end reads extending outside the bait sequence during sequencing to fill the gap. This design has been found to have a high percentage of off-target enrichment (Clark *et al.* 2011), which reduces its target efficiency compared with the other platforms. This kit detects more single-nucleotide variants in the untranslated regions (UTRs) than the other platforms (Clark *et al.* 2011), although comparisons of performance with NimbleGen and Agilent's "+UTR" kits have yet to be performed. Following filtering for duplicates, multiple mappers, improper pairs, and off-target reads,Chilamakuri *et al.* (2014) found that this platform retained fewer reads (54.8%) than NimbleGen (66%) or Agilent (71.7%). At high read counts (>50M), this platform outperformed Agilent's SureSelect in downstream identification of indels (Clark *et al.*2011).

Illumina's Nextera Rapid Capture Exome and Expanded Exome kits are similar to the TruSeq kit in their probe design. They differ from the other kits in that they use transposomes to fragment the genomic DNA, whereas the other platforms use ultrasonication. These kits have not been extensively compared with the other platforms, with only Chilamakuri *et al.* (2014) having included Nextera in a comparison study. At the time of the study there was only one Nextera kit, which was the Expanded Exome version: the kit with the larger target region of the two. The Expanded Exome kit shares a target region with the TruSeq kit, which includes UTRs and miRNAs. Chilamakuri *et al.* (2014) found that the Nextera kit had increased coverage of high GC content areas as the result of altered bias in the transposome technology used during fragmentation, decreasing its overall uniformity; however, recent changes to the protocol in the new versions may have improved this. They also found that out of all platforms tested, the Nextera platform retained the fewest reads after filtering for duplicates, multiple mappers, improper pairs, and off-target reads, at 40.1%.

Chilamakuri *et al.* (2014) compared the target regions of the human kits from three providers (NimbleGen SeqCap EZ v3.0 – 64.1Mb; Agilent SureSelect V4 – 51.1Mb; and Illumina TruSeq and Illumina Nextera original version and protocol – 62.08 Mb). Despite the fact that all of the platforms are targeting the human exome, there is surprisingly little overlap between the three designs, with just 26.2 Mb covered by all three target regions. NimbleGen's and Agilent's coverage are more similar to one another than either are to Illumina's; this is likely attributable to the large amount of UTRs covered in Illumina's target region. Illumina has 22.5Mb of targets unique to their platforms, and 21.8Mb of these are UTRs. NimbleGen and Agilent have 16.1 Mb and 7 Mb of unique targets, respectively. It would be interesting to see how NimbleGen and Agilent's "+UTR" kits compare with Illumina in single-nucleotide variant calling, particularly in UTRs.

Currently, in addition to human kits, NimbleGen offers capture kits for maize, barley, wheat, soy, mouse, and pig exomes, and Agilent offers capture kits for mouse, cattle, and zebrafish exomes. Both providers also offer the opportunity to design custom kits for other species. The kits for nonhuman species use similar bait designs and protocols to the providers' human kits. Both manufacturers offer a flexible design process that allows for modifications to improve coverage for specific regions and purposes.

USES IN HUMANS

The first successful use of WES to diagnose and inform subsequent treatment in a human patient was in the identification of the causal variant of a rare form of inflammatory bowel disease in an infant (Worthey *et al.* 2011). In this case, conventional diagnostics had failed to find an explanation for the patient's severe symptoms, and doctors needed to understand the underlying cause of the symptoms before they could decide how to treat the child. A multidisciplinary team combined clinical phenotyping, exome sequencing, bioinformatics, and functional studies, eventually finding the causative mutation that influenced the future treatment of the child. The analysis of the exome data was hampered by the relative lack of software designed for this purpose at the time, and researchers had to manually inspect more than 2000 variants. Through filtering and manual inspection, the candidate pool was reduced to 70 genes exhibiting hemi- or homozygous variants. Of these, only eight variants were novel and predicted to be damaging to protein function. Analysis of evolutionary conservation identified two variants that were located in highly conserved sequences and one of these had a high null genotype frequency. This left a single hemizygous, nonsynonymous variant in the *X-linked inhibitor of apoptosis (XIAP)* gene. The child was diagnosed with X-linked lymphoproliferative disease 2, exhibiting a novel inflammatory bowel disease–like manifestation caused by loss of tolerance to commensal organisms in the digestive system. This diagnosis allowed for effective treatment through allogeneic hematopoietic progenitor cell transplant. The unusual manifestation of the condition meant that the patient was unlikely to have been diagnosed without the exome sequence data.

After the success of this initial diagnosis, exome sequencing has been used extensively to diagnose novel diseases and find novel causative mutations for known disease phenotypes. Exome sequencing is useful in human medicine for diagnosis of particularly difficult-to-diagnose patients, diagnosis of young patients who may not yet exhibit a full spectrum of symptoms (Iglesias *et al.* 2014), prenatal diagnosis (Iglesias *et al.* 2014; Xu*et al.* 2014), and early diagnosis of debilitating disease (Bras and Singleton 2011; Sassi*et al.* 2014). In addition to reaching a diagnosis, finding the causative mutation can allow for alteration of treatment, prevention of further invasive testing, accurate prognoses, and confirmed diagnoses, which are essential for eligibility for benefits and access to clinical trials (Grossmann *et al.* 2011; Rabbani *et al.* 2012; Taneri *et al.* 2012; Iglesias *et al.*2014) and in the future may allow for more targeted treatment.

Exome sequencing in human medicine benefits from the availability of large databases of known SNPs, known pathogenic variants, and control genomes; during analysis variants found in these databases can generally be excluded when one is looking for novel variants, significantly reducing the variant pool. The Exome Aggregation Consortium (ExAC) has created a user-friendly database containing the exome sequences of more than 60,000 unrelated individuals, which is freely available (Exome Aggregation Consortium (ExAC) 2014). The exomes have been analyzed using a uniform bioinformatic pipeline and are from individuals with adult-onset diseases. This provides researchers with a large set of reference exomes that should be free from homozygous variants that cause childhood-onset Mendelian diseases. The database provides a wealth of information such as depth of coverage, genotype quality, allele frequency, and variant consequences. The filtering capabilities and large number of exomes will simplify the process of prioritizing variants when using exome sequencing as a diagnostic tool, particularly in children.

Where many unrelated, affected individuals are available, an "overlap" strategy can be used to simplify analysis and search for common variants likely to affect gene function (Johansson *et al.* 2012). If there is a known inheritance pattern this can be used to search for specific genotype zygosity. Sequencing of multiple affected related individuals also can increase the power of the analysis (Johnson *et al.* 2010; Shi *et al.* 2011). In rare diseases, case–parent trios can be used to exclude nonpathogenic variants found in the parents (Smith *et al.* 2014). However, single-patient sequencing may be sufficient to identify the causative mutation (Wang *et al.* 2011; Worthey *et al.* 2011; Xu *et al.* 2014). Analysis examining how conserved an amino acid sequence is through evolution and between genes in a family can help to increase the confidence of a mutation having a deleterious effect (Wang *et al.* 2011; Johansson *et al.* 2012; Sassi *et al.* 2014; Smith *et al.*2014; Xu *et al.* 2014).

Examples of diseases for which exome sequencing has been used to detect a causative variant include Leber congenital amaurosis (Wang *et al.* 2011), Alzheimer disease (Sassi*et al.* 2014), maturity-onset diabetes of the young (Johansson *et al.* 2012), high myopia (Shi *et al.* 2011), autosomal recessive polycystic kidney disease (Xu *et al.* 2014), amyotrophic lateral sclerosis (Johnson *et al.* 2010), immunodeficiency leading to infection with human herpes virus 8 causing Kaposi Sarcoma (Byun *et al.* 2010), acromelic frontonasal dystois (Smith *et al.* 2014), and a number of cancer predisposition mutations (*e.g.*, Yan *et al.* 2011; Greif *et al.* 2012; Snape *et al.* 2012; Kiiski *et al.* 2014;Cai *et al.* 2015).

The Deciphering Developmental Disorders project aims to develop a scalable exome sequencing workflow to facilitate the translation of the method from the research environment to a clinical environment. So far more than 1000 children with undiagnosed developmental disorders and their parents have been sequenced with plans to increase this number to 12,000 patients. With diagnostic yields up to 31% in the children sequenced thus far, this project demonstrates the future potential of the method in a clinical setting (Gecz and Corbett 2015; Wright *et al.* 2015; The Deciphering Developmental Disorders 2015).

USES IN OTHER SPECIES

Variants Discovery for Agricultural Improvement

Not all protein-altering variants cause disease (Macarthur and Tyler-Smith 2010). To understand the potential of exome sequencing in areas other than disease variant discovery, which is the main focus in human research, we can look to studies using the technique to sequence agricultural species.

Plant genomes can be extremely complex, repetitive, and are often polyploid; as a result, some of the most economically important crops are not well suited for genome re-sequencing studied. For example, Bread Wheat (*Triticum aestivum*) has an allohexaploid (AABBDD) genome around 17Gb in size (Brenchley *et al.* 2012; The International Wheat Genome Sequencing Consortium 2014). This is likely too large for subsequent WGS studies at the current price of sequencing and data storage. Wheat is an extremely important crop for both human and livestock consumption and genetic improvement is slow; with the human population increasing and additional challenges from environmental changes, it is essential to gain a better understanding of the crop to improve it. An exome capture kit has been designed for wheat based on the accumulated transcriptome data (Winfield *et al.* 2012). The capture region for this kit is 56.5 Mb, which is around the lower estimated size of one diploid wheat exome and because of similarity between the three genomes may be sufficient to capture most of the exome data from the whole allohexaploid genome. This kit has been used to identify induced mutations in the genome to aid studies investigating gene function, a use that also was applied to the rice (*Oryza sativa*) exome in the same study (Henry *et al.* 2014) and the soybean (*Glycine max*) exome in a separate study (Bolon *et al.* 2011). WES in soybean has also been used to identify unwanted intracultivar genetic heterogeneity in the exome that may affect the plant's phenotype (Haun *et al.* 2011).

The genome of barley (*Hordeum vulgare* L.) has not been fully sequenced. Barley's genome is smaller than wheat's at around 5 Gb, but is still larger than is practical for routine whole-genome sequence analysis and contains repetitive elements that complicate the genome's assembly. A gene space assembly has been produced (The International Barley Genome Sequencing Consortium 2012) and a barley exome capture kit has been developed based on this assembly (Mascher *et al.* 2013). The kit has since been used to identify a mutation involved in early maturation, a trait relevant to production (Pankin *et al.* 2014).

Exome capture in barley has also been used to identify a gene causative of many-noded dwarfism using mapping-by-sequencing (Mascher *et al.* 2014). The many-noded dwarfism phenotype is a shorter plant with more, narrower leaves than the wild type. The mutant in this study was created using X-ray mutagenesis, a technique that often causes large deletions. An F_2 population between mutant and wild-type phenotypes was created and 18 mutant individuals and 30 wild-type individuals were exome sequenced. From these sequences SNPs were identified and allele frequencies of these were used to identify an allele overrepresented in the mutant group. Researchers queried the sequencing reads for exome targets that were present in the wild type but not the mutant. This lead to the identification of a candidate gene (MLOC_64838.2, now HvMND), which has a homolog known to play a role in a similar phenotype in rice. Screening of other mutants showing this phenotype found a variety of null mutations in this gene. The family to which this gene belongs is known to have effects on important production traits and may include good selection targets to improve production.

Identification of New Genetic Markers

Robert *et al.* (2014) have designed exome capture probes for the pig *Sus scrofa*, used these to sequence the exomes of 96 healthy pigs, and identified potentially deleterious variants. Bioinformatic analysis identified 236,608 high-confidence predicted variants and 28,115 predicted indels in the target region. This work revealed notable gaps in the current Ensembl *S.scrofa* genome annotation and identified a large number of potential protein truncating variants. As the pigs tested were healthy, it is possible that some of these protein truncating variants have phenotypic effects on traits other than those relating to the health of the pigs or to the production traits currently under selection. This work is an important step in identifying phenotype altering variants in the pig: a production animal and medical model.

The barley exome kit has been used to differentiate between markers of *H. vulgare* L. and *H. bulbosum* L.; *H. bulbosum* L. is a wild species that has superior pathogen resistance and tolerance compared to the domestic species and the two can be crossed to improve the domesticated crop, however negative linkage drag on production traits has hampered its use in elite barley lines. Using exome sequencing to identify specific markers can allow selective crossing to be used to incorporate the beneficial variants without incorporating linked variants that are detrimental to production (Wendler *et al.*2014). Similarly, the wheat exome kit has allowed for discovery of previously unidentified markers in the genome which can be used in future genetic studies and marker assisted selection (Allen *et al.* 2013).

Black cottonwood (*Populus trichocarpa*) has had an exome capture kit designed (Zhou and Holliday 2012). *P. trichocarpa* is a model organism and was the first tree to have its whole genome sequenced (Tuskan *et al.* 2006). The tree is used in lumber production and in cosmetics. The tree has experienced a whole genome duplication fairly recently and the exome study found that this does not appear to have had an effect on SNP detection through exome sequencing (Zhou and Holliday 2012). There is potential to use the identified markers to improve production in this species.

Health Traits

As with exome sequencing in humans, exome sequencing has been used in other mammals to discover variants associated with health traits. WES has been used in conjunction with a genome wide association study to identify a frameshift mutation causing blindness in Phalène dogs (Ahonen *et al.* 2013). In cattle (*Bos taurus*), WES has been used successfully to identify strong candidate variants for haplotypes relating to reduced fertility rates in Holsteins which can be used to selectively breed against these detrimental haplotypes (McClure *et al.* 2014).

Special Considerations for Nonhuman Species

For many species, the reference genome is not completed to the same standard as the human genome; many species have only a draft genome or, as discussed for barley, no reference genome. This is an important consideration when using WES in nonhuman species. Poor annotation of genomes mean that in the design of capture probes, causative genes may be missed because they are not annotated whereas with whole genome sequencing, the data will be there whether the gene is annotated or not. Where available, including predicted genes identified from RNA-sequencing data may be beneficial to maximize coverage of functional elements in poorly annotated genomes. For example,Robert *et al.* (2014) added an additional 14Mb of data to the capture region in pigs using EST evidence. Additionally, errors in the reference genome will greatly increase the number of false-positive variant calls in these species. This increases computational burden and forces more stringent filtering which may inadvertently discard causative variants.

WES may be especially useful for model organisms, particularly where the sequences of large numbers of individuals are needed. The expense of WGS on large cohorts makes it less feasible in animal studies than human studies, particularly in species with fairly large genomes such as mice (~3.5 Gb) and zeberafish (~1.5 Gb), both common model organisms. Sequencing a smaller portion of the genome, particularly if candidate regions are known and can be targeted, would be more cost-effective, allow deeper coverage and potentially increase the number of individuals that can be used in these studies.

The genomes of animals have different levels of linkage disequilibrium (LD), which are often high in domestic species and model organisms with limited effective population sizes. The level of LD also may vary by breed, as demonstrated in the domestic dog (Stern *et al.* 2013). When using exome sequencing to identify causal variants, high LD may lead to the identification of a benign variant that is in LD with the causative variant. In this case, the causative variant may have been missed in variant calling or lie outside the sequence space; it is therefore important to consider the known function of the element the candidate variant is associated with and the predicted effect of the variant on function. Additionally, efforts to sample individuals that are as outbred as possible may help to reduce this problem. Another consideration with high LD is that it may be possible to use this information in imputation, as has been done with other technologies such as imputing genotypes from SNP arrays (Hickey *et al.* 2012), from WGS data (Deelen *et al.* 2014; Gudbjartsson *et al.* 2015) and from human exome data (Auer *et al.*2012). The larger haplotypes associated with high LD increase the accuracy and reduce the computational burden associated with imputation. Imputation may be less accurate than sequencing; however it allows for a larger numbers of individuals to be used at lower coverage and may help to reduce problems associated with variable coverage in exome sequencing.

EXOME CAPTURE TRANSFERABILITY BETWEEN SPECIES

The Neanderthal exome was successfully sequenced by Burbano *et al.* (2010) using an exome capture kit designed for the human exome. The study compared the Neanderthal exome with human exomes and found 88 fixed substitutions in 83 genes in the human exomes; these substitutions did not appear to be a result of positive selection and may be a result of accelerated genetic drift from reduced effective population size in humans following historical bottlenecks or reduced purifying selection.

The kit designed for cattle also can be applied to other bovid species; Cosart *et al.* (2011) demonstrated that the kit could be successfully used to capture the exomes of, and identify SNPs in, zebu (*Bos indicus*) and American Bison (*Bison bison*). This transferability of exome capture kits, as also demonstrated in studies involving the sequencing of Neanderthals and nonhuman primates using human capture kits (Burbano*et al.* 2010; Vallender 2011), is possible because despite millions of years of divergence, functional elements tend to be highly conserved.

WHY NOT USE GENOME SEQUENCING?

With the price of sequencing decreasing as rapidly as it has during the past decade, questions have been raised concerning WES's usefulness in the era of affordable WGS. WGS, however, is still more expensive than WES. The costs of WES consist of the cost of the capture plus the cost of sequencing, whereas WGS consists only of the sequencing costs. If we assume that the cost of capture remains fixed, then as the costs of sequencing decrease, the cost of WGS will approach the cost of WES. However, at present that is not the case, and it would be unwise to assume that the cost of sequence capture will not decrease.

Human genomes are a special case in that the HiSeq X platform offers a cost-per-Gb far less than other platforms yet is limited (contractually, rather than technically) to 30X WGS human genomes (Watson 2014). However, even given that advantage, the $1000 price tag for a 30X human genome is (in our estimate) two to three times the cost of a 40X human exome (depending on scale). It may be advantageous to sequence more samples using WES, and gain statistical power, than to sequence more of the genome. In other species, the price difference is even greater, for example, in pigs (a similar sized genome to human but currently without the benefit of access to the HiSeq X platform) we estimate the cost of WGS to be 9-10 times the cost of WES (Robert *et al.* 2014).

Although WGS does have benefits over WES, the cost of this technology is more than simply the price of sequencing. Although sequencing technology has been improving at a much faster rate than would be predicted by Moore's law (a prediction of improvement in computing hardware, but often also applied to other technologies), the technology for storing and analyzing the data has not seen a matching acceleration in improvement (Mardis 2010; Sboner *et al.* 2011). WGS produces approximately one hundred times the data that WES does at the same coverage. The infrastructure needed to store, manage, and analyze data significantly increases the costs of WGS. WGS produces a much larger number of variants than WES does, not only because of the size of the sequencing space, but because regions outside the exome are less well conserved; although this number might include a variant of interest, the larger data set significantly increases the computational burden for analysis. Additionally variation in noncoding regions is less well understood than variation in the coding region (Mu *et al.* 2011; Maurano *et al.*2012; Ward and Kellis 2012), making it more difficult to predict which variants might be relevant to a trait of interest in WGS datasets. The majority of causative variants identified so far in Mendelian disease have been found in coding regions (Botstein and Risch 2003), although ascertainment bias is likely to play a role in this conclusion. A study investigating functional noncoding variants based on WGS data from 1092 human genomes showed that functionally deleterious non-coding mutations were under strong negative selection, in a similar way to that of loss-of-function variants in protein-coding regions (Khurana *et al.* 2013). The authors developed a tool to prioritize non-coding variants in disease studies, which was used to identify noncoding candidate drivers in tumor genomes.

With the continuous decrease in sequencing cost, new studies making use of WGS to investigate causative variants will lead to the discovery of additional mutations in regulatory elements that contribute to the pool of disease-associated variants. In that context, the sampling bias currently observed toward coding variants is likely to be reduced by WGS investigations of non-coding genomic regions.

However, cost is not the only consideration. WGS covers the whole genome at more consistent coverage than WES, can provide more accurate detection of structural variants, and does not have reference sequence bias caused by probe sequences in WES (Majewski *et al.* 2011; Meynert *et al.* 2014; Belkadi *et al.* 2015). Recent studies have highlighted and suggested roles for promoters (The Fantom Consortium and the RIKEN PMI and CLST 2014) and enhancers (Andersson *et al.* 2014) in a range of different cell types, and these are not traditionally captured by exome sequencing. Importantly, WES requires previous knowledge of the location and sequence of features to target them, whereas WGS covers the entire genome. This means that exome sequencing relies on the accuracy of the genome annotation, so phenotype altering variants may be missed in poorly or incompletely annotated genomes. It is therefore important to take care when designing the capture region or purchasing a commercial kit to ensure any specific regions of interest are included. Exome sequencing also invariably fails to successfully capture the entire target region (Asan *et al.* 2011; Bodi *et al.* 2013; Chilamakuri *et al.*2014; Robert *et al.* 2014), causing even properly annotated regions to be missed. Another consideration is that exome sequencing involves a PCR stage that is known to reduce coverage of GC-rich regions (Kozarewa *et al.* 2009; Veal *et al.* 2012). New sequencing technology has allowed for sequencing of DNA without the need for DNA amplification, generating sequencing from single molecules. This technology can produce accurate, longer reads without the artifacts and biases associated with the amplification process in other sequencing methods and the exome capture stage of WES (Shin *et al.* 2013). The longer read length in these third-generation sequencers is particularly beneficial for detecting structural variants and for resolving repetition in assemblies and copy number variable regions (Roberts *et al.* 2013). However, for now the price of this long-read sequencing is still prohibitively expensive and is not commonly in use for analysis of genetic variation. In the future as these sequencers improve and prices come down they may make WGS more attractive to researchers even if their analysis focuses solely on the protein-coding region and known functional elements.

The power of WGS on a large cohort for variant discovery already has been demonstrated by Gudbjartsson *et al.* (2015), who sequenced the whole genomes of 2636 Icelanders to 20X and imputed the sequence variants into 101,584 further chip-genotyped and phased individuals from the same population. This allowed for discovery of 6795 null mutations in 4924 genes that may play a role in disease. They also found evidence of lethal mutations that are not found in the homozygous state, which likely impact on fertility in heterozygous couples. Imputation reduces the amount of sequencing and data storage required, though the study found that fewer null mutations were detected in the imputed data set suggesting that some mutations were missed in these individuals.

WGS will eventually take a leading role in genome interrogation; however, it will likely have to wait for data storage and analysis to improve before its full potential can be realized. In the meantime, WES provides many of the benefits of WGS with lower storage requirements and computational burden, at an affordable price. This is particularly useful in large scale studies, for example, a recent study sequenced the exomes of more than 9000 people (Schick *et al.* 2015). It is also useful for the sharing of information such as in the ExAC database where there are more than 60,000 exomes stored. WES will likely also remain the method of choice in species with exceptionally large genomes, for example in some polyploid plant species.

Exome sequencing is a technology that allows interrogation of the most well-understood portion of the genome: the protein-coding sequence and functional elements. Variants of interest don't necessarily fall within the exome, but so far most of the known variants responsible for Mendelian disease have been found in the coding region and the target region can be extended to include other regions of interest. Although the decreasing price of sequencing may soon make genome sequencing more attractive, the additional costs of data handling and downstream analysis cannot be ignored. The applications of variant discovery, particularly in disease gene identification, cannot afford to wait for data storage and processing technology to catch up to sequencing improvements. The amount of data produced by WES is far more manageable than WGS, particularly for small research groups and groups studying organisms with large genomes. WES has established itself as an important method in disease gene identification in humans and increasingly in domestic species. The applications of WES in crop research are allowing genomic techniques to be used in species with complex genomes, potentially identifying variants important for production that can be incorporated into marker-assisted selection. At present, WES is a useful and powerful method for variant discovery within coding regions offering most of the benefits of WGS while allowing for easier analysis and storage of the data produced.

However, WGS will eventually be the NGS technology of choice with regard to the investigation of genomic variations due to the more uniform coverage achieved with WGS *vs.* WES, including coverage within the exome (Belkadi *et al.* 2015) and the more uniform distribution of sequencing quality parameters (*e.g.*, coverage depth, genotype quality) observed with WGS *vs.* WES (Meynert *et al.* 2014; Belkadi *et al.* 2015). In addition, WGS extends the variation search space to the whole genome, allowing for the additional detection of noncoding variants (The 1000 Genomes Project Consortium 2012), the functional impact of which is becoming easier to interpret with the systematic annotation of functional noncoding elements (The Encode Project Consortium 2012;Ward and Kellis 2012; Andersson *et al.* 2014; The Fantom Consortium and the RIKEN PMI and CLST 2014).

Ideally, regardless of sequencing costs and storage limitations, integrated approaches combining the advantages of both WES and WGS would be beneficial for variant discovery studies with the addition of WES-/WGS-exclusive variants with WES providing additional variants missed in low-coverage dataset (The 1000 Genomes Project Consortium 2012). In addition, both technologies' usefulness depend on the quality of the reference assembly that the sequenced reads are mapped to; poor quality references increase the number of false-positive variants identified in the analysis, which inevitably leads to more stringent filtering, increasing the potential for discarding a variant of interest. In the future, the improvement of reference assemblies, bioinformatic tools and sequencing technology will be necessary to improve the power of variant discovery techniques.

The term "exome" may no longer be appropriate for a technique that is simply a subset of the more generic technique of sequence capture. It is already possible to extend the exome capture region beyond protein coding genes to capture noncoding genes, and regulatory elements such as promoters and enhancers. In this way, we may sequence all of the functional areas of a genome without the cost of sequencing everything. Indeed, as clinically important variants are discovered, the paradigm may change to sequencing small panels of genes that are known to be relevant to disease, as is common in cancer genomics.

LITERATURE CITED

1. Ahonen S. J., Arumilli M., Lohi H., 2013. A CNGB1 frameshift mutation in Papillon and Phalene dogs with progressive retinal atrophy. PLoS One 8: e72122.
2. Albert T. J., Molla M. N., Muzny D. M., Nazareth L., Wheeler D., et al., 2007. Direct selection of human genomic loci by microarray hybridization. Nat. Methods 4: 903–905.
3. Allen A. M., Barker G. L., Wilkinson P., Burridge A., Winfield M., et al., 2013. Discovery and development of exome-based, co-dominant single nucleotide polymorphism markers in hexaploid wheat (Triticum aestivum L.). Plant Biotechnol. J. 11: 279–295.
4. Andersson R., Gebhard C., Miguel-Escalada I., Hoof I., Bornholdt J., et al., 2014. An atlas of active enhancers across human cell types and tissues. Nature 507: 455–461.
5. Asan Y., Xu H., Jiang C., Tyler-Smith Y., Xue, et al., 2011. Comprehensive comparison of three commercial human whole-exome capture platforms. Genome Biol. 12: R95.
6. Auer P. L., Johnsen J. M., Johnson A. D., Logsdon B. A., Lange L. A., et al., 2012. Imputation of Exome Sequence Variants into Population- Based Samples and Blood-Cell-Trait-Associated Loci in African Americans: NHLBI GO Exome Sequencing Project. Am. J. Hum. Genet. 91: 794–808.
7. Belkadi A., Bolze A., Itan Y., Cobat A., Vincent Q. B., et al., 2015. Whole-genome sequencing is more powerful than whole-exome sequencing for detecting exome variants. Proc. Natl. Acad. Sci. USA 112: 5473–5478.

8. Bodi K., Perera A. G., Adams P. S., Bintzler D., Dewar K., et al., 2013. Comparison of commercially available target enrichment methods for next-generation sequencing.J. Biomol. Tech. 24: 73–86.

9. Bolon Y. T., Haun W. J., Xu W. W., Grant D., Stacey M. G., et al., 2011. Phenotypic and genomic analyses of a fast neutron mutant population resource in soybean. Plant Physiol. 156: 240–253.

10. Botstein D., Risch N., 2003. Discovering genotypes underlying human phenotypes: past successes for Mendelian disease, future approaches for complex disease. Nat. Genet. 33: 228–237.

11. Bras J. M., Singleton A. B., 2011. Exome sequencing in Parkinson's disease. Clin. Genet. 80: 104–109.

12. Brenchley R., Spannagl M., Pfeifer M., Barker G. L. A., D'Amore R., et al., 2012. Analysis of the bread wheat genome using whole-genome shotgun sequencing. Nature 491: 705–710.

13. Burbano H. A., Hodges E., Green R. E., Briggs A. W., Krause J., et al., 2010. Targeted investigation of the Neandertal genome by array-based sequence capture. Science 328: 723–725.

14. Byun M., Abhyankar A., Lelarge V., Plancoulaine S., Palanduz A., et al., 2010. Whole-exome sequencing-based discovery of STIM1 deficiency in a child with fatal classic Kaposi sarcoma. J. Exp. Med. 207: 2307–2312.

15. Cai J., Li L., Ye L., Jiang X., Shen L., et al., 2015. Exome sequencing reveals mutant genes with low penetrance involved in MEN2A-associated tumorigenesis. Endocr. Relat. Cancer 22: 23–33.

16. Chilamakuri C. S. R., Lorenz S., Madoui M.-A., Vodák D., Sun J., et al., 2014. Performance comparison of four exome capture systems for deep sequencing.BMC Genomics 15: 449.

17. Clark M. J., Chen R., Lam H. Y., Karczewski K. J., Euskirchen G., et al., 2011. Performance comparison of exome DNA sequencing technologies. Nat. Biotechnol. 29: 908–914.

18. Cosart T., Beja-Pereira A., Chen S., Ng S. B., Shendure J., et al., 2011. Exome-wide DNA capture and next generation sequencing in domestic and wild species. BMC Genomics 12: 347.

19. Deelen P., Menelaou A., van Leeuwen E. M., Kanterakis A., van Dijk F., et al., 2014. Improved imputation quality of low-frequency and rare variants in European samples using the 'Genome of The Netherlands'. Eur. J. Hum. Genet. 22: 1321–1326.

20. Exome Aggregation Consortium (ExAC), 2014 ExAC Browser. Cambridge, MA. Available at: http://exac.broadinstitute.org. Accessed April 2015.

21. Gecz J., Corbett M., 2015. Developmental disorders: deciphering exomes on a grand scale. Lancet 385: 1266–1267.

22. Greif P. A., Dufour A., Konstandin N. P., Ksienzyk B., Zellmeier E., et al., 2012. GATA2 zinc finger 1 mutations associated with biallelic CEBPA mutations define a unique genetic entity of acute myeloid leukemia. Blood 120: 395–403.

23. Grossmann V., Tiacci E., Holmes A. B., Kohlmann A., Martelli M. P., et al., 2011. Whole-exome sequencing identifies somatic mutations of BCOR in acute myeloid leukemia with normal karyotype. Blood 118: 6153–6163.

24. Gudbjartsson D. F., Helgason H., Gudjonsson S. A., Zink F., Oddson A., et al., 2015. Large-scale whole-genome sequencing of the Icelandic population. Nat. Genet. 47: 435–444.

25. Haun W. J., Hyten D. L., Xu W. W., Gerhardt D. J., Albert T. J., et al., 2011. The composition and origins of genomic variation among individuals of the soybean reference cultivar Williams 82. Plant Physiol. 155: 645–655.

26. Hedges D. J., Guettouche T., Yang S., Bademci G., Diaz A., et al., 2011. Comparison of three targeted enrichment strategies on the SOLiD sequencing platform. PLoS One6: e18595.

27. Henry I. M., Nagalakshmi U., Lieberman M. C., Ngo K. J., Krasileva K. V., et al., 2014. Efficient genome-wide detection and cataloging of EMS-induced mutations using exome capture and next-generation sequencing. Plant Cell 26: 1382–1397.

28. Hickey J. M., Kinghorn B. P., Tier B., van der Werf J. H. J., Cleveland M. A., 2012. A phasing and imputation method for pedigreed populations that results in a single-stage genomic evaluation. Genet. Sel. Evol. 44: 9.

29. Iglesias A., Anyane-Yeboa K., Wynn J., Wilson A., Truitt Cho M., et al., 2014. The usefulness of whole-exome sequencing in routine clinical practice. Genet. Med. 16: 922–931.

30. International Human Genome Sequencing Consortium, 2004. Finishing the euchromatic sequence of the human genome. Nature 431: 931–945.

31. Johansson S., Irgens H., Chudasama K. K., Molnes J., Aerts J., et al., 2012. Exome sequencing and genetic testing for MODY. PLoS One 7: e38050.

32. Johnson J. O., Mandrioli J., Benatar M., Abramzon Y., Van Deerlin V. M., et al., 2010. Exome sequencing reveals VCP mutations as a cause of familial ALS. Neuron 68: 857–864.

33. Khurana E., Fu Y., Colonna V., Mu X. J., Kang H. M., et al., 2013. Integrative annotation of variants from 1092 humans: application to cancer genomics. Science342: 1235587.

34. Kiiski J. I., Pelttari L. M., Khan S., Freysteinsdottir E. S., Reynisdottir I., et al., 2014. Exome sequencing identifies FANCM as a susceptibility gene for triple-negative breast cancer. Proc. Natl. Acad. Sci. USA 111: 15172–15177.

35. Kozarewa I., Ning Z., Quail M. A., Sanders M. J., Berriman M., et al., 2009. Amplification-free Illumina sequencing-library preparation facilitates improved mapping and assembly of GC-biased genomes. Nat. Methods 6: 291–295.

36. MacArthur D. G., Tyler-Smith C., 2010. Loss-of-function variants in the genomes of healthy humans. Hum. Mol. Genet. 19: R125–R130.

37. Majewski J., Schwartzentruber J., Lalonde E., Montpetit A., Jabado N., 2011. What can exome sequencing do for you? J. Med. Genet. 48: 580–589.
38. Mardis E. R., 2010. The $1,000 genome, the $100,000 analysis? Genome Med. 2: 84.
39. Mascher M., Richmond T. A., Gerhardt D. J., Himmelbach A., Clissold L., et al., 2013. Barley whole exome capture: a tool for genomic research in the genus Hordeum and beyond. Plant J. 76: 494–505.
40. Mascher M., Jost M., Kuon J.-E., Himmelbach A., Aßfalg A., et al., 2014. Mapping-by-sequencing accelerates forward genetics in barley. Genome Biol. 15: R78.
41. Maurano M. T., Humbert R., Rynes E., Thurman R. E., Haugen E., et al., 2012. Systematic localization of common disease-associated variation in regulatory DNA. Science 337: 1190–1195.
42. McClure M. C., Bickhart D., Null D., Vanraden P., Xu L., et al., 2014. Bovine exome sequence analysis and targeted SNP genotyping of recessive fertility defects BH1, HH2, and HH3 reveal a putative causative mutation in SMC2 for HH3. PLoS One 9: e92769.
43. Meynert A. M., Ansari M., FitzPatrick D. R., Taylor M. S., 2014. Variant detection sensitivity and biases in whole genome and exome sequencing. BMC Bioinformatics15: 247.
44. Mu X. J., Lu Z. J., Kong Y., Lam H. Y. K., Gerstein M. B., 2011. Analysis of genomic variation in non-coding elements using population-scale sequencing data from the 1000 Genomes Project. Nucleic Acids Res. 39: 7058–7076.
45. Ng S. B., Turner E. H., Robertson P. D., Flygare S. D., Bigham A. W., et al., 2009. Targeted capture and massively parallel sequencing of twelve human exomes. Nature 461: 272–276.
46. Pankin A., Campoli C., Dong X., Kilian B., Sharma R., et al., 2014. Mapping-by-sequencing identifies HvPHYTOCHROME C as a candidate gene for the early maturity 5 locus modulating the circadian clock and photoperiodic flowering in barley. Genetics 198: 383–396.
47. Rabbani B., Mahdieh N., Hosomichi K., Nakaoka H., Inoue I., 2012. Next-generation sequencing: impact of exome sequencing in characterizing Mendelian disorders. J. Hum. Genet. 57: 621–632.
48. Robert C., Fuentes-Utrilla P., Troup K., Loecherbach J., Turner F., et al., 2014. Design and development of exome capture sequencing for the domestic pig (Sus scrofa).BMC Genomics 15: 9.
49. Roberts R. J., Carneiro M. O., Schatz M. C., 2013. The advantages of SMRT sequencing. Genome Biol. 14: 405.
50. Sassi C., Guerreiro R., Gibbs R., Ding J., Lupton M. K., et al., 2014. Exome sequencing identifies 2 novel presenilin 1 mutations (p.L166V and p.S230R) in British early-onset Alzheimer's disease. Neurobiol Aging 35: 2422.e13-6.
51. Sboner A., Mu X. J., Greenbaum D., Auerbach R. K., Gerstein M. B., 2011. The real cost of sequencing: higher than you think! Genome Biol. 12: 125–125.
52. Schick U. M., Auer P. L., Bis J. C., Lin H., Wei P., et al., 2015. Association of exome sequences with plasma C-reactive protein levels in >9000 participants. Hum. Mol. Genet. 24: 559–571.
53. Shi Y., Li Y., Zhang D., Zhang H., Li Y., et al., 2011. Exome sequencing identifies ZNF644 mutations in high myopia. PLoS Genet. 7: e1002084.
54. Shin S. C., Ahn D. H., Kim S. J., Lee H., Oh T.-J., et al., 2013. Advantages of single-molecule real-time sequencing in high-gc content genomes. PLoS One 8: e68824.
55. Smith J. D., Hing A. V., Clarke C. M., Johnson N. M., Perez F. A., et al., 2014. Exome sequencing identifies a recurrent de novo ZSWIM6 mutation associated with acromelic frontonasal dysostosis. Am. J. Hum. Genet. 95: 235–240.
56. Snape K., Ruark E., Tarpey P., Renwick A., Turnbull C., et al., 2012. Predisposition gene identification in common cancers by exome sequencing: insights from familial breast cancer. Breast Cancer Res. Treat. 134: 429–433.
57. Stern J., White S., Meurs K., 2013. Extent of linkage disequilibrium in large-breed dogs: chromosomal and breed variation. Mamm. Genome 24: 409–415.
58. Sulonen A. M., Ellonen P., Almusa H., Lepisto M., Eldfors S., et al., 2011. Comparison of solution-based exome capture methods for next generation sequencing. Genome Biol. 12: R94.
59. Taneri B., Asilmaz E., Gaasterland T., 2012. Biomedical impact of splicing mutations revealed through exome sequencing. Mol. Med. 18: 314–319.
60. The 1000 Genomes Project Consortium, 2012. An integrated map of genetic variation from 1,092 human genomes. Nature 491: 56–65.
61. The Deciphering Developmental Disorders Study, 2015. Large-scale discovery of novel genetic causes of developmental disorders. Nature 519: 223–228.
62. The ENCODE Project Consortium, 2012. An integrated encyclopedia of DNA elements in the human genome. Nature 489: 57–74.
63. The Fantom Consortium and the RIKEN PMI and CLST, 2014. A promoter-level mammalian expression atlas. Nature 507: 462–470.
64. The International Barley Genome Sequencing Consortium, 2012. A physical, genetic and functional sequence assembly of the barley genome. Nature 491: 711–716.
65. The International Wheat Genome Sequencing Consortium, 2014. A chromosome-based draft sequence of the hexaploid bread wheat (Triticum aestivum) genome.Science 345: 1251788.
66. Tuskan G. A., DiFazio S., Jansson S., Bohlmann J., Grigoriev I., et al., 2006. The genome of Black Cottonwood, Populus trichocarpa (Torr. & Gray). Science 313: 1596–1604.
67. Vallender E. J., 2011. Expanding whole exome resequencing into non-human primates. Genome Biol. 12: R87.

68. Veal C., Freeman P., Jacobs K., Lancaster O., Jamain S., et al., 2012. A mechanistic basis for amplification differences between samples and between genome regions.BMC Genomics 13: 1–10.

69. Wang H., Chen X., Dudinsky L., Patenia C., Chen Y., et al., 2011. Exome capture sequencing identifies a novel mutation in BBS4. Mol. Vis. 17: 3529–3540.

70. Ward L. D., Kellis M., 2012. Interpreting non-coding variation in complex disease genetics. Nat. Biotechnol. 30: 1095–1106.

71. Watson M., 2014. Illuminating the future of DNA sequencing. Genome Biol. 15: 108.

72. Wendler N., Mascher M., Noh C., Himmelbach A., Scholz U., et al., 2014. Unlocking the secondary gene-pool of barley with next-generation sequencing. Plant Biotechnol. J. 12: 1122–1131.

73. Winfield M. O., Wilkinson P. A., Allen A. M., Barker G. L., Coghill J. A., et al., 2012. Targeted re-sequencing of the allohexaploid wheat exome. Plant Biotechnol. J. 10: 733–742.

74. Worthey E. A., Mayer A. N., Syverson G. D., Helbling D., Bonacci B. B., et al., 2011. Making a definitive diagnosis: successful clinical application of whole exome sequencing in a child with intractable inflammatory bowel disease. Genet. Med. 13: 255–262.

75. Wright C. F., Fitzgerald T. W., Jones W. D., Clayton S., McRae J. F., et al., 2015. Genetic diagnosis of developmental disorders in the DDD study: a scalable analysis of genome-wide research data. Lancet 385: 1305–1314.

76. Xu Y., Xiao B., Jiang W. T., Wang L., Gen H. Q., et al., 2014. A novel mutation identified in PKHD1 by targeted exome sequencing: guiding prenatal diagnosis for an ARPKD family. Gene 551: 33–38.

77. Yan X. J., Xu J., Gu Z. H., Pan C. M., Lu G., et al., 2011. Exome sequencing identifies somatic mutations of DNA methyltransferase gene DNMT3A in acute monocytic leukemia. Nat. Genet. 43: 309–315.

78. Zhou L., Holliday J. A., 2012. Targeted enrichment of the black cottonwood (Populus trichocarpa) gene space using sequence capture. BMC Genomics 13: 703.

CHAPTER 11

NETWORK ANALYSIS OF HUMAN GENES INFLUENCING SUSCEPTIBILITY TO MYCOBACTERIAL INFECTIONS

Ettie M. Lipner[1,2], Benjamin J. Garcia[1,2], and Michael Strong[1,2]

[1]Center for Genes, Environment, and Health, National Jewish Health, Denver, Colorado, United States of America

[2]Computational Bioscience Program, University of Colorado Denver, School of Medicine, Aurora, Colorado, United States of America

ABSTRACT

Tuberculosis and nontuberculous mycobacterial infections constitute a high burden of pulmonary disease in humans, resulting in over 1.5 million deaths per year. Building on the premise that genetic factors influence the instance, progression, and defense of infectious disease, we undertook a systems biology approach to investigate relationships among genetic factors that may play a role in increased susceptibility or control of mycobacterial infections. We combined literature and database mining with network analysis and pathway enrichment analysis to examine genes, pathways, and networks, involved in the human response to Mycobacterium tuberculosis and nontuberculous mycobacterial infections. This approach allowed us to examine functional relationships among reported genes, and to identify novel genes and enriched pathways that may play a role in mycobacterial susceptibility or control. Our findings suggest that the primary pathways and genes influencing mycobacterial infection control involve an interplay between innate and adaptive immune proteins and pathways. Signaling pathways involved in autoimmune disease were significantly enriched as revealed in our networks. Mycobacterial disease susceptibility networks were also examined within the context of gene-chemical relationships, in order to identify putative drugs and nutrients with potential beneficial immunomodulatory or anti-mycobacterial effects.

INTRODUCTION

Tuberculosis (TB), an airborne infectious disease caused by the bacterium *Mycobacterium tuberculosis* (MTB), is an ongoing global health crisis [1,2] resulting in over 9 million illnesses and 1.5 million deaths each year [3]. In contrast, nontuberculous mycobacterial (NTM) disease, caused by phylogenetically related environmental mycobacteria [4], has emerged as an increasingly prevalent infectious disease, particularly over the last two decades [5–8].

Although exposure to TB is common in certain regions of the world, a relatively small proportion of exposed people progress to develop active pulmonary disease. As an example, one third of the world's population is latently infected with MTB, but only 10% of those individuals will ever progress to become ill with active TB. Similarly, many individuals come into contact with NTM through soil or municipal water sources [4], but few develop pulmonary NTM disease. Certain clinical conditions, including immunodeficiencies and individuals with compromised lungs, increase susceptibility, but most TB and NTM disease occur in otherwise healthy people [3,9].

Citation: Lipner EM, Garcia BJ, Strong M (2016) Network Analysis of Human Genes Influencing Susceptibility to Mycobacterial Infections. PLoS ONE 11(1): e0146585. doi:10.1371/journal.pone.0146585

We hypothesized that a systems biology approach would help reveal critical human pathways involved in mycobacterial susceptibility, and help elucidate why some individuals progress to active disease while most do not.

Accumulating evidence suggests that host genetic factors influence the susceptibility to MTB and NTM infection. Research studies utilizing twin design [10,11], linkage analysis [12,13], candidate gene association [14–17], genome-wide association analysis [18–21], and fine mapping studies [22] have implicated numerous human genetic markers as contributing factors to the susceptibility of MTB infection. Although fewer studies have examined the human genetic contribution to NTM susceptibility, familial clustering of pulmonary NTM [23] and candidate gene association studies have implicated certain genetic factors [24–26] and support the hypothesis of a genetic predisposition to NTM in some individuals.

In this study, we examine genes critical to the human response to TB and NTM infection, as well as highlight enriched biological pathways and networks that may play a critical role in mycobacterial susceptibility. We explore whether there is any commonality between TB and NTM susceptibility genes and the functional implications of these shared genes and pathways. Shared susceptibility genes or pathways may suggest related mechanisms for the response or control of TB and NTM disease. Furthermore, we examined the resulting networks in order to identify drugs and nutrients with potential immunomodulatory or anti-mycobacterial effects.

MATERIALS AND METHODS

Data Sources & Gene Selection

We identified genes associated with TB and NTM utilizing three publically available databases: the Online Mendelian Inheritance in Man (OMIM) [27, 28] database, the Comparative Toxicogenomics Database (CTD) [29], and the Human Genome Epidemiology encyclopedia (HuGE Navigator) [30]. The Online Mendelian Inheritance in Man (OMIM) database [27] is considered to be the best curated resource of genotype-phenotype relationships [28]. The Comparative Toxicogenomics Database (CTD) [29] curates relationships between chemicals, genes and human diseases, and is unique because it integrates chemical and gene/protein-disease relationships with the goal of understanding the effects of environmental chemicals on human health. In CTD, disease-gene associations are reported as curated or inferred. We selected only curated associations due to a higher confidence than inferred associations. Lastly, we used the Human Genome Epidemiology encyclopedia (HuGE Navigator) [30] which mines the scientific literature on human gene-disease associations and maintains a comprehensive database of population-based epidemiologic studies of human genes [31]. We selected these databases because of their unique approach, breadth, and depth to cataloguing human disease-gene associations.

TB Key Word Search

- For searches of OMIM, we used the key words: "*Mycobacterium tuberculosis*, susceptibility to" which resulted in 11 genes associated with *Mycobacterium tuberculosis* susceptibility or protection.
- For searches of CTD, we used the key words "*Mycobacterium tuberculosis*, susceptibility to infection by", which resulted in 15 genes associated with*Mycobacterium tuberculosis* susceptibility.
- For searches of the HuGE Navigator, we searched using key words "mycobacterium infections". Tuberculosis was defined by disease phenotypes, such as, "Tuberculosis, Gastrointestinal", "Tuberculosis, Pleural", or "Tuberculosis, Pulmonary". We therefore chose to use "Tuberculosis, Pulmonary" as our search term, since our focus for TB and NTM disease in this study is related to lung disease. We excluded genes that have been implicated in hepatotoxicity and other adverse reactions, rather than susceptibility to infection. None of these excluded genes were later identified in the network analyses. In the HuGE database, we found 154 genes that were associated with pulmonary tuberculosis. We further refined this list and selected only genes with at least 3 references. This restricted our list to 42 genes that were associated with pulmonary tuberculosis. Three of the excluded genes were identified from other databases and thus included in the overall gene list.

NTM Key Word Search

- For searches of OMIM, we searched using the key words "mycobacterium infection, nontuberculous, familial". Three separate phenotypes, "Atypical mycobacteriosis, familial", Atypical mycobacteriosis, familial, x-linked", "Atypical mycobacteriosis, familial, x-linked 2", were associated with 7 genes.
- For searches of CTD, we searched using key words "mycobacterium infections, nontuberculous", identifying 6 genes.
- For searches of the HuGE Navigator, we searched using key words "mycobacterium infections", and then selected "Mycobacterium avium-intracellulare infection" identifying 10 genes and "Mycobacterium Infections, Atypical" identifying 10 genes.

From each of the three databases, we compiled lists of genes that were associated with susceptibility to TB and NTM infection. All subsequent analyses included 50 TB-associated genes and 15 NTM-associated genes (Fig 1). In the overall TB list, ten genes were identified in more than 1 database, and seven genes in the overall NTM list were identified in more than 1 database.

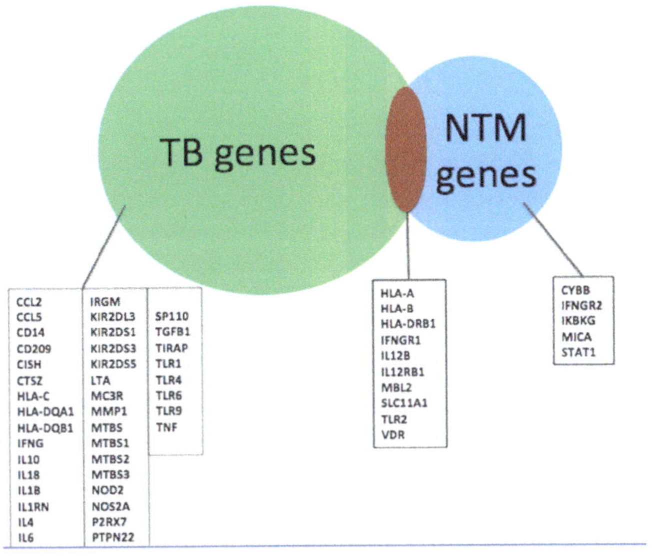

Figure 1. Database-derived TB and NTM-associated genes.

Pathway and Network Analysis

We utilized the Ingenuity Pathway Analysis (IPA), a web-based software application (Ingenuity Systems, www.ingenuity.com), to investigate gene-gene relationships within the context of networks and pathway enrichment. IPA uses a database of human, mouse, and rat genes/proteins, as well as other biological and chemical targets of interest, to find interactions between genes, proteins, related networks, functions/diseases, and canonical pathways. IPA uses a manually curated database, mined from over 300 top journals using the full text, from over 3,600 journals using abstracts, as well as interaction data from third party databases such as GNF, IntAct, BIND, DIP, MINT, MIPS, BIOGRID, COGNIA, GNF, GO, Entrez Gene, OMIM, RefSeq, ClinVar, COSMIC, GWAS Database, HMDB, clinicaltrials.gov, TarBase, TargetScan, miRecords, DrugBank, HSDB, CCRIS, (www.ingenuity.com), [32]. IPA's data analysis allows for understanding the significance of genes or gene products of interest within a larger biological system.

We generated TB and NTM-associated gene networks with the IPA Core Analysis function using 50 TB focus genes and 15 NTM focus genes (Fig 1). Focus genes are genes that were identified in our initial OMIM, CTD, and HuGE database queries. Focus genes represented in networks must interact with at least one other gene in the database. Interactions between genes are represented as nodes connected by edges during network generation. Non-focus genes are genes that were not in our initial database derived search, but are connected in the resulting networks. We also refer to these as "network-suggested genes". We trimmed the resulting gene networks such that nodes with four or fewer edges were removed to facilitate visualization of the most highly connected nodes and networks.

Gene-Chemical Network

Our list of input genes for the gene-chemical network analysis was obtained by combining genes from the union of networks depicted in Figs 2 & 3. We defined our threshold for gene inclusion as the top 50% of all connections between genes. This corresponded to 9 or more edges per node in the union network 1, and 5 or more edges per node in the union network 2. The complete list of input genes included: *BCR (complex)*, *CCL2*, *CCL5*, *IKBKG*, *IL10*,*IL12 (complex)*, *IL1B*, *IL1RN*, *IL4*, *IL6*, *Immunoglobulin*, *MMP1*, *NFkB (complex)*, *TLR1*,*TLR2*, *TLR 4*, *TLR 9*, *TNF*, *Vegf*. The final list of genes corresponds to only those that mapped in CTD (this excluded: BCR, IL12, Immunoglobulin, and TNF). The chemical-gene network was constructed using data from CTD [29] and DrugBank [33]. Chemicals listed in CTD were excluded if they did not have a synonymous drug name in DrugBank. Gene-gene interactions and gene-chemical interactions were extracted from CTD based on our gene list. In this network analysis, gene-gene interactions represent a variety of interaction or relational types, including genes encoding proteins that physically interact or participate in sequential steps of a biochemical pathway. This analysis builds upon our previous work on disease-gene-chemical networks [34].

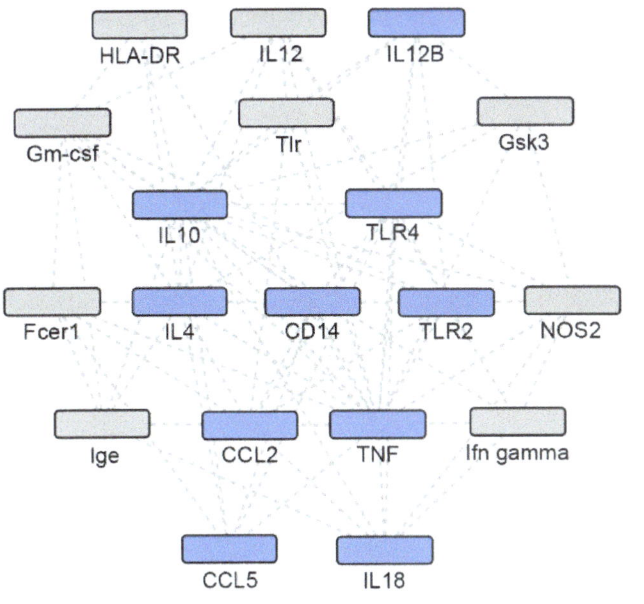

Figure 2. Union of TB and NTM gene sets: network 1. Grey: Network identified genes; Blue: TB or NTM focus genes.

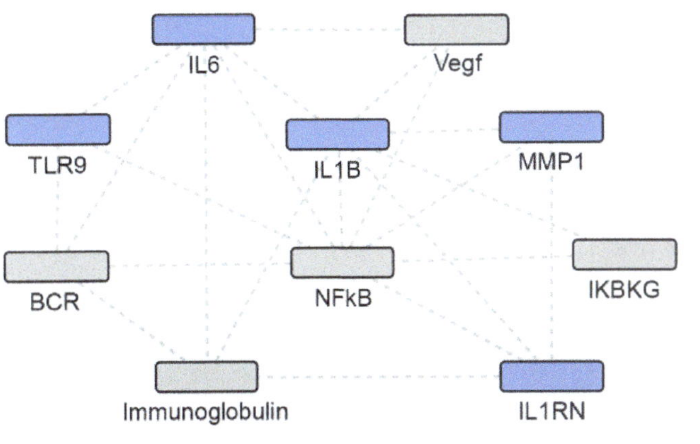

Figure 3. Union of TB and NTM gene sets: network 2. Grey: Network identified genes; Blue: TB or NTM focus genes.

Chemical exclusion criteria were based upon literature evaluation in that any intervention deemed potentially harmful to a patient would not be considered in these networks. Therefore, we excluded: known toxins (ie. alcohols and arsenic), drugs with obvious negative side effects (ie. anticonvulsants, warfarin, opiods/opiates, mood stabilizers, narcotics), drugs with known negative side effects (ie. NSAIDs, immune suppressing steroids/hormones), non-concise chemicals that are generally seen as harmful (ie. dust, CTD definition: "Earth or other dry matter in fine, dry particles"), anti-retrovirals due to frequent HIV-TB co-infection, their use in HIV treatment and concern for development of drug resistance, as well as antibiotics that are effective against the bacteria but detrimental to the patient. The remaining chemicals included: statins/fibrates, select group of antibiotics, and vitamins. Macrolides and Gemfibrozil were grouped together because of their potential to positively alter the human immunological response to infection as well as both being pharmaceutical drugs. Gene networks and gene-chemical networks were recreated using Cytoscape [35].

RESULTS AND DISCUSSION

Many studies using conventional genetic methods have identified single genes/loci influencing the susceptibility to TB or NTM infection and control of disease [12–22,24–26]. Despite these reports, little has been done to examine these reported loci within the context of biological networks. Since the progression to active TB and NTM disease are influenced by multiple genetic factors, genes identified through single gene approaches, in and of themselves, may provide an incomplete etiological story [36,37]. As a result, we undertook a comprehensive network and pathway analysis of human genes associated with TB and NTM infection susceptibility. Using this approach we identify new candidate genes, as well as investigate the overlap of genes and pathways involved in both TB and NTM susceptibility and control.

Based on our database search, we identified 50 TB-associated focus genes and 15 NTM-associated focus genes (Fig 1). We generated 10 TB-associated networks, 1 NTM-associated network, and 1 network using the intersection of TB- and NTM-associated focus genes. Five networks were generated using the union of TB- and NTM-associated genes (Tables 1–4). In the following text, we discuss the most statistically significant networks from each gene set.

Table 1. Network associated with TB.

Molecules in Network	Score*	Focus Molecules	Top Diseases and Functions
BCR (complex), caspase, **CISH**, Fcger3, **HLA-A, HLA-B, HLA-C, HLA-DQB1, HLA-DRB1,** IFN Beta, **IFNG**, Iga, Ikb, IL12 (family), **IL12RB1, IL1B, IL1RN,** Immunoglobulin, Interferon alpha, KIR, **KIR2DL1/KIR2DL3, KIR2DS4, LTA, MBL2, MMP1,** NFkB (complex), **P2RX7, PTPN22,** Ras, SAA, TCR, **TIRAP, TLR1, TLR9**, Vegf	42	20	Cell-To-Cell Signaling and Interaction, Immunological Disease, Connective Tissue Disorders

Bold: TB focus genes.

*Score: -log(Fisher's Exact p-value)

doi:10.1371/journal.pone.0146585.t001

Table 2. Network associated with NTM.

Molecules in Network	Score*	Focus Molecules	Top Diseases and Functions
CLEC11A, **CYBB**, DUOX2, ERK1/2, FXN, GRM1, Gsk3, **HLA-A, HLA-B, HLA-DRB1,** Ifn gamma, **IFNGR1, IFNGR2, IKBKG, IL12B, IL12RB1,** IL36A, Immunoglobulin, IRAK, **MBL2, MICA,** NEU1, NFkB (complex), NLRP2, P38 MAPK, PASK, SH3GLB2, **STAT1,** SUMO4, TCR, TFG, **TLR2,** TLR10, USP21, **VDR**	36	14	Cancer, Cell Death and Survival, Cellular Compromise

Bold: NTM focus genes.

*Score: -log(Fisher's Exact p-value)

doi:10.1371/journal.pone.0146585.t002

Table 3. Network associated with the intersection of TB and NTM genes.

Molecules in Network	Score*	Focus Molecules	Top Diseases and Functions
CCL8, CCL26, CD6, CLEC11A, ERK1/2, FXN, **HLA-A, HLA-B, HLA-DRB1, IFNGR1**, IL25, IL10RA, **IL12B, IL12RB1,** IL36A, Immunoglobulin, IRAK, KIR, LILRB1, MAP3K10, **MBL2,** MED28, NEU1, NFkB (complex), NOD1, P2RY6, P38 MAPK, PTPN12, SH3GLB2, SUMO4, **TLR2,** TLR10, TNIP3, UNC93B1, **VDR**	22	9	Inflammatory Response, Hematological System Development and Function, Tissue Morphology

Bold: TB and NTM focus genes.

*Score: -log(Fisher's Exact p-value)

doi:10.1371/journal.pone.0146585.t003

Table 4. Networks associated with the union of TB and NTM genes.

Molecules in Network	Score*	Focus Molecules	Top Diseases and Functions
CCL2, CCL5 CD14, CTSZ, CYBB, Fcer1, Fcgr2, Gm-csf, Gsk3, HLA-DQ, HLA-DR, Ifn gamma, **IFNGR1, IFNGR2,** Ige, IgG1, **IL4, IL10, IL18,** IL23, IL12 (complex), **IL12B,** lymphotoxin-alpha1-beta2, MHC Class II (complex), **NOD2, NOS2,** Nr1h, SCAVENGER receptor CLASS A, Sod, Tlr, **TLR2, TLR4, TLR6, TNF,** U1 snRNP	33	17	Cellular Function and Maintenance, Hematological System Development and Function, Infectious Disease
26s Proteasome, BCR (complex), caspase, **CD209,** Cdk, Collagen type 1, Eotaxin, Fcgr3, **HLA-DQB1,** Hsp27, Iga, Ikb, **IKBKG,** IL1, **IL6,** IL12 (family), **IL1B, IL1RN,** Immunoglobulin, JK, **LTA, MBL2, MICA, MMP1,** N-cor, NFkB(complex), **P2RX7, PTPN22,** Ras homolog, SAA, **TIRAP, TLR1, TLR9, VDR,** Vegf	33	16	Cell-To-Cell Signaling and Interaction, Hematological System Development and Function, Immunological Disease

Bold: TB and NTM focus genes.
*Score: -log(Fisher's Exact p-value)

doi:10.1371/journal.pone.0146585.t004

Gene Network Analyses

Using the IPA enrichment algorithm, five networks were generated using the union of TB and NTM genes. The most statistically significant network ($p = 10^{-33}$) included 17 focus genes implicated in either infection (Table 4, Fig 2). The second network ($p = 10^{-30}$) included 16 focus genes in a 35 gene network (Table 4, Fig 3).

The top TB network was highly significant ($p = 10^{-42}$) and included 20 focus genes implicated in susceptibility to *Mycobacteria tuberculosis* infection in a 35 gene network (Table 1, Fig 4).

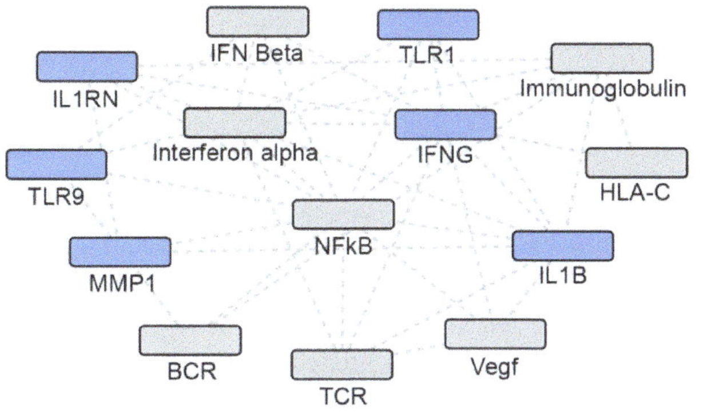

Figure 4. TB network. Grey: Network identified genes; Blue: TB focus genes.

The top NTM network was also highly significant ($p = 10^{-36}$). This 35 gene network included 14 focus genes, which have been previously implicated in susceptibility to NTM infection (Table 2, Fig 5).

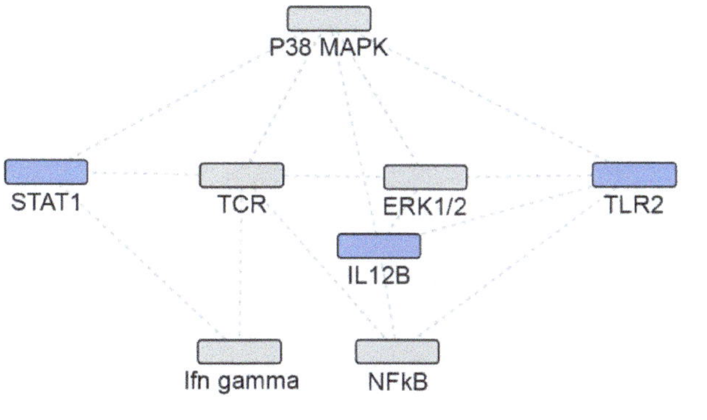

Figure 5. NTM network. Grey: Network identified genes; Blue: NTM focus genes.

Using the intersection of TB- and NTM-associated genes, one network was generated ($p = 10^{-22}$). From the 10 focus genes common to both infections, 9 genes were implicated in this 35 gene network (Table 3, Fig 6).

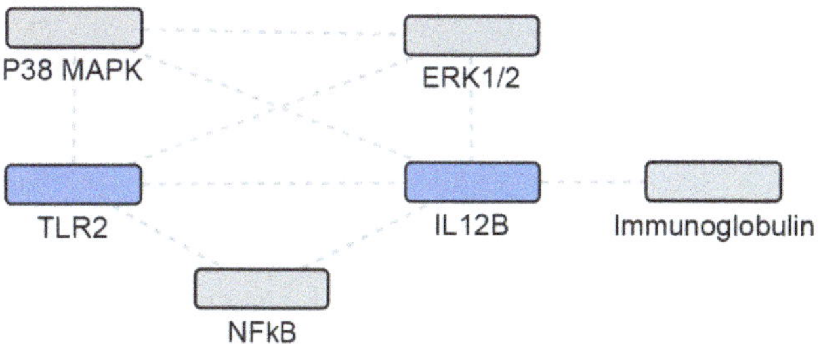

Figure 6. Intersection of TB and NTM gene sets. Grey: Network identified genes; Blue: TB and NTM focus genes.

Discussion of Network-Central Genes

Networks Using the Union of TB and NTM Gene Sets

The two most statistically significant networks demonstrated that NTM-associated focus genes were the most peripheral, while the TB-associated focus genes and the focus genes common to both TB and NTM infection were the more central and connected nodes in the network. When comparing the two networks (Figs 2 and 3), different genes appear to be most central and interconnected. In network 1 (Fig 2), the most central and highly connected genes are the TB- associated focus genes (blue), *IL4*, *IL10*, *TNF*, *CCL2*, *CCL5*, and *TLR4* as well as the focus genes common to both TB and NTM (blue) *IL12B*, and *TLR2*. While in network 2 (Fig 3), different TB- associated focus genes (blue), *IL6* and *IL1B*, are the most central and highly connected, while genes common to both TB and NTM infection are not central or highly connected nodes. Among the genes putatively associated with TB or NTM identified via network analysis, *Gm-csf* (Fig 2) and *NFkB* complex (Fig 3) are the most central and highly connected nodes. Overall, NTM-associated genes appear the most peripheral, while the TB-associated focus genes seem to play a more central role in these networks.

Focus Genes

TLR2: *TLR2* stands out as being one of the most central and highly interconnected nodes in all the networks. Toll Like Receptors (TLRs) can recognize molecular patterns of MTB and initiate signaling pathways to activate the innate and adaptive immune responses [38]. The major receptors for MTB and those most critical in the recognition process are TLR-1, -2, and -6. Because of the TLR's essential role in MTB recognition, many studies have examined associations of TLR polymorphisms and TB, however results have been inconsistent. In a recent meta-analysis [38], authors reported a positive association between a *TLR2* polymorphism and TB risk, especially among Asians and Europeans. Yim et al. demonstrated that a GT repeat microsatellite polymorphism in intron II of *TLR2* contributes to the development of NTM lung disease, especially for *M. avium-intracellulare* complex (MAC) infection [39], although another study from Korea did not find an association between TLR2 and NTM susceptibility [40]. Supported by previous studies, our network analyses highlight the important role that *TLR2* may play in the immune response to both TB and NTM infection.

IL12B: The *IL12B* gene encodes the p40 subunit of *IL-12* and *IL-23* cytokines, which play important roles in bridging the innate and adaptive immune systems. Interleukin 12 (*IL-12*) is a proinflammatory cytokine and acts as a key regulator in determining the T helper 1 or 2 immune response [41]. *IL12B* polymorphisms have been associated with increased susceptibility to TB [42], as well as to *Mycobacterium leprae* [43], but findings have been inconsistent [44,45]. *IL12B* has also been implicated in the development of autoimmune disease. A recent meta-analysis demonstrated a significant association between *IL12B*polymorphisms and risk of psoriasis [46]. *IL12B* has also been implicated [47] as a susceptibility gene for leprosy and inflammatory bowel disease. This may imply a shared genetic susceptibility to inflammation and infectious disease.

IFNG: Interferon-gamma (*IFNG*) binds to its own receptor made of transmembrane proteins,*IFNGR1* and *IFNGR2*, to induce antimicrobial mechanisms and upregulates antigen processing and presentation pathways [48]. Consequently, *IFNG* is central to innate and adaptive cell-mediated immunity against intracellular pathogens and thus crucial to controlling MTB replication. Studies have shown that low production of *IFNG* has been associated with active TB [49–51]. A recent meta-analysis reported a statistically significant protective association of an *IFNG* polymorphism and pulmonary/extrapulmonary TB in different ethnicities [52]. Polymorphisms in this gene may be good genetic markers for TB resistance.

Network-Suggested Genes

NFkB complex: Among the putatively associated genes identified via network analysis, the nuclear factor-kB (*NFkB*) complex stands out as one of the most highly central and interconnected nodes in all our networks. The *NFkB* family of transcription factors plays a key role in the regulation of genes involved in immune and inflammatory responses [53,54]. A wide range of stimuli, including viral and bacterial products, can lead to the activation of*NFkB*. It regulates gene expression in many different cell types during development in response to injury and infection, thus it is often referred to as a central mediator of the immune response [53–55]. Research has demonstrated that activation of *NFkB* enhances immunity against certain microbial pathogens [56,57], however some *NFkB* pathways can be exploited to promote pathogen survival [56]. A recent study demonstrated that inhibiting*NFkB* reduced MTB survival in human macrophages [58], emphasizing the complex role of*NFkB* in MTB infection. While the human immune response against NTM infection is not well understood, many studies have shown that the role of the *NFkB* is a crucial step in the regulation of genes involved in the killing of NTM. One study [59] focused on the role of*NFkB* in the innate immune response of macrophages. The authors reported that *M.smegmatis*, a nonpathogenic mycobacteria, induces *NFkB* activation and is killed by macrophages, while *M. avium*, a pathogenic mycobacteria, represses *NFkB* activation and survives within macrophages. Additional cell, mouse and bovine studies [60–63] have also examined the influence of *NFkB* on the response to infection. These studies demonstrated that *NFkB* is involved in the initiation of a proinflammatory cytokine response in the macrophage and is rapidly activated by the interaction of host cell and bacterium [60,63]. However, the mechanisms and sequence of events underlying *NFkB* activation and cytokine response to NTM infection remains unclear.

ERK1/2 & p38 MAPK: In response to invading pathogens, innate immune cells, such as macrophages/monocytes and dendritic cells, use pattern recognition receptors (PRRs) to recognize bacteria, which in turn, leads to the activation of signaling pathways, such as the mitogen-activated protein (MAP) kinase pathway [64–66]. The NTM gene network (Fig 5) and the network using the intersection of TB and NTM gene sets (Fig 6) identify *ERK1/2* and*p38 MAPK*, both members of the MAP kinases, as highly central and interconnected nodes. Depending on the stimulus, activation of *ERK1/2* together with *p38* can have contrasting functions. A greater amount of *ERK* activity relative to *p38* can promote cell proliferation and survival, whereas a greater amount of *p38* activity relative to *ERK* can trigger cell death and apoptosis [67,68]. In an *in vitro* study using primary human monocytes, authors reported that *Mycobacterium avium-intracellulare* (MAI) infection activates both *ERK* and *p38*. While*ERK* appears to regulate pathogenic MAI replication in human monocytes, *p38* influences cytokine release more than *ERK* does [69].

VEGF: Vascular endothelial growth factor (*VEGF*) is a key regulator of normal angiogenesis. It promotes endothelial cell survival, growth and migration, in addition to being implicated in pathological angiogenesis associated with tumor growth [70]. *VEGF* has been also studied as a prognostic biomarker for infectious diseases but studies have yielded inconsistent results. In sepsis infection, some studies found higher *VEGF* levels in survivors compared with non-survivors, while other studies found the opposite [71,72]. Some studies examining dengue infection have reported significantly higher *VEGF* levels in patients with dengue hemorrhagic fever compared with dengue fever [73,74], other studies have reported a lack of association between *VEGF* and illness severity [75]. Similarly, several TB studies have reported elevated plasma VEGF levels in active TB [76–78]. And in a separate TB study, authors found that plasma VEGF concentrations were significantly reduced upon TB treatment and could potentially represent a surrogate marker to monitor sputum culture conversion [79].

Pathway Enrichment Analysis

Canonical Pathway Analysis using IPA

For TB, the top two canonical pathways most significantly associated with this gene set were "altered T cell and B cell signaling in rheumatoid arthritis" and "communication between innate and adaptive immune cells" (Table 5). For NTM, the top two pathways most significantly associated with this gene set were "type 1 diabetes mellitus signaling" and "T Helper cell differentiation" (Table 6). Examining the intersection of the TB and NTM gene sets, pathway analysis revealed that the top two canonical pathways were "communication between Innate and Adaptive Immune Cells" and "type 1 diabetes mellitus signaling" (Table 7). Examining the union of TB and NTM gene sets, pathway analysis revealed that the top two canonical pathways were "altered T cell and B cell signaling in rheumatoid arthritis" and "communication between innate and adaptive immune cells" (Table 8).

Table 5. Top canonical pathways associated with TB gene set.

Pathways	p-value	Ratio*	Molecules in pathway
Altered T Cell and B Cell Signaling in Rheumatoid Arthritis	1.14E-33	19/81 (0.235)	HLA-DQA1, HLA-DQB1, HLA-DRB1, IFNG, IL4, IL6, IL10, IL18, IL12B, IL1B, IL1RN, LTA, TGFB1, TLR1, TLR2, TLR4, TLR6, TLR9, TNF
Communication between Innate and Adaptive Immune Cells	1.48E-33	19/82 (0.232)	CCL5, HLA-A, HLA-B, HLA-C, HLA-DRB1, IFNG, IL4, IL6, IL10, IL18, IL12B, IL1B, IL1RN, TLR1, TLR2, TLR4, TLR6, TLR9, TNF

*Ratio: number of genes from our dataset that map to a canonical pathway divided by the total number of genes in that pathway.

doi:10.1371/journal.pone.0146585.t005

Table 6. Top canonical pathways associated with NTM gene set.

Pathways	p-value	Ratio*	Molecules in pathway
Type 1 Diabetes Mellitus Signaling	9.08E-15	8/106 (0.075)	HLA-A, HLA-B, HLA-DRB1, IFNGR1, IFNGR2, IKBKG, IL12B, STAT1
T Helper Cell Differentiation	1.27E-11	6/67 (0.09)	HLA-DRB1, IFNGR1, IFNGR2, IL12B, IL12RB1, STAT1

*Ratio: number of genes from our dataset that map to a canonical pathway divided by the total number of genes in that pathway.

doi:10.1371/journal.pone.0146585.t006

Table 7. Top canonical pathways associated with the intersection of TB and NTM gene sets.

Pathways	p-value	Ratio*	Molecules in pathway
Communication between Innate and Adaptive Immune Cells	5.11E-10	5/82 (0.047)	HLA-A, HLA-B, HLA-DRB1, IFNGR1, IL12B,
Type 1 Diabetes Mellitus Signaling	1.89E-09	5/106 (0.047)	HLA-A, HLA-B, HLA-DRB1, IFNGR1, IL12B,

*Ratio: number of genes from our dataset that map to a canonical pathway divided by the total number of genes in that pathway.

doi:10.1371/journal.pone.0146585.t007

Table 8. Top canonical pathways associated with the union of TB and NTM gene sets.

Pathways	p-value	Ratio*	Molecules in pathway
Altered T Cell and B Cell Signaling in Rheumatoid Arthritis	6.25E-33	19/81 (0.235)	HLA-DQA1, HLA-DQB1, HLA-DRB1, IFNG, IL4, IL6, IL10, IL18, IL12B, IL1B, IL1RN, LTA, TGFB1, TLR1, TLR2, TLR4, TLR6, TLR9, TNF
Communication between Innate and Adaptive Immune Cells	8.13E-33	19/82 (0.232)	CCL5, HLA-A, HLA-B, HLA-C, HLA-DRB1, IFNG, IL4, IL6, IL10, IL18, IL12B, IL1B, IL1RN, TLR1, TLR2, TLR4, TLR6, TLR9, TNF

*Ratio: number of genes from our dataset that map to a canonical pathway divided by the total number of genes in that pathway.

doi:10.1371/journal.pone.0146585.t008

Signaling pathways for autoimmune disease and communication or differentiation of immune cells are overrepresented in our pathway enrichment analyses. This may indicate similar susceptibility or pathogenic mechanisms for TB and NTM infections.

The top canonical pathway for the TB gene set and the union of TB and NTM gene sets includes "altered T cell and B cell signaling in rheumatoid arthritis" (RA). Our analysis has identified a signaling pathway specific to RA, which has been confirmed in the literature. The relationship between mycobacterial infection and autoimmune disease, in particular, Inflammatory Bowel Disease (IBD) and RA, has been reported previously [80–84]. Studies have shown that persons with RA are at increased risk for TB and NTM disease, independent of immunosuppressive medications used in RA treatment [85,86]. The link between specific genes, for example SLC11A1 (NRAMP1) and susceptibility to autoimmune disease and infectious disease has been widely explored. Shaw et al (1996) demonstrated genetic linkage of NRAMP1 and RA. Searle & Blackwell (1999) found that high expression of NRAMP1 allele 3 contributes to autoimmune susceptibility, specifically to RA [80]. This association was also found in diabetes patients with a first or second degree relative with RA [87]. Conversely, low expression of NRAMP1 allele 2 was found to contribute to infectious disease, specifically tuberculosis [88,89]. Interestingly, Mobley (2004) demonstrated similarity in the epidemiology of RA and the epidemiology of tuberculosis deaths from 2 centuries ago [90]. The author suggested the possibility that genetic factors influencing tuberculosis survival may now be influencing susceptibility to the development of RA. Other genetic associations have reported between mycobacterial infection and chronic inflammation. IL12B, for example, has been identified as a susceptibility gene for leprosy, psoriasis and IBD [46,47]. Similarly, the IRGM autophagy gene, a strong mediator of inflammation, has also been associated with an increased risk to leprosy [91], Crohn's disease and IBD [92,93].

"Type 1 diabetes mellitus signaling" was also ranked among the top canonical pathways for both the NTM gene set and the intersection of TB and NTM gene sets. Diabetes mellitus (DM) has been associated with increased risk of progression to TB, increased TB severity and with poor TB treatment outcomes [94,95]. While some studies hypothesize that DM impairs the immune responses necessary to control mycobacterial replication [96], the exact mechanisms by which DM increases TB risk have not yet been elucidated.

Network-Identified Immune-Modulating Chemicals

Pro-immune altering chemicals remain an understudied and underutilized area of disease treatment. Within our gene-chemical network, there were three macrolides (Azithromycin, Clarithromycin, and Roxithromycin) and a fibrate (Gemfibrozil) that are known to have immune altering effects (Fig 7). Although macrolides play a central role in the treatment of NTM infection [97,98], they are not commonly used to treat TB. There is conflicting evidence regarding the implications of potential macrolide induced alterations in macrophage activation [99–101] and immune cell composition [102] during the course of infection in humans. In mice models, it has been hypothesized that azithromycin may contribute to NTM infection by decreasing autophagy [103], despite evidence showing azithromycin successfully treats NTM infections in mice with chronic pulmonary infection [104]. In humans with COPD, azithromycin increases phagocytosis [105,106], which suggests a beneficial effect on macrophages-mediated degradation of microorganisms. It has been shown that TB can inhibit phagosome maturation by using mannose receptors to mediate transport to phagosomes with limited fusion capabilities [107], the same receptors that are greatly increased by azithromycin treatment [106]. However, given that mycobacteria often reside in inactivated macrophages [108] and that azithromycin promotes altered activation of macrophages [99], this altered immune response may be beneficial to treating TB [34], even though direct macrolide effects are inconsistently seen *in vitro* [109–114]. Gemfibrozil represents a novel drug for both NTM and TB treatment. This drug inhibited 27 strains of MTB grown in macrophages by decreasing their ability to acquire fatty acids from macrophages [115], leading to novel mechanisms for aiding the host in combating active TB.

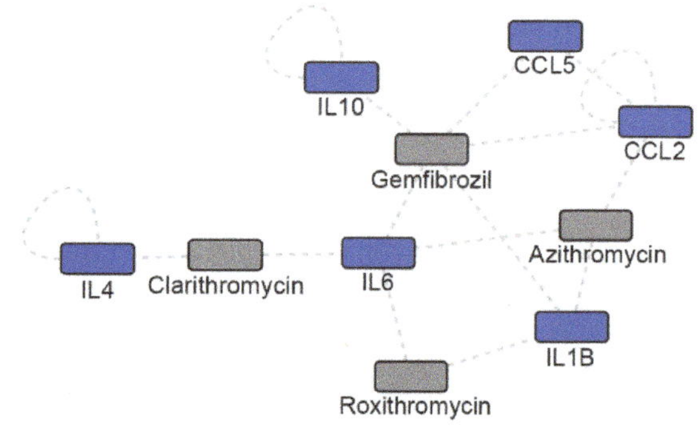

Figure 7. Network-identified immune-modulating chemicals. Grey: Network identified chemicals; Blue: Network identified genes.

Network-Identified Nutrients

Nutrients play an important role in both basic immune health and pathogen virulence. There is a correlation between malnutrition and active TB; malnutrition increases the likelihood of progressing to TB and TB causes a decrease in available nutrients [116]. There are guidelines for providing nutrient supplementation in malnourished individuals and pregnant women; however, there are no guidelines for additional supplementation in individuals receiving adequate food intake [116]. The overall chemical-gene network contains a variety of vitamins (A, B, C, D, E, K, lycopene), suggesting the importance of micronutrients in both immune regulation and mycobacterial infection (Fig 8). While there is clear support for nutrient supplementation in malnourished individuals, particularly in treating malaria [117], there is some evidence that nutrient modulation in healthy individuals may improve response to infection. Vitamin A mediates anti-microbial activity, which may be beneficial in MTB treatment [118]. MTB is highly sensitive to vitamin C induced Fenton reaction [119]. This reaction generates hydroxyl radicals, chemicals that promote eradication of actively growing MTB [120]. Vitamin D supplementation improved treatment response among patients with specific vitamin D receptor mutations who were infected with MTB [121]. Vitamin D deficiency has also been associated with susceptibility to NTM infection [122]. However, based on the evidence from randomized controlled trials, vitamin D supplementation has not shown benefit in the general population of TB patients [121,123]. Further, a Cochrane Review concludes that "there is currently no reliable evidence that routinely supplementing at or above recommended daily amounts has clinical benefits" [124].

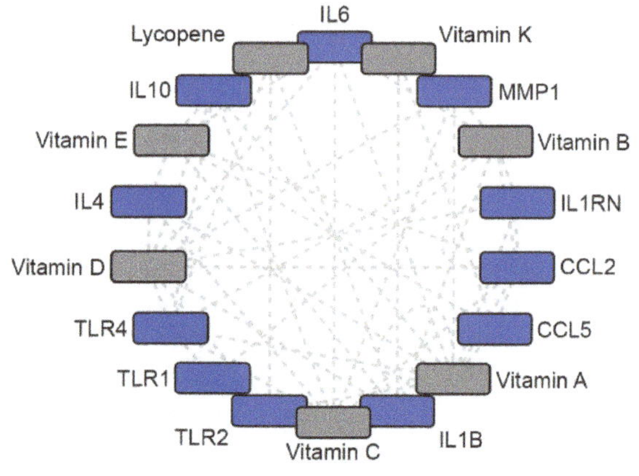

Figure 8. Network-identified nutrients. Grey: Network identified vitamins; Blue: Network identified genes.

Just as nutrients can help augment host response, nutrients also represent key chemicals for regulating the pathogen. As iron is a necessary molecule for NTM and TB survival and virulence [125], the human host is the main source of iron acquisition during infection [125,126]. The heme acquisition system offers a possible target for drugs [126] by limiting mycobacterial nutrient uptake and preserving these essential nutrients for the host. The folate pathway represents another potential target for therapeutics against mycobacterium [127]. Folate cofactors are an essential carbon donor used for amino acid and DNA synthesis [127] and inhibition can possibly disrupt bacterial growth and replication.

CONCLUSION

Our findings suggest that the genetic contribution to MTB and NTM infection operates through similar genes and pathways, providing insight into the underlying pathogenesis and human immune response to mycobacterial disease. Genes involved in bridging the innate and adaptive immune responses are central to both TB and NTM infection susceptibility and control. Genes in these processes are essential to host protection, thus providing an important basis for further research. TB and NTM disease gene sets were also overrepresented in autoimmune signaling pathways. This implies that overlapping physiological mechanisms and pathways influence susceptibility to mycobacterial infection and the development of autoimmune disease, suggesting another avenue for future research.

We identified drugs and nutrients via network analysis that interact with genes of interest, and suggest potential therapeutics with immune modulating effects. Our network findings suggest that three well-known macrolides and a fibrate may target our genes of interest and may boost human immune response to infection. While research examining the effectiveness of these treatments on mycobacterial disease is inconclusive, our findings suggest that further research may be warranted to expand the repertoire of treatment options for mycobacterial disease. Lastly, we identified nutrients that target our genes of interest, which may improve response to infection even in healthy individuals. If vitamins A, B, C, D, E, K, and lycopene positively impact response to infection, then nutrient supplementation may be a simple intervention to reduce the incidence and prevalence of mycobacterial disease worldwide.

Although literature dependent network analyses are often limited by an incomplete knowledge base, by relying on what has been published in the literature, network inferences can suggest important pathways involved in disease and lead to novel hypotheses. In our study, for example, our TB input gene list was larger compared with the NTM gene list, as a result of TB being a more extensively studied disease. The fact that a gene is associated with risk of TB but not NTM may either be due to a true functional difference, or alternatively, may result as a byproduct of a smaller pool of published NTM literature. Nonetheless, our network analysis examines two important categories of mycobacterial disease, TB and NTM, and the resulting networks provide a visually intuitive and statistically sound methodology for data interpretation and examination. The resulting network and pathway analysis offers a powerful and complementary approach to other methods, to help identify underlying mechanisms and pathways involved in complex diseases where multiple genes and gene products interact.

In summary, this analysis explores the connectivity between TB and NTM-associated genes, susceptibility to infection, and possible therapeutics. It also provides a foundation for further examination of these target genes among infected and uninfected individuals.

ACKNOWLEDGMENTS

We thank Dr. Nicholas Walter and Dr. Lynn Petukhova for thorough reviews and feedback.

AUTHOR CONTRIBUTIONS

Conceived and designed the experiments: EML MS. Performed the experiments: EML BJG. Analyzed the data: EML BJG. Contributed reagents/materials/analysis tools: EML BJG. Wrote the paper: EML BJG MS.

REFERENCES

1. Donoghue HD, Spigelman M, Greenblatt CL, Lev-Maor G, Bar-Gal GK, Matheson C, et al. Tuberculosis: from prehistory to Robert Koch, as revealed by ancient DNA. Lancet Infect Dis. 2004 Sep; 4(9): 584–92. pmid: 15336226 doi: 10. 1016/s1473-3099(04)01133-8
2. Zink AR, Grabner W, Reischl U, Wolf H, Nerlich AG. Molecular study on human tuberculosis in three geographically distinct and time delineated populations from ancient Egypt. Epidemiol Infect 2003 Mar; 130(2): 239–49. pmid: 12729192 doi: 10. 1017/ s0950268802008257
3. World Health Organization. Global tuberculosis report 2014. World Health Organization; 2014.
4. Falkinham JO. Surrounded by mycobacteria: nontuberculous mycobacteria in the human environment. J Appl Microbiol. 2009 Aug; 107(2): 356–67. doi: 10. 1111/j. 1365-2672. 2009. 04161. x. pmid: 19228258
5. Kennedy TP, Weber DJ. Nontuberculous mycobacteria. An underappreciated cause of geriatric lung disease. Am J Respir Crit Care Med. 1994 Jun; 149(6): 1654–8 pmid: 8004325 doi: 10. 1164/ajrccm. 149. 6. 8004325
6. Adjemian J, Olivier KN, Seitz AE, Holland SM, Prevots DR. Prevalence of Nontuberculous Mycobacterial Lung Disease in U. S. Medicare Beneficiaries. Am J Respir Crit Care Med. 2012 Apr 15; 185(8): 881–6. doi: 10. 1164/rccm. 201111-2016OC. pmid: 22312016
7. Prevots DR, Shaw PA, Strickland D, Jackson LA, Raebel MA, Blosky MA, et al. Nontuberculous mycobacterial lung disease prevalence at four integrated health care delivery systems. Am J Respir Crit Care Med 2010 Oct 1; 182(7): 970–6. doi: 10. 1164/rccm. 201002-0310OC. pmid: 20538958
8. Kendall B, Winthrop K. Update on the Epidemiology of Pulmonary Nontuberculous Mycobacterial Infections. Semin Respir Crit Care Med. 2013 Mar 4; 34(01): 087–94. doi: 10. 1055/s-0033-1333567
9. Wagner D, Young LS. Nontuberculous Mycobacterial Infections: A Clinical Review. Infection. 2004. Oct;

32(5): 257–70. pmid: 15624889 doi: 10. 1007/s15010-004-4001-4

10. Simonds B. Twin research in tuberculosis. Eugen Rev. 1957 Apr; 49(1): 25–32. pmid: 21260731
11. Jepson A, Fowler A, Banya W, Singh M, Bennett S, Whittle H, et al. Genetic Regulation of Acquired Immune Responses to Antigens of Mycobacterium tuberculosis: a Study of Twins in West Africa. Infection and Immunity. 2001 Jun 1; 69(6): 3989–94. pmid: 11349068 doi: 10. 1128/iai. 69. 6. 3989-3994. 2001
12. Bellamy R, Beyers N, McAdam KP, Ruwende C, Gie R, Samaai P, et al. Genetic susceptibility to tuberculosis in Africans: A genome-wide scan. Proceedings of the National Academy of Sciences. 2000 Jun 20; 97(14): 8005–9. doi: 10. 1073/pnas. 140201897
13. Cooke GS, Campbell SJ, Bennett S, Lienhardt C, McAdam KP, Sirugo G, et al. Mapping of a Novel Susceptibility Locus Suggests a Role for MC3R and CTSZ in Human Tuberculosis. Am J Respir Crit Care Med. 2008 Jul 15; 178(2): 203–7. doi: 10. 1164/rccm. 200710-1554OC. pmid: 18420963
14. Hoal EG, Lewis L-A, Jamieson SE, Tanzer F, Rossouw M, Victor T, et al. SLC11A1 (NRAMP1) but not SLC11A2 (NRAMP2) polymorphisms are associated with susceptibility to tuberculosis in a high-incidence community in South Africa. Int J Tuberc Lung Dis. 2004 Dec; 8(12): 1464–71 pmid: 15636493
15. Möller M, Nebel A, Valentonyte R, van Helden PD, Schreiber S, Hoal EG. Investigation of chromosome 17 candidate genes in susceptibility to TB in a South African population. Tuberculosis. 2009 Mar; 89(2): 189–94. doi: 10. 1016/j. tube. 2008. 10. 001. pmid: 19147409
16. Rossouw M, Nel HJ, Cooke GS, van Helden PD, Hoal EG. Association between tuberculosis and a polymorphic NFkappaB binding site in the interferon gamma gene. Lancet. 2003 May 31; 361(9372): 1871–2. pmid: 12788577 doi: 10. 1016/s0140-6736(03)13491-5
17. Pan H, Dai Y, Tang S, Wang J. Polymorphisms of NOD2 and the risk of tuberculosis: a validation study in the Chinese population. Int J Immunogenet. 2012 Jun; 39(3): 233–40. doi: 10. 1111/j. 1744-313X. 2011. 01079. x. pmid: 22212192
18. Thye T, Vannberg FO, Wong SH, Owusu-Dabo E, Osei I, Gyapong J, et al. Genome-wide association analyses identifies a susceptibility locus for tuberculosis on chromosome 18q11. 2. Nat Genet. 2010 Sep; 42(9): 739–41. doi: 10. 1038/ng. 639. pmid: 20694014
19. Thye T, Owusu-Dabo E, Vannberg FO, van Crevel R, Curtis J, Sahiratmadja E, et al. Common variants at 11p13 are associated with susceptibility to tuberculosis. Nat Genet. 2012 Mar; 44(3): 257–9. doi: 10. 1038/ng. 1080. pmid: 22306650
20. Png E, Alisjahbana B, Sahiratmadja E, Marzuki S, Nelwan R, Balabanova Y, et al. A genome wide association study of pulmonary tuberculosis susceptibility in Indonesians. BMC Medical Genetics. 2012 Jan 13; 13: 5. doi: 10. 1186/1471-2350-13-5. pmid: 22239941
21. Mahasirimongkol S, Yanai H, Mushiroda T, Promphittayarat W, Wattanapokayakit S, Phromjai J, et al. Genome-wide association studies of tuberculosis in Asians identify distinct at-risk locus for young tuberculosis. J Hum Genet. 2012 Jun; 57(6): 363–7. doi: 10. 1038/jhg. 2012. 35. pmid: 22551897
22. Cervino AC, Lakiss S, Sow O, Bellamy R, Beyers N, Hoal-van Helden E, et al. Fine mapping of a putative tuberculosis-susceptibility locus on chromosome 15q11-13 in African families. Hum Mol Genet. 2002 Jul 1; 11(14): 1599–603. pmid: 12075004 doi: 10. 1093/hmg/11. 14. 1599
23. Colombo RE, Hill SC, Claypool RJ, Holland SM, Olivier KN. Familial clustering of pulmonary nontuberculous mycobacterial disease. Chest. 2010 Mar; 137(3): 629–34. doi: 10. 1378/chest. 09-1173. pmid: 19858235
24. Um S-W, Ki C-S, Kwon OJ, Koh W-J. HLA Antigens and Nontuberculous Mycobacterial Lung Disease in Korean Patients. Lung. 2009 Feb 14; 187(2): 136–40. doi: 10. 1007/s00408-009-9136-8. pmid: 19219503
25. Koh W-J, Kwon OJ, Kim EJ, Lee KS, Ki C-S, Kim JW. NRAMP1 Gene Polymorphism and Susceptibility to Nontuberculous Mycobacterial Lung Diseases. Chest. 2005 Jul 1; 128(1): 94–101. pmid: 16002921 doi: 10. 1378/chest. 128. 1. 94
26. Hwang JH, Kim EJ, Koh W-J, Kim SY, Lee S-H, Suh GY, et al. Polymorphisms of interferon-gamma and interferon-gamma receptor 1 genes and non-tuberculous mycobacterial lung diseases. Tuberculosis. 2007 Mar; 87(2): 166–7. pmid: 17023216 doi: 10. 1016/j. tube. 2006. 07. 003
27. Hamosh A. Online Mendelian Inheritance in Man (OMIM), a knowledgebase of human genes and genetic disorders. Nucleic Acids Research. 2004 Dec 17; 33(Database issue): D514–7. doi: 10. 1093/nar/gki033
28. Amberger JS, Bocchini CA, Schiettecatte F, Scott AF, Hamosh A. OMIM. org: Online Mendelian Inheritance in Man (OMIM®), an online catalog of human genes and genetic disorders. Nucleic Acids Research. 2015 Jan; 43: D789–D798. doi: 10. 1093/nar/gku1205. pmid: 25428349
29. Davis AP, Murphy CG, Johnson R, Lay JM, Lennon-Hopkins K, Saraceni-Richards C, et al. The Comparative Toxicogenomics Database: update 2013. Nucleic Acids Research. 2013 Jan; 41(D1): D1104–14. doi: 10. 1093/nar/gks994
30. Yu W, Gwinn M, Clyne M, Yesupriya A, Khoury MJ. A navigator for human genome epidemiology. Nat Genet. 2008 Feb; 40(2): 124–5. doi: 10. 1038/ng0208-124. pmid: 18227866
31. Yesupriya A, Evangelou E, Kavvoura FK, Patsopoulos NA, Clyne M, Walsh MC, et al. Reporting of human genome epidemiology (HuGE) association studies: an empirical assessment. BMC Med Res Methodol. 2008 May 20; 8: 31. doi: 10. 1186/1471-2288-8-31. pmid: 18492284
32. Thomas S, Bonchev D. A survey of current software for network analysis in molecular biology. Hum Genomics. 2010 Jun; 4(5): 353–60. pmid: 20650822 doi: 10. 1186/1479-7364-4-5-353

33. Law V, Knox C, Djoumbou Y, Jewison T, Guo AC, Liu Y, et al. DrugBank 4. 0: shedding new light on drug metabolism. Nucleic Acids Res. 2014 Jan; 42(Database issue): D1091–7. doi: 10. 1093/nar/gkt1068. pmid: 24203711

34. Garcia B, Datta G, Cosgrove GP, Strong M. Network and matrix analysis of the respiratory disease interactome. BMC Syst Biol. 2014; 8(1): 34. doi: 10. 1186/1752-0509-8-34

35. Smoot ME, Ono K, Ruscheinski J, Wang P- L, Ideker T. Cytoscape 2. 8: new features for data integration and network visualization. Bioinformatics. 2011 Feb 1; 27(3): 431–2. doi: 10. 1093/bioinformatics/btq675. pmid: 21149340

36. de Wit E, van der Merwe L, van Helden PD, Hoal EG. Gene-gene interaction between tuberculosis candidate genes in a South African population. Mamm Genome. 2010 Aug 27; 22(1–2): 100–10. doi: 10. 1007/s00335-010-9280-8. pmid: 20799037

37. Möller M, Hoal EG. Current findings, challenges and novel approaches in human genetic susceptibility to tuberculosis. Tuberculosis. 2010 Mar; 90(2): 71–83 doi: 10. 1016/j. tube. 2010. 02. 002. pmid: 20206579

38. Zhang Y, Jiang T, Yang X, Xue Y, Wang C, Liu J, et al. Toll-like receptor -1, -2, and -6 polymorphisms and pulmonary tuberculosis susceptibility: a systematic review and meta-analysis. PLoS ONE. 2013; 8(5): e63357. doi: 10. 1371/journal. pone. 0063357. pmid: 23691034

39. Yim J-J, Kim HJ, Kwon OJ, Koh W-J. Association between microsatellite polymorphisms in intron II of the human Toll-like receptor 2 gene and nontuberculous mycobacterial lung disease in a Korean population. Hum Immunol. 2008 Sep; 69(9): 572–6. doi: 10. 1016/j. humimm. 2008. 06. 003. pmid: 18602432

40. Ryu YJ, Kim EJ, Koh W-J, Kim H, Kwon OJ, Chang JH. Toll-like receptor 2 polymorphisms and nontuberculous mycobacterial lung diseases. Clin Vaccine Immunol. 2006 Jul; 13(7): 818–9. pmid: 16829623 doi: 10. 1128/cvi. 00025-06

41. Hall MA, McGlinn E, Coakley G, Fisher SA, Boki K, Middleton D, et al. Genetic polymorphism of IL-12 p40 gene in immune-mediated disease. Genes Immun. 2000 Feb; 1(3): 219–24. pmid: 11196715 doi: 10. 1038/sj. gene. 6363661

42. Tso HW, Lau YL, Tam CM, Wong HS, Chiang AK. Associations between IL12B polymorphisms and tuberculosis in the Hong Kong Chinese population. The Journal of infectious diseases. 2004 Sep 1; 190(5): 913–9. pmid: 15295696 doi: 10. 1086/422693

43. Ali S, Srivastava AK, Chopra R, Aggarwal S, Garg VK, Bhattacharya SN, et al. IL12B SNPs and copy number variation in IL23R gene associated with susceptibility to leprosy. Journal of Medical Genetics. 2013 Jan; 50(1): 34–42. doi: 10. 1136/jmedgenet-2012-101214. pmid: 23240095

44. Morahan G, Kaur G, Singh M, Rapthap CC, Kumar N, Katoch K, et al. Association of variants in the IL12B gene with leprosy and tuberculosis. Tissue Antigens. 2007 Apr; 69 Suppl 1: 234–6. pmid: 17445208 doi: 10. 1111/j. 1399-0039. 2006. 773_3. x

45. Ma X, Reich RA, Gonzalez O, Pan X, Fothergill AK, Starke JR, et al. No evidence for association between the polymorphism in the 3› untranslated region of interleukin-12B and human susceptibility to tuberculosis. The Journal of Infectious Diseases. 2003 Oct 15; 188(8): 1116–8. pmid: 14551880 doi: 10. 1086/378674

46. Zhu K-J, Zhu C-Y, Shi G, Fan Y-M. Meta-analysis of IL12B polymorphisms (rs3212227, rs6887695) with psoriasis and psoriatic arthritis. Rheumatol Int. 2013 Jul; 33(7): 1785–90. doi: 10. 1007/s00296-012-2637-4. pmid: 23297015

47. Liu H, Irwanto A, Tian H, Fu X, Yu Y, Yu G, et al. Identification of IL18RAP/IL18R1 and IL12B as leprosy risk genes demonstrates shared pathogenesis between inflammation and infectious diseases. American Journal of Human Genetics. 2012 Nov 2; 91(5): 935–41. doi: 10. 1016/j. ajhg. 2012. 09. 010. pmid: 23103228

48. Manry J, Laval G, Patin E, Fornarino S, Tichit M, Bouchier C, et al. Evolutionary genetics evidence of an essential, nonredundant role of the IFN-γ pathway in protective immunity. Hum Mutat. 2011 Jun; 32(6): 633–42. doi: 10. 1002/humu. 21484. pmid: 21448974

49. Zhang M, Lin Y, Iyer DV, Gong J, Abrams JS, Barnes PF. T-cell cytokine responses in human infection with Mycobacterium tuberculosis. Infect Immun. 1995 Aug; 63(8)3231–4. pmid: 7622255

50. Vallinoto ACR, Graça ES, Araújo MS, Azevedo VN, Cayres-Vallinoto I, Machado LF, et al. IFNG +874T/A polymorphism and cytokine plasma levels are associated with susceptibility to Mycobacterium tuberculosis infection and clinical manifestation of tuberculosis. Hum Immunol. 2010 Jul; 71(7): 692–6. doi: 10. 1016/j. humimm. 2010. 03. 008. pmid: 20353805

51. Onwubalili JK, Scott GM, Robinson JA. Deficient immune interferon production in tuberculosis. Clin Exp Immunol. 1985 Feb; 59(2): 405–13. pmid: 2579755

52. Pacheco AG, Cardoso CC, Moraes MO. IFNG +874T/A, IL10 -1082G/A and TNF -308G/A polymorphisms in association with tuberculosis susceptibility: a meta-analysis study. Hum Genet. 2008 Jun; 123(5): 477–84. doi: 10. 1007/s00439-008-0497-5. pmid: 18414898

53. Li X. NFκB-dependent signaling pathways. Experimental Hematology. 2002 Apr; 30(4): 285–96. pmid: 11937262 doi: 10. 1016/s0301-472x(02)00777-4

54. Sha WC. Regulation of immune responses by NF-kappa B/Rel transcription factor. J Exp Med. 1998 Jan 19; 187(2): 143–6. pmid: 9432972 doi: 10. 1084/jem. 187. 2. 143

55. Bowie A, O'Neill LA. Oxidative stress and nuclear factor-κB activation. Biochemical Pharmacology. 2000 Jan; 59(1): 13–23. pmid: 10605930 doi: 10. 1016/s0006-2952(99)00296-8

56. Tato CM, Hunter CA. Host-Pathogen Interactions: Subversion and Utilization of the NF- B Pathway during Infection. Infection and Immunity. 2002 Jul 1; 70(7): 3311–7. pmid: 12065467 doi: 10. 1128/iai. 70. 7. 3311-3317. 2002

57. Yamada H, Mizuno S, Reza-Gholizadeh M, Sugawara I. Relative Importance of NF-κB p50 in Mycobacterial Infection. Infect Immun. 2001 Nov; 69(11): 7100–5. pmid: 11598086 doi: 10. 1128/iai. 69. 11. 7100-7105. 2001
58. Bai X, Feldman NE, Chmura K, Ovrutsky AR, Su W-L, Griffin L, et al. Inhibition of nuclear factor-kappa B activation decreases survival of Mycobacterium tuberculosis in human macrophages. PLoS ONE. 2013; 8(4): e61925. doi: 10. 1371/journal. pone. 0061925. pmid: 23634218
59. Gutierrez MG, Mishra BB, Jordao L, Elliott E, Anes E, Griffiths G, et al. NF- B Activation Controls Phagolysosome Fusion-Mediated Killing of Mycobacteria by Macrophages. J Immunol. 2008 Aug 15; 181(4)2651–63. pmid: 18684956 doi: 10. 4049/ jimmunol. 181. 4. 2651
60. Giri DK, Mehta RT, Kansal RG, Aggarwal BB. Mycobacterium avium-intracellulare complex activates nuclear transcription factor-kappaB in different cell types through reactive oxygen intermediates. J Immunol. 1998 Nov 1; 161(9): 4834–4841. pmid: 9794416
61. Kim K-H, Kim T-S, Lee JG, Park J-K, Yang M, Kim JM, et al. Characterization of Proinflammatory Responses and Innate Signaling Activation in Macrophages Infected with Mycobacterium scrofulaceum. Immune Netw. 2014 Dec; 14(6): 307–320. doi: 10. 4110/in. 2014. 14. 6. 307. pmid: 25550697
62. Weiss DJ, Souza CD, Evanson OA. Effects of nuclear factor-kappaB on regulation of cytokine expression and apoptosis in bovine monocytes exposed to Mycobacterium avium subsp paratuberculosis. Am J Vet Res. 2008 Jun; 69(6): 804–810. doi: 10. 2460/ajvr. 69. 6. 804. pmid: 18518662
63. Ghassemi M, Andersen BR, Roebuck KA, Rabbi MF, Plate JM, Novak RM. Mycobacterium avium Complex Activates Nuclear Factor κB via Induction of Inflammatory Cytokines. Cellular Immunology. 1999 Feb 1; 191(2)117–23. pmid: 9973533 doi: 10. 1006/ cimm. 1998. 1419
64. Kawai T, Akira S. Signaling to NF-κB by Toll-like receptors. Trends in Molecular Medicine. 2007 Nov; 13(11): 460–9. pmid: 18029230 doi: 10. 1016/j. molmed. 2007. 09. 002
65. Kawai T, Akira S. Toll-like Receptors and Their Crosstalk with Other Innate Receptors in Infection and Immunity. Immunity. 2011 May; 34(5): 637–50. doi: 10. 1016/j. immuni. 2011. 05. 006. pmid: 21616434
66. Schorey JS, Cooper AM. Macrophage signalling upon mycobacterial infection: the MAP kinases lead the way. Cell Microbiol. 2003 Mar; 5(3): 133–42. pmid: 12614457 doi: 10. 1046/j. 1462-5822. 2003. 00263. x
67. Xia Z, Dickens M, Raingeaud JL, Davis RJ, Greenberg ME. Opposing Effects of ERK and JNK-p38 MAP Kinases on Apoptosis. Science. 1995 Nov 24; 270(5240): 1326–31. pmid: 7481820 doi: 10. 1126/science. 270. 5240. 1326
68. Pearson G, Robinson F, Beers Gibson T, Xu B-E, Karandikar M, Berman K, et al. Mitogen-Activated Protein (MAP) Kinase Pathways: Regulation and Physiological Functions. Endocrine Reviews. 2001 Apr; 22(2): 153–83. pmid: 11294822 doi: 10. 1210/edrv. 22. 2. 0428
69. Shiratsuchi H, Ellner JJ, Basson MD. Extracellular-regulated kinase activation regulates replication of Mycobacterium avium intracellularly in primary human monocytes. Cell Tissue Res. 2008 May; 332(2): 237–44. doi: 10. 1007/s00441-008-0594-8. pmid: 18335241
70. Ferrara N, Gerber H-P, LeCouter J. The biology of VEGF and its receptors. Nat Med. 2003 Jun; 9(6): 669–76. pmid: 12778165 doi: 10. 1038/nm0603-669
71. van der Flier M, van Leeuwen HJ, van Kessel KP, Kimpen JL, Hoepelman AI, Geelen SP. Plasma Vascular Endothelial Growth Factor in Severe Sepsis. Shock. 2005 Jan 1; 23(1): 35. pmid: 15614129 doi: 10. 1097/01. shk. 0000150728. 91155. 41
72. Karlsson S, Pettilä V, Tenhunen J, Lund V, Hovilehto S, Ruokonen E, et al. Vascular endothelial growth factor in severe sepsis and septic shock. Anesth Analg. 2008 Jun; 106(6): 1820–6. doi: 10. 1213/ane. 0b013e31816a643f. pmid: 18499616
73. Srikiatkhachorn A, Ajariyakhajorn C, Endy TP, Kalayanarooj S, Libraty DH, Green S, et al. (2007) Virus-induced decline in soluble vascular endothelial growth receptor 2 is associated with plasma leakage in dengue hemorrhagic Fever. J Virol. 2007 Feb; 81(4): 1592–600. pmid: 17151115 doi: 10. 1128/jvi. 01642-06
74. Tseng C-S, Lo H-W, Teng H-C, Lo W-C, Ker C-G. Elevated levels of plasma VEGF in patients with dengue hemorrhagic fever. FEMS Immunol Med Microbiol. 2005 Jan 1; 43(1): 99–102. pmid: 15607642 doi: 10. 1016/j. femsim. 2004. 10. 004
75. Sathupan P, Khongphattanayothin A, Srisai J, Srikaew K, Poovorawan Y. The role of vascular endothelial growth factor leading to vascular leakage in children with dengue virus infection. Ann Trop Paediatr. 2007 Sep; 27(3): 179–84. pmid: 17716445 doi: 10. 1179/146532807x220280
76. Matsuyama W, Hashiguchi T, Matsumuro K, Iwami F, Hirotsu Y, Kawabata M, et al. Increased serum level of vascular endothelial growth factor in pulmonary tuberculosis. Am J Respir Crit Care Med. 2000 Sep; 162(3 Pt 1): 1120–2. pmid: 10988140 doi: 10. 1164/ ajrccm. 162. 3. 9911010
77. Abe Y, Nakamura M, Oshika Y, Hatanaka H, Tokunaga T, Ohkubo Y, et al. Serum levels of vascular endothelial growth factor and cavity formation in active pulmonary tuberculosis. Respiration. 2001; 68(5): 496–500. pmid: 11694812 doi: 10. 1159/000050557
78. Alatas F, Alatas O, Metintas M, Özarslan A, Erginel S, Yildirim H. Vascular Endothelial Growth Factor Levels in Active Pulmonary Tuberculosis. Chest. 2004 Jun 1; 125(6): 2156–9. pmid: 15189936 doi: 10. 1378/chest. 125. 6. 2156
79. Riou C, Perez Peixoto B, Roberts L, Ronacher K, Walzl G, Manca C, et al. Effect of standard tuberculosis treatment on plasma cytokine levels in patients with active pulmonary tuberculosis. PLoS ONE. 2012; 7(5):

e36886. doi: 10. 1371/journal. pone. 0036886. pmid: 22606304

80. Shaw MA, Clayton D, Atkinson SE, Williams H, Miller N, Sibthorpe D, et al. Linkage of rheumatoid arthritis to the candidate gene NRAMP1 on 2q35. Journal of Medical Genetics. 1996; 33: 672–677. pmid: 8863160 doi: 10. 1136/jmg. 33. 8. 672

81. Searle S, Blackwell JM. Evidence for a functional repeat polymorphism in the promoter of the human NRAMP1 gene that correlates with autoimmune versus infectious disease susceptibility. Journal of Medical Genetics. 1999; 36: 295–299. pmid: 10227396

82. Blackwell JM, Searle S, Mohamed H, White JK. Divalent cation transport and susceptibility to infectious and autoimmune disease: continuation of the Ity/Lsh/Bcg/Nramp1/Slc11a1 gene story. Immunol Lett. 2003 Jan 22; 85(2): 197–203. pmid: 12527228 doi: 10. 1016/s0165-2478(02)00231-6

83. Ates O, Dalyan L, Müsellim B, Hatemi G, Türker H, Ongen G, et al. NRAMP1 (SLC11A1) gene polymorphisms that correlate with autoimmune versus infectious disease susceptibility in tuberculosis and rheumatoid arthritis. Int J Immunogenet. 2009 Feb; 36(1): 15–9. doi: 10. 1111/j. 1744-313X. 2008. 00814. x. pmid: 19055603

84. Jostins L, Ripke S, Weersma RK, Duerr RH, McGovern DP, Hui KY, et al. Host-microbe interactions have shaped the genetic architecture of inflammatory bowel disease. Nature. 2012 Nov 1; 491(7422): 119–24. doi: 10. 1038/nature11582. pmid: 23128233

85. Winthrop KL, Iseman M. Bedfellows: mycobacteria and rheumatoid arthritis in the era of biologic therapy. Nat Rev Rheumatol. 2013 Jun 25; 9(9): 524–31. doi: 10. 1038/nrrheum. 2013. 82. pmid: 23797309

86. Yeh J-J, Wang Y-C, Sung F-C, Kao C-H. Rheumatoid arthritis increases the risk of nontuberculosis mycobacterial disease and active pulmonary tuberculosis. PLoS ONE. 2014; 9(10): e110922. doi: 10. 1371/journal. pone. 0110922. pmid: 25337995

87. Esposito L, Hill NJ, Pritchard LE, Cucca F. Genetic analysis of chromosome 2 in type 1 diabetes: analysis of putative loci IDDM7, IDDM12, and IDDM13 and candidate genes NRAMP1 and IA-2 and the interleukin-1 gene cluster. IMDIAB Group. Diabetes. 1998 Nov; 47(11): 1797–9. pmid: 9792551 doi: 10. 2337/diabetes. 47. 11. 1797

88. Bellamy R, Ruwende C, Corrah T, McAdam KP, Whittle HC, Hill AV. Variations in the NRAMP1 gene and susceptibility to tuberculosis in West Africans. N Engl J Med. 1998 Mar 5; 338(10): 640–4. pmid: 9486992 doi: 10. 1056/nejm199803053381002

89. Blackwell JM, Black GF, Peacock CS, Miller EN, Sibthorpe D, Gnanandha D, et al. Immunogenetics of leishmanial and mycobacterial infections: the Belem Family Study. Philos Trans R Soc Lond B Biol Sci. 1997 Sep 29; 352(1359): 1331–45. pmid: 9355125 doi: 10. 1098/rstb. 1997. 0118

90. Mobley JL. Is rheumatoid arthritis a consequence of natural selection for enhanced tuberculosis resistance? Med Hypotheses. 2004; 62(5): 839–43. pmid: 15082117 doi: 10. 1016/j. mehy. 2003. 12. 006

91. Yang D, Chen J, Zhang L, Cha Z, Han S, Shi W, et al. Mycobacterium leprae upregulates IRGM expression in monocytes and monocyte-derived macrophages. Inflammation. 2014 Aug; 37(4): 1028–34. doi: 10. 1007/s10753-014-9825-1. pmid: 24469081

92. Zhang H, Massey D, Tremelling M, Parkes M. Genetics of inflammatory bowel disease: clues to pathogenesis. Br Med Bull. 2008; 87: 17–30. doi: 10. 1093/bmb/ldn031. pmid: 18753178

93. Lu XC, Tao Y, Wu C, Zhao PL, Li K, Zheng JY, et al. Association between variants of the autophagy related gene—IRGM and susceptibility to Crohn›s disease and ulcerative colitis: a meta-analysis. PLoS ONE. 2013 Nov 13; 8(11): e80602. doi: 10. 1371/journal. pone. 0080602. pmid: 24232856

94. Baker MA, Harries AD, Jeon CY, Hart JE, Kapur A, Lonnroth K, et al. The impact of diabetes on tuberculosis treatment outcomes: A systematic review. BMC Med. 2011; 9(1): 81. doi: 10. 1186/1741-7015-9-81

95. Dooley KE, Tang T, Golub JE, Dorman SE, Cronin W. Impact of diabetes mellitus on treatment outcomes of patients with active tuberculosis. Am J Trop Med Hyg. 2009 Apr; 80(4): 634–9. pmid: 19346391

96. Dooley KE, Chaisson RE. Tuberculosis and diabetes mellitus: convergence of two epidemics. Lancet Infect Dis. 2009 Dec; 9(12): 737–46. doi: 10. 1016/S1473-3099(09)70282-8. pmid: 19926034

97. Griffith DE, Brown-Elliott BA, Langsjoen B, Zhang Y, Pan X, Girard W, et al. Clinical and Molecular Analysis of Macrolide Resistance in Mycobacterium avium Complex Lung Disease. Am J Respir Crit Care Med. 2006 Oct 15; 174(8): 928–34. pmid: 16858014 doi: 10. 1164/rccm. 200603-450oc

98. Griffith DE, Brown BA, Girard WM, Griffith BE, Couch LA, Wallace RJ Jr. Azithromycin-Containing Regimens for Treatment of Mycobacterium avium Complex Lung Disease. Clin Infect Dis. 2001 Jun 1; 32(11): 1547–53. pmid: 11340525 doi: 10. 1086/320512

99. Murphy BS, Sundareshan V, Cory TJ, Hayes D, Anstead MI, Feola DJ. Azithromycin alters macrophage phenotype. Journal of Antimicrobial Chemotherapy. 2008 Feb 4; 61(3): 554–60. doi: 10. 1093/jac/dkn007. pmid: 18230686

100. Xu G, Fujita J, Negayama K, Ohnishi T, Miyawaki H, Hojo S, Takigawa K, et al. Effect of clarithromycin on macrophage functions. Kansenshogaku Zasshi. 1995 Aug 1; 69(8): 864–72. pmid: 7594777 doi: 10. 11150/kansenshogakuzasshi1970. 69. 864

101. Kumar V, Harjai K, Chhibber S. Effect of Clarithromycin on Lung Inflammation and Alveolar Macrophage Function in Klebsiella pneumoniaeB5055-Induced Acute Lung Infection in BALB/c mice. Journal of Chemotherapy. 2008 Oct; 20(5): 609–14. pmid: 19028625 doi: 10. 1179/joc. 2008. 20. 5. 609

102. Feola DJ, Garvy BA, Cory TJ, Birket SE, Hoy H, Hayes D Jr, et al. Azithromycin Alters Macrophage Phenotype and Pulmonary Compartmentalization during Lung

Infection with Pseudomonas. Antimicrobial Agents and Chemotherapy. 2010 May 17; 54(6): 2437–47. doi: 10. 1128/AAC. 01424-09. pmid: 20231397

103. Renna M, Schaffner C, Brown K, Shang S, Tamayo MH, Hegyi K, et al. Azithromycin blocks autophagy and may predispose cystic fibrosis patients to mycobacterial infection. J Clin Invest. 2011 Sep; 121(9): 3554–63. doi: 10. 1172/JCI46095. pmid: 21804191

104. De Groote MA, Johnson L, Podell B, Brooks E, Basaraba R, Gonzalez-Juarrero M. GM-CSF knockout mice for preclinical testing of agents with antimicrobial activity against Mycobacterium abscessus. Journal of Antimicrobial Chemotherapy. 2014 Mar 17; 69(4): 1057–64. doi: 10. 1093/jac/dkt451. pmid: 24222613

105. Hodge S, Hodge G, Jersmann H, Matthews G, Ahern J, Holmes M, et al. Azithromycin improves macrophage phagocytic function and expression of mannose receptor in chronic obstructive pulmonary disease. Am J Respir Crit Care Med. 2008 Jul 15; 178(2): 139–48. doi: 10. 1164/rccm. 200711-1666OC. pmid: 18420960

106. Hodge S, Reynolds PN. Low-dose azithromycin improves phagocytosis of bacteria by both alveolar and monocyte-derived macrophagesin chronic obstructive pulmonary disease subjects. Respirology. 2012 Jul; 17(5): 802–7. doi: 10. 1111/j. 1440-1843. 2012. 02135. x. pmid: 22288725

107. Kang PB. The human macrophage mannose receptor directs Mycobacterium tuberculosis lipoarabinomannan-mediated phagosome biogenesis. Journal of Experimental Medicine. 2005 Oct 3; 202(7): 987–99. pmid: 16203868 doi: 10. 1084/jem. 20051239

108. Leemans JC, Thepen T, Weijer S, Florquin S, van Rooijen N, van de Winkel JG, et al. Macrophages Play a Dual Role during Pulmonary Tuberculosis in Mice. J Infect Dis. 2005 Jan 1; 191(1): 65–74. pmid: 15593005 doi: 10. 1086/426395

109. Watt B, Rayner A, Harris G. Comparative activity of azithromycin against clinical isolates of mycobacteria. J Antimicrob Chemother. 1996 Sep; 38(3): 539–42. pmid: 8889727 doi: 10. 1093/jac/38. 3. 539

110. Prammananan T, Arjratanakool W, Chaiprasert A, Tingtoy N, Leechawengwong M, Asawapokee N, et al. Second-line drug susceptibilities of Thai multidrug-resistant Mycobacterium tuberculosis isolates. Int J Tuberc Lung Dis. 2005 Feb; 9(2)216–9. pmid: 15732744 doi: 10. 5588/ijtld. 13. 0197

111. van Halsema C, Humphreys S, Bonington A. Extensively drug-resistant tuberculosis: early access to bedaquiline for a UK patient. European Respiratory Journal. 2013 Dec 31; 43(1): 292–4. doi: 10. 1183/09031936. 00128613. pmid: 23988770

112. Luna-Herrera J, Reddy VM, Daneluzzi D, Gangadharam PR. Antituberculosis activity of clarithromycin. Antimicrobial Agents and Chemotherapy. 1995 Dec 1; 39(12): 2692–5. pmid: 8593004 doi: 10. 1128/aac. 39. 12. 2692

113. Bolhuis MS, van Altena R, van Soolingen D, de Lange WC, Uges DR, van der Werf TS, et al. Clarithromycin increases linezolid exposure in multidrug-resistant tuberculosis patients. European Respiratory Journal. 2013 Dec; 42(6): 1614–21. doi: 10. 1183/09031936. 00001913. pmid: 23520311

114. Bosne-David S, Barros V, Verde SC, Portugal C, David HL. Intrinsic resistance of Mycobacterium tuberculosis to clarithromycin is effectively reversed by subinhibitory concentrations of cell wall inhibitors. J Antimicrob Chemother. 2000 Sep; 46(3): 391–5. pmid: 10980165 doi: 10. 1093/jac/46. 3. 391

115. Reich-Slotky R, Kabbash CA, Della-Latta P, Blanchard JS, Feinmark SJ, Freeman S, et al. Gemfibrozil Inhibits Legionella pneumophila and Mycobacterium tuberculosis Enoyl Coenzyme A Reductases and Blocks Intracellular Growth of These Bacteria in Macrophages. Journal of Bacteriology. 2009 Jul 27; 191(16): 5262–71. doi: 10. 1128/JB. 00175-09. pmid: 19429621

116. Guideline: Nutritional Care and Support for Patients with Tuberculosis. Geneva: World Health Organization; 2013.

117. Shankar AH. Nutritional modulation of malaria morbidity and mortality. The Journal of Infectious Diseases. 2000 Sep; 182 Suppl 1: S37–53. pmid: 10944483 doi: 10. 1086/315906

118. Wheelwright M, Kim EW, Inkeles MS, De Leon A, Pellegrini M, Krutzik SR, et al. All-trans retinoic acid-triggered antimicrobial activity against Mycobacterium tuberculosis is dependent on NPC2. J Immunol. 2014 Mar 1; 192(5): 2280–90. doi: 10. 4049/jimmunol. 1301686. pmid: 24501203

119. Vilchèze C, Hartman T, Weinrick B, Jacobs WR. Mycobacterium tuberculosis is extraordinarily sensitive to killing by a vitamin C-induced Fenton reaction. Nat Comms. 2013; 4: 1881. doi: 10. 1038/ncomms2898

120. Grant SS, Kaufmann BB, Chand NS, Haseley N, Hung DT. Eradication of bacterial persisters with antibiotic-generated hydroxyl radicals. Proc Natl Acad Sci USA. 2012 Jul 24; 109(30): 12147–52. doi: 10. 1073/pnas. 1203735109. pmid: 22778419

121. Martineau AR, Timms PM, Bothamley GH, Hanifa Y, Islam K, Claxton AP, et al. High-dose vitamin D(3) during intensive-phase antimicrobial treatment of pulmonary tuberculosis: a double-blind randomised controlled trial. Lancet. 2011 Jan 15; 377(9761): 242–50. doi: 10. 1016/S0140-6736(10)61889-2. pmid: 21215445

122. Jeon K, Kim S- Y, Jeong B- H, Chang B, Shin SJ, Koh WJ. Severe vitamin D deficiency is associated with non-tuberculous mycobacterial lung disease: a case-control study. Respirology. 2013 Aug; 18(6): 983–8. doi: 10. 1111/resp. 12109. pmid: 23627546

123. Wejse C, Gomes VF, Rabna P, Gustafson P, Aaby P, Lisse IM, et al. Vitamin D as supplementary treatment for tuberculosis: a double-blind, randomized, placebo-controlled trial. Am J Respir Crit Care Med. 2009 May 1; 179(9): 843–50. doi: 10. 1164/rccm. 200804-567OC. pmid: 19179490

124. Sinclair D, Abba K, Grobler L. Nutritional supplements for people being treated for active tuberculosis. Cochrane Database Syst Rev. 2011 Nov 9; (11): CD006086. doi: 10. 1002/14651858. CD006086. pub3. pmid: 22071828

125. Ratledge C. Iron, mycobacteria and tuberculosis. Tuberculosis. 2004; 84(1–2): 110–30. pmid: 14670352 doi: 10. 1016/j. tube. 2003. 08. 012

126. Tullius MV, Harmston CA, Owens CP, Chim N, Morse RP, McMath LM, et al. Discovery and characterization of a unique mycobacterial heme acquisition system. Proceedings of the National Academy of Sciences. 2011 Mar 22; 108(12): 5051–6. doi: 10. 1073/pnas. 1009516108

127. Nixon MR, Saionz KW, Koo M-S, Szymonifka MJ, Jung H, Roberts JP, et al. Folate Pathway Disruption Leads to Critical Disruption of Methionine Derivatives in Mycobacterium tuberculosis. Chemistry & Biology. 2014 Jul; 21(7): 819–30. doi: 10. 1016/j. chembiol. 2014. 04. 009

CHAPTER 12

LOCAL DNA DYNAMICS SHAPE MUTATIONAL PATTERNS OF MONONUCLEOTIDE REPEATS IN HUMAN GENOMES

Albino Bacolla[1], Xiao Zhu[2], Hanning Chen[3], Katy Howells[4], David N. Cooper[4], and Karen M. Vasquez[1]

[1]Division of Pharmacology and Toxicology, College of Pharmacy, Dell Pediatric Research Institute, The University of Texas at Austin, 1400 Barbara Jordan Boulevard, Austin, TX 78723, USA

[2]Texas Advanced Computing Center, Austin, TX 78758-4497, USA

[3]Department of Chemistry, George Washington University, 725 21st Street, NW, Washington, DC 20052, USA

[4]Institute of Medical Genetics, School of Medicine, Cardiff University, Cardiff CF14 4XN, UK

ABSTRACT

Single base substitutions (SBSs) and insertions/deletions are critical for generating population diversity and can lead both to inherited disease and cancer. Whereas on a genome-wide scale SBSs are influenced by cellular factors, on a fine scale SBSs are influenced by the local DNA sequence-context, although the role of flanking sequence is often unclear. Herein, we used bioinformatics, molecular dynamics and hybrid quantum mechanics/molecular mechanics to analyze sequence context-dependent mutagenesis at mononucleotide repeats (A-tracts and G-tracts) in human population variation and in cancer genomes. SBSs and insertions/deletions occur predominantly at the first and last base-pairs of A-tracts, whereas they are concentrated at the second and third base-pairs in G-tracts. These positions correspond to the most flexible sites along A-tracts, and to sites where a 'hole', generated by the loss of an electron through oxidation, is most likely to be localized in G-tracts. For A-tracts, most SBSs occur in the direction of the base-pair flanking the tracts. We conclude that intrinsic features of local DNA structure, i.e. base-pair flexibility and charge transfer, render specific nucleotides along mononucleotide runs susceptible to base modification, which then yields mutations. Thus, local DNA dynamics contributes to phenotypic variation and disease in the human population.

INTRODUCTION

Changes in human genomic DNA in the form of base substitutions and insertions/deletions (indels) are essential to ensure population diversity, adaptation to the environment, defense from pathogens and self-recognition; they are also a critical source of human inherited disease and cancer. On a genome-wide scale, base substitutions result from the combined action of several factors, including replication fidelity, lagging versus leading strand DNA synthesis, repair, recombination, replication timing, transcription, nucleosome occupancy, etc., both in the germline and in cancer (1–4). On a much finer scale [(over a few base pairs (bp)], rates of base substitutions may be strongly influenced by interrelationships between base–protein and base–base interactions. For example, the mutator role of activation-induced deaminase (AID) in B-cells during class-switch recombination and somatic hypermutation (5) targets preferentially cytosines within WRC (W: A|T; R: A|G) sequences (6), whereas apolipoprotein B mRNA editing enzyme, catalytic polypeptide-like (APOBEC) overexpression displays a preference for base substitutions at cytosines in TCW contexts (7). Other examples, such as the induction of C→T

Citation: Albino Bacolla, Xiao Zhu, Hanning Chen, Katy Howells, David N. Cooper, and Karen M. Vasquez. Local DNA dynamics shape mutational patterns of mononucleotide repeats in human genomes, Nucleic Acids Research (26 May 2015), Volume 43, Issue 10, Pp. 5065-5080. DOI: 10.1093/nar/gkv364

transitions at CG:CG dinucleotides by cytosine-5-methylation and the role of UV light in promoting base substitutions at pyrimidine dimers have been well documented (reviewed in (4,8)). More recently, complex patterns of base substitution at guanosines in cancer genomes have been found to correlate with changes in guanosine ionization potentials as a result of electronic interactions with flanking bases (9), suggesting a role for electron transfer and oxidation reactions in sequence-dependent mutagenesis. However, despite these advances, the increasing number of sequence-dependent patterns of mutation noted in genome-wide sequencing studies has met with a lack of understanding of most of the underlying mechanisms (10). Thus, a picture is emerging in which mutations are often heavily dependent on sequence-context, but for which our comprehension is limited.

Mononucleotide repeats comprise blocks of identical base pairs (A|T or C|G; hereafter referred to as A-tracts and G-tracts) and display distinct features: they are abundant in vertebrate genomes; mutations within the tracts occur more frequently than the genome-wide average; mutations generally increase with increasing tract length; length instability is a hallmark of mismatch repair-deficiency in cancers; and sequence polymorphism within the general population has been linked to phenotypic diversity (11–15). Thus, mononucleotide repeats appear ideal for addressing the question of sequence-dependent mutagenesis since base pairs within the tracts are flanked by identical neighbors. Both historic and recent investigations concur with the conclusion that a major source of mononucleotide repeat polymorphism is the occurrence of slippage (i.e. repeat misalignment) during semiconservative DNA replication, which gives rise to the addition or deletion of repeat units (11,12). An additional and equally important source of mutation has recently been suggested to arise from errors in DNA replication by translesion synthesis DNA polymerases, such as pol η and pol κ (13), also on slipped intermediates, leading to single base substitutions.

A key question that remains unanswered in these studies and which is relevant to the issue of sequence context-dependent mutagenesis is whether all base pairs within mononucleotide repeats display identical susceptibility to single base changes and whether indels (which are consequent to DNA breakage) occur randomly within the tracts.

Herein, we combine bioinformatics analyses on mononucleotide repeat variants from the 1000 Genomes Project and cancer genomes with molecular dynamics simulations and hybrid quantum mechanics/molecular mechanics calculations to address the question of sequence-dependent mutagenesis within these tracts. We show that mutations along both A-tracts and G-tracts are highly non-uniform. Specifically, both base substitutions and indels occur preferentially at the first and last bp of A-tracts, whereas they are concentrated between the second and third G:C base pairs in G-tracts. These positions coincide with the most flexible base pairs for A-tracts and with the preferential localization of a 'hole' that results when one electron is lost due to an oxidation reaction anywhere along G-tracts. Thus, despite the uniformity of sequence composition, mutations occur in a sequence-dependent context at homopolymeric runs according to a hierarchy that is imposed by both local DNA structural features and long-range base–base interactions. We also show that the repair processes leading to base substitution must differ between A- and G-tracts, since in the former, but not in the latter, base substitutions occur predominantly in the direction of the base immediately flanking the tracts. Additional sequence-dependent patterns of mutation are likely to arise from studies of more heterogeneous sequence combinations, possibly involving other aspects intrinsic to the structure of DNA.

MATERIALS AND METHODS

Datasets

A- and G-tracts were retrieved from the hg19 human genome assembly (or from exons) available at UCSC (http://hgdownload.soe.ucsc.edu) using custom scripts. The 1KGP SNV and indel data were obtained from version 3 of the integrated variant call set atftp://ftp.1000genomes.ebi.ac.uk/vol1/ftp/release/20110521. For the Neanderthal genomes, ~18.5M contigs sequenced from bone of three Neanderthal individuals were downloaded fromftp://hgdownload.cse.ucsc.edu/gbdb/hg18/neandertal/seqAlis/. Approximately half of the numbers of tracts present in the hg19 assembly were found in the three Neanderthal samples. For gene enrichment analyses, genes with A- and G-tracts were retrieved fromhttp://hgdownload.cse.ucsc.edu/goldenpath/hg19/database/ and the resulting NM_* entries were used as input to DAVID (http://david.abcc.ncifcrf.gov) for gene enrichment analyses. The charts generated were used to obtain two types of information: (i) the terms with a Benjamini-corrected P-value <E-10 for A-tracts and <E-05 for G-tracts; and (ii) the terms with a Benjamini-corrected P-value <0.05 that were in common between the A- and G-tracts. Nucleotides from A- and G-tracts intersecting transcription factor-bound chromatin immunoprecipitated genomic segments were obtained fromhttp://hgdownload.cse.ucsc.edu/goldenpath/hg19/encodeDCC/wgEncodeRegTfbsClustered/. For A- and G-tracts along well-positioned nucleosomes, file GSE36979_mnase_mids_combined_147.wig.gz deposited in GEO with accession number GSE36979 (16) was downloaded fromhttp://www.ncbi.nlm.nih.gov/geo/query/acc.cgi?acc=GSE36979. We mapped all tract coordinates matching those covered by the GSE36979_mnase_mids_combined_147 file intervals. Cancer SBSs were downloaded from ftp://data.dcc.icgc.org/,ftp://ftp.sanger.ac.uk/pub/cancer/AlexandrovEtAl and https://tcga-data.nci.nih.gov/tcga/findArchives.htm, and sorted to exclude duplicate SBSs. HGMD data were obtained from HGMD® Professional release 2014.2 (http://www.biobase-international.com/product/hgmd). SBSs (n = 3805), 1-bp deletions (n = 1540) and 1-bp insertions (n = 1324) occurring in A- and G-tracts of 4–13 bp were extracted using the 'Advanced' data analysis tool.

MD Simulations

MD simulations were performed as described (17). Briefly, we describe the interactions between DNA molecules and the environment (e.g. water and counter ions) by the AMBER parm99 force field with parmbsc0 nucleotide modifications (parmbsc0). We solvated the DNA fragments (G[6] and G[9] capped by CGC and A[6] and A[9] capped by CGCGCG at both ends to mitigate end effects) in a cubic water box using the SOLVATE module implemented in Visual Molecular Dynamics (VMD) (18). Each side of the box extended 10

Å away from any solute atom and water molecules within a distance of 2.6 Å of any solute atom were deleted. Sodium ions were randomly added to the water box in each case to neutralize the net charge. Periodic boundary conditions were used and the periodic box was filled with water molecules based on the modified TIP3P potential. MD simulations were carried out with the molecular dynamics package NAMD (19) at 298K. A 2-ns MD simulation at 298K with an isothermal-isobaric (NPT) ensemble was performed to equilibrate the system. During the equilibration, the position of DNA atoms was kept fixed with harmonic constraints, which were released stepwise in the final 1 ns. A second equilibration was then carried out with a NPT ensemble for 4 ns at 298K without constraints, followed by 100 ns production runs for G-tracts and 32 ns for A-tracts. Langevin dynamics was employed with a damping coefficient $\gamma = 5\ ps^{-1}$ (20). Pressure was maintained at 1 atm by the Langevin piston method, with a piston period of 200 fs, a damping time constant of 100 fs and a piston temperature of 298K (20). Non-bonded interactions were calculated with a group-based cutoff and a switching function at each time step; full electrostatics was employed using the particle-mesh Ewald method, with 1 Å grid width at every other step. Covalent bonds involving hydrogen atoms were held rigid with the SHAKE algorithm, allowing a 2 fs time step.

DFT Simulations

Vertical ionization potentials of double-stranded G[6] and G[9] DNA fragments were calculated by comparing the energies before and after removing an electron from the static solvated structures. To account for structural fluctuation and thus to ensure statistical reliability, a total of 40 solvated DNA structures for both G[6] and G[9] were randomly selected from the 100-ns MD trajectories. A hybrid quantum mechanics/molecular mechanics (QM/MM) approach was employed (21) with the Gaussian electrostatic coupling scheme (22) to attain a balance between accuracy and efficiency. Specifically, the DNA molecules were modeled by the density functional theory (DFT) (23), whereas water molecules and sodium counterions were described by the modified TIP3P and AMBER parm99 potentials, respectively. All DFT simulations were performed using the CP2K software (24) with the Goedecker-Teter-Hutter (GTH) pseudopotential (25), the Perdew-Burke-Ernzehof (PBE) exchange-correlation functional (26) and the polarized-valence-double-ζ (PVDZ) basis set (27). Charge redistribution upon vertical ionization was examined by tracking the differences on the density-derived atomic partial charges (DDAPC) (28), grouped by DNA base pairs. The calculated charge difference is always negative since the charge per bp after electron loss is subtracted from that of the non-perturbed state.

RESULTS

Mononucleotide Repeat Variation is defined by Tract Length and Flanking Base Composition

We define mononucleotide repeats in the GRCh37/hg19 (hg19) human genome assembly as uninterrupted runs of A:T and G:C base pairs (hereafter referred to as A-tracts and G-tracts, respectively) from 4 to 13 base pairs in length (Figure 1A). We retrieved a total of 48,767,945 A-tracts and 13,633,781 G-tracts, both of which displayed a biphasic distribution with an inflection point between tract lengths of 8 and 9 (bp) and with the number of runs declining with length more dramatically for G-tracts than for A-tracts (Figure 1B), as noted previously (29). Both the number of short tracts and the extent of decline varied with flanking base composition, TA[n]T runs being two- to three-fold more abundant than CA[n]Cs and AG[n]As declining the most rapidly. Thus, mononucleotide runs exist as a collection of separate pools of sequences in extant human genomes, each maintained at distinctive rates of sequence stability, as determined by factors such as bp composition (A:T versus G:C), tract length and flanking sequence composition.

We examined the extent of sequence variation in the human population by mapping 38,878,546 single nucleotide variants (SNVs) from 1092 haplotype-resolved genomes (the 1000 Genomes Project, 1KGP) (30) to the hg19 A- and G-tracts. The normalized fractions of polymorphic tracts (F SNV) were greater for G-tracts than A-tracts and both displayed Gaussian-type distributions, with maxima of 0.067 for G-tracts of length 8 and 0.017 for A-tracts of length 9 (Figure 1C). CA[n]C and AG[n]A runs displayed the highest F SNV values for A- and G-tracts, respectively, with F SNV values for AG[n]As attaining ~0.10 at length 8. We conclude that flanking base composition influences the rates of SNV within mononucleotide runs and, as a consequence, their representation in the reference human genome.

F SNV values at the flanking 5′ and 3′ bp were similar between A- and G-tracts, except for minor differences for the least represented (i.e. longest) tracts and did not exceed 0.02. These fractions are expected to be greater than at more distant positions from the tracts, based on previous data (29). SNVs at G-tracts, but not at A-tracts, were more frequent than at flanking base pairs. F SNVs for base pairs flanking short (≤8 bp) tracts were at least twice as high as those flanking long tracts; F SNVs also displayed distinct sequence preference with most (~0.1) variants occurring at Ts 3′ of G-tracts (Figure 1D and E). In summary, SNVs at mononucleotide runs do not increase monotonically with length but peak at 8–9 bp. This behavior mirrors the genomic distributions, both with respect to the total number of tracts (Figure 1B) and the subsets flanked by specific-sequence combinations. Variation at flanking base pairs also displayed a biphasic pattern centered at a length of 8–9 bp, with a greater chance of variation adjacent to G- than A-tracts and with characteristic sequence preferences.

A

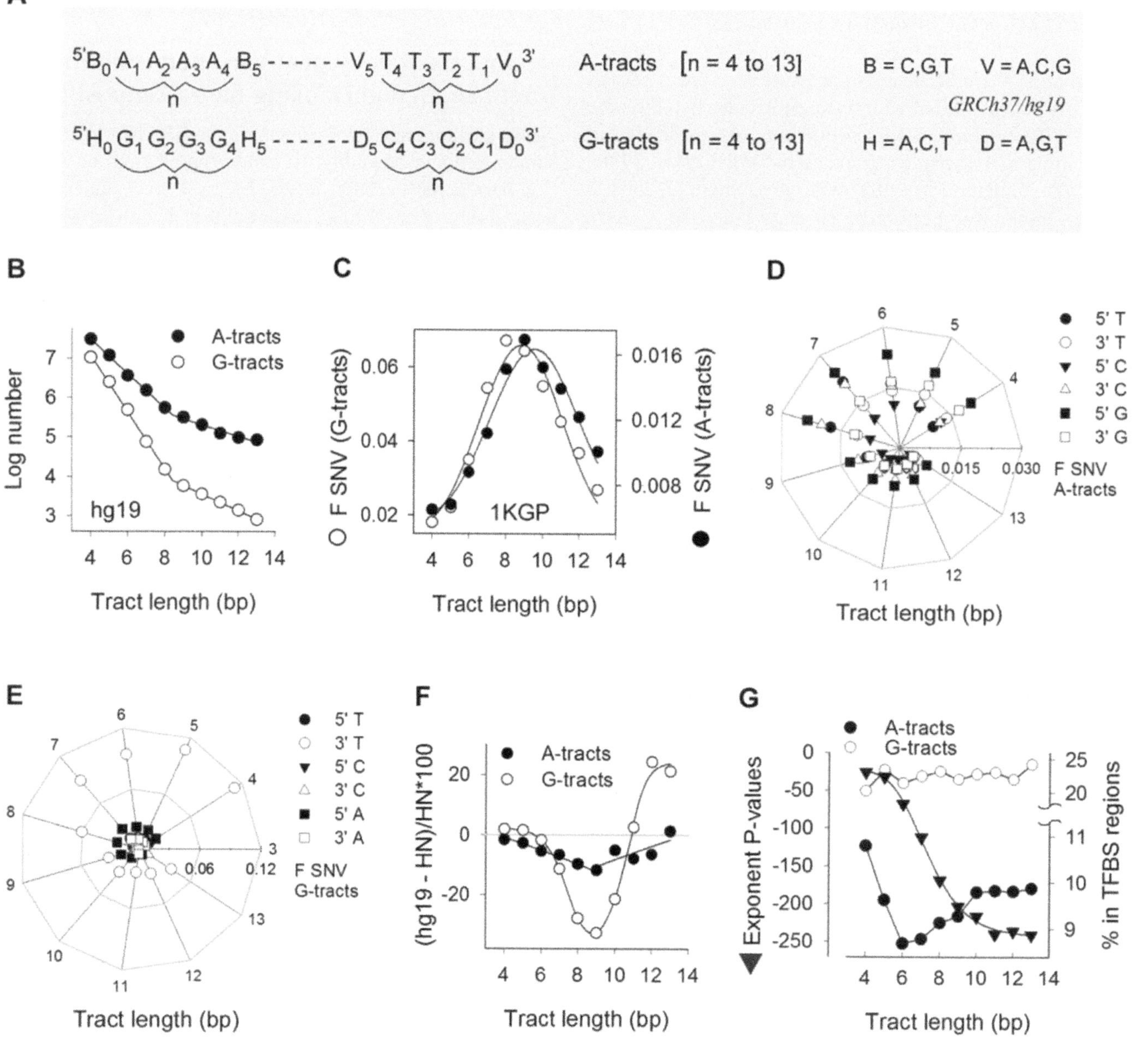

H

Category	Term
CGAP_SAGE_QUARTILE	Brain 3rd
GNF_U133A_QUARTILE	Appendix 3rd; Dorsal root ganglia 3rd; Fetal liver 3rd; Occipital lobe 3rd; Olfactory bulb 3rd
SP_PIR_KEYWORDS	Alternative splicing; Phosphoprotein
UNIGENE_EST_QUARTILE	Adipose tissue normal 3rd; Adrenal tumor disease 3rd; Breast cancer disease 3rd Colorectal tumor disease 3rd; Ear normal 3rd; Eye normal 3rd; Laryngeal cancer disease 3rd Larynx normal 3rd; Mammary gland normal 3rd; Mixed (normal and tumor) disease 3rd Muscle normal 3rd; Pharyngeal tumor disease 3rd; Pharynx normal 3rd; Thyroid normal 3rd Thyroid tumor disease 3rd; Tongue normal 3rd
UP_TISSUE	Brain

Figure 1. Mononucleotide repeat variation, evolutionary conservation and association with transcription. (A) The search algorithm was designed to retrieve runs of As or Ts (A-tracts) and Gs or Cs (G-tracts) length n (n = 4 to 13), along with their 5′ (n= 0) and 3′ (n = n + 1) nearest neighbors from hg19. Tract bases were numbered 5′ to 3′ with respect to the purine-rich sequence. The panel exemplifies the nomenclature for A- and G-tracts of length 4. (B) Logarithmic plot of the number of A-tracts (closed circles) and G-tracts (open circles) in hg19 as a function of length. (C) Normalized fractions of polymorphic tracts (F SNV) (number of SNVs divided by both hg19 number of tracts and n) from the 1KGP for A-tracts (closed circles) and G-tracts (open circles). (D) Radial plot of SNVs in the 1KGP at the 5′ and 3′ nearest neighbors of A-tracts. Periphery, tract length; horizontal axis, scale for the fraction of SNVs (F SNV). (E) Radial plot of SNVs in the 1KGP at the 5′ and 3′ nearest neighbors of G-tracts. (F) Percent difference in the numbers of A-tracts (closed circles) and G-tracts (open circles) between syntenic regions of hg19 and HN genomes. (G) The exponents of Benjamini-corrected P-values for A-tract-containing genes enriched in transcription-factor binding sites plotted as a function of A-tract length (triangles); each value represents the median of the top 11 USCS_TFBS terms. The percent A-tracts (closed circles) and G-tracts (open circles) intersecting genomic regions pulled-down by chromatin immunoprecipitation using antibodies against transcription factors are plotted as a function of tract length. (H) List of gene enrichment terms with a Benjamini-corrected P-value of <0.05 in common between genes containing A- and G-tracts of lengths 4–13, excluding the UCSC_TFBS terms.

Long Tracts Are Evolutionarily Conserved and Associated With High Transcription

To assess whether more variable monosatellite runs (Figure 1C) might have undergone a greater reduction in number in extant humans relative to extinct hominids, we compared the number of A- and G-tracts between syntenic regions of five individuals comprising hg19 and three Neanderthal (HN) specimens (31). The difference between hg19 and HN was very small (<±2%) for the short tracts, but it displayed more negative values in hg19 with increasing tract length, which reached a maximum of −11.8 and −32.7% for A- and G-tracts, respectively, of length 9. Beyond this threshold, the numbers of tracts converged for A-tracts, whereas they were more abundant in hg19 for G-tracts >11 bp (Figure 1F). In summary, the largest difference in the number of mononucleotide runs between hg19 and HN sequences was centered at 9 bp for both A- and G-tracts, suggesting that the length distributions reflect distinct rates of evolutionary gains and losses due to differential sequence mutability (Figure 1C) as a function of length and flanking sequence composition (12).

The fact that long (>9 bp) mononucleotide runs display low variability in the human population (Figure 1C) and sequence conservation during evolutionary divergence (Figure 1F) raises the possibility that they might serve functional roles. Through gene enrichment analyses, we found that genes containing A- and G-tracts were enriched for genes associated with the term 'UCSC_TFBS', which pertains to transcripts harboring frequent transcription factor binding sites (32,33). For A-tract-containing genes, the median P-values for the top 11 UCSC_TFBS terms decreased from 2.95E-26 for tracts of length 4 to 5.22E-241 for tracts of length 13 (Figure 1G). The percent of A-tracts intersecting genomic fragments amplified from chromatin immunoprecipitation using transcription-factor binding antibodies (32,33) also increased from 8.7 to 9.9 from length 6 to 13, whereas it was constant (mean ± SD, 22.4 ± 1.1) for G-tracts (Figure1G). For gene classes excluding 'UCSC_TFBS', a search for categories enriched at $P < 0.05$ and common to all A- and G-tract-containing genes returned a set of 25 terms, 22 of which were associated with high levels of tissue-specific gene expression (Figure 1H). In summary, these analyses extend prior work (14) supporting a role for mononucleotide tracts in enhancing gene expression, a function that for A-tracts appears to increase with increasing tract length.

Repeat Variability Is Highly Skewed

Next we addressed whether bp along A- and G-tracts display equal probability and type of variation. In the 1KGP dataset, the number of SNVs at each position along both A- and G-tracts of length 4 was within a two-fold difference (144,000–240,000); for both types of sequence, transitions (i.e. A→G and G→A) were the predominant (51–78%) type of base substitution. However, with increasing length, the number of SNVs decreased up to 30-fold more drastically for G-tracts than for A-tracts, with increasing numbers of transversions (A→T and G→C|T) being predominant. Normalizing the data for the number of tracts genome-wide revealed that the extent of SNV varied by up to 10-fold, depending upon tract length and bp position. Specifically, the highest degree of variation was observed at the first and last A within the A-tracts (i.e. A_1 and A_n), which underwent up to 61% A→T and 43% A→C transversions, respectively, at length 9 (Figure 2A). Likewise, for G-tracts, the most polymorphic sites were G_3, followed by G_2, for mid-size tracts of 8–10 bp, with 44% G→C transversions at G_3 for tracts of length 8 (Figure2B). Thus, the extent of SNV at mononucleotide runs is grossly skewed in human genomes, both along the sequence itself and across tract length, which must account for the bell-shape behavior in F SNV for the tracts as a whole (Figure 1C).

We assessed whether SNV hypervariability was associated with specific combinations of nearest neighbors. For A-tracts flanked 5′ by a T, C or G, the highest percentage of SNVs was observed at A_1 when preceded by a T, which reached 7.9% for TA[n] tracts of length 9. By contrast, for 3′ T, C or G, the greatest effect was elicited by a C, with the highest percentage (7.1%) of SNVs at A_n for A[n]C tracts of length 9. Therefore, flanking base pairs play a critical role both in the spectra and frequencies of SNVs at A-tracts. More detailed plots along A-tracts either preceded, followed or preceded and followed (Figure 2C) by a T, C or G revealed the dramatic and long-range (up to 9–10 bp for the longest tracts, higher than the value of 4 bp predicted by mathematical models of slippage (11)) influence of flanking base pairs on variation spectra, in which up to 95% of the changes were in the direction of the base flanking the tract. Because the number of A-tracts preceded or followed by a specific base varies by up to three-fold, we conclude that for A-tracts, the overall mutation fractions and spectra are the result of at least three variables; length, position along the tract, and base composition of the 5′ and 3′ nearest-neighbors.

For G-tracts flanked 5′ by a T, C or A, high percentages (10–12%) of SNVs were observed at G_1 for tracts preceded by a C, an effect that decreased with increasing tract length. This result, together with an exceedingly low number of G→A transitions at G_1 for tracts not preceded by a C relative to all tracts, is consistent with the known high mutability of CG:CG dinucleotides as a result of cytosine-5 methylation (9). The hypermutability at G_2 was observed preferentially for tracts preceded by an A, and to a lesser extent T, whereas that at G_3 was insensitive to flanking sequence composition. Likewise, G-tracts flanked 3′ by a T, C or A did not display marked sequence-dependent effects. Detailed plots of the SNV spectra along G-tracts either preceded, followed, or preceded and followed (Figure 2D) by a T, C or A revealed a noticeable effect only for 5′ T in association with G→T substitutions at G_1 for tracts of length ≥8. Thus, despite a consistent over-representation of G-tracts flanked 5′ by a T, which must account for the high absolute number of SNVs at G_1 for TG[n] relative to AG[n] and CG[n], nearest-neighbor base composition seems to play a lesser role in SNV spectra at G-tracts than at A-tracts.

With respect to SNVs at the flanking 5′ and 3′ nearest positions, no B→A or H→G substitutions (Figure 1A) were found above a length threshold of 9 for A-tracts and 8 for G-tracts (Figure 2E, gray shading) out of 5969 SNVs, implying that tract expansion by recruiting flanking base pairs is disfavored at these lengths. In summary, base substitution along mononucleotide repeats is strongly skewed towards the edges of A-tracts and within the 5′ half of G-tracts, with frequencies that peak at midsize lengths (8–9 bp). For A-tracts ≥7 bp, base substitution occurred almost exclusively in the direction of the flanking nearest-neighbors. Finally, base substitution at flanking bases did not contribute to tract expansion for mononucleotide runs longer than 8–9 bp.

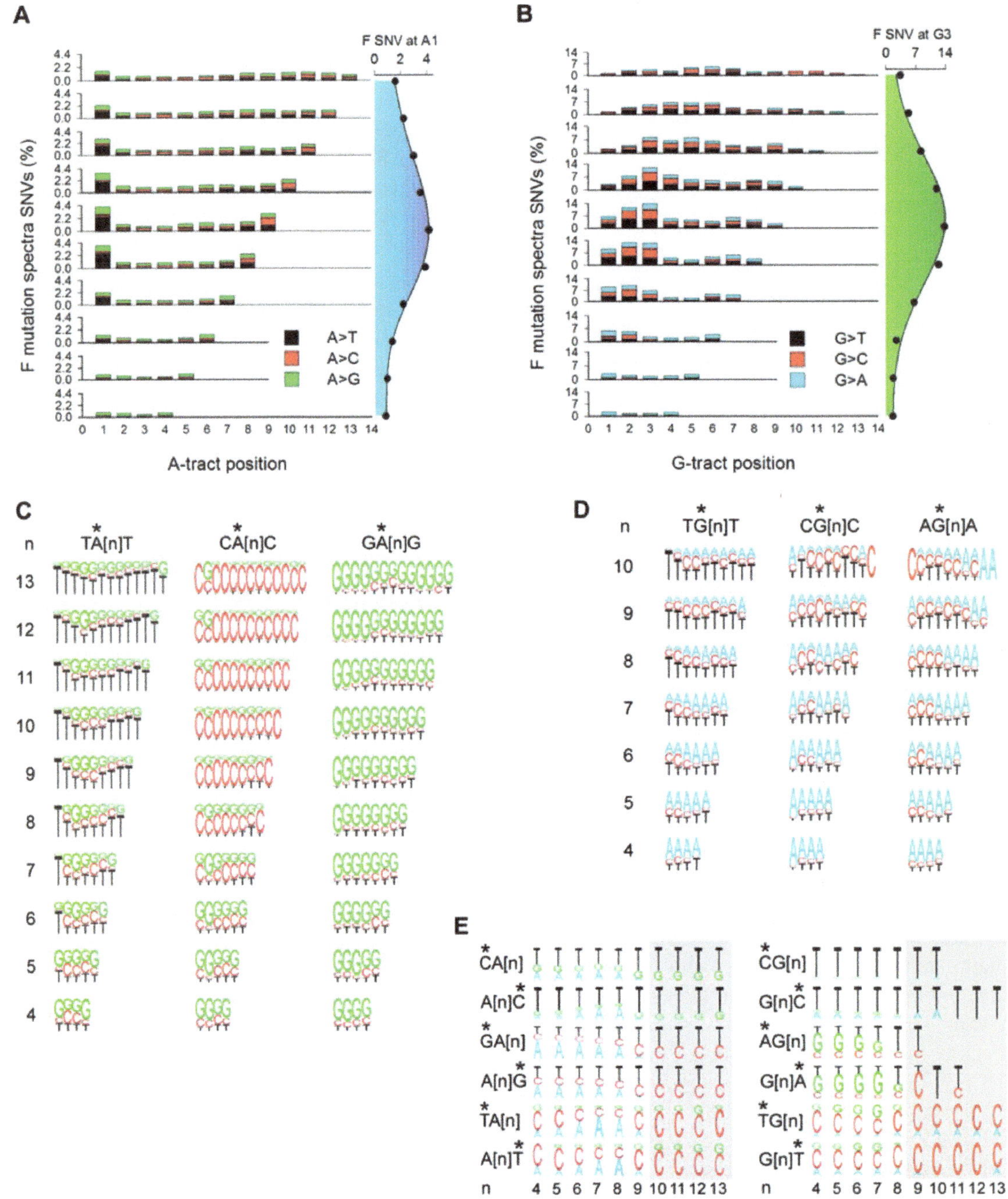

Figure 2. Population variation spectra. (A) Variation spectra of A-tracts. Percent (number of SNVs at each position divided by the number of tracts in hg19 × 100) of A→T (black), A→C (red) and A→G (green) SNVs in the 1KGP dataset (left). Percent SNVs at A_1 as a function of tract length (right). (B) Variation spectra of G-tracts. As in panel A with G→T (black), G→C (red) and G→A (cyan) (left). Percent SNVs at G_3 as a function of tract length (right). (C) Percent A→T, A→C and A→G transitions at each position along A-tracts (stars) preceded and followed by a T (TA[n]T, left), C (CA[n]C), center) and G (GA[n]G, right) as a function of tract length. (D) Percent G→T, G→C and G→A transitions at each position along G-tracts (stars) preceded and followed by a T (TG[n]T, left), C (CG[n]C), center) and A (AG[n]A, right) as a function of tract length. (E) Percent transitions at base pairs (stars) preceding or following A-tracts (left) and G-tracts (right) as a function of tract length (n). *, mutated position.

Insertions and Deletions Display Length and Positional Preference

In addition to SNVs, mononucleotide runs are polymorphic in length as a result of indels. Herein, we consider separately two types of indels: one in which tract length changes by ±1 and flanking bp composition is not altered (slippage); the other comprising all other cases involving the addition or removal of 1–200 bp (indels). Slippage is a widely accepted mutational mechanism (11–12,34), whereby DNA replication errors at reiterated DNA motifs cause changes in the number of motifs (most often +/−1). The normalized fractions of slippage in the 1KGP dataset peaked at lengths of 8 bp for A-tracts and 9 bp for G-tracts (Figure 3A), generating bell-shaped curves similar to those observed for SNVs (Figure 1C) and with no differences in the highest fraction of 'slipped' tracts, which peaked at ~0.02. By contrast, +1 slippage occurred more frequently than −1 slippage at A-tracts (Figure 3B). These results support recent studies on microsatellite repeats (12) and contrast with previous conclusions that slippage increases monotonically with tract length, and that the extent of slippage differs between A- and G-tracts (35,36).

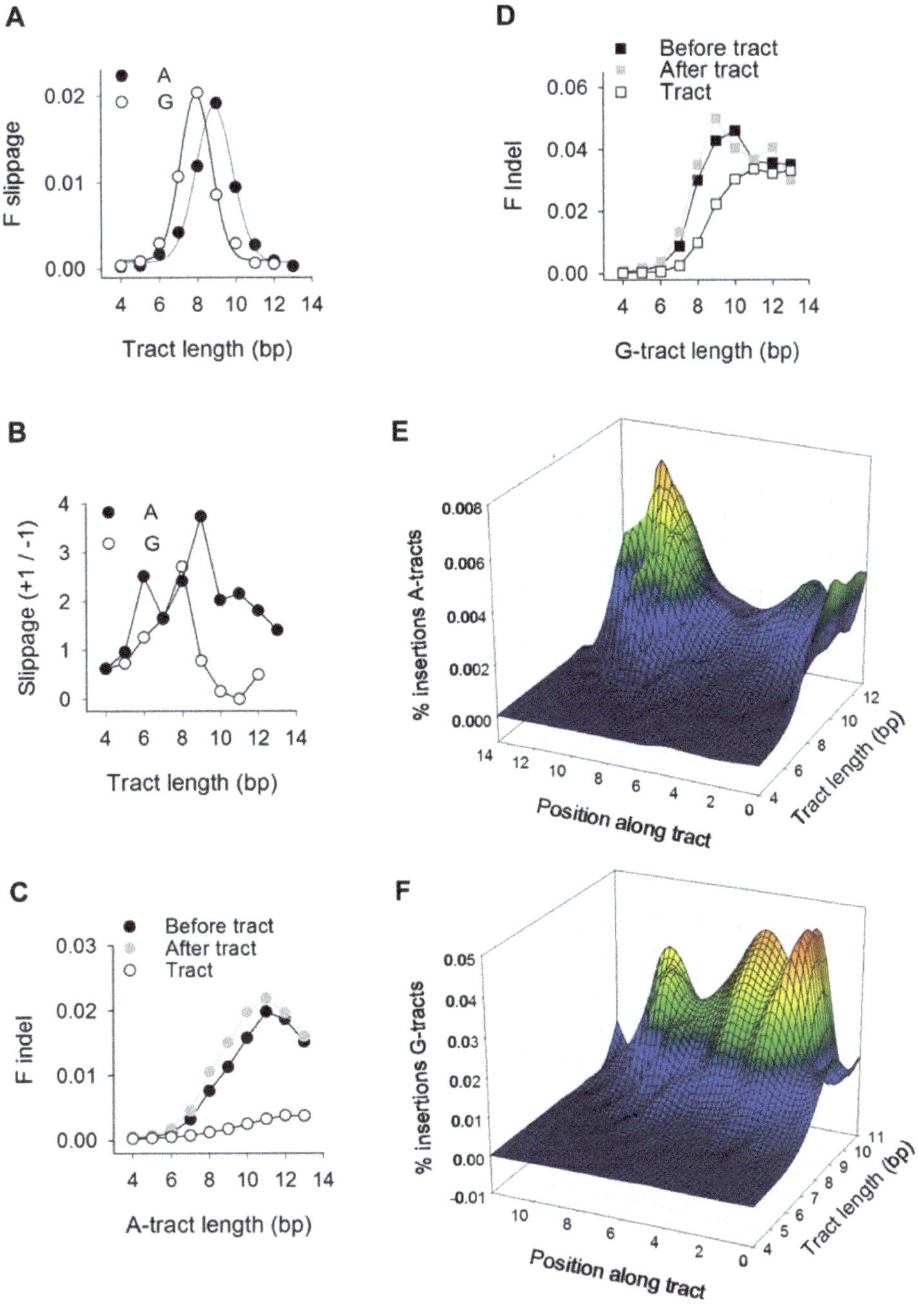

Figure 3. Population insertions and deletions. (A) Normalized fractions of A-tracts (closed circles) and G-tracts (open circles) displaying +/−1 bp slippage in the 1KGP dataset as a function of tract length. Data were obtained by dividing the number of events by both the number of hg19 tracts and tract length (n). (B) Ratio of the number of +1 to −1 slippage for A-tracts (closed circles) and G-tracts (open circles). (C) Indels at A-tracts. For positions along the tracts ('Tract'), 'F Indel' is the ratio between the number of indels and the number of tracts in hg19 multiplied by tract length. For the positions immediately flanking the tracts genomic coordinates ('Before tract' and 'After tract'), 'F Indel' is the ratio between the number of indels and the number of tracts in hg19. (D) Indels at G-tracts, calculated as described in panel C. (E) Heatmap representation of insertions along A-tracts. The percent insertions (i.e. the number of insertions at each position divided by the number of tracts in hg19) (y-axis) plotted as a function of location (x-axis) from position 0 (insertion between the bp 5′ to the tract and the first bp of the tract) to position n + 1 (insertion between the bp 3′ to the last bp of the tract and the following bp) (see Figure 1A) and as a function of tract length (z-axis). (F) Heatmap representation of insertions along G-tracts.

With respect to indels, the normalized fractions were low ($<1 \times 10^{-3}$) along short (4–6 bp) A- and G-tracts, but rose to a plateau for longer tracts as reported earlier (11); this plateau was 10-fold higher for G-tracts (~0.03) than for A-tracts (~0.003) (Figure 3C and D). Indels also occurred more frequently (up to six-fold for A-tracts of length 11) at nearest-neighboring base pairs ('Before tract' and 'After tract' in Figure 3C and D) than along the tracts. Thus, contrary to SNVs and slippage, indels increased to a plateau with mononucleotide tract length.

We analyzed in detail the locations of insertions along the tracts and the flanking positions with respect to the 5′ to 3′ orientation of the tracts (Figure 1A). The normalized fractions demonstrated that insertions peaked at the 3′, and to a lesser extent 5′, ends of the longest A-tracts (Figure3E), but remained low. For G-tracts, insertions occurred most efficiently at two locations ($G_{2\text{-}3}$ and G_5) (Figure 3F), they increased with tract length (up to ~0.04), and attained ~10-fold higher values than for A-tracts. In conclusion, insertion sites at A- and G-tracts followed the patterns observed for SNVs (Figure 2A and B), suggesting that factors associated with local DNA dynamics sensitize specific bases along the tracts to genetic alteration, inducing both SBS and indels.

Base Pair Flexibility and Charge Localization Map to Sites of Sequence Changes

To elucidate elements of intrinsic DNA dynamics that may be responsible for the biases in SNV and insertion sites, we performed molecular dynamics (MD) and hybrid quantum mechanics/molecular mechanics (QM/MM) simulations on model A[6], A[9], G[6] and G[9] duplex DNA fragments. We focused on water bridge coordination (Figure 4A), bp step flexibility, and for the G[6] and G[9], charge localization, as these properties are known to impact the susceptibility of DNA to base damage, repair and mutation. The fractions of one water coordination increased along the A[9] and A[6] structures in a 5′ to 3′ direction, irrespective of flanking sequence composition, in concert with a decrease in minor groove width (Figure 4B) as predicted (37). V_{step}, a measure of bp structural fluctuation, displayed a prominent peak of ~40 Å^{3}deg^3 at the 5′-TA-3′ step for both structures (Figure 4C), which together with low water occupancy points to 5′-TA-3′ being a preferred location for base modification and mutation. In the G[9] and G[6] structures water coordination involved mostly two-water bridges due to wide (~14 Å) minor grooves (Figure 4D), whereas flexibility was modest (~20–22 Å^{3}deg^3, Figure 4E). Thus, bp dynamics are likely to impact mutations at A-tracts to a greater extent than at G-tracts. Guanine has the lowest ionization potential (IP) of all four bases and IP further decreases at guanine runs, rendering them targets for electron loss, charge localization, oxidation and eventually mutation (4,38). Because after electron loss the ensuing charge (hole) can migrate along the DNA double-helix and relocalize at specific guanines, we addressed whether the preferred sites of mutation along G-tracts, i.e. G_{2-3} and G_5, would also be preferred sites for charge localization. The QM/MM determinations indicated that whereas for the short G[6] fragment the difference in the density-derived atomic partial charges (DDAPC) (i.e. the hole) localized most often (~50%) to the first position (Figure 4F), for the long G[9] fragment charge localization shifted downstream (mostly to the second, but also to positions 6–7, Figure 4G). Importantly, the charge was found exclusively around the guanine rings (Figure 4H). Thus, the two main sites of sequence change along G-tracts, i.e. G_{2-3} and G_5, coincide with positions where charge localization and hence one-electron oxidation reactions is predicted to occur most frequently. In summary, bp flexibility at A-tracts and charge transfer at G-tracts likely represent intrinsic DNA features underlying the bias in SNV and insertions at mononucleotide runs in human genomes.

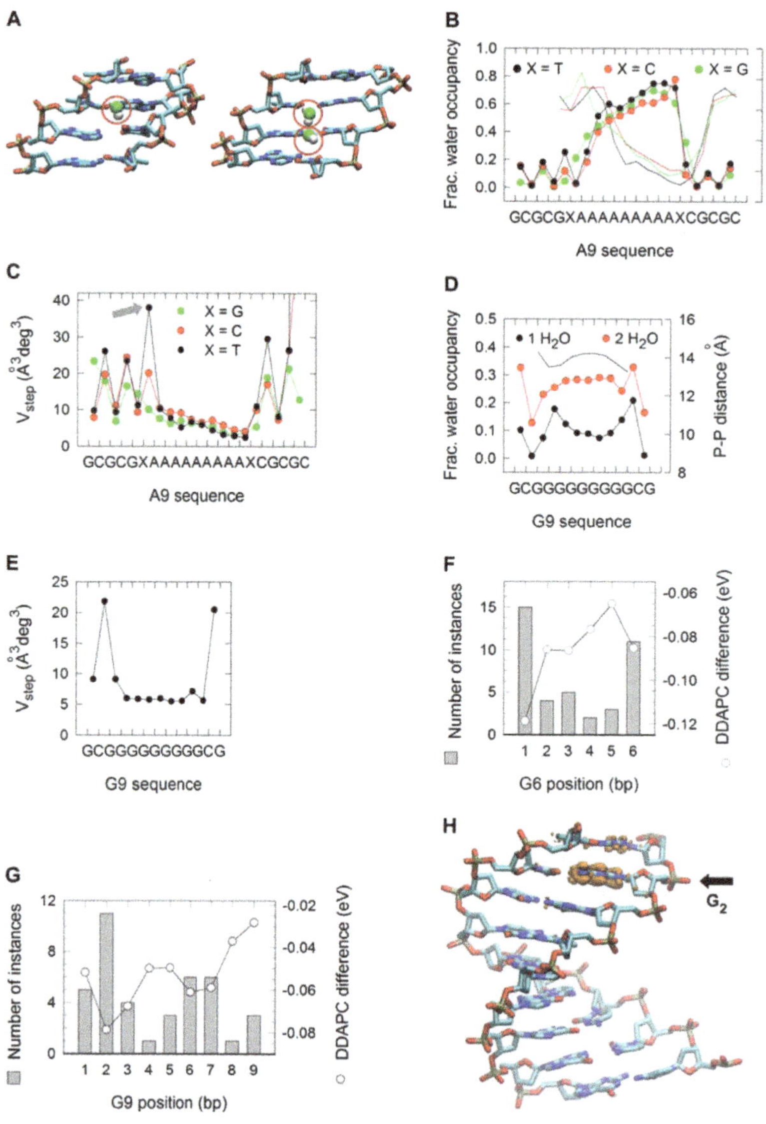

Figure 4. MD and QM/MM simulations. (A) Molecular modeling of one (left) and two (right) minor groove water bridge coordination. (B) Fraction of one-water bridge occupancy (left axis) at A[9] DNA sequences flanked 5′ and 3′ by a T (black circles), C (red circles) or G

(green circles). Minor groove widths (right axis), as determined from intrastrand phosphate-to-phosphate distances. (C) V_{step} for A[9] DNA sequences, determined as the product of the square root of the eigenvalues (λ_i) described by the six bp step parameters shift, slide, rise, tilt, roll and twist; i.e. Vstep=∏6i=1λi−−√Vstep=∏i=16λi. (D) Fraction of one- (black circles) and two-water (red circles) bridge occupancy (left axis) at G[9] DNA sequences. Minor groove widths (right axis), as assessed from intrastrand phosphate-to-phosphate distances. (E) V_{step} for G9 DNA sequences. (F) Average charge redistribution (open circles and right axis) for G[6] DNA structures upon vertical ionization, examined by calculating the difference on the density-derived atomic partial charges (DDAPC) for the neutral and negatively charged states. Histogram of the number of instances (left axis) in which the largest charge redistribution occurred at a specific position along the G[6] structures. (G) DDAPC for G[9] DNA structures (open circles and right axis) and histogram of the number of instances (left axis) in which the largest charge redistribution occurred at a specific position. (H) VMD rendering of a G[9] DNA structure displaying hole localization at G_2. Capped base pairs were removed for clarity.

Position and Orientation along Nucleosome Core Particles Modulate Sequence Variation

DNA wrapped around histones in nucleosomes is subject to local deformation (39), which may impact mutation. Thus, we analyzed the 1KGP SNVs at A- and G-tracts predicted to overlap with well-positioned nucleosome core particles (NCPs) (16). In hg19, the percentage of tracts that overlap with NCPs decreased moderately from ~90% at length of 4 to 81% and 71% for A- and G-tracts of length 13, respectively (Figure 5A), suggesting that mononucleotide runs are not depleted in NCPs in human genomes as previously proposed (40). A-tracts of lengths 4–8 base pairs displayed distinctive peaks along the NCP surface in phase with the helical repeat of DNA (10.5 bp) and with minor grooves facing toward the inner protein core (lengths 4–5) (16) (Figure 5B). A-tracts of length of 9–13 bp exhibited only half (six) the peaks evident for the shorter tracts. For the G-tracts, only small peaks with no clear minor groove-inward-facing regions were detected.

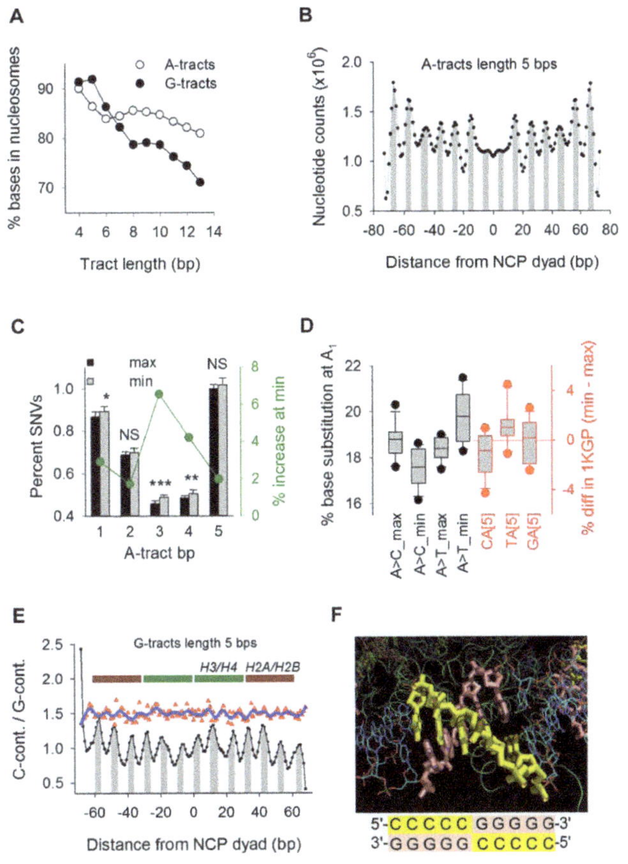

Figure 5. Positioning along nucleosome core particles. (A) Percent of A-tract (open circles) and G-tract (closed circles) base pairs in hg19 overlapping with well-positioned NCP genomic coordinates as a function of tract length. (B) Counts of base pairs in hg19 A-tracts of length 5 overlapping with NCPs genomic regions as a function of distance from the histone octamer dyad axis. Minor groove-inward-facing regions (gray) were derived from the X-ray crystal structure of NCP147 (41). (C) Percent SNVs in the 1KGP dataset (left axis) at every bp along A-tracts of length 5 for tracts centered at maxima (black) and minima (gray) along NCPs (Figure 5B). Percent increase (right axis) of SNVs at minima relative to maxima (green). P-values for paired t-tests: 0.013 (*), 0.002 (**) and 4.7×10^{-6} (***). (D) Whisker plots of%SNVs (left axis) at A_1 for A-tracts of length 5 centered at maxima and minima (black) along NCPs (Figure 5B). Percent difference (right axis) in the number of A-tracts of length 5 in hg19 preceded by C, T or G (red) between those centered at minima and those centered at maxima (Figure5B). (E) C-containing/G-containing ratios (see text) for G-tracts of length 5 in hg19 as a function of distance from the NCP dyad axis (black) and location of core histones (maroon and green). Peaks correspond to negative iSAT (i.e. tilt parameters multiplied by the corresponding sin θ) values (gray) (39). Ratios of%SNV at G_1 (upshifted by 0.5 for clarity) between C-containing (5′-CCCCCG-3′ sequences on the hg19 forward strand) and G-containing (5′-CGGGGG-3′ sequences on the hg19 forward strand) (Figure 1A) CG[5] tracts mapping NCP Chip-seq genomic intervals (red) fitted by a non-parametric local regression (loess; sampling proportion, 0.100; polynomial degree, 3). (F) VMD rendering (top) of TATTT residues 34–38 (yellow) and the complementary AAATA residues 672–753 (pink) from the 1EQZ pdb nucleosomal crystal structure, corresponding to peak area from −40 to −36 in Figure 5E. The switch in G-tract (lengths of 5 and 7) orientation along NCPs (bottom) serves to position the C-containing strand on the outside (yellow) and, correspondingly, the G-containing strand on the inside (pink).

To assess if tract-positioning along NCPs influences SNVs, we selected A-tracts of lengths 5, 7 and 9 bp and G-tracts of lengths 5 and 7 bp whose central positions coincided with either the maxima or minima (41) (Figure 5B) and conducted pairwiset-tests (330 total) between permutations of 'categories', including 'tracts centered at maxima versus minima', 'position along the tracts', 'flanking sequence composition', 'specific NCP locations' and 'tract orientation'. For A-tracts, 79/207 (38%) significant pairs were found, 68 (86%) of which were related to differences between tracts centered at maxima versus minima, with a preponderance (63%) of tests displaying increased %SNVs at minima. For example, %SNVs at length 5 bp were greater at minima than at maxima at each position along the A-tracts (Figure 5C). A→C substitutions at A_1 were more abundant at maxima than at minima (mean ± SD, 18.7 ± 0.7% at max and 17.6 ± 0.8% at min; P-value 0.001), whereas A→T substitutions at the same position displayed the opposite trend (mean ± SD, 18.4 ± 0.5% at max and 19.8 ± 1.1% at min; P-value 0.0005) (Figure 5D). A-tracts of length 7 also exhibited a similar pattern at A_7. The percentages of CA[5] and A[7]C tracts in hg19 centered at maxima were greater than at minima and the reverse was observed for the TA[5] and A[7T] tracts (Figure 5D). Thus, we conclude that positioning along the NCP surface of both the double-helical grooves and junctions with flanking base pairs influence SNVs along A-tracts. However, this influence is complex and for the most part, difficult to predict.

For G-tracts, most pairwise comparisons (18/34, 53%) indicated SNV variation according to sequence orientation. In hg19, the ratio of the numbers of G-tracts of lengths 5 and 7 for which the C-containing strand coincided with the forward sequence (downstream example sequence in Figure 1A) to the numbers of G-tracts for which the G-containing strand coincided with the forward sequence (upstream example sequence in Figure 1A) (C-containing/G-containing ratios) displayed a prominent 10.5-bp oscillation in phase with iSAT (Figure 5E), a measure of 'inside' and 'outside' bases, according to the bp step tilt parameter (39). Analysis of the helical path of a 146-bp DNA fragment wrapped around histones showed that the oscillation in the C-containing/G-containing ratios corresponds to a preference for guanine bases to face the protein core (Figure 5F). We analyzed the subset of G-tracts preceded by a 5′ C (i.e. CG[5]) to assess whether SNVs at G_1, the position known to be mutable due to CpG methylation also oscillated with the C-containing/G-containing ratios. Oscillation in SNV-C-containing/SNV-G-containing values was evident, with peaks aligning to the hg19 troughs (Figure 5E) implying that the cytosines facing the protein surface harbor more variants than those facing away. We conclude that A- and G-tracts display preferential positioning (the former) and orientation (the latter) along NCPs, which in turn modulate the rate of sequence variation.

Mutations Associated with Human Disease

Knowing that the first and last As of long A-tracts and G_{2-3} in G-tracts are the major sites of SNV in the human population, we addressed whether these features are also discernible in mutated mononucleotide tracts associated with human genetic disease. We collected 9,450,456 unique SBSs (both SBSs and SNVs refer to single base changes) from sequenced cancer genomes and normalized the percent mutations along A- and G-tracts to enable a direct comparison with the 1KGP dataset. For A-tracts (Figure 6A), SBSs displayed the same trend as the 1KGP data (Figure 2A) with respect to the bell-shape increase in mutations at A_1 and A_n and the mutation spectra, although the susceptibility to mutation as a function of tract length attained greater values (6.36% for length 11 in cancer versus 4.15% for length 9 in the 1KGP datasets at A_1). The first and last 3 bp also harbored more SBSs than in the 1KGP dataset for tracts >7 bp, a feature that we found to be due exclusively to a large cancer dataset (42) containing high-level microsatellite instability (MSI) samples, which are known to result from mismatch-repair deficiency (15). Thus, A-tracts display similar patterns of base substitution between the germline and somatic cancer tissues. For G-tracts, mutation spectra were characterized by G→T transversions at tract lengths >7, particularly at G_1, the most frequently mutated position for tracts lengths up to 11 bp (Figure 6B). This trend persisted even when the high rates of methylation-mediated deamination mutations at the CG dinucleotide were removed. Thus, mutation patterns in cancer genomes contrast with those observed in the germline, both with respect to the most mutable position (G_1 versus G_{2-3}) and the types of base substitution (G→T in cancer genomes versus G→T and G→C in the germline).

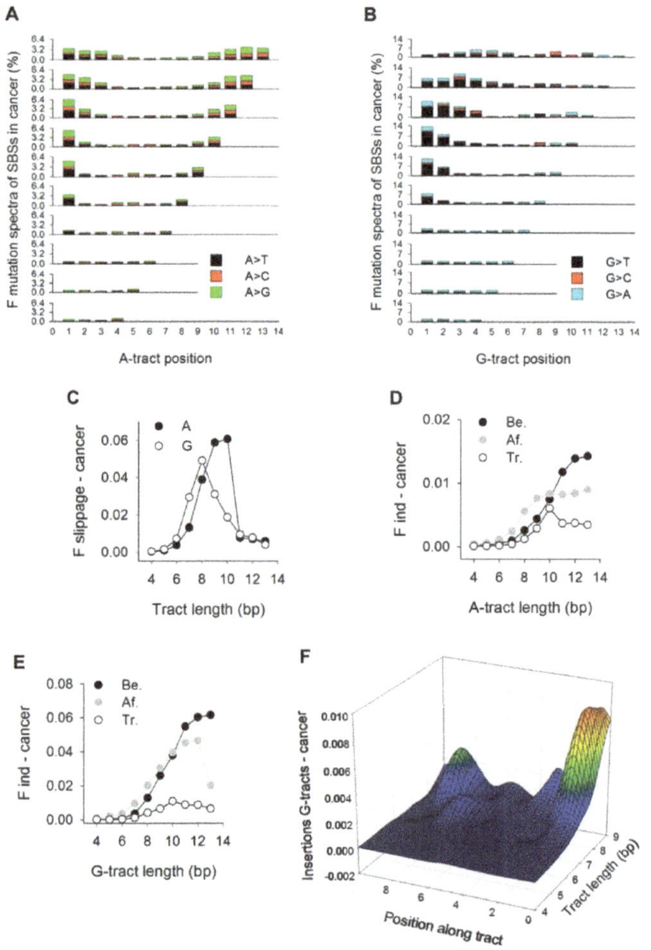

Figure 6. Mutation patterns in cancer genomes. (A) Mutation spectra for SBSs at A-tracts. Percent values were obtained by dividing the total number of SBSs at each position by the number of tracts in hg19 and then multiplying by 3.2516 to equalize the percentage of A-tracts of length 4 between the cancer genomes and the 1KGP datasets. (B) Mutation spectra for SBSs at G-tracts in cancer genomes. Percent values were obtained as in (A) using a multiplication factor of 3.7419. (C) Normalized fractions of A-tracts (closed circles) and

G-tracts (open circles) displaying +/−1 bp slippage, obtained by dividing the number of events by both the number of tracts in hg19 and tract length. (D) Indels at A-tracts, calculated as described in Figure 3C. (E) Indels at G-tracts, calculated as described in Figure3C. (F) Heatmap representation of insertions along G-tracts, as described in Figure 3E.

With respect to slippage, the fractions for A-tracts elicited an excess at lengths 9 and 10 bp relative to the 1KGP dataset, which was also due to the MSI-containing dataset. For G-tracts, the fractions peaked at length 8, as for the 1KGP dataset (Figures 3A and 6C), implying that the propensity to undergo slippage is indistinguishable between the germline and soma. Indels were also more abundant at flanking base pairs than along the tracts (Figure 6D and E), particularly for G-tracts of length >7, similar to the 1KGP dataset (Figure 3C and D). Detailed analyses of insertions revealed that both G_1 and the preceding position were the most significant sites of mutation (F-values up to 0.08 at G_1 for tracts of length 8) (Figure6F). Thus, the 5′ end of long G-tracts is the most susceptible site for both SBSs and insertions in cancer genomes, in contrast to the germline where these occur within the runs, typically at G_{2-3}.

We also extracted the mutated A- and G-tracts from the Human Gene Mutation Database (HGMD), a collection of >150,000 germline gene mutations associated with human inherited disease. A total of 1519 genes were mutated at A- or G-tracts out of a total of 3972 (38%); 3480 SBSs and 2866 slippage events were noted within these tracts, 85 and 46% of which were predicted to be disease-causing, respectively (Figure 7A). Ranking genes by the number of literature reports indicated that among the top 10 entries three were associated with cancer (BRCA1, BRCA2 and APC), two with hemophilia (F8 and F9), four with debilitating lesions of the skin (COL71A), muscle (DMD), lung (CFTR) and kidney (PKD1), with one causing hypercholesterolemia (LDLR) (Figure7B). Thus, mutations within A- and G-tracts carry a high social burden by contributing to some of the most common human pathological conditions.

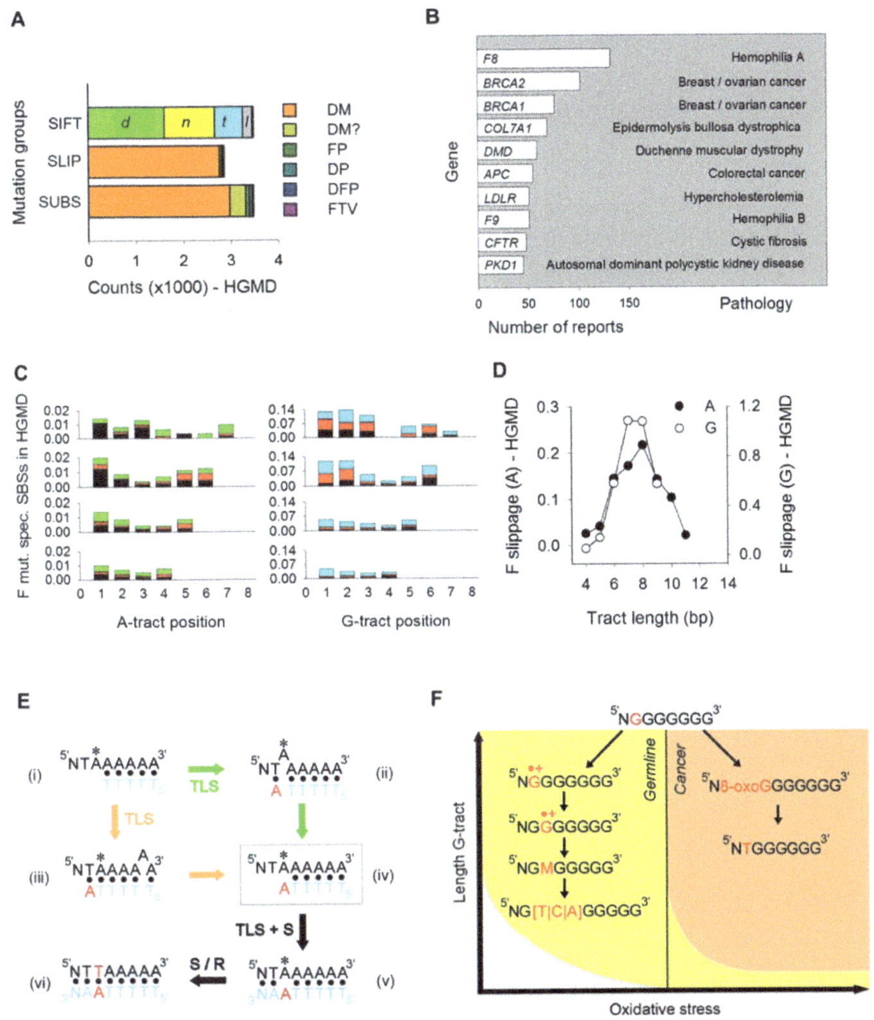

Figure 7. Mutation patterns in HGMD and model for sequence context-dependent changes. (A) Number of germline SBSs and slippage events (Slip.) at A- and G-tracts in HGMD. Gene alterations were classified as disease-causing mutation (DM), likely disease-causing mutation (DM?), disease-associated and putatively functional polymorphism (DFP), disease-associated polymorphism with additional supporting functional evidence (DP) and invitro/laboratory or invivofunctional polymorphism (FP). Codon changes (SIFT predictor) were classified as damaging (d), null (n), tolerated (t) and low-confidence prediction (l). (B) The 10 most commonly reported genes in HGMD with mutations at A- and G-tracts. Various mutated tracts were generally reported for the same gene in different reports. (C) Mutation spectra for SBSs at A- (left) and G-tracts (right) in HGMD. Percent values were obtained by dividing the total number of SBSs at each position by the number of tracts in hg19 exons. A|G→T (black), A|G→C (red), A→G (green), G→A (cyan). (D) Normalized fractions of A-tracts (closed circles) and G-tracts (open circles) displaying +/−1 bp slippage, obtained by dividing the total number of events by the number of tracts in hg19 exons and by tract length. (E) Model for sequence context-dependent changes at A-tracts (left) and G-tracts (right). *, site of base modification.

For both A- and G-tracts, SBSs occurred mostly at tract lengths of 4–7, with patterns more similar to those in the 1KGP than in the cancer datasets, both with respect to the location of the most mutable positions (first and last As and first/second Gs) and the types of base substitution (A→T and G→H) (Figure 7C). Likewise, slippage events peaked at tract lengths of 7–9 as observed in the 1KGP dataset (Figure 7D). In summary, the patterns of both SBSs and slippage in the HGMD dataset followed the trend observed in the 1KGP dataset, suggesting that germline variants at mononucleotide repeats leading to either population variation or human inherited disease may have arisen through similar mechanisms.

DISCUSSION

Why are specific A:T and G:C base pairs within A- and G-tracts more susceptible to sequence changes than their identical neighbors? For A-tracts, bp flexibility may play a role. Chemical damage to DNA, such as by hydroxyl radicals has been shown to be proportional to the geometrical solvent-accessible surface of the atomic groups, which increases with DNA flexibility (43). Along A-tracts flexibility is restricted, but it is high at both the 5′ and 3′ junctions. Thus, the fact that the highest rates of mutation coincide with the highest degree of flexibility at the 5′-TA-3′ bp step is consistent with the view that this position may be susceptible to DNA damage as a result of flexibility. Other sources of DNA dynamics are also likely to be relevant, such as sugar flexibility at the junctions, which increases with tract length (44). Chemical modification at these junctions may then lead to base substitution and indels, the latter as a result of strand breaks.

With respect to SNV mutation spectra, these were found mostly in the direction of flanking base composition above a length of 7–8 bp. We interpret this behavior in terms of DNA slippage along A-tracts when attempts are made during translesion synthesis (TLS) to bypass a damaged site (Figure 7Ei). Two scenarios may be considered to account for A→T transitions at A_1. In the first, the last tract-template base would loop out into the polymerase active site permitting base-pairing and strand elongation (Figure 7Eii) using the tract-flanking base as a template (34,45–46). In the second (Figure 7Eiii), slippage would occur behind the polymerase, prompting extension past the newly created A*:T mispair generated by primer/template misalignment. Either pathway would yield a common intermediate (Figure 7Eiv) that contains the base complementary to the junction across from the damaged site upon slippage resolution (34). Following DNA synthesis (S) and/or repair (R) (Figure 7Ev and vi), this mispair will generate a base change that is always identical to the tract-flanking base.

For G-tracts, the high rates of G→T transversions at G_1 in cancer genomes are also consistent with preferred chemical attack at this site due to high flexibility (Figure 7F top). Direct chemical attack at a guanine is known to result in stable products, such as 8-oxo-G and Fapy-G, both of which are known to yield G→T transversions (47–50). Thus, G_1 may be the most susceptible site for such reactions for G-tracts of lengths ≥7 (Figure 7Fright), which in cancer genomes would become a mutation hotspot. In the germline, SNVs peaked inside G-tract base pairs, while mutational spectra were insensitive to flanking base composition; these events are inconsistent with a role for template misalignment and slippage as noted for A-tracts. Rather, the correspondence between hotspot mutations at G_{2-3} and G_5 and the QM/MM simulations suggest a role for charge transfer. A large body of work during the past 20 years using computational, theoretical chemistry and biophysical techniques on short oligonucleotides, has shown that guanine is the most easily oxidizable base in DNA and that indeed a guanine radical cation can be generated through long-range hole transfer from an oxidant via one-electron oxidation mechanisms (51–55). GGG triplets were found to act as the most effective traps in hole transfer by both experimental and theoretical work (56–59), demonstrating that the resulting guanine radical cation (or its neutral deprotonated form) became rather delocalized, but it preferentially centered at the first and second G. These well-established patterns of chemical reactivity are consistent with our experimental observation of high mutation frequencies at G_1 for short G-tracts and the results from QM/MM simulations on G6. For longer tracts, the downstream shift in mutation hotspots, i.e., G_{2-3} and G_5, also correlate well with the charge localization predicted from QM/MM simulations, which explicitly included solvent effects and structural fluctuations. Thus, in conjunction with the constrained density functional theory (60), both the neutral and oxidized forms of a guanine nucleobase can be reliably constructed to infer the accurate determination of mutational patterns of mononucleotide repeats in human genomic DNA.

The compact organization of the sperm genome (61), and presumably low levels of oxidative stress in the germline, may enable guanine oxidization through one-electron oxidation reactions rather than by direct chemical attack, thereby favoring the formation of radical cations. A charge injected at G_1 by electron loss would then migrate to neighboring guanines and localize at sites of low IP, such as G_2 (Figure 7F left). Guanine radical cations are known to readily undergo further chemical modification leading to products such as 8-oxo-G, oxazolone, imidazolone, guanidinohydantoin, and spiroiminodyhydantoin (62) (M in Figure 7F), to yield G→T, G→C and G→A substitutions (4,63). Our model is in line with recent observations in which mutations at guanines within short G-runs (1–4 bp) correlate with sequence-dependent IPs at the target guanine in cancer genomes (9). Interestingly, these correlations were not observed in the germline (9). We interpret these composite observations as follows. The IP values for G-runs have been shown to decrease asymptotically with tract length, although the absolute values vary according to the methods and assumptions used (we obtained a value of 5.43 eV for both G[6] and G[9]) (64,65). We suggest that short G-runs with high IPs undergo one-electron oxidation reactions in the oxidative environment of cancer cells but would be refractory to such a mechanism in the germline (Figure 7Fright yellow and left white sectors). As length increases and IP values fall, G-runs would be attacked directly by oxidants abundant in tumor cells (Figure 7F orange sector), whereas oxidation will be limited to electron loss in the germline environment (Figure 7F left yellow sector).

These models (template misalignment for A-tracts and charge transfer for G-tracts) suggest a more complex scenario for mechanisms underlying mononucleotide repeat polymorphism in the human population than recently proposed (13), in which nucleotide misincorporation by error-prone polymerases is proposed as a primary source of mutations at both A- and G-tracts. As already stated, the directionality of

SNVs toward tract-flanking bases in A-tracts and the hotspot mutations at G_{2-3}, supports multiple and distinct mechanisms of base substitution at mononucleotide repeats.

Our analyses highlight additional information, including the lack of mutations in the direction of tract-base composition for base pairs flanking long tracts, the association with gene expression and the preference of guanines for the inner NCP surface, and extend prior observations (12) such as the bell-shape character of base substitution and slippage, whose mechanisms remain to be fully clarified. Finally, we document the contribution of mononucleotide mutagenesis to key aspects of human pathology beyond the well-established MSI instability in cancer (15), including hemophilia and tissue degeneration. Our collective work supports the conclusion that as the human genome undergoes evolutionary diversification and along the way suffers disease-associated mutations, oxidation reactions including charge transfer may play a prominent role.

REFERENCES

1. Lujan S. A., Clausen A. R., Clark A. B., MacAlpine H. K., MacAlpine D. M., Malc E. P., Mieczkowski P. A., Burkholder A. B., Fargo D. C., Gordenin D. A., et al. Heterogeneous polymerase fidelity and mismatch repair bias genome variation and composition. Genome Res. 2014; 24: 1751-1764.
2. Roberts S. A., Gordenin D. A. Hypermutation in human cancer genomes: footprints and mechanisms. Nat. Rev. Cancer 2014; 14: 786-800.
3. Helleday T., Eshtad S., Nik-Zainal S. Mechanisms underlying mutational signatures in human cancers. Nat. Rev. Genet. 2014; 15: 585-598.
4. Bacolla A., Cooper D. N., Vasquez K. M. Mechanisms of base substitution mutagenesis in cancer genomes. Genes 2014; 5: 108-146.
5. Keim C., Kazadi D., Rothschild G., Basu U. Regulation of AID, the B-cell genome mutator. Genes Dev. 2013; 27: 1-17.
6. Carpenter M. A., Rajagurubandara E., Wijesinghe P., Bhagwat A. S. Determinants of sequence-specificity within human AID and APOBEC3G. DNA Repair 2010; 9: 579-587.
7. Roberts S. A., Lawrence M. S., Klimczak L. J., Grimm S. A., Fargo D., Stojanov P., Kiezun A., Kryukov G. V., Carter S. L., Saksena G., et al. An APOBEC cytidine deaminase mutagenesis pattern is widespread in human cancers. Nat. Genet. 2013; 45: 970-976.
8. Pfeifer G. P., Besaratinia A. UV wavelength-dependent DNA damage and human non-melanoma and melanoma skin cancer. Photochem. Photobiol. Sci. 2012; 11: 90-97.
9. Bacolla A., Temiz N. A., Yi M., Ivanic J., Cer R. Z., Donohue D. E., Ball E. V., Mudunuri U. S., Wang G., Jain A., et al. Guanine holes are prominent targets for mutation in cancer and inherited disease. PLoS Genet. 2013; 9: e1003816.
10. Alexandrov L. B., Nik-Zainal S., Wedge D. C., Aparicio S. A., Behjati S., BiankinA. V., Bignell G. R., Bolli N., Borg A., Borresen-Dale A. L., et al. Signatures of mutational processes in human cancer. Nature 2013; 500: 415-421.
11. Montgomery S. B., Goode D. L., Kvikstad E., Albers C. A., Zhang Z. D., Mu X. J., Ananda G., Howie B., Karczewski K. J., Smith K. S., et al. The origin, evolution, and functional impact of short insertion-deletion variants identified in 179 human genomes. Genome Res. 2013; 23: 749-761.
12. Ananda G., Walsh E., Jacob K. D., Krasilnikova M., Eckert K. A., Chiaromonte F., Makova K. D. Distinct mutational behaviors differentiate short tandem repeats from microsatellites in the human genome. Genome Biol. Evol. 2013; 5: 606-620.
13. Ananda G., Hile S. E., Breski A., Wang Y., Kelkar Y., Makova K. D., Eckert K. A. Microsatellite interruptions stabilize primate genomes and exist as population-specific single nucleotide polymorphisms within individual human genomes. PLoS Genet. 2014; 10: e1004498.
14. Sawaya S., Bagshaw A., Buschiazzo E., Kumar P., Chowdhury S., Black M. A., Gemmell N. Microsatellite tandem repeats are abundant in human promoters and are associated with regulatory elements. PLoS One 2013; 8: e54710.
15. Kim T. M., Laird P. W., Park P. J. The landscape of microsatellite instability in colorectal and endometrial cancer genomes. Cell 2013; 155: 858-868.
16. Gaffney D. J., McVicker G., Pai A. A., Fondufe-Mittendorf Y. N., Lewellen N., Michelini K., Widom J., Gilad Y., Pritchard J. K. Controls of nucleosome positioning in the human genome. PLoS Genet. 2012; 8: e1003036.
17. Zhu X., Schatz G. C. Molecular dynamics study of the role of the spine of hydration in DNA A-tracts in determining nucleosome occupancy. J. Phys. Chem. B2012; 116: 13672-13681.
18. Humphrey W., Dalke A., Schulten K. VMD: visual molecular dynamics. J. Mol. Graph. Model. 1996; 14: 33-38.
19. Kale L., Skeel R., Bhandarkar M., Brunner R., Gursoy A., Krawetz N., Phillips J., Shinozaki A., Varadarajan K., Schulten K. NAMD2: greater scalability for parallel molecular dynamics. J. Comput. Phys. 1999; 151: 283-312.
20. Martyna G. J., Tobias D. J., Klein M. L. Constant-pressure molecular-dynamics algorithms. J. Chem. Phys. 1994; 101: 4177-4189.
21. Warshel A., Levitt M. Theoretical studies of enzymic reactions: dielectric, electrostatic and steric stabilization of the carbonium ion in the reaction of lysozyme. J. Mol. Biol. 1976; 103: 227-249.
22. Laino T., Mohamed F., Laio A., Parrinello M. An efficient real space multigrid QM/MM electrostatic coupling. J. Chem. Theory Comput. 2005; 1: 1176-1184.
23. Kohn W., Sham L. J. Self-consistent equations including exchange and correlation effects. Phys. Rev. 1965; 140: A1133-A1138.
24. VandeVondele J., Krack M., Mohamed F., Parrinello M., Chassaing T., Hutter J. Quickstep: Fast and accurate density functional calculations using a mixed Gaussian

and plane waves approach. Comput. Phys. Commun. 2005; 167: 103-128.

25. Goedecker S., Teter M., Hutter J. Separable dual-space Gaussian pseudopotentials. Phys. Rev. B 1996; 54: 1703-1710.

26. Perdew J. P., Burke K., Ernzerhof M. Generalized gradient approximation made simple. Phys. Rev. Lett. 1996; 77: 3865-3868.

27. Woon D. E., Dunning T. H. Gaussian basis sets for use in correlated molecular calculations. IV. Calculation of static electrical response properties. J. Chem. Phys. 1994; 100: 2975-2988.

28. Blöchl P. E. Electrostatic decoupling of periodic images of plane-wave-expanded densities and derived atomic point charges. J. Chem. Phys. 1995; 103: 7422-7428.

29. Siddle K. J., Goodship J. A., Keavney B., Santibanez-Koref M. F. Bases adjacent to mononucleotide repeats show an increased single nucleotide polymorphism frequency in the human genome. Bioinformatics 2011; 27: 895-898.

30. Abecasis G. R., Auton A., Brooks L. D., DePristo M. A., Durbin R. M., HandsakerR. E., Kang H. M., Marth G. T., McVean G. A. An integrated map of genetic variation from 1, 092 human genomes. Nature 2012; 491: 56-65.

31. Green R. E., Krause J., Briggs A. W., Maricic T., Stenzel U., Kircher M., PattersonN., Li H., Zhai W., Fritz M. H., et al. A draft sequence of the Neandertal genome. Science 2010; 328: 710-722.

32. Gerstein M. B., Kundaje A., Hariharan M., Landt S. G., Yan K. K., Cheng C., MuX. J., Khurana E., Rozowsky J., Alexander R., et al. Architecture of the human regulatory network derived from ENCODE data. Nature 2012; 489: 91-100.

33. Wang J., Zhuang J., Iyer S., Lin X. Y., Greven M. C., Kim B. H., Moore J., PierceB. G., Dong X., Virgil D., et al. Factorbook. org: a Wiki-based database for transcription factor-binding data generated by the ENCODE consortium. Nucleic Acids Res. 2013; 41: D171-D176.

34. Mukherjee P., Lahiri I., Pata J. D. Human polymerase kappa uses a template-slippage deletion mechanism, but can realign the slipped strands to favour base substitution mutations over deletions. Nucleic Acids Res. 2013; 41: 5024-5035.

35. Campregher C., Scharl T., Nemeth M., Honeder C., Jascur T., Boland C. R., Gasche C. The nucleotide composition of microsatellites impacts both replication fidelity and mismatch repair in human colorectal cells. Hum. Mol. Genet. 2010; 19: 2648-2657.

36. Ellegren H. Microsatellites: simple sequences with complex evolution. Nat. Rev. Genet. 2004; 5: 435-445.

37. Haran T. E., Mohanty U. The unique structure of A-tracts and intrinsic DNA bending. Q. Rev. Biophys. 2009; 42: 41-81.

38. Giese B. Long-distance electron transfer through DNA. Annu. Rev. Biochem. 2002; 71: 51-70.

39. Richmond T. J., Davey C. A. The structure of DNA in the nucleosome core. Nature 2003; 423: 145-150.

40. Segal E., Widom J. Poly(dA: dT) tracts: major determinants of nucleosome organization. Curr. Opin. Struct. Biol. 2009; 19: 65-71.

41. Chua E. Y., Vasudevan D., Davey G. E., Wu B., Davey C. A. The mechanics behind DNA sequence-dependent properties of the nucleosome. Nucleic Acids Res. 2012; 40: 6338-6352.

42. Wang K., Yuen S. T., Xu J., Lee S. P., Yan H. H., Shi S. T., Siu H. C., Deng S., Chu K. M., Law S., et al. Whole-genome sequencing and comprehensive molecular profiling identify new driver mutations in gastric cancer. Nat. Genet. 2014; 46: 573-582.

43. Viduna D., Hinsen K., Kneller G. Influence of molecular flexibility on DNA radiosensitivity: a simulation study. Phys. Rev. E 2000; 62: 3986-3990.

44. Nikolova E. N., Bascom G. D., Andricioaei I., Al-Hashimi H. M. Probing sequence-specific DNA flexibility in A-tracts and pyrimidine-purine steps by nuclear magnetic resonance (13)C relaxation and molecular dynamics simulations. Biochemistry2012; 51: 8654-8664.

45. Brenlla A., Markiewicz R. P., Rueda D., Romano L. J. Nucleotide selection by the Y-family DNA polymerase Dpo4 involves template translocation and misalignment. Nucleic Acids Res. 2014; 42: 2555-2563.

46. Ling H., Boudsocq F., Woodgate R., Yang W. Crystal structure of a Y-family DNA polymerase in action: a mechanism for error-prone and lesion-bypass replication. Cell 2001; 107: 91-102.

47. Margolin Y., Shafirovich V., Geacintov N. E., DeMott M. S., Dedon P. C. DNA sequence context as a determinant of the quantity and chemistry of guanine oxidation produced by hydroxyl radicals and one-electron oxidants. J. Biol. Chem. 2008; 283: 35569-35578.

48. Mesaros C., Arora J. S., Wholer A., Vachani A., Blair I. A. 8-Oxo-2′-deoxyguanosine as a biomarker of tobacco-smoking-induced oxidative stress. Free Radic. Biol. Med. 2012; 53: 610-617.

49. Mangal D., Vudathala D., Park J. H., Lee S. H., Penning T. M., Blair I. A. Analysis of 7, 8-dihydro-8-oxo-2′-deoxyguanosine in cellular DNA during oxidative stress. Chem. Res. Toxicol. 2009; 22: 788-797.

50. Delaney S., Jarem D. A., Volle C. B., Yennie C. J. Chemical and biological consequences of oxidatively damaged guanine in DNA. Free Radic. Res. 2012; 46: 420-441.

51. Erkkila K. E., Odom D. T., Barton J. K. Recognition and reaction of metallointercalators with DNA. Chem. Rev. 1999; 99: 2777-2796.

52. Kanvah S., Joseph J., Schuster G. B., Barnett R. N., Cleveland C. L., Landman U. Oxidation of DNA: damage to nucleobases. Acc. Chem. Res. 2010; 43: 280-287.

53. Sugiyama H., Saito I. Theoretical studies of GG-specific photocleavage of DNA via electron transfer: significant lowering of ionization potential and 5′-localization of HOMO of stacked GG bases in B-form DNA. J. Am. Chem. Soc. 1996; 118: 7063-7068.

54. Meggers E., Michel-Beyerle M. E., Giese B. Sequence dependent long range hole transport in DNA. J. Am. Chem. Soc. 1998; 120: 12950-12955.

55. Giese B., Wessely S., Spormann M., Lindemann U., Meggers E., Michel-BeyerleM. E. On the mechanism of long-range electron transfer through DNA. Angew. Chem. Int. Ed. Engl. 1999; 38: 996-998.

56. Saito I., Takayama M., Sugiyama H., Nakatani K., Tsuchida A., Yamamoto M. Photoinduced DNA cleavage via electron transfer: demonstration that guanine residues located 5′ to guanine are the most electron-donating sites. J. Am. Chem. Soc. 1995; 117: 6406-6407.

57. Saito I., Nakamura T., Nakatani K., Yoshioka Y., Yamaguchi K., Sugiyama H. Mapping of the hot spots for DNA damage by one-electron oxidation: efficacy of GG doublets and GGG triplets as a trap in long-range hole migration. J. Am. Chem. Soc. 1998; 120: 12686-12687.

58. Yoshioka Y., Kitagawa Y., Takano Y., Yamaguchi K., Nakamura T., Saito I. Experimental and theoretical studies on the selectivity of GGG triplets toward one-electron oxidation in B-form DNA. J. Am. Chem. Soc. 1999; 121: 8712-8719.

59. Voityuk A. A., Jortner J., Bixon M., Rosch N. Energetics of hole transfer in DNA. Chem. Phys. Lett. 2000; 324: 430-434.

60. Wu Q., Van Voorhis T. Direct optimization method to study constrained systems within density-functional theory. Phys. Rev. A. 2005; 72: 024502.

61. Noblanc A., Damon-Soubeyrand C., Karrich B., Henry-Berger J., Cadet R., SaezF., Guiton R., Janny L., Pons-Rejraji H., Alvarez J. G., et al. DNA oxidative damage in mammalian spermatozoa: where and why is the male nucleus affected? Free Radic. Biol. Med. 2013; 65: 719-723.

62. Lim K. S., Taghizadeh K., Wishnok J. S., Babu I. R., Shafirovich V., Geacintov N. E., Dedon P. C. Sequence-dependent variation in the reactivity of 8-Oxo-7, 8-dihydro-2′-deoxyguanosine toward oxidation. Chem. Res. Toxicol. 2012; 25: 366-373.

63. Suzuki M., Kino K., Morikawa M., Kobayashi T., Komori R., Miyazawa H. Calculation of the stabilization energies of oxidatively damaged guanine base pairs with guanine. Molecules 2012; 17: 6705-6715.

64. Kubar T., Elstner M. Coarse-grained time-dependent density functional simulation of charge transfer in complex systems: application to hole transfer in DNA. J. Phys. Chem. B 2010; 114: 11221-11240.

65. Yokojima S., Yoshiki N., Yanoi W., Okada A. Solvent effects on ionization potentials of guanine runs and chemically modified guanine in duplex DNA: effect of electrostatic interaction and its reduction due to solvent. J. Phys. Chem. B2009; 113: 16384-16392.

CHAPTER 13

REVIEW OF CURRENT METHODS, APPLICATIONS, AND DATA MANAGEMENT FOR THE BIOINFORMATICS ANALYSIS OF WHOLE EXOME SEQUENCING

Riyue Bao,[1] Lei Huang,[1] Jorge Andrade,[1] Wei Tan,[2] Warren A Kibbe,[3] Hongmei Jiang,[4] and Gang Feng[3]

[1]Center for Research Informatics, The University of Chicago, Chicago, IL, USA

[2]IBM Thomas J. Watson Research Center, Yorktown Heights, New York, USA

[3]Biomedical Informatics Center (NUBIC), Clinical and Translational Sciences Institute (NUCATS), Northwestern University, Chicago, IL, USA

[4]Department of Statistics, Northwestern University, Evanston, IL, USA

ABSTRACT

The advent of next-generation sequencing technologies has greatly promoted advances in the study of human diseases at the genomic, transcriptomic, and epigenetic levels. Exome sequencing, where the coding region of the genome is captured and sequenced at a deep level, has proven to be a cost-effective method to detect disease-causing variants and discover gene targets. In this review, we outline the general framework of whole exome sequence data analysis. We focus on established bioinformatics tools and applications that support five analytical steps: raw data quality assessment, pre-processing, alignment, post-processing, and variant analysis (detection, annotation, and prioritization). We evaluate the performance of open-source alignment programs and variant calling tools using simulated and benchmark datasets, and highlight the challenges posed by the lack of concordance among variant detection tools. Based on these results, we recommend adopting multiple tools and resources to reduce false positives and increase the sensitivity of variant calling. In addition, we briefly discuss the current status and solutions for big data management, analysis, and summarization in the field of bioinformatics.

INTRODUCTION

High-throughput "-omics" technologies, such as microarray, sequencing, and quantitative real-time polymerase chain reaction (PCR), have been widely applied in basic biological and biomedical research for more than a decade. Advances in these techniques have enabled a broad spectrum of applications in genomic, transcriptomic, proteomic, metagenomic, and metabolomic studies.[1–8]

Two important challenges accompany these technologies: (1) How do we best manage the enormous amount of "-omics" data? (2) What are the most appropriate choices among the available computational methods and analysis tools[9]? In this review, we first present an overview of next-generation sequencing (NGS) technologies, highlight some of the issues associated with NGS data analysis, and then survey the current bioinformatics strategies and computational tools for whole exome variant detection. Based on our evaluation of the major analysis tools, we present our recommendations for the selection of variant analysis tools for specific research tasks. We will also discuss challenges in large-scale NGS data analysis and management.

First-generation sequencing such as Sanger and shotgun sequencing techniques were successfully employed for the

Citation: Riyue Bao, Lei Huang, Jorge Andrade, Wei Tan, Warren A Kibbe, Hongmei Jiang, and Gang Feng. Review of Current Methods, Applications, and Data Management for the Bioinformatics Analysis of Whole Exome Sequencing. Cancer Informatics. 2014; 13(Suppl 2):67-82. DOI: 10.4137/CIN.S13779.

creation of the human genome project (HGP). The HGP required 3.4 billion USD, 13 years, and collaboration of hundreds of international labs to complete the first human genome assembly. Since the introduction of NGS technologies in 2007, the price for sequencing a whole human genome has dropped rapidly. In early 2014, Illumina Inc. introduced the HiSeq X platform into the market, claiming a 1,000 USD personal genome price (experimental cost). With this new machine, 16 human genomes can be sequenced in 3 days at a depth of 30×.[10] Several new initiatives have been established to take advantage of this dramatic price drop to sequence thousands of human genomes. For instance, the Genome England project will sequence 100,000 genomes by 2017 and attach it to the medical record data as part of the UK medical system. In early March, Human Longevity Inc., founded by J. Craig Venter and two other partners, announced that it will build the world largest human genome sequencing center with the capacity of sequencing up to 40,000 human genomes per year.[11] The drop in per-base sequencing price is expected to drive the generation of immense amount of NGS data, creating a big data challenge in bioinformatics.

NGS sequencing experiments produce millions to billions of short sequence reads at a high speed. Current NGS platforms including Illumina, Ion Torrent/Life Technologies, Pacific Bioscience, Nanopore, and GenapSys can generate reads of 100–10,000 bases long,[12,13] allowing better coverage of the genome at lower cost. However, these platforms also generate huge amounts of raw data. For example, the raw data produced by Illumina HiSeq2500 platform add up to 1TB per run, covering 150–180 human whole exome sequencing (WES) samples at a depth of 50× or higher (Illumina Inc.). For tumor samples, a sequencing depth of at least 125× is recommended accounting for intra-tumor heterogeneity.[14,15]

Common NGS applications include DNA-seq, RNA-seq, ChIP-seq, and methyl-seq. DNA-seq can be applied to the whole genome sequencing (WGS), WES, or a specific targeted region of the genome. In general, the goal of DNA sequencing is to discover genomic variations in the form of single nucleotide variants (SNVs), small DNA insertions or deletions (indels), copy number variations (CNVs), or other structural variants (SVs), with the ultimate goal of associating those variations to human disease. RNA-seq that measures gene expression changes can be used to discover new transcripts including noncoding RNAs and detect transcript splicing or gene fusion events. ChIP-seq detects genome-wide transcription factor binding sites and chromatin-associated modifications. Methyl-seq is used to profile various types of DNA methylation such as 5-methylcytosine and 5-hydroxymethylcytosine at single nucleotide resolution. In this review, we focus primarily on WES techniques and data analysis. This evaluation of established bioinformatics tools covers the variant analysis workflow from the quality control (QC) of raw reads to prioritization of biologically meaningful or clinically relevant variants. We also explore current solutions for the management of big data and its applications in bioinformatics.

A typical workflow of WES analysis consists of the following steps: raw data QC, preprocessing, mapping, post-alignment processing, variant calling, annotation, and prioritization (Fig. 1).

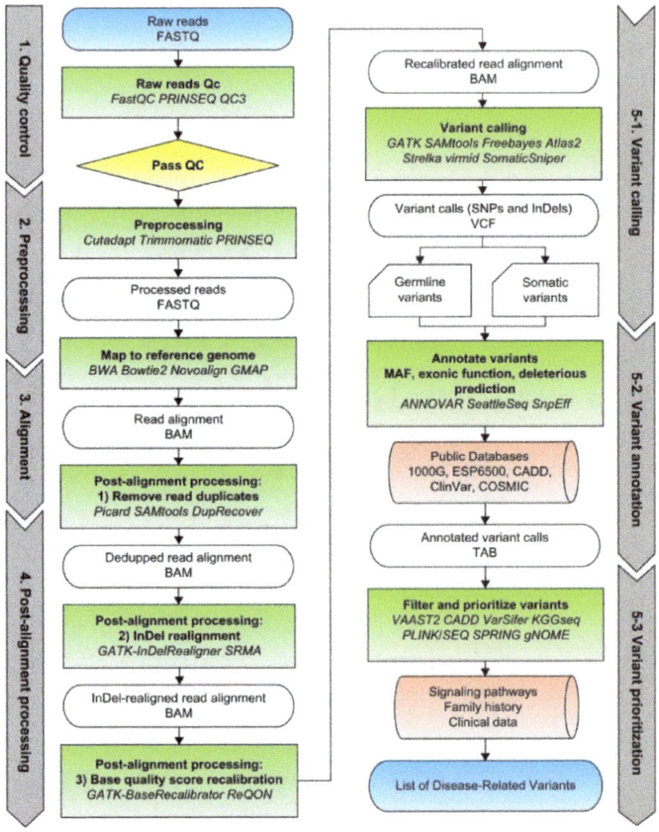

Figure 1. A general framework of WES data analysis. Five major steps are shown: raw reads QC, preprocessing, alignment, post-processing, and variant analysis (variant calling, annotation, and prioritization).

RAW DATA QC AND PREPROCESSING

FASTQ and FASTA are standard formats for representing biological sequence data. The FASTA format is a text-based representation of sequences, which begins with the sequence name followed by lines of single-letter coded nucleotides or amino acids. FASTQ format was developed to incorporate the Phred-scaled base quality scores to facilitate the assessment of sequence quality. It is widely accepted as the standard file format for NGS raw data.

Several tools have been developed to assess the quality of raw NGS data. Some of the commonly used ones include FastQC,[16] FastQ Screen,[17] FASTX-Toolkit,[18] NGS QC Toolkit,[19] PRINSEQ,[20] QC-Chain,[21] and recently published QC3.[22] FastQC is a java application that generates many useful data diagnosis and plots such as Phred score distribution along the reads, GC content distribution, read length distribution, and sequence duplication level. It also detects over-represented sequences that may be an indication of primer or adaptor contamination. With a comprehensive raw reads QC report generated by FastQC, researchers are able to determine whether any preprocessing steps such as base trimming, read filtering, or adaptor clipping are necessary prior to alignment.

Standard preprocessing procedure includes 3′ end adapter removal and trimming of low quality bases at the ends of the reads. Depending on the study design and use of the data, redundant reads and undesired sequences such as contamination from primers, adaptors, or other species may be removed at this point. Several tools are available to perform those tasks, such as Cutadapt[23] and Trimmomatic.[24] PRINSEQ[20] and QC3,[22] on the other hand, provide both QC and preprocessing functions as a suite. In addition to generic preprocessing functions as listed above, each program is equipped with their own custom features. For example, Cutadapt allows detection of adaptor sequences anywhere in the read and performs clipping afterwards. Trimmomatic is a java application that provides useful functions for handling paired-end reads. QC3 offers multi-perspective evaluation of data quality from the raw reads, mapped reads, and detected variants, with the unique feature to detect batch effects and cross-contamination.

SEQUENCE ALIGNMENT

After raw data QC and preprocessing, the next step is to map the reads to the reference genome and with high efficiency and accuracy. Alignment mapping is a classical "string match" task in computer science. For example, most web browsers and text editors provide a "Find" function to search for the perfect matching string with a given query. However, finding the optimal alignment for a sequence read requires an alignment algorithm that is tolerant to imperfect matches, where genomic variations may occur. Moreover, the algorithm needs to be able to align millions of reads at a reasonable speed. As a first step to address this challenge, the reference genome is usually indexed in a hash table for efficient querying.

Many different tools have been developed for short reads mapping. They use Burrows–Wheeler Transformation (BWT) compression techniques, Smith–Waterman (SW) Dynamic programing algorithm or the combination of both in order to find the optimal alignment match within an acceptable computational time.

Bowtie2[25] and BWA[26] are two well-known short reads alignment tools that implement BWT algorithm. SW is a score-based dynamic programming algorithm that provides at least one optimal local alignment even though the solution might not be unique. This algorithm is tolerant to mismatches and gaps at the expense of increased computational time. MOSAIK,[27] SHRiMP2,[28] and Novoalign (http://www.novocraft.com) are implementations of SW algorithms with increased alignment accuracy. Multithreading and/or MPI implementations are employed in those mapping tools allowing significant reduction in the runtime.

Evaluation of Short-Read Aligners

Using simulated datasets of 5 million 100-bp reads, we evaluated four commonly used alignment tools, Bowtie2, BWA, Novoalign V3, and genomic mapping and alignment program (GMAP).[25,29] Those reads were randomly generated from the human genome assembly hg19 with various types of genomic variations introduced. These include 1–5 bp SNVs, insertions only, deletions only, and mixed insertions and deletions (indels), as well as a mixture of SNVs and indels. If a read is aligned to the expected target region, we consider this as a true alignment. Sensitivity was calculated as the percentage of true alignments out of the total 5 million reads, and precision as the ratio of the number of true alignments to the total number of aligned reads.

Figure 2A and 2B shows the sensitivity and precision of the four tools with 1–5 bp variants introduced in each read. Bowtie2, Novoalign, and GMAP show high accuracy for all simulated genomic variations and are tolerant to various numbers of errors in a read. Bowtie2 and Novoalign have a similar level of sensitivity and precision between 94% and 97%. Compared with Bowtie2 and Novoalign, a 2–5% increased specificity and 4–10% reduced sensitivity were observed in GMAP alignment. The sensitivity of Novoalign drops to between 77% and 85% when a read contains 5 bp deletions. On the other hand, BWA performs well when there is at most one error in a read, and its sensitivity and specificity are reduced by 10% to 66% when attempting to align reads with more than one error.

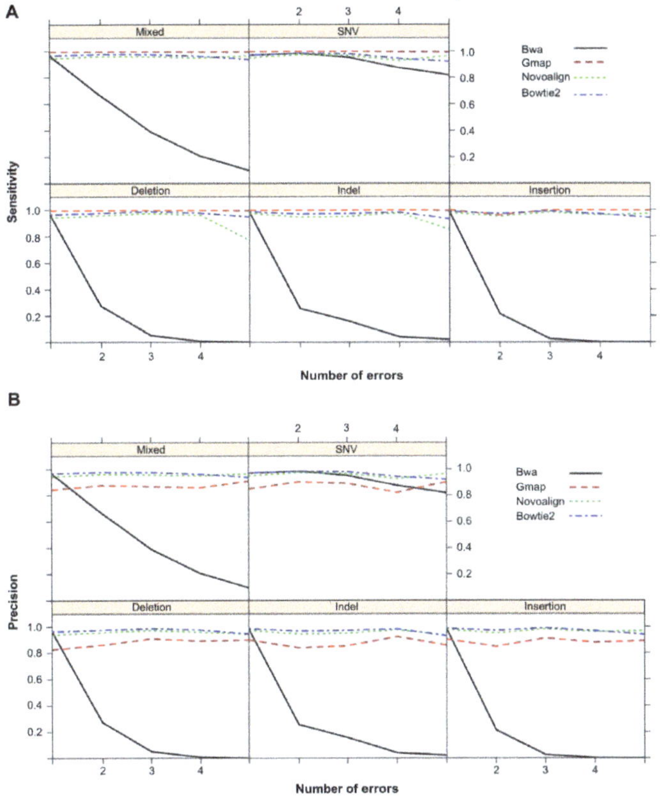

Figure 2. Comparison of alignment tools in terms of sensitivity (**A**) and precision (**B**) with 1–5 bp genomic variations per simulated read. Five sets of alignment are shown with introduction of errors categorized by types of errors (deletions only, insertions only, insertions and deletions (indels), SNVs, and a mixture of indels and SNVs (mixed)).

As for the running time of each tool, Bowtie2 and BWA are dramatically faster than the other two. Bowtie2 took approximately 8–25 minutes to finish aligning 5 million reads in each alignment, which is two to four times faster than BWA. A 5- to 10-fold increase in runtime was observed for GMAP compared to Bowtie2 and BWA. Novoalign took almost 3 hours to finish, which is probably due to its implementation of SW alignment algorithm as it usually takes longer time than BWT-based aligners. All tools were tested using one single processor on Northwestern University's Quest high-performance computing cluster.

POST-ALIGNMENT PROCESSING

After mapping reads to the reference genome, a three-step post-alignment processing procedure is recommended to minimize the artifacts that may affect the quality of downstream variant calling. It consists of read duplicate removal, indel realignment, and base quality score recalibration (BQSR).

In the alignment, reads aligned with exact mapping coordinates are considered "read duplicates," which represent either true DNA materials or PCR artifacts. The two cases, however, cannot be distinguished solely based on sequence or alignment information. Before sequencing, a library of DNA fragments from genomic regions of interest is prepared. Those fragments are amplified via certain amount of PCR cycles to provide a sufficient amount of DNA materials for sequencing, while limiting the duplication level of templates introduced by rounds of amplifications. For WES analysis, it is recommended to remove duplicates before variant calling, with the purpose of eliminating PCR-introduced bias due to uneven amplification of DNA fragments. Programs such as Picard MarkDuplicates (http://picard.sourceforge.net) and SAM-tools[26] determine read duplicates based on their identical 5′ mapping coordinates and orientation on the genome. 3′ Coordinates are usually not considered due to two reasons. First, the quality of bases generated by sequencers tends to drop down toward the 3′ end of a read; thus its alignment is less reliable compared to the 5′ bases. Second, if reads are trimmed at 3′ low-quality bases before alignment, they will have different read lengths resulting in different 3′ mapping coordinates. It should also be noted that read sequence information is not taken into account during duplicate removal. PCR duplicates do not necessarily have the identical sequence due to errors introduced in PCR amplification or sequencing processes. However, this may introduce bias in the calculation of variant frequency with de-duped read alignment. The bias becomes more severe in ultra-deep sequenced tumor samples (500× or higher), where removing large amounts of duplicates may affect allele frequency-based tumor subclone discovery and CNV detection. To address this issue, Zhou et al.[27] proposed a quantitative solution, DupRecover, which systematically estimates the degree of PCR-introduced bias and corrects for allele fractions.

After duplicate removal, the second step is to identify genomic regions that contain indels and improve the alignment quality in the target region. Compared to reads that contain only SNVs, mapping reads composed of indels requires gapped alignment which is more prone to noise. When aligning reads to the genome (discussed in the previous section), most short-read aligners walk through the reads one by one and the optimal alignment is determined for each read independently. As a result, the introduction of gaps in each alignment may be different among overlapping reads. The quality of the resulting gapped alignment can be improved by considering all aligned reads in the same region after mapping. Two algorithms have been developed to achieve this task: (1) local realignment of gapped reads to the reference genome or alternative candidate haplotypes; (2) local de novo assembly of the reads aligned around the target region followed by construction of a consensus sequence for indel discovery. Programs that implement one algorithm or a mixture of the two include SRMA[28] and IndelRealigner from the Genome Analysis Toolkit (GATK).[30] Some aligners and variant callers incorporate indel alignment improvement as part of the mapping or variant calling procedure. For example, Novoalign internally performs local SW alignments in addition to the hash-table based "seed" string searching to determine the optimal alignment for each read. The newly published GATK HaplotypeCaller program conducts a local de novo assembly of aligned reads prior to indel calling, which demonstrates to greatly improve the quality of indel calls.[31] Dindel implements a Bayesian approach to detect indels by calculating the posterior possibility of a haplotype after realigning reads to a number of candidate haplotypes constructed from the target region.[32] It is designed to accurately call indels correcting for mapping errors, base-calling errors, and sequencing errors in homopolymer-rich regions.

In the sequencing reads, each base is assigned with a Phred-scaled quality score generated by the sequencer, which represents the confidence of a base call. Base quality is a critical factor for accurate variant detection in the downstream analysis. However, the machine-generated scores are often inaccurate and systematically biased.[33]Therefore, BQSR is recommended to improve the accuracy of confidence scores before variant calling. It takes into account all reads per lane and analyzes covariation among the raw quality score, machine cycle, and dinucleotide content of adjacent bases. A corrected Phred-scaled quality score is reported for each base in the read alignment, assuming that all observed differences between the aligned reads and the reference genome are sequencing errors. One of the most commonly used BQSR programs is BaseRecalibrator from the GATK suite, which takes alignment files and recalibrates base scores across multiple sequencing runs. After base score recalibration, the variant calling accuracy was shown to be significantly improved and the bias was greatly reduced (GATK Best Practices recommendations[31,34]). Other well-established programs include Recab from the NGSUtils suite,[35] which provides similar functions as GATK BaseRecalibrator,[30] and the Bioconductor package ReQON,[36] which uses logistic regression for recalibration of the base quality scores. In addition, ReQON outputs a set of diagnostic data and plots before and after recalibration to illustrate the improved accuracy. Some aligners, such as Novoalign, implement BQSR as one of the internal post-alignment processing options and the output contains aligned reads with improved base score accuracy. It is important to exclude known variants before score recalibration, as those represent true genomic variations and should not be considered as sequencing errors. Most programs take a list of known variants in addition to the alignment files for recalibration. For genomes without known variants available, it is recommended to run the variant calling without BQSR first to generate a list of variants, filter for high-quality ones, and then re-run BQSR with the list of high-confidence variants as the known genomic variations. Furthermore, in case of targeted sequencing where only a small region of the genome is sequenced, BQSR is not recommended, as it will not be able to accurately estimate the errors with limited coverage of the genome.

VARIANT ANALYSIS

Variant analysis consists of genotyping, variant calling, annotation, and prioritization. Different types of genomic variants including SNVs, indels, CNVs, and large SVs can be detected from the sample by comparing the aligned reads to the reference genome. In cancer studies, it is important to distinguish somatic from germline variants as the two classes of variants often play distinct roles in tumor development. Germline variants are inherited mutations present in the germ cells, which are related to patient family history. Somatic variants are mutations that are present only in somatic cells and can be tissue-specific.

Variant Calling

With a reasonable number of samples available, multiple sample variant calling is usually recommended. By taking into account all reads from one genomic region across multiple samples, it reduces the possibility of calling randomly presented sequencing errors and increases the possibility of calling alleles of low frequency or low coverage in a single sample. As a result, the accuracy and sensitivity of multi-sample variant calling are in general much higher than single sample variant calling.[37–39] In some circumstances, however, multi-sample calling becomes less practical. If the sample size is large, the requirement of computational resources and time for multi-sample variant calling increases dramatically. Furthermore, if one project is executed at multiple stages where a subset of the samples are sequenced each time, multi-sample calling would require re-running the variant calling step when new samples are added. Therefore, it may be more feasible to conduct single sample variant calling under those circumstances. Alternatively, large amounts of samples may be pooled into smaller groups and called per group. It should be noted that the results may be different from calling all the samples together.

Programs available for germline variant calling include GATK [38],[30,31] SAMtools,[26]FreeBayes,[40] and Atlas2.[41] GATK is a NGS analysis suite that employs a MapReduce framework to accelerate the processing of large amounts of sequence alignment files in parallel. It implements a simple Bayesian model to estimate the likelihood of genotype in the sample based on the observed sequence reads that cover the specified locus.[30,31,34]GATK consists of two main variant calling programs, UnifiedGenotyper and HaplotypeCaller. UnifiedGenotyper calls SNVs and indels separately with the assumption that each variant locus is independent. HaplotypeCaller simultaneously detects SNVs, indels, and some SVs with increased accuracy by performing a local de novo assembly of the aligned reads (discussed in the previous section).

SAMtools contains a set of utilities for the manipulation of aligned sequence reads in the SAM/BAM format.[26] One of the available utilities, mpileup, scans every position along the covered genome, computes possible genotypes from the aligned reads, and calculates the likelihood that each of these genotypes is truly present in the sample. Another tool, bcftools, then uses the genotype likelihoods to call the SNVs and indels. The main difference of the variant calling models between SAMtools and GATK is the estimation of the genotype likelihood of SNVs and indels. The differences also lie in the filtering step, where SAMtools uses predefined filters while GATK learns the filters from the data.

FreeBayes is a haplotype-based short polymorphism caller that can simultaneously detect SNVs, indels, multi-base mismatches, polyallelic sites, polyploidy, as well as CNVs in a single sample, pooled samples, or mixed populations.[40] It is built on a Bayesian statistical framework.

Compared to the three tools discussed above, Atlas2 is quite different in the implementation of variant calling algorithms.[41] For data generated by SOLiD™platform, it uses logistic regression models trained on validated WES data to detected SNVs and indels. For Illumina data, it uses logistic regression models for calling indels and a mixture of logistic regression and a Bayesian model for SNV detection. Atlas2 consists of two separate programs, Atlas2-SNV and Atlas2-InDel.

Somatic Variant Detection

A major application of NGS variant analysis in cancer research is to distinguish somatic mutations in tumor cells from germline polymorphisms present in normal tissue. However, sequencing errors, insufficient variant coverage, sample contamination, and misclassification of germline variations often pose significant challenges in the detection of somatic variants. A number of tools have been developed to identify somatic mutations with paired tumor–normal samples, with their algorithms classified in two categories.

The first type of algorithms treats both samples as the same type, performs multi-sample variant calling on all samples, applies genotype-based subtraction methods with integration of sample pair information, and retrieves variants that are only present in tumor samples (somatic) or in both samples (germline). This approach is prone to false positives if a germline variant is not called due to low frequency in the normal sample and false negatives if the tumor samples possess low mutation levels that cannot be distinguished from sequencing error. However, this type of caller has been well established and has demonstrated high accuracy and sensitivity in calling variants, such as GATK,[30] SAMtools mpileup,[42] and Isaac variant caller.[43]

The other type of algorithm treats tumor and normal as paired samples from the onset and detects variants simultaneously on both samples using joint diploid genotype likelihoods or shared allele frequency between the samples. A number of somatic mutation tools have emerged in the past 2 years, including deepSNV,[44] Strelka,[45]MutationSeq,[46] MutTect,[47] QuadGT (http://www.iro.umontreal.ca/~csuros/quadgt), Seurat,[48] Shimmer[49] and SolSNP (http://sourceforge.net/projects/solsnp), jointSNVMix,[50] SomaticSniper,[51] VarScan2,[51] and Virmid.[52] Each of them is equipped with unique features and applications. For example, deepSNV[44] was specially designed for detection of subclonal variants in ultra-deep sequenced tumor samples. Most of the above tools require matching tumor–normal samples as mandatory input. We discuss five of them in detail.

JointSNVMix implements a probabilistic model for somatic mutation discovery. It utilizes the joint genotype information across the paired tumor/normal samples and treats them as being conditionally independent. This leads to an increase in specificity while the sensitivity stays nearly unchanged.[50]

Strelka is built on a novel Bayesian model where the normal sample is represented as a mixture of diploid germline variation with noises and the tumor sample is represented as a mixture of the normal sample with somatic variations through allele frequencies.[45] A score derived from the joint probability of a somatic variant and a specific genotype in the normal sample is used to make the variant call. This algorithm computes allele frequency variation in samples at any level without requiring an estimation of tumor purity.

SomaticSniper is a tool that compares the diploid genotype likelihood in the tumor and normal pair for the somatic variant calling.[51] The likelihood is computed using the germline genotyping algorithm adopted from the MAQ program with the consideration of the dependency between the tumor and normal genotypes from the sample patient.[42] A set of somatic detection filters is applied to the calls from the MAQ genotyping. The current tool does not take into account tumor purity or copy number state.

Varscan2 applies a heuristic and statistical algorithm to identify variants as germline or somatic and detects loss of heterozygosity events from the variant calls.[53] The genotype for each sample is first determined from mpileup calling of the variants, and then one-sided Fisher's exact test is applied to call somatic or germline variants. Tumor purity can be added to the program explicitly. This provides the robustness when the sample is contaminated or alternate ploidy exists in the normal sample.

Virmid uses the level of impurity in the sample to improve the somatic variation detection.[52] The algorithm estimates the sample contamination level and the disease genotypes using the maximum likelihood estimation. A Bayesian model utilizes the joint genotype probability of tumor and normal samples to call the variant. The unique design of the algorithm allows the program to recalibrate the genotype probabilities with varied contamination levels. This procedure reduces the running time and greatly improves the accuracy. The method also takes into account other types of noise such as sequencing and mapping errors, mapping bias, and CNV stage.

Evaluation of Variant Callers

Previous studies reported that the accuracy and robustness vary among variant calling methods.[54,55] Liu et al.[38] evaluated the performance of four variant callers, SAMtools, GATK, glftools, and Atlas2. They recommended GATK for general-purpose variant analysis. Xu et al.[56] compared the performance of five somatic SNV calling methods (GATK UnifiedGenotyper followed by simple subtraction, MuTect, Strelka, SomaticSniper, and VarScan2) for matched tumor–normal sequencing data. They used the NIST-GIAB gold standard dataset to demonstrate that the sensitivities of these methods vary in regard to the allelic fraction of the mutation in the tumor sample. Roberts et al.[57] conducted a comparison between VarScan, Somatic-Sniper, JointSNVMix2, and Strelka. Their results revealed substantial discordance among the called variants. To facilitate the method evaluation process, Talwalkar et al.[58] proposed a benchmarking methodology for the evaluation of the human genome variant callers. Their SMASH toolkit consists of three components: (1) short reads from NGS experiments; (2) reference genome; and (3) the validation data in standard VCF format are used to measure the algorithm performance. A set of evaluation metrics is defined to measure the accuracy of the variant calls and the computational performance of the algorithm.

We evaluated the performance of four variant callers (GATK UnifiedGenotyper, SAMtools mpileup, Atlas2, and FreeBayes) with read alignment generated by three aligners (BWA, Bowtie2, and Novoalign V3) using the NIST-GIAB benchmark genotype calls.[59] It contains 2,915,731 high-confidence SNVs, indels, and homozygous reference genotype calls for NA12878 (version 2.18) after integrating 14 datasets from five sequencing platforms. For the tool evaluation, we downloaded 100× NA12878 WES data generated by University of Washington (SRX079575). A total of 170,987,444 50 bp paired-end reads were preprocessed and mapped to the hg19 reference genome using the three aligners. On average, over 85% of the reads were mapped with mapping quality higher than 30.[42] The read alignment was post-processed for duplicate removal using Picard and indel realignment and base score recalibration using GATK.

Twelve sets of variants were generated from the three aligners and four callers, with low-quality calls removed. Variants not located in the targeted regions and/or with read depth lower than 6× were excluded from further consideration. The variants were further filtered to exclude those located within genome regions where no confident calls could be made.[59] A total of 21,661 calls were kept and compared to the benchmark set, which consists of 23,294 on-target calls covering approximately 31.9 Mb of the human exome. 95.3% of the variants were detected by at least two callers in reads mapped by at least two aligners (referred to as filter "2Aligner×2Caller"), yielding a final variant list with high sensitivity (99.04% for SNVs and 87.28% for small indels), specificity (>99.99%), and precision rate (>99%) with overall performance better than any single algorithm (Fig. 3). When comparing variants called from the alignment generated by each of the three aligners, close to 80% of the united variant set is concordantly detected by all four callers in each alignment. Furthermore, 83–93% of the united variant set was concordantly detected in all three alignments. When comparing SNVs with indels in between callers, indels showed a lower fraction of consistently called variants, which may be due to difficulties in gapped alignment and noises from variants present in the neighborhood regions (Fig. 4A and 4B).

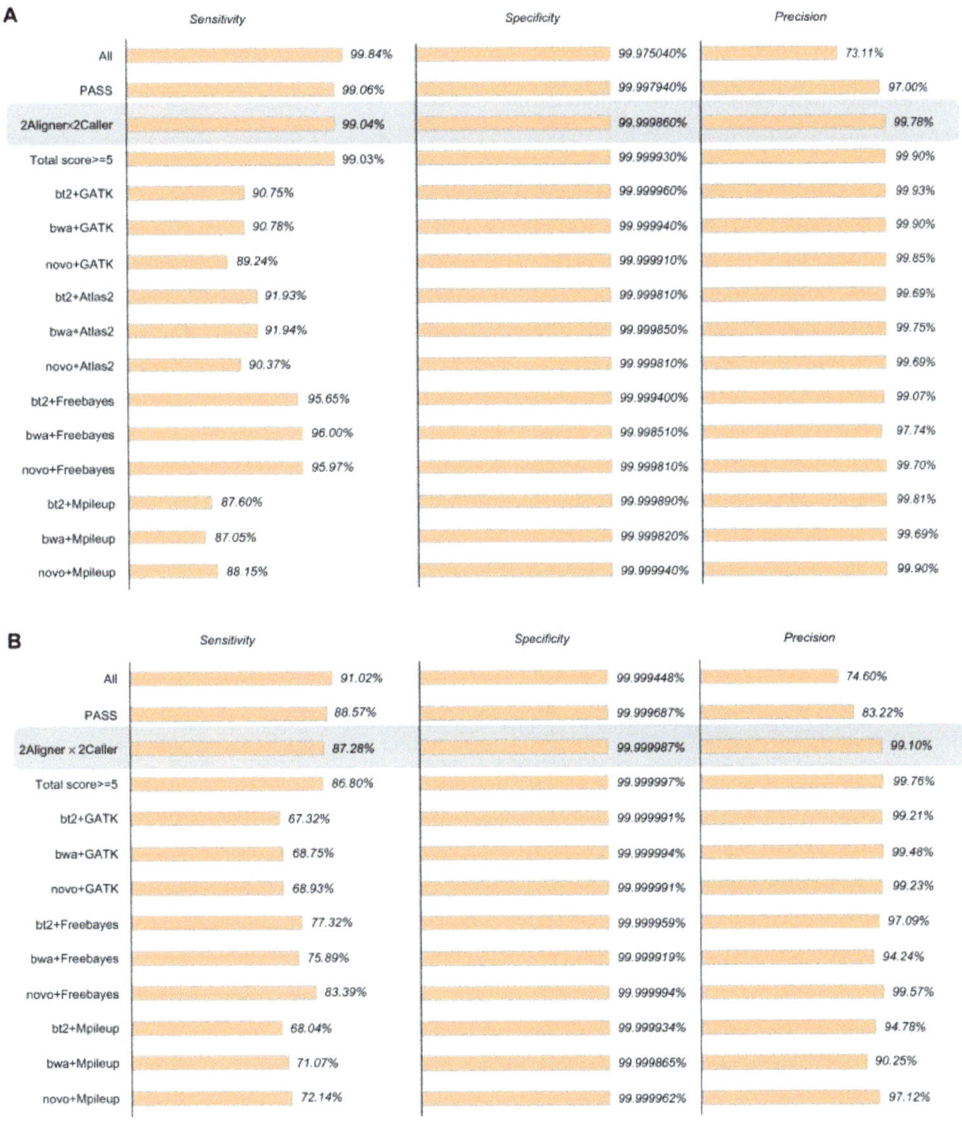

Figure 3. Evaluation results of variant callers with alignment generated by three aligners for SNVs (**A**) and indels (**B**). Aligners used for mapping the reads to the genome include Bowtie2 (bt2), BWA, and Novoalign (novo).

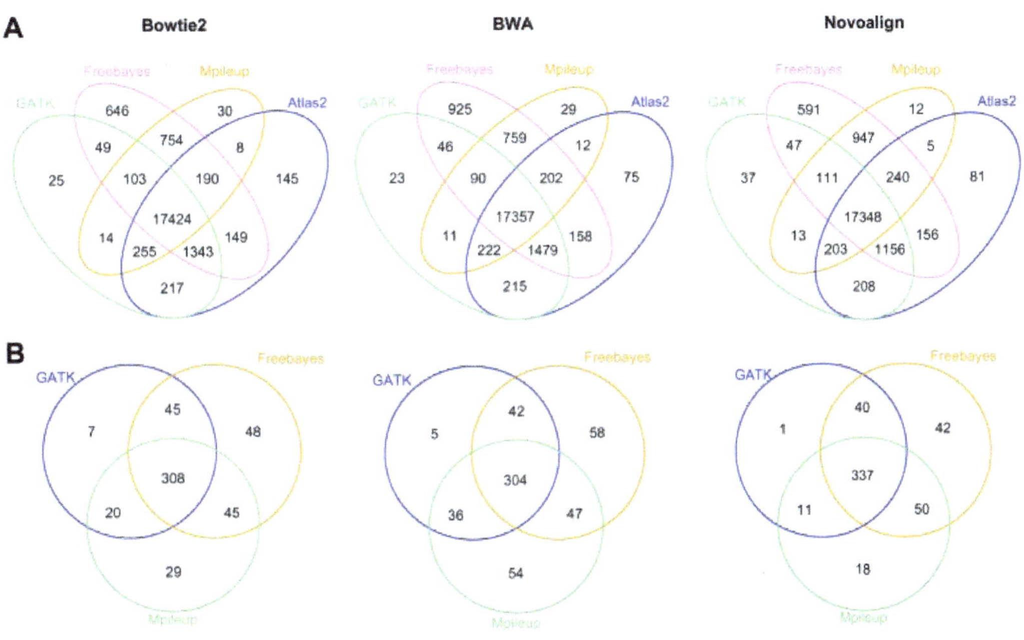

Figure 4. Counts of variants detected by four callers with alignment generated by three aligners as shown in Venn diagram of (**A**) SNVs and (**B**) indels.

In summary, these results suggest that by integrating multiple aligners and callers, variants of high confidence and high sensitivity are obtained. Among all 12 variant sets, variants called by FreeBayes from Novoalign-mapped reads are associated with the highest sensitivity (SNVs: 95.97% and indels: 83.39%) and precision rate (SNVs: 99.70% and indels: 99.57%). Therefore, if only one aligner and one caller to be chosen, a combination of Novoalign and FreeBayes is recommended for data analysis.

Variant Annotation

Even with the increasing amount of sequencing data, identification of disease-causing mutations from a background of random errors and polymorphisms remains challenging. After variants are detected, annotation attributes such as genomic feature, gene symbol, exonic function, and amino acid change can be attached to the variant list. Most studies focus on the non-synonymous SNVs and indels in the protein-coding regions, which account for 85% of the discovered disease-causing mutations in Mendelian disorders[60,61] and many disease-associated mutations in complex diseases. Synonymous SNVs are important for estimating the background mutation rate in the genome. For example, the background mutation rate in the melanoma patients was discovered to be approximately 10–20 mutations per MB of human exome, indicating a low mutation rate in the melanoma genome.[62,63]

In addition to the basic annotation discussed above, many programs have been developed to integrate public databases for additional information of the variants, including minor allele frequency (MAF) in normal populations, experimental evidence from clinical assays, deleterious prediction of variant function, and collection of variant and genes in known cancer or genetic disease studies. Detailed genomic contents, including tissue-specific expression, transcription factor binding sites, histone modifications, DNase I hypersensitive sites, enhancers, and eQTLs can be retrieved from the ENCODE project[64]and public databases such as RegulomeDB[65] and HaploReg[66] and added to variant annotation. While most of those tools only provide a small number of annotations, a few programs were developed to combine the annotations from numerous sources. One of the most commonly used variant annotation programs is ANNOVAR, which provides three annotation modes, gene-based, region-based, and filter-based, with a collection of over 4,000 public databases for annotation.[67] It integrates dbSNP, 1000Genomes, ESP6500, Complete Genomics personal genomes and NCI-60 human tumor cell line panel exome sequencing data for accessing MAF information, seven programs from the LJB23 database[68] plus Combined Annotation Dependent Depletion (CADD) database[69] for deleterious prediction, Phylop[70] and Genomic Evolutionary Rate Profiling (GERP)[71]score for indicating conservation of the mutated site across 29 mammalian species, as well as experimental evidence in pathogenesis of the variant from disease variant databases such as COSMIC[72] and ClinVar.[73] CADD uses a unified score for the potential deleteriousness of all 8.6 billion possible human mutations by comparing variants that survived natural selection with simulated mutations.[69] The database combines 63 annotations through a machine-learning framework. Those annotations consist of prediction scores from GERP, PolyPhen, and other programs. The simple metric provides a straightforward approach to interpret high-penetrance mutations in Mendelian disease and low-penetrance variants found in genome-wide association studies. SeattleSeq, a web-based variant annotation system, offers direct upload of variant lists through the website and annotates with public databases such as dbSNP and deleterious prediction programs such as PolyPhen.[74] It integrates databases of KEGG Pathways, CpG islands, transcription binding sites, and protein–protein interactions and provides useful information regarding the gene regulatory network and regulation of the mutated genes. Other stand-alone variant programs include VariantAnnotator from the GATK[30] and SnpEff.[75] They are equipped with various read filters or unique features such as gene set enrichment analysis for downstream analysis.

Variant Filtration and Prioritization

Exome sequencing of human samples with at least 100× coverage was estimated to detect approximately 20,000–30,000 SNV and indel calls on average.[54] The number of candidate variants is reduced using a three-step filtration and prioritization strategy to generate a short candidate mutation list for experimental validation.

The first step is to remove less reliable variant calls. This includes variants with low coverage, low quality, strand biased, located in SNV clusters, and/or supported by low-confidence read alignment.[76] Depending on the quality settings, this step may reduce the variant list by 1.3× to 1.5× (unpublished data). Moreover, in the evaluation of family trios, variants detected in a child that could not have been inherited from both parents (Mendelian errors) can be identified, most of which are likely to result from sequencing artifacts.[77] Improvement in accuracy of rare or novel variant calls was observed when multiple family members are called simultaneously with knowledge of pedigree structure. Of note, the accuracy of error detection and variant identification increases with the number of relatives and generations sequenced per family.[78,79] Filtering out these errors should be done with extra caution as this class of variants also includes de novo mutations in the child.

The second step is to restrict variants to those of relatively low population frequency, assuming that common variants are less likely to cause disease than rare ones. The MAF threshold should be carefully chosen based on disease model of the studies. For germline variants, usually over 70% of the variants are removed at MAF <1% (unpublished data). The filter does not have as dramatic effect as on somatic variants because most of those are novel and patient specific. Detection of compound heterozygous mutations is more complicated, in which case, each of the heterozygous variants does not cause observable phenotypic changes (recessive alleles), yet together they knock out both copies of the same gene and lead to disease outcome. In addition to rare variants, common variants may also show compound heterozygous effects with other variants in the same gene. For example, Heresbach et al.[80] reported that compound heterozygotes of three common variants in *NOD2/CARD15* gene increases the risk of Crohn's disease, which is greater than any single heterozygote. Furthermore, combinatorial effect of a common variant together with a novel mutation in the *MTHFR* gene was demonstrated to cause more severe symptoms in MTHFR deficiency disorders such as hyperh omocysteinaemia.[81]Publicly accessible tools that can detect compound heterozygous mutations in familial diseases include the GeneTalk Suite, which provides a web-based interface for customized variant filtration.[82] Stand-alone programs designed for identification of recessively acting variants (homozygotes or compound heterozygotes) include the SCOREASSOC program, which detects the association between variants based on derivation from the Hardy–Weinberg equilibrium in outbred populations.[83]

The third step is to prioritize variants relative to the disease. In general, SNVs can be ordered by their coding effect, in which case splice mutations (SNVs that occurred at splice donor or receptor sites) and nonsense mutations are in general more damaging than missense mutations. Indels, on the other hand, can be ordered based on whether they cause splice disruption or frameshift of the coding sequence. Furthermore, candidate disease-causing mutations are identified using discovery-based and hypothesis-based approaches. If little is known about the disease, the discovery-based approach is usually employed to filter and prioritize variants based on the mutation frequency across patient groups if the patients are genetically unrelated or on the inheritance pattern of mutations with the presence of a pedigree. Other criteria, including MAF values, deleterious function prediction, experimental evidence from published studies, and pathway information are also important factors for variant prioritization. In cancer genomics, computational approaches have been developed to distinguish driver mutations from passenger ones based on high mutation frequency in the patient cohort and highly damaging amino acid changes on the protein function or structure.[84]Moreover, variants that predispose an individual toward cancer may be identified from the germline variants and interpreted together with the pattern of somatic mutations.[85] In the hypothesis-based approach, a disease model is drawn based on the family history or known studies, and the discovery of candidate pathogenic mutations is driven by the proposed hypothesis. For Mendelian disorders, this may be recessive, dominant or compound heterozygous mutation model, or a combination of the models for the explanation of the inherited traits.[86] Moreover, pathways discovered in previous studies may guide discovery of new mutations in the known pathways, yet it is limited by the prior knowledge of the disease. It should be noted that new pathways can be established by linking newly identified mutations to known diseases. For example, six novel mutations in *C5ORF42* were reported to be causative in the development of Joubert syndrome in French Canadian population, although the pathogenic mechanism remains unknown.[87] In most cases, more than one approach is employed to identify the most interesting candidate variants.

Several tools have been developed to systematically filter, evaluate, and prioritize thousands of variants all at once, taking into account the annotation results of the variants, patient family information, as well as phenotypes and disease subtype information. VAAST2[88] implements an aggregative variant test combining the amino acid changes, allele frequencies, and phylogenetic conservation. The program generates a ranking list of variants sorted by its importance for the disease, which is useful for analysis of complex genetic diseases and rare Mendelian disorders. Other tools that are publicly available include VarSifer,[89] KGGseq,[90] PLINK/SEQ,[91] and SPRING.[92] A newly developed GUI tool, gNOME, allows direct upload of the variant file and performs streamline variant annotation, filtration, and prioritization, with its output summarized at variant, gene, or genome level.[93] One unique feature of gNOME is that it takes group information and detects disease-related genes or variants enriched in cases but not controls. One commercial variant analysis software, Ingenuity® Variant Analysis™ (http://www.ingenuity.com/variants), utilizes public databases such as those derived from the 1000 Genomes Project and the NHLBI Exome Sequencing Project (ESP6500), as well as experimental data collected from the literature and signaling pathways to filter and prioritize variants driven by data or with a specified disease model.

The above filter-based approaches have a higher success rate in identification of inherited or novel deleterious mutations in familial diseases, whereas with unrelated subjects, its power becomes limited. For rare diseases, one solution is to study unrelated probands with the same syndrome and focus on rare mutations that occurred in multiple individuals. The approach is known to be effective if the variant is clearly deleterious and disease related. Using this approach, Hood et al.[94] discovered truncating mutations in the C-terminal of SNF2-related CREBBP activator protein as the causative mutations for Floating-Harbour syndrome in five unrelated patients. In other cases, statistical tests are often necessary to discover mutations or target genes that contribute to a disease.[86]This is particularly important for addressing the effects of multiple rare variants that cause functional damage in a combinatorial manner. Previous studies identified rare predisposing variants that are significantly associated with complex traits such as colorectal adenomas,[95] high-density lipoprotein cholesterol,[96] and schizophrenia.[97]There are two major types of variant association tests: (1) burden tests, which include collapsing methods such as CAST,[98] CMC,[99] RareCover,[100] aSum,[101] and aggregation methods such as WSS,[102] KBAC, and[103] RBT[104]) (2) non-burden test based methods such as VT,[105] C-alpha,[106] EREC,[107] and SKAT.[108] For example, Wu et al.[108] applied SKAT to Dallas Heart Study data on 93 variants in three genes to test the association between log-transformed serum triglyceride levels and rare variants. The results showed that SKAT was a very powerful test for the dichotomous trait and performed comparably with burden-test-based methods for continuous trait. It should be noted that SKAT could be applied to both common and rare variants. Furthermore, the combination of genetic linkage, association analysis, and WES can serve as a useful approach to reduce the search space for rare variants in complex diseases with increased discovery power.[109]

With all the tools available and new ones emerging monthly, variant filtration and prioritization are becoming more automated like other parts of variant analysis such as the detection and annotation. Regardless, a deep understanding of the biological questions being asked and the etiology of the disease being studied is crucial for properly choosing tools and parameters that suit a study the best. Integration with clinical data such as patient diagnosis and family history is often helpful for identifying the variants that are responsible for the symptoms.[110]

NGS DATA MANAGEMENT

The technology evolution in molecular biology, especially in NGS, has moved biology into the big data era. For example, European Bioinformatics Institute (EBI) currently stores 20 petabytes of data, 2 petabyte of which is genomics related.[111] While the capacity of computing hardware doubles every 18 months, new biological data are doubling every 9 months. With this trend, the challenges faced by life scientists have been shifted from data acquisition to data management, processing, and knowledge extraction. While many studies have recognized the big data challenge, few systematically present approaches to tackle it. In this section, we propose a framework to address the big data challenge faced by biological scientists which consists of data, computation, workflow, and knowledge.

Data

As the volume of biological data generated by NGS instrumentation grows from hundreds of terabytes to petabytes, it has moved beyond the storage capabilities typically handled by individual scientists. Only a few organizations, such as the EBI and NIH NCBI have the capacity and the mandate to store large datasets and provide public access. The sustainability of this model is not clear as we move into the 100,000 genomes world. Big data hosted in the cloud seems to be a promising solution, one that many institutions are exploring. As an example, Amazon Simple Storage Service (Amazon S3) offers a cloud-based file system, with virtually unlimited capacity and charged by usage. Currently, NCBI stores a subset of the human genomics 1000 data (about 200 TB) in S3 (http://aws.amazon.com/1000 genomes).

Commercial vendors are coming into this space too. In the USA, Illumina's NGS environment is cloud based with data hosted by Amazon AWS. Illumina also has established Clinical Services Laboratory to provide genome-based testing to quickly access genomic aberrations and assist medical diagnosis. Another example is Genewiz, a DNA services company based in New Jersey, which provides similar services. International examples also exist, including Beijing Genomic Institute (BGI, Shenzhen), who acquired Complete Genomics last year, that has built solid connections with many healthcare providers. BGI also has set up two data centers, Bio-Data Centre (CliMB) and Biol-Cloud Computing centre (BGI Cloud), offering cloud-based data service, sequence alignment, and many other features. Meanwhile, healthcare providers are also launching genome-sequencing programs, such as Massachusetts General Hospital, Geisinger Health System, Scripps Health, and Inova Health System. For instance, the Falls Church, Virgina-based Inova Health System, has performed "1,500 complete DNA (whole genome) sequences as part of the first clinical study aimed at unraveling the mysteries of pre-term birth" in its Inova Translational Medicine Institute. In the next 2 years, they plan to produce 20,000 family-based whole genome sequences. In the foreseeable future, we expect to see more public or private biology data stored in cloud.

Despite the massive capacity provided by cloud computing, storing the huge amount of genome sequencing data is still challenging. To make the storage economic, various data compression techniques have been developed, including naive bit encoding, dictionary-based, statistical, and referential approaches.[112]

Computation

A key to reduce the latency of analyzing big data is to move computation to the data. This has two meanings in a cloud environment. First, computation should occur in the cloud, rather than moving data out of the cloud in order to compute on it. Cloud providers such as Amazon usually offer computation clusters collocated with the storage clusters and intra-cloud data movement is usually fast and free of charge. Second, when data reach the computation cluster, the computation framework should address the issue of data locality. Amazon AWS has offered a comprehensive suite of tools to process large volume of genomics data (http://aws.amazon.com/genomics/). For another example, Hadoop is an open-source software framework that parallelizes the computation and makes it easier to co-localize data and parallel computation. It has been well adopted in the field of bioinformatics,[113] including a few sequence alignment tools such as CloudBurst, Crossbow, SeqMapReduce, and CloudAligner.[114]

Workflow

New findings in biological sciences usually come out of multi-step data pipelines, also known as work-flows. Biologists already used a few workflow tools in the pre-cloud and pre-big-data era. Synapse (http://aws.amazon.com/swf/testimonials/swfsagebio) and Galaxy[115,116] are two cloud-based workflow tools dealing with big data. However, it is still necessary to globally optimize the data flow in an overall multi-step workflow in order to eliminate unnecessary data movement and redundant computation.

Knowledge embedded in big data, as well as the routines and workflow used to derive it, needs to be captured properly. New journals such as *Nature Protocols* are promoting the documentation of "recipe" style of step-by-step procedures leading to a discovery. In the bioinformatics domain, some scientists are promoting the idea of embedding a provenance workflow inside a publication to preserve both the data and the routine (http://tridentworkflow.codeplex.com). Since scientific publications are the carriers of new knowledge, we expect that knowledge-embedded data and workflows should be an integral part of future scientific publications.

CONCLUSION

The unprecedented reduction in the cost of high-throughput sequencing has made it possible to conduct ever-larger studies on human diseases. The bottleneck of NGS has shifted from producing sequence data to data management, analysis, and summarization. In this review, we examined bioinformatics software available for whole exome data analysis, including data preprocessing, alignment, post-alignment processing, variant calling, annotation, and prioritization tools. We compared the performance of alignment tools and variant calling programs using simulated and benchmark datasets. Along the way, we attempted to highlight frequent considerations and procedures necessary for the identification of causal variants of the disease.

Despite the large number of data analysis options, the complexity of the human genome and the lack of concordant results from different variant detection tools[54] highlight the urgency for developing community standardized protocols, tools, and benchmarks. As one of the efforts, The Cancer Genome Atlas (TCGA) and the International Cancer Genomics Consortium (ICGC) launched the ICGC-TCGA DREAM Somatic Mutation Calling Challenge, a competition of the best tools/pipelines for the detection of the cancer genome mutations using NGS data.[117] The organizers hope the challenge will accelerate the adoption of the best somatic variant identification techniques and help answer other key questions in cancer genome research.

The rapid growth of NGS technologies affords new opportunities to conduct large-scale patient sequencing projects, which facilitates discovery of tumor-specific mutations as potential targets for development of precision medicine.[45] As an example of the power of genomic techniques in understanding human disease and identifying new treatment options, J. Carpten's group discovered a unique pattern in the tumor mutation profile of 14 metastatic triple-negative breast cancer (TNBC) patients based on whole genome and transcriptome sequence data. They suggested that TNBC should be treated as a genetically different disease instead of a subtype of breast cancer.[118] They identified somatic mutations in the RAS/RAF/MEK and PI3K/AKT/mTOR signaling pathways, which have led to a clinical trial combining agents targeting of MEK and mTOR genes with encouraging results. Furthermore, by performing WES on a metastatic, castration-resistant prostate cancer patient, Van Allen et al. discovered somatic genomic mutations in the PI3K pathway and a *BRCA2* germline variant that may predispose individuals to cancer.[119] An increasing number of studies employ whole genome and exome sequencing to successfully identify disease-causing mutations that provide attractive treatment targets using gene therapy, therapeutic drugs, or transplantation.[120–122]

In the next five to twenty years, whole exome/genome analysis may be adopted as a routine procedure as part of the clinical laboratory for disease treatment. Universities and institutions including University of Pennsylvania (http://www.pennmedicine.org/personalized-diagnostics), Emory University (http://genetics.emory.edu/egl), and University of Washington (http://depts.washington.edu/labweb/Divisions/MolDiag/MolDiagGen/index.htm) have already established clinical laboratories that offer genetic testing that employ streamline NGS technologies and data analysis, with the goal of efficient molecular diagnosis of human diseases. Two years ago, the NGS: Standardization of Clinical Testing (Nex-StoCT) workgroup, led by The US Centers for Disease Control and Prevention, published general principles and guidelines for quality practices of NGS data used in clinical testing.[123] The next challenge will be data integration on millions of genomic variants, clinical records, and patient information, allowing novel discovery of variants contributing to disease, rapid retrieval of this information, and user-friendly visualization and decision support for care givers.

Establishment of disease-specific variant databases and web servers to store, retrieve, and view the genome-wide contents is an ongoing task at universities and institutions, including the integration of visualization tools such as GBrowse[124] and Bioconductor packages such as Gviz[125] and Shiny.[126] International efforts to build a federated cancer genome resource where genomics and clinical data can be easily searched and visualized via a web-based interface include the ICGC Cancer Genome Portal (http://icgc.org), which was established based on the ICGC cancer genomes and Oncomine,[127] which integrates both microarray and NGS data with a large collection of pre-analyzed results. NCI also has the cancer Genomics Data Commons effort (http://ocg.cancer.gov/news/genomics-data-commons) focused on TCGA and TARGET and an affiliated effort to build cloud computing capabilities around TCGA data (http://cbiit.nci.nih.gov/ncip/nci-cancer-genomics-cloud-pilots). Because of protected clinical information, some of the data in these resources require permission to access. Standards for patient consent, what is freely sharable, and what is available as limited datasets are under active debate. The upcoming changes to the USA Common Rule (http://www.hhs.gov/ohrp/humansubjects/anprm2011page.html) and some of the proposed activities by the Global Alliance for Genomics and Health (http://genomicsandhealth.org/) are examples of activities that are attempting to streamline and standardized data access for clinically derived genomic data. At some point in the near future, systems will be designed that enable the full processing of NGS data, from QA through to variant calling. Such a system might even incorporate pieces or concepts from existing frameworks like the Galaxy project.[115,128,129]

As whole exome and genome sequencing are now starting to be integrated into research and clinical practice, some ethical and legal issues have arisen regarding the resulting genomic data. Particularly, what procedures are needed to protect patient's privacy, how do we interpret the causality of identified variants in human disease and return the reports to the research participants or patients, how should we report incidental findings of pathologic mutations that were not originally ordered by the clinician.

For privacy-related issues, careful control has to be exerted in order to prevent the identity of individuals (de-identified datasets). Informed consent should be acquired to address return of individual research results. However, like other health information data, technological solutions cannot completely resolve confidentiality problems. Even if data are anonymized, individuals can be re-identified if phenotype and genotype data are combined.[130] To increase the confidence of identified disease-causing variants, standards of data interpretation need to be imposed. Clinician should convey the patient with any information about variants that is clinically important but not necessarily with information of variants with unknown significance.[131] Recently, MacArthur et al.[132]proposed a list of guidelines for interpreting and reporting sequence variants in human disease. For incidental findings, the American College of Medical Genetics and Genomics (ACMG) has published a policy statement that provides recommendations by an appointed working group for reporting incidental findings in clinical sequencing.[133]The working group developed a minimum list of 56 genes (23 cancer susceptibility genes) which should be reported as incidental findings. Due to the limitations of current technology, the disorders associated with the gene list are restricted to those caused by SNVs and small indels. The list will be refined and updated at least annually as the technologies evolve. The ACMG recommendation also describes the responsibility of the ordering clinician who provides comprehensive pre- and post-test genetic counseling to the patient.

As the costs of NGS continue to fall, we expect WGS will eventually overtake WES as the mainstream tool for human genomics and disease studies. The principles outlined in this review are generally applicable to WGS as well.

AUTHOR CONTRIBUTIONS

Conceived and designed the experiments: GF, HJ, JA, RB, LH. Analyzed the data: RB, LH, GF, HJ. Wrote the first draft of the manuscript: RB, LH, JA, WT, WAK, HJ, GF. Contributed to the writing of the manuscript: RB, LH, JA, WT, WAK, HJ, GF. Agree with manuscript results and conclusions: RB, LH, JA, WT, WAK, HJ, GF. Jointly developed the structure and arguments for the paper: RB, LH, JA, WT, HJ, GF. Made critical revisions and approved final version: RB, LH, JA, WT, HJ, GF. All authors reviewed and approved of the final manuscript.

REFERENCES

1. Golub TR, Slonim DK, Tamayo P, et al. Molecular classification of cancer: class discovery and class prediction by gene expression monitoring. Science. 1999; 286: 531–7.
2. Cancer Genome Atlas Research Network. Integrated genomic analyses of ovarian carcinoma. Nature. 2011; 474: 609–15.
3. Fertig EJ, Slebos R, Chung CH. Application of genomic and proteomic technologies in biomarker discovery. Am Soc Clin Oncol Educ Book. 2012; 32: 377–82.
4. Ansari NA, Bao R, Voichita C, Draghici S. Detecting phenotype-specific interactions between biological processes from microarray data and annotations. IEEE/ACM Trans Comput Biol Bioinform. 2012; 9: 1399–409.
5. Huang L, Zhao S, Frasor JM, Dai Y. An integrated bioinformatics approach identifies elevated cyclin E2 expression and E2F activity as distinct features of tamoxifen resistant breast tumors. PLoS One. 2011; 6: e22274.
6. Shi Y, Sha G, Sun X. Genome-wide study of NAGNAG alternative splicing in.Arabidopsis Planta. 2014; 239: 127–38.
7. Sheppard S, Lawson ND, Zhu LJ. Accurate identification of polyadenylation sites from 3' end deep sequencing using a naive Bayes classifier. Bioinformatics.2013; 29: 2564–71.
8. Jiang H, An L, Lin SM, Feng G, Qiu Y. A statistical framework for accurate taxonomic assignment of metagenomic sequencing reads. PLoS One. 2012; 7: e46450.
9. Pabinger S, Dander A, Fischer M, et al. A survey of tools for variant analysis of next-generation genome sequencing data. Brief Bioinform. 2013: bbs086.
10. Hayden E. [Is the $1,000 genome for real? January 15, 2014]; Nature. doi: 10.1038/nature.2014.14530.
11. Craig Venter Wants to Build the World's Biggest Database of Genome and Physiological Information. Mar 11, 2014. Available athttp: //www.technologyreview.com/news/525416/microbes-and-metabolites-fuel-an-ambitious-aging-project/
12. Kowalczyk SW, Wells DB, Aksimentiev A, Dekker C. Slowing down DNA translocation through a nanopore in lithium chloride. Nano Lett. 2012; 12: 1038–44.
13. Esfandyarpour H. Genapsys 100X solution: label-free fully-integrated "personal genomixer. J Biomol Tech. 2012; 23(suppl): S9.
14. Wagle N, Allen EMV, Treacy DJ, et al. MAP kinase pathway alterations in BRAF-mutant melanoma patients with acquired resistance to combined RAF/MEK inhibition.Cancer Discov. 2013; 4(1): 1–8.
15. Londin ER, Keller MA, D'Andrea MR, et al. Whole-exome sequencing of DNA from peripheral blood mononuclear cells (PBMC) and EBV-transformed lymphocytes from the same donor. BMC Genomics. 2011; 12: 464.
16. Andrews S. FastQC. Babraham Bioinformatics; Cambridge, UK: 2012.
17. Babraham Bioinformatics. FastQ Screen. Babraham Bioinformatics; Cambridge, UK: 2013.
18. Hannon Lab. FASTX-Toolkit. Hannon Lab; Cold Spring Harbor, NY: 2010.
19. Patel RK, Jain M. NGS QC toolkit: a toolkit for quality control of next generation sequencing data. PLoS One. 2012; 7: e30619.
20. Schmieder R, Edwards R. Quality control and preprocessing of metagenomic datasets. Bioinformatics. 2011; 27: 863–4.
21. Zhou Q, Su X, Wang A, Xu J, Ning K. QC-chain: fast and holistic quality control method for next-generation sequencing data. PLoS One. 2013; 8: e60234.
22. Guo Y, Zhao S, Sheng Q, et al. Multi-perspective quality control of Illumina exome sequencing data using QC3. Genomics. 2014; 103(5–6): 323–8.
23. Martin M. Cutadapt removes adapter sequences from high-throughput sequencing reads. EMBnet J. 2011; 17: 10–2.
24. Bolger AM, Lohse M, Usadel B. Trimmomatic: a flexible trimmer for Illumina sequence data. Bioinformatics. 2014; 30(15): 2114–20.
25. Wu TD, Nacu S. Fast and SNP-tolerant detection of complex variants and splicing in short reads. Bioinformatics. 2010; 26: 873–81.
26. Li H, Handsaker B, Wysoker A, et al. The sequence alignment/map format and SAMtools. Bioinformatics. 2009; 25: 2078–9.
27. Zhou W, Chen T, Zhao H, et al. Bias from removing read duplication in ultra-deep sequencing experiments. Bioinformatics. 2014; 30(8): 1073–80.
28. Homer N, Nelson SF. Improved variant discovery through local re-alignment of short-read next-generation sequencing data using SRMA. Genome Biol. 2010; 11: R99.
29. Wu TD, Watanabe CK. GMAP: a genomic mapping and alignment program for mRNA and EST sequences. Bioinformatics. 2005; 21: 1859–75.
30. McKenna A, Hanna M, Banks E, et al. The genome analysis toolkit: a MapReduce framework for analyzing next-generation DNA sequencing data. Genome Res.2010; 20: 1297–303.

31. Van der Auwera GA, Carneiro MO, Hartl C, et al. From FastQ data to high-confidence variant calls: the genome analysis toolkit best practices pipeline. Current Protocols in Bioinformatics. Curr Protoc Bioinform. 2013; 43: 11.10.1–33.
32. Albers CA, Lunter G, MacArthur DG, McVean G, Ouwehand WH, Durbin R. Dindel: accurate indel calls from short-read data. Genome Res. 2011; 21: 961–73.
33. Minoche AE, Dohm JC, Himmelbauer H. Evaluation of genomic high-throughput sequencing data generated on Illumina HiSeq and genome analyzer systems. Genome Biol. 2011; 12: R112.
34. DePristo MA, Banks E, Poplin R, et al. A framework for variation discovery and genotyping using next-generation DNA sequencing data. Nat Genet. 2011; 43: 491–8.
35. Breese MR, Liu Y. NGSUtils: a software suite for analyzing and manipulating next-generation sequencing datasets. Bioinformatics. 2013; 29: 494–6.
36. Cabanski CR, Cavin K, Bizon C, et al. ReQON: a bioconductor package for recalibrating quality scores from next-generation sequencing data. BMC Bioinformatics.2012; 13: 221.
37. Cheng AY, Teo YY, Ong RT. Assessing single nucleotide variant detection and genotype calling on whole-genome sequenced individuals. Bioinformatics.2014; 30: 1707–13.
38. Liu X, Han S, Wang Z, Gelernter J, Yang BZ. Variant callers for next-generation sequencing data: a comparison study. PLoS One. 2013; 8: e75619.
39. Liu Q, Guo Y, Li J, Long J, Zhang B, Shyr Y. Steps to ensure accuracy in genotype and SNP calling from Illumina sequencing data. BMC Genomics. 2012; 13(suppl 8): S8.
40. Garrison E, Marth G. Haplotype-based variant detection from short-read sequencing.*ArXiv*12073907 *Q-Bio*. 2012
41. Challis D, Yu J, Evani US, et al. An integrative variant analysis suite for whole exome next-generation sequencing data. BMC Bioinformatics. 2012; 13: 8.
42. Li H, Ruan J, Durbin R. Mapping short DNA sequencing reads and calling variants using mapping quality scores. Genome Res. 2008; 18: 1851–8.
43. Raczy C, Petrovski R, Saunders CT, et al. Isaac: ultra-fast whole-genome secondary analysis on Illumina sequencing platforms. Bioinformatics. 2013; 29: 2041–3.
44. Gerstung M, Beisel C, Rechsteiner M, et al. Reliable detection of subclonal single-nucleotide variants in tumour cell populations. Nat Commun. 2012; 3: 811.
45. Saunders CT, Wong WSW, Swamy S, Becq J, Murray LJ, Cheetham RK. Strelka: accurate somatic small-variant calling from sequenced tumor–normal sample pairs.Bioinformatics. 2012; 28: 1811–7.
46. Ding J, Bashashati A, Roth A, et al. Feature-based classifiers for somatic mutation detection in tumour–normal paired sequencing data. Bioinformatics. 2012; 28: 167–75.
47. Cibulskis K, Lawrence MS, Carter SL, et al. Sensitive detection of somatic point mutations in impure and heterogeneous cancer samples. Nat Biotechnol. 2013; 31: 213–9.
48. Christoforides A, Carpten JD, Weiss GJ, Demeure MJ, Hoff DDV, Craig DW. Identification of somatic mutations in cancer through Bayesian-based analysis of sequenced genome pairs. BMC Genomics. 2013; 14: 302.
49. Hansen NF, Gartner JJ, Mei L, Samuels Y, Mullikin JC. Shimmer: detection of genetic alterations in tumors using next-generation sequence data. Bioinformatics.2013; 29: 1498–503.
50. Roth A, Ding J, Morin R, et al. JointSNVMix: a probabilistic model for accurate detection of somatic mutations in normal/tumour paired next-generation sequencing data.Bioinformatics. 2012; 28: 907–13.
51. Larson DE, Harris CC, Chen K, et al. SomaticSniper: identification of somatic point mutations in whole genome sequencing data. Bioinformatics. 2012; 28: 311–7.
52. Kim S, Jeong K, Bhutani K, et al. Virmid: accurate detection of somatic mutations with sample impurity inference. Genome Biol. 2013; 14: R90.
53. Koboldt DC, Zhang Q, Larson DE, et al. VarScan 2: Somatic mutation and copy number alteration discovery in cancer by exome sequencing. Genome Res. 2012; 22: 568–76.
54. O'Rawe J, Jiang T, Sun G, et al. Low concordance of multiple variant-calling pipelines: practical implications for exome and genome sequencing. Genome Med.2013; 5: 28.
55. Alkan C, Sajjadian S, Eichler EE. Limitations of next-generation genome sequence assembly. Nat Methods. 2011; 8: 61–5.
56. Xu H, DiCarlo J, Satya RV, Peng Q, Wang Y. Comparison of somatic mutation calling methods in amplicon and whole exome sequence data. BMC Genomics.2014; 15: 244.
57. Roberts ND, Kortschak RD, Parker WT, et al. A comparative analysis of algorithms for somatic SNV detection in cancer. Bioinformatics. 2013; 29: 2223–30.
58. Talwalkar A, Liptrap J, Newcomb J, et al. SMaSH: a benchmarking toolkit for human genome variant calling. *ArXiv*13108420 *Q-Bio*. 2013
59. Zook JM, Chapman B, Wang J, et al. Integrating human sequence data sets provides a resource of benchmark SNP and indel genotype calls. Nat Biotechnol. 2014; 32: 246–51.
60. Botstein D, Risch N. Discovering genotypes underlying human phenotypes: past successes for Mendelian disease, future approaches for complex disease. Nat Genet.2003; 33: 228–37.
61. Rabbani B, Tekin M, Mahdieh N. The promise of whole-exome sequencing in medical genetics. J Hum Genet. 2014; 59: 5–15.
62. Shull AY, Latham-Schwark A, Ramasamy P, et al. Novel somatic mutations to PI3 K pathway genes in metastatic melanoma. PLoS One. 2012; 7: e43369.

63. Wei X, Walia V, Lin JC, et al. Exome sequencing identifies GRIN2 A as frequently mutated in melanoma. Nat Genet. 2011; 43: 442–6.

64. The ENCODE Project Consortium. An integrated encyclopedia of DNA elements in the human genome. Nature. 2012; 489: 57–74.

65. Boyle AP, Hong EL, Hariharan M, et al. Annotation of functional variation in personal genomes using RegulomeDB. Genome Res. 2012; 22: 1790–7.

66. Ward LD, Kellis M. HaploReg: a resource for exploring chromatin states, conservation, and regulatory motif alterations within sets of genetically linked variants. Nucleic Acids Res. 2012; 40: D930–4.

67. Wang K, Li M, Hakonarson H. ANNOVAR: functional annotation of genetic variants from high-throughput sequencing data. Nucleic Acids Res. 2010; 38: e164–e164.

68. Liu X, Jian X, Boerwinkle E. dbNSFP v2.0: a database of human non-synonymous SNVs and their functional predictions and annotations. Hum Mutat. 2013; 34: E2393–402.

69. Kircher M, Witten DM, Jain P, O'Roak BJ, Cooper GM, Shendure J. A general framework for estimating the relative pathogenicity of human genetic variants. Nat Genet. 2014; 46: 310–5.

70. Pollard KS, Hubisz MJ, Rosenbloom KR, Siepel A. Detection of nonneutral substitution rates on mammalian phylogenies. Genome Res. 2010; 20: 110–21.

71. Davydov EV, Goode DL, Sirota M, Cooper GM, Sidow A, Batzoglou S. Identifying a high fraction of the human genome to be under selective constraint using GERP++PLoS Comput Biol. 2010; 6: e1001025.

72. Forbes SA, Bindal N, Bamford S, et al. COSMIC: mining complete cancer genomes in the catalogue of somatic mutations in cancer. Nucleic Acids Res. 2011; 39(suppl 1): D945–50.

73. Landrum MJ, Lee JM, Riley GR, et al. ClinVar: public archive of relationships among sequence variation and human phenotype. Nucleic Acids Res. 2014; 42: D980–85.

74. Ng SB, Turner EH, Robertson PD, et al. Targeted capture and massively parallel sequencing of 12 human exomes. Nature. 2009; 461: 272–6.

75. Cingolani P, Platts A, Wang LL, et al. A program for annotating and predicting the effects of single nucleotide polymorphisms, SnpEff: SNPs in the genome of *Drosophila melanogaster* strain w1118; iso–2; iso–3. Fly (Austin) 2012; 6: 80–92.

76. Li H. Toward better understanding of artifacts in variant calling from high-coverage samples. Bioinformatics. 2014: btu356.

77. Patel ZH, Kottyan LC, Lazaro S, et al. The struggle to find reliable results in exome sequencing data: filtering out Mendelian errors. Appl Genet Epidemiol. 2014; 5: 16.

78. Li B, Chen W, Zhan X, et al. A likelihood-based framework for variant calling and de novo mutation detection in families. PLoS Genet. 2012; 8: e1002944.

79. Peng G, Fan Y, Palculict TB, et al. Rare variant detection using family-based sequencing analysis. Proc Natl Acad Sci U S A. 2013; 110: 3985–90.

80. Heresbach D, Gicquel-Douabin V, Birebent B, et al. NOD2/CARD15 gene polymorphisms in Crohn's disease: a genotype- phenotype analysis. Eur J Gastroenterol Hepatol. 2004; 16: 55–62.

81. Rummel T, Suormala T, Häberle J, et al. Intermediate hyperhomocysteinaemia and compound heterozygosity for the common variant c.677C > T and a MTHFR gene mutation. J Inherit Metab Dis. 2007; 30: 401.

82. Kamphans T, Sabri P, Zhu N, et al. Filtering for compound heterozygous sequence variants in non-consanguineous pedigrees. PLoS One. 2013; 8: e70151.

83. Curtis D. Approaches to the detection of recessive effects using next generation sequencing data from outbred populations. Adv Appl Bioinform Chem. 2013; 6: 29–35.

84. Raphael BJ, Dobson JR, Oesper L, Vandin F. Identifying driver mutations in sequenced cancer genomes: computational approaches to enable precision medicine. Genome Med. 2014; 6: 5.

85. Frank SA. Genetic predisposition to cancer–insights from population genetics. Nat Rev Genet. 2004; 5: 764–72.

86. Ionita-Laza I, Makarov V, Yoon S, et al. Finding disease variants in Mendelian disorders by using sequence data: methods and applications. Am J Hum Genet.2011; 89: 701–12.

87. Srour M, Schwartzentruber J, Hamdan FF, et al. Mutations in C5ORF42 cause Joubert syndrome in the French Canadian population. Am J Hum Genet. 2012; 90: 693–700.

88. Hu H, Huff CD, Moore B, Flygare S, Reese MG, Yandell M. VAAST 2.0: improved variant classification and disease-gene identification using a conservation-controlled amino acid substitution matrix. Genet Epidemiol. 2013; 37: 622–34.

89. Teer JK, Green ED, Mullikin JC, Biesecker LG. VarSifter: visualizing and analyzing exome-scale sequence variation data on a desktop computer. Bioinformatics.2012; 28: 599–600.

90. Li M-X, Gui H-S, Kwan JSH, Bao S-Y, Sham PC. A comprehensive framework for prioritizing variants in exome sequencing studies of Mendelian diseases. Nucleic Acids Res. 2012; 40: e53–e53.

91. ATGU. PLINK/SEQ. Massachusetts General Hospital; 2012.

92. Wu J, Li Y, Jiang R. Integrating multiple genomic data to predict disease-causing nonsynonymous single nucleotide variants in exome sequencing studies. PLoS Genet.2014; 10: e1004237.

93. Lee I-H, Lee K, Hsing M, et al. Prioritizing disease-linked variants, genes, and pathways with an interactive whole-genome analysis pipeline. Hum Mutat.2014; 35(5): 537–47.

94. Hood RL, Lines MA, Nikkel SM, et al. Mutations in SRCAP, encoding SNF2-related CREBBP activator

protein, cause floating-harbor syndrome. Am J Hum Genet.2012; 90: 308–13.

95. Azzopardi D, Dallosso AR, Eliason K, et al. Multiple rare nonsynonymous variants in the adenomatous polyposis coli gene predispose to colorectal adenomas. Cancer Res.2008; 68: 358–63.

96. Cohen JC, Kiss RS, Pertsemlidis A, Marcel YL, McPherson R, Hobbs HH. Multiple rare alleles contribute to low plasma levels of HDL cholesterol. Science. 2004; 305: 869–72.

97. Walsh T, McClellan JM, McCarthy SE, et al. Rare structural variants disrupt multiple genes in neurodevelopmental pathways in schizophrenia. Science. 2008; 320: 539–43.

98. Morgenthaler S, Thilly WG. A strategy to discover genes that carry multi-allelic or mono-allelic risk for common diseases: a cohort allelic sums test (CAST) Mutat Res.2007; 615: 28–56.

99. Li B, Leal SM. Methods for detecting associations with rare variants for common diseases: application to analysis of sequence data. Am J Hum Genet. 2008; 83: 311–21.

100. Bhatia G, Bansal V, Harismendy O, et al. A covering method for detecting genetic associations between rare variants and common phenotypes. PLoS Comput Biol.2010; 6: e1000954.

101. Han F, Pan W. A data-adaptive sum test for disease association with multiple common or rare variants. Hum Hered. 2010; 70: 42–54.

102. Madsen BE, Browning SR. A groupwise association test for rare mutations using a weighted sum statistic. PLoS Genet. 2009; 5: e1000384.

103. Liu DJ, Leal SM. A novel adaptive method for the analysis of next-generation sequencing data to detect complex trait associations with rare variants due to gene main effects and interactions. PLoS Genet. 2010; 6: e1001156.

104. Ionita-Laza I, Buxbaum JD, Laird NM, Lange C. A new testing strategy to identify rare variants with either risk or protective effect on disease. PLoS Genet.2011; 7: e1001289.

105. Price AL, Kryukov GV, de Bakker PIW, et al. Pooled association tests for rare variants in exon-resequencing studies. Am J Hum Genet. 2010; 86: 832–8.

106. Neale BM, Rivas MA, Voight BF, et al. Testing for an unusual distribution of rare variants. PLoS Genet. 2011; 7: e1001322.

107. Lin D-Y, Tang ZZ. A general framework for detecting disease associations with rare variants in sequencing studies. Am J Hum Genet. 2011; 89: 354–67.

108. Wu MC, Lee S, Cai T, Li Y, Boehnke M, Lin X. Rare-variant association testing for sequencing data with the sequence kernel association test. Am J Hum Genet.2011; 89: 82–93.

109. Rosenthal EA, Ranchalis J, Crosslin DR, et al. Joint linkage and association analysis with exome sequence data implicates SLC25 A40 in hypertriglyceridemia. Am J Hum Genet. 2013; 93: 1035–45.

110. Van El CG, Cornel MC, Borry P, et al. Whole-genome sequencing in health care.Eur J Hum Genet. 2013; 21: 580–4.

111. Marx V. Biology: the big challenges of big data. Nature. 2013; 498: 255–60.

112. Wandelt S, Bux M, Leser U. Trends in genome compression. Curr Bioinf. 2013; 9: 1–12.

113. Russell J. Hadoop's rise in life sciences. BioIT World. 2012

114. Wandelt S, Rheinländer A, Bux M, Thalheim L, Haldemann B, Leser U. Data management challenges in next generation sequencing. Datenbank-Spektrum.2012; 12: 161–71.

115. Goecks J, Nekrutenko A, Taylor J. Galaxy: a comprehensive approach for supporting accessible, reproducible, and transparent computational research in the life sciences. Genome Biol. 2010; 11: R86.

116. Afgan E, Baker D, Coraor N, et al. Harnessing cloud computing with galaxy cloud.Nat Biotechnol. 2011; 29: 972–4.

117. Boutros PC, Ewing AD, Ellrott K, et al. Global optimization of somatic variant identification in cancer genomes with a global community challenge. Nat Genet.2014; 46: 318–9.

118. Craig DW, O'Shaughnessy JA, Kiefer JA, et al. Genome and transcriptome sequencing in prospective metastatic triple-negative breast cancer uncovers therapeutic vulnerabilities. Mol Cancer Ther. 2013; 12: 104–16.

119. Van Allen EM, Foye A, Wagle N, et al. Successful whole-exome sequencing from a prostate cancer bone metastasis biopsy. Prostate Cancer Prostatic Dis. 2014; 17: 23–7.

120. Chiang P-W, Wang J, Chen Y, et al. Exome sequencing identifies NMNAT1 mutations as a cause of Leber congenital amaurosis. Nat Genet. 2012; 44: 972–4.

121. Goh V, Helbling D, Biank V, Jarzembowski J, Dimmock D. Next-generation sequencing facilitates the diagnosis in a child with twinkle mutations causing cholestatic liver failure. J Pediatr Gastroenterol Nutr. 2012; 54: 291–4.

122. Worthey EA, Mayer AN, Syverson GD, et al. Making a definitive diagnosis: successful clinical application of whole exome sequencing in a child with intractable inflammatory bowel disease. Genet Med. 2011; 13: 255–62.

123. Gargis AS, Kalman L, Berry MW, et al. Assuring the quality of next-generation sequencing in clinical laboratory practice. Nat Biotechnol. 2012; 30: 1033–6.

124. Stein LD, Mungall C, Shu S, et al. The generic genome browser: a building block for a model organism system database. Genome Res. 2002; 12: 1599–610.

125. Hahne F, Durinck S, Ivanek R, Mueller A, Lianoglou S, Tan G. Gviz: Plotting Data and Annotation Information along Genomic Coordinates. 2014

126. RStudio Inc Shiny: Web Application Framework for R. 2014.

127. Rhodes DR, Kalyana-Sundaram S, Mahavisno V, et al. Oncomine 3.0: genes, pathways, and networks in a collection of 1 8,000 cancer gene expression profiles. Neoplasia. 2007; 9: 166–80.

128. Blankenberg D, Von Kuster G, Coraor N, et al. Galaxy: a web-based genome analysis tool for experimentalists. Curr Protoc Mol Biol. 2010 Chapter 19: Unit 19.10.1–21.

129. Giardine B, Riemer C, Hardison RC, et al. Galaxy: a platform for interactive large-scale genome analysis. Genome Res. 2005; 15: 1451–5.

130. Gymrek M, McGuire AL, Golan D, Halperin E, Erlich Y. Identifying personal genomes by surname inference. Science. 2013; 339: 321–4.

131. Institute of Medicine. Integrating Large-Scale Genomic Information into Clinical Practice: Workshop Summary. Washington, DC: The National Academies Press; 2012.

132. MacArthur DG, Manolio TA, Dimmock DP, et al. Guidelines for investigating causality of sequence variants in human disease. Nature. 2014; 508: 469–76.

133. Green RC, Berg JS, Grody WW, et al. ACMG recommendations for reporting of incidental findings in clinical exome and genome sequencing. Genet Med. 2013; 15: 565–74.

CHAPTER 14

THE EFFECTS OF POLYMORPHISMS ON HUMAN GENE TARGETING

David R. Deyle,[1] Li B. Li,[1] Gaoying Ren,[1] and David W. Russell[1,2]

[1]Department of Medicine, University of Washington, Seattle, WA, 98195, USA

[2]Department of Biochemistry, University of Washington, Seattle, WA, 98195, USA

ABSTRACT

DNA mismatches that occur between vector homology arms and chromosomal target sequences reduce gene targeting frequencies in several species; however, this has not been reported in human cells. Here we demonstrate that even a single mismatched base pair can significantly decrease human gene targeting frequencies. In addition, we show that homology arm polymorphisms can be used to direct allele-specific targeting or to improve unfavorable vector designs that introduce deletions.

INTRODUCTION

Gene targeting has revolutionized genetics and allowed for site-specific manipulation of the mammalian genome. Increasingly, it is now used to engineer human cell lines and it may be used therapeutically in the future. The human genome contains multiple types of genetic variations, with single-nucleotide polymorphisms (SNPs) occurring on average every 1000–2000 bp when comparing haplotypes (1). Because the homology arms present in targeting vectors are frequently several killobases in length, these polymorphisms could impact homologous pairing and recombination.

Prior studies in mouse embryonic stem cells showed that gene targeting frequencies were lower when vectors contained as few as 0.6% DNA mismatches, which has led to the routine use of isogenic DNA for the preparation of targeting constructs (2–4). Surprisingly, in the case of human cells, a comparison of targeting frequencies at eight different loci in seven human cell lines concluded that isogenic DNA was not advantageous and that human recombination was tolerant of mismatches (5). However, in these human cell experiments, it was not established if the targeting constructs included polymorphisms not present in the chromosomal target loci. Given the importance of sequence homology on recombination in other systems, we examined this issue again in human cells. Here we show that the number and position of SNPs affect gene targeting, and that the inclusion of polymorphisms in vectors can enhance targeting at a specific allele and the generation of a deletion mutation.

MATERIALS AND METHODS

Plasmids and Vectors

The plasmids pA2HSN5′ (6), pA2-APPe3ITKNA (7), pA2HPe3 (8) and pHPe2/3 (8) have been described. Plasmids pLHSNΔ53O and pLHSNins4O are based on pLHSNO (9), but contain a 53-bp deletion at base pair 63 or a 4-bp insertion at base pair 64 of the*neo* reading frame, respectively. Plasmid

Citation: Deyle D.R., Li L.B., Ren G, Russell D.W. The effects of polymorphisms on human gene targeting. Nucleic Acids Research. 2014; 42(5):3119-3124. DOI: 10.1093/nar/gkt1303.

pLHSNΔ53O-SNP6 is identical to pLHSNΔ53O except that G to A transitions were introduced by sequential site-directed mutagenesis into the 5′ homology arm at −2, −87, −304, −700, −999, −1503 bp relative to the *neo* translation start site. Plasmids pLHSNΔ53O−2; pLHSNΔ53O−87, −304, −700; pLHSNΔ53O−999, and −1503; and pLHSNΔ53O−1503 contain only the specified SNPs. Plasmids pLHSΔ1Nins4O, pLHSΔ4Nins4O, pLHSΔ1Nins4Δ1O and pLHSΔ4Nins4Δ4O are identical to pLHSNins4O except that 1 or 4 bp deletions were engineered at base pair −5 and +103 relative to the *neo* start site. Plasmids pA2HPe3(i2i3+1) and pHPe2/3(i2i3+1) are identical to pA2HPe3 and pHPe2/3, respectively, except for 1 bp insertions located 15 bp upstream and 20 bp downstream of*HPRT* exon 3. All plasmid sequences are available on request.

Retroviral vectors were produced by calcium phosphate transfection of PG13 packaging cells (10) with pLHSNΔ53O, pLHSNins4O or their derivatives, collection of medium 2 days later and passage through 0.45-μm filters. Adeno-associated virus (AAV) vectors AAV-HSN5′, AAV-APPe3ITKNA, AAV-HPe3 and AAV-HPe3(i2i3+1) were prepared by transient cotransfection of helper and vector plasmids as described (11). All AAV vector stocks were serotype 2, purified on iodixanol gradients, and titered by quantifying full-length vector genomes on alkaline Southern blots (12). To ensure that an equivalent number of AAV genomes were used to transduce HT-1080 subclones in the *HPRT*experiments, AAV-HPe3 and AAV-HPe3(i2i3+1) titers were determined on the same alkaline Southern blot.

Cell Culture

HT-1080 human fibrosarcoma cells (13) were cultured at 37°C in Dulbecco's modified Eagle's medium with 10% heat-inactivated fetal bovine serum (Hyclone), 100 U/ml of penicillin and 100 μg/ml streptomycin. Moloney murine leukemia virus (MLV) provirus target sites were introduced into HT-1080 cells and selected with hygromycin to produce polyclonal transduced populations derived from $>10^4$ independent transduction events, as determined by plating dilutions of the transduced cells in selective medium as described (9). Down syndrome induced pluripotent stem cells (iPSCs) were derived from Down syndrome fibroblasts (AG06872; Coriell Institute for Medical Research) as described previously (7). iPSCs were grown on irradiated mouse embryonic fibroblasts as described (14).

Gene Targeting

SNPs were engineered into the MLV target sites rather than the AAV targeting vectors to allow for the rapid generation of genetic variants and the use of a single gene targeting vector to minimize experimental variation. AAV-HSN5′ was used to correct *neo* mutations as described (9). *APP* locus gene targeting was conducted as described (7). In Figure 4, independent HT-1080 subclones Δ4 c1 and Δ4 c2 were engineered to contain a 4-bp deletion in exon 3 of *HPRT*, and independent subclones +4 c1 and +4 c2 harbor a 4-bp insertion in exon 3 of *HPRT* as previously described (8). When correcting *HPRT* mutations with AAV vectors, HT-1080 *HPRT* subclones were plated at 5.5×10^5 cells per 6 cm dish on day 1 and infected with AAV-HPe3 or AAV-HPe3(i2i3+1) at an multiplicity of infection of 10^4 genome-containing particles on day 2. On day 3, cells were treated with trypsin and replated at 0.008 and 99% dilutions in 10-cm dishes and 15-cm dishes, respectively. On day 4, the cells in 99% dishes were grown in HAT medium (contains hypoxanthine, aminopterin, and thymidine) and the 0.008% dishes were grown without selection. After 7–10 days, colonies were counted. Gene targeting with transfected linearized plasmids pHPe2/3 and pHPe2/3(i2i3+1) was done by transfecting with Superfect reagent (Qiagen) as described (8). Briefly, HT-1080 subclones with *HPRT* mutations were plated at 5×10^5 cells per 10-cm dish (42 dishes for each subclone) on day 1, and 20 dishes were each transfected with 10 μg of linearized pHPe2/3 and 0.1 μg of pCMVβ, or 10 μg of linearized pHPe2/3(i2i3+1) and 0.1 μg of pCMVβ on day 2. pCMVβ, which expresses the *lacZ* gene from a cytomegalovirus (CMV) promoter (Clontech) was used as a transfection efficiency control. On day 3, the transfected cells were treated with trypsin and pooled, 5×10^4cells were plated in a well of a 6-well plate for β-galactosidase staining, and the remaining cells were counted and plated in 20 15-cm dishes. On day 4, the six-well plate was stained for β-galactosidase expression and the number of positive foci determined. On day 5, the remaining dishes were switched to HAT medium for selection, and the surviving colonies were counted 10–14 days later.

DNA Analysis and Plasmid Rescue

Genomic DNA was isolated by the Puregene DNA purification protocol (Gentra Systems/Qiagen). To determine whether mutations were introduced into the target loci during retroviral production or cell expansion, genomic DNA was isolated from HT-1080 polyclonal populations containing LHSNΔ53O and LHSNΔ53O-SNP6 and the MLV sequences homologous to AAV-HSN5′ were amplified by polymerase chain reaction (PCR) using primers LHSNO-for (ACCTGAGGAAGGGAGTCGAT) and LHSNO-rev (CGCTATGTCCTGATAGCGGT). Ten independent PCR products for both LHSNΔ53O and LHSNΔ53O-SNP6 were cloned into the TA cloning vector pGEM T-easy (Promega), sequenced and all were found to lack mutations. Integrated MLV provirus target loci were rescued as described (15) with the following modifications: 20 μg of genomic DNA containing a corrected MLV site was digested with 80 units of Kpn I, extracted with phenol and chloroform and precipitated with ethanol. The resulting DNA fragments were resuspended and circularized with 2000 U of T4 DNA ligase in 400 μl at 16°C overnight. The DNA was precipitated, resuspended in 10 μl of H_2O and 2 μg was electroporated into supercompetent *Escherichia coli* DH10B cells (Invitrogen). Targeted *APP* alleles were determined by PCR amplification of the 5′ homology region as described (7).

Statistical Analysis

In all cases, statistical significance was determined using Student's *t*-test to compare gene targeting frequencies. $P < 0.05$ were considered significant.

RESULTS

Homology Arm SNPs Influence Gene Targeting Frequencies

We designed a SNP-dependent targeting assay in which a MLV vector that confers hygromycin resistance is used to deliver a mutated *neomycin phosphotransferase* (*neo*) gene, which is then corrected by an AAV gene targeting vector to generate G418-resistant colonies. AAV-mediated gene targeting is efficient in human cells (16), so targeting frequencies can be measured accurately with this system. The MLV vectors LHSNΔ53O and LHSNΔ53O-SNP6 used to deliver the target loci both have a 53-bp deletion in the *neo* gene, but LHSNΔ53O-SNP6 also contains six G to A SNPs in the 5′ homology arm located at −2, −87, −304, −700, −999 and −1503 bp relative to the *neo*start site. The targeting vector AAV-HSN5′ contains 3149 bp of sequence homology to LHSNΔ53O, with a truncated *neo* gene that lacks the 53 bp deletion.

HT-1080 human fibrosarcoma cells were transduced with LHSNΔ53O or LHSNΔ53O-SNP6 to generate polyclonal populations consisting of at least 10^4 independent target proviruses. This allowed us to avoid genomic position effects, which can influence targeting frequencies significantly (6). To ensure the fidelity of MLV target loci in these polyclonal populations, integrated proviral DNA was amplified by PCR and the region homologous to the AAV targeting vector was sequenced. We analyzed 10 independent PCR products from each polyclonal population and did not identify any mutations. Both populations were infected with AAV-HSN5′ and the number of G418-resistant colonies determined (Figure 1A). The six SNPs present in LHSNΔ53O-SNP6 decreased targeting ~5-fold. Next, we introduced additional MLV vectors with different combinations of SNPs in the homology arm region to determine the effects of SNP position and number on gene targeting (Figure 1A). A single SNP reduced targeting when it was present 64 bp from the *neo* deletion (LHSNΔ53O−2 target), but SNPs located ~1 kb or more from the deletion had little or no measurable effect (LHSNΔ53O−999,1503, for example). In general, the decrease in targeting frequencies due to SNP mismatches was greatest when they were closest to the mutation being corrected, and the effect of multiple SNPs was additive.

Mapping the extent of chromosomal sequence changes

We used this system to determine the extent of vector homology arm sequence introduced into the human genome during gene targeting by tracking which SNPs were present in targeted loci. The LHSNΔ53O-SNP6 vector contains a bacterial promoter and replication origin that allows for the rescue of integrated proviruses as circularized bacterial plasmids. Sixty independent gene-targeted clones were isolated, and their LHSNΔ53O-SNP6 target loci were rescued from genomic DNA and sequenced. Based on the presence of SNPs, 72% of targeting events incorporated all of the identifiable vector homology arm sequence into the chromosome, extending to the distal 1016 bp of the 5′ homology arm over 1.5 kb from the 53-bp deletion mutation (Figure 1B). This could be due to preferential recombination near the vector inverted terminal repeats, which may form a recombinogenic hairpin structure (17), or extended tracts of mismatch repair occurring on vector: chromosome heteroduplexes. The next most frequent region that limited the extent of vector SNP incorporation during targeting was in the SV40 promoter located 366–762 bp from the *neo* deletion, which could be due to genomic instability associated with these sequences (18). Three clones had a discontinuous SNP pattern that could not be explained by a single recombination or gene conversion event. In these cases, segmented DNA mismatch repair could have led to incomplete inheritance of the SNPs as a contiguous block, or there may have been four recombination crossover events to account for the discontinuity. None of the targeted loci corrected the *neo* mutation without an accompanying A to G conversion.

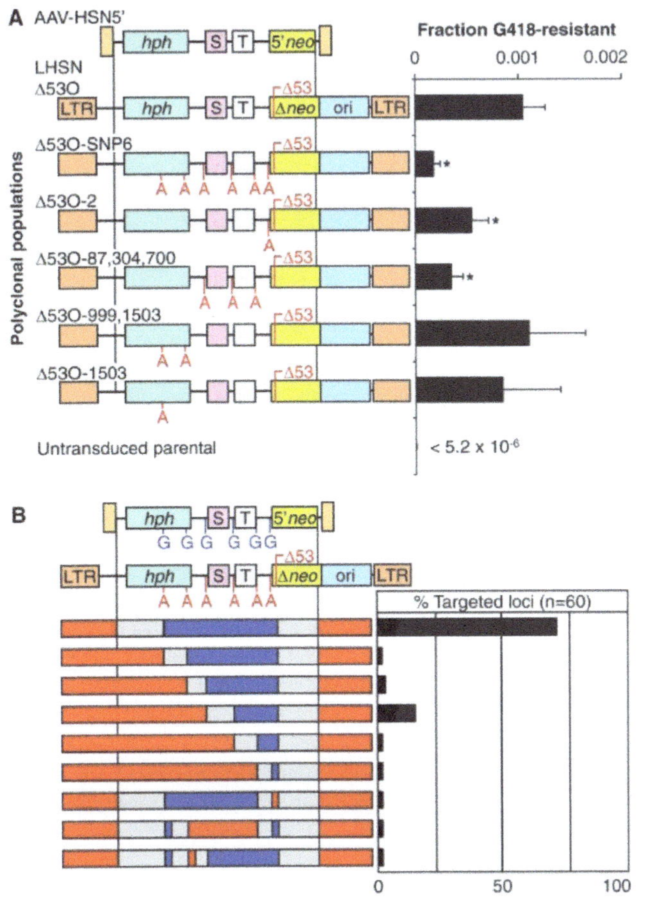

Figure 1. Gene targeting at MLV target loci with SNPs. (**A**) Schematic of the AAV-HSN5′ targeting vector and six MLV provirus loci with the location of the 53 bp *neo* gene deletion (Δ53) and the G to A SNPs shown. Targeting frequencies were measured as the fraction of G418-resistant colonies obtained after infecting polyclonal HT-1080 populations containing each indicated MLV provirus with AAV-HSN5′. LTR, long terminal repeat; *hph*, hygromycin phosphotransferase; S, SV40 promoter; T, Tn5 bacterial promoter, *5′neo*, truncated neo gene; ori, p15A plasmid origin; *$P < 0.05$ versus LHSNΔ53O. (**B**) Graphic illustration of SNPs present in the recovered targeted loci. Colors indicate the presence of MLV target SNPs (red), AAV vector SNPs (blue) or segments between identifiable SNPs (gray).

SNPs Affect Gene Targeting at an Endogenous Locus

Based on these results, we reasoned that SNPs should also impact targeting at an endogenous chromosomal locus. The amyloid precursor protein (*APP*) gene on chromosome 21 contains two SNPs located 339- and 389-bp 5′ of exon 3. We transduced a Down syndrome–iPSC line harboring three different *APP* SNP haplotypes (GT, GG and CG) with an AAV targeting vector that contained the GT SNPs in the 5′ homology arm, and was designed to introduce an internal ribosome entry site (IRES) and *TKNeo*gene into exon 3 (7) (Figure 2A). Twelve G418-resistant, targeted clones were isolated, and the targeted and untargeted alleles were amplified by PCR and sequenced to determine which allele was targeted (Figure 2B). The allele lacking SNP mismatches was targeted in nine of the clones, while those with one and two mismatches were targeted in two and one clones, respectively (Figure 2C). These data corroborate our MLV provirus experiments, and they show that SNPs can also decrease targeting at a chromosomal locus in a human stem cell.

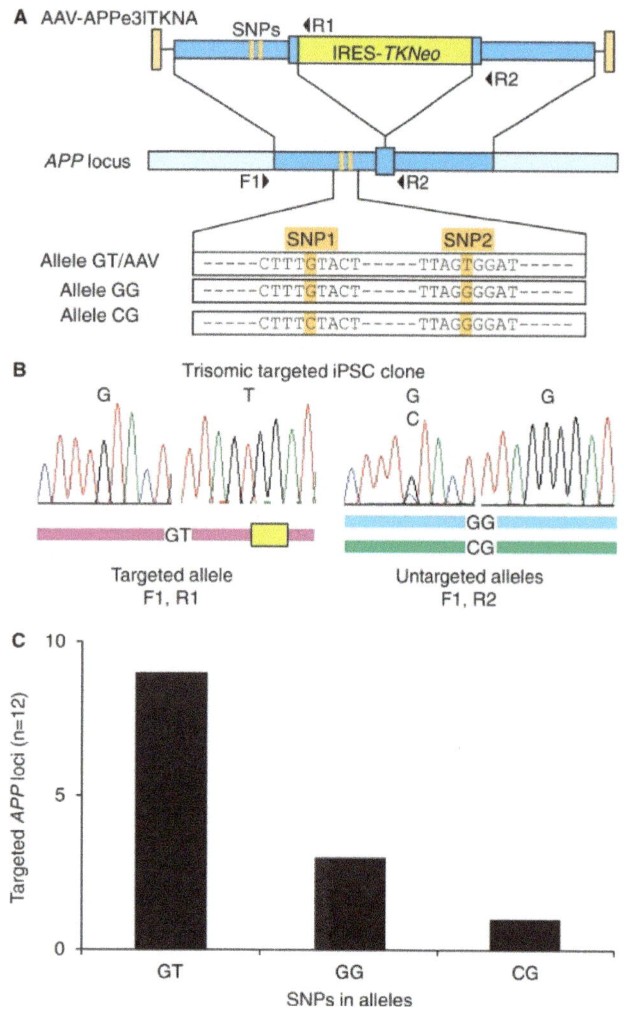

Figure 2. Targeting in trisomic iPSC cells containing three different SNP haplotypes. (**A**) *APP* locus showing AAV-APPe3ITKNA targeting vector, forward (F1) and reverse (R1, R2) PCR primers, SNP locations and sequences found in each of three *APP* alleles. (**B**) Representative sequence reads are shown demonstrating targeting at the GT allele in a Down syndrome iPSC clone. (**C**) The percentage of targeted *APP* alleles with each SNP haplotype.

Insertion Polymorphisms can Increase Gene Targeting with Deletion Vectors

We previously found that introducing an insertion mutation by gene targeting was ~10 times more efficient than introducing a deletion, suggesting that cellular DNA repair and recombination mechanisms preferentially preserve the unpaired vector insertions that arise in vector: chromosome heteroduplexes (8). We reasoned that insertion polymorphisms in the vector homology arms might increase targeting by a related mechanism. To test this, we redesigned the MLV-LHSNO provirus targeting system so that a 4-bp *neo* gene insertion would be corrected by introducing an unfavorable deletion with the AAV targeting vector. When this target backbone was specifically modified to include 1- or 4-bp silent flanking deletions that could be corrected by favorable insertions in the AAV vector, targeting frequencies increased 4-fold, with the best results obtained when 4-bp insertions were present on both sides of the *neo* mutation (Figure 3). Similar results were obtained when correcting mutations in the endogenous X-linked*HPRT* locus. HT-1080 subclones engineered to contain a 4-bp deletion in exon 3 of *HPRT* (8) were corrected by the AAV-HPe3 targeting vector containing wild-type sequence at 45–93 times the frequency of subclones containing a 4-bp insertion in*HPRT*. In comparison, the AAV-HPe3(i2i3+1) targeting vector containing silent 1-bp insertions in the flanking introns increased targeting 3- to 4-fold in the subclones with an unfavorable insertion in exon 3 (Figure 4A and B).

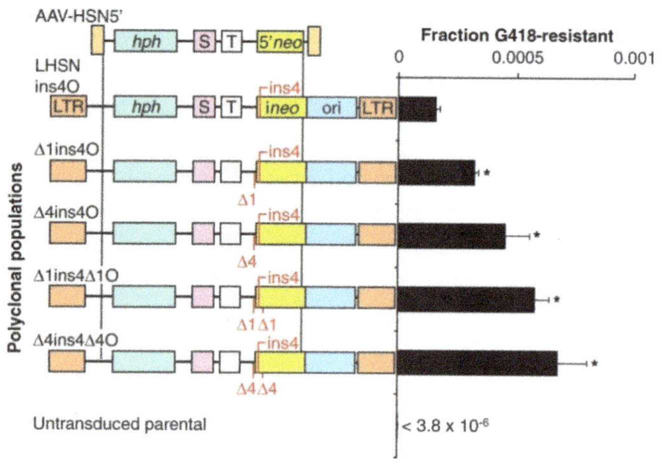

Figure 3. Gene targeting with a deletion vector. Illustration of MLV provirus targets with a 4-bp *neo* insertion (ins4) and different flanking deletions (Δ1, Δ4), with their targeting frequencies as shown in Figure 1. *$P < 0.05$ versus LHSNins4O.

Silent insertions also increased targeting when transfecting plasmid-based targeting constructs (Figure 4C), demonstrating that the effects of sequence polymorphisms on gene targeting are not limited to AAV vectors. Although we did not calculate statistical significance in these experiments, the results were consistent in two pairs of clones.

DISCUSSION

These experiments show that DNA polymorphisms have a significant impact on human gene targeting, as demonstrated at multiple target loci in normal and transformed human cells. Similar results were obtained when using single-stranded AAV gene targeting vectors or double-stranded plasmid constructs. Based on these findings, human targeting vectors should contain isogenic DNA to maximize targeting frequencies, as even a single polymorphism can significantly reduce targeting. This can complicate vector design and require the construction of multiple vector stocks, especially when preparing for clinical gene targeting applications in genetically diverse human populations.

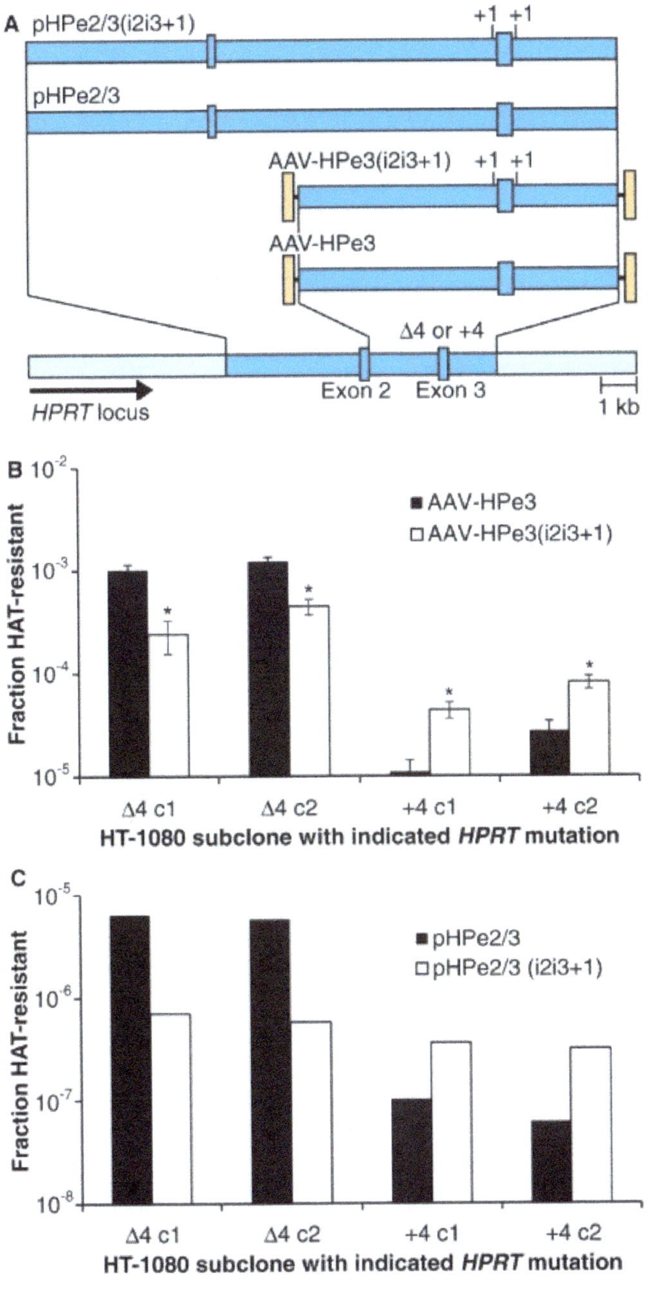

Figure 4. Correction of *HPRT* mutations with AAV and plasmid vectors. (**A**) Structure of the *HPRT* locus containing either a 4-bp deletion or insertion in exon 3, the AAV-HPe3 and AAV-Hpe3(i2i3+1) targeting vectors used and the analogous plasmid-based targeting constructs pHPe2/3 and pHPe2/3(i2i3+1). Gene targeting frequencies are shown as the fraction of HAT-resistant colonies obtained after transducing HT-1080 subclones harboring *HPRT* mutations with either AAV targeting vectors (*$P < 0.05$ versus AAV-HPe3) (**B**) or linearized plasmid targeting constructs (**C**).

Our results show that human cells are like mouse cells, where sequence polymorphisms reduce homologous recombination frequencies (2–4). This result stands in contrast to a prior analysis of human gene targeting (5), which concluded that isogenic DNA did not enhance targeting. However, this prior study did not sequence the target loci to demonstrate if chromosomal SNPs were also present in the plasmid-based targeting constructs, so there may not have been any sequence heterologies. In addition, we found that sequence polymorphism effects were maximal when they were located close to the sequence change being introduced, which may not be the case with many targeting constructs.

We focused on AAV vectors because their high targeting frequencies resulted in many targeted clones and accurate measurements of SNP effects. As shown in Figure 4, the targeting frequencies of transfected plasmid-based constructs were 100- to 1000-fold lower than those of AAV vectors, raising the possibility that unique features of AAV may limit the general applicability of our results. One major difference is that the single-stranded form of the AAV vector genome appears to be the substrate for targeting. This is supported by the lack of targeting observed with AAV vectors containing double-stranded encapsidated genomes (19), and by strand-specific differences in targeting frequencies obtained with related parvoviral vectors that package only one vector strand (20). The AAV capsid could also influence targeting, as shown by microinjection experiments demonstrating that purified AAV vector genomes do not target efficiently (21). Despite these differences, AAV and plasmid-based targeting also have similarities, including stimulation by double-strand breaks (22,23), the involvement of similar recombination proteins (24) and a preference for introducing insertions over deletions (8). Here we show that AAV and plasmid-based targeting are both inhibited by homology arm sequence heterologies, suggesting another shared aspect of their mechanism. Because homologous recombination requires that the plasmid constructs unwind and form heteroduplexes with the chromosome, there is still the opportunity for mismatched bases to reduce homologous pairing or influence mismatch repair, just as with single-stranded AAV vector genomes.

Our results also show that SNPs can be advantageous in certain situations. They can be used to direct allele-specific targeting, which could be useful when correcting distinct mutations in compound heterozygotes, or when inactivating dominant mutations in a single allele, or to avoid recombination with a previously targeted allele when knocking out multiple alleles. Another application of vector sequence heterologies is their potential for enhancing the introduction of unfavorable deletion mutations (8). Although the basis for the reduced targeting frequencies of deletion vectors is not understood, this effect can be minimized by including favorable, silent, flanking insertions in the homology arms.

ACKNOWLEDGEMENTS

The authors would like to thank Roli K. Hirata, Harkirat Sohi and Raisa Stolitenko for valuable technical advice and assistance.

REFERENCES

1. Sachidanandam R, Weissman D, Schmidt SC, Kakol JM, Stein LD, Marth G, Sherry S, Mullikin JC, Mortimore BJ, Willey DL, et al. A map of human genome sequence variation containing 1.42 million single nucleotide polymorphisms. Nature.2001; 409: 928–933.
2. Deng C, Capecchi MR. Reexamination of gene targeting frequency as a function of the extent of homology between the targeting vector and the target locus. Mol. Cell. Biol.1992; 12: 3365–3371.

3. van Deursen J, Wieringa B. Targeting of the creatine kinase M gene in embryonic stem cells using isogenic and nonisogenic vectors. Nucleic Acids Res. 1992; 20: 3815–3820.
4. te Riele H, Maandag ER, Berns A. Highly efficient gene targeting in embryonic stem cells through homologous recombination with isogenic DNA constructs. Proc. Natl Acad. Sci. USA. 1992; 89: 5128–5132.
5. Sedivy JM, Vogelstein B, Liber HL, Hendrickson EA, Rosmarin A. Gene targeting in human cells without isogenic DNA. Science. 1999; 283: 9a.
6. Cornea AM, Russell DW. Chromosomal position effects on AAV-mediated gene targeting. Nucleic Acids Res. 2010; 38: 3582–3594.
7. Li LB, Chang KH, Wang PR, Hirata RK, Papayannopoulou T, Russell DW. Trisomy correction in down syndrome induced pluripotent stem cells. Cell Stem Cell.2012; 11: 615–619.
8. Russell DW, Hirata RK. Human gene targeting favors insertions over deletions. Hum. Gene Ther. 2008; 19: 907–914.
9. Inoue N, Hirata RK, Russell DW. High-fidelity correction of mutations at multiple chromosomal positions by adeno-associated virus vectors. J.Virol. 1999; 73: 7376–7380.
10. Miller AD, Garcia JV, von Suhr N, Lynch CM, Wilson C, Eiden MV. Construction and properties of retrovirus packaging cells based on gibbon ape leukemia virus. J. Virol.1991; 65: 2220–2224.
11. Zolotukhin S, Byrne BJ, Mason E, Zolotukhin I, Potter M, Chesnut K, Summerford C, Samulski RJ, Muzyczka N. Recombinant adeno-associated virus purification using novel methods improves infectious titer and yield. Gene Ther. 1999; 6: 973–985.
12. Inoue N, Russell DW. Packaging cells based on inducible gene amplification for the production of adeno-associated virus vectors. J. Virol. 1998; 72: 7024–7031.
13. Graham FL, Smiley J, Russell WC, Nairn R. Characteristics of a human cell line transformed by DNA from human adenovirus type 5. J. Gen. Virol. 1977; 36: 59–72.
14. Yu J, Vodyanik MA, Smuga-Otto K, Antosiewicz-Bourget J, Frane JL, Tian S, Nie J, Jonsdottir GA, Ruotti V, Stewart R, et al. Induced pluripotent stem cell lines derived from human somatic cells. Science. 2007; 318: 1917–1920.
15. Rutledge EA, Russell DW. Adeno-associated virus vector integration junctions. J. Virol. 1997; 71: 8429–8436.
16. Khan IF, Hirata RK, Russell DW. AAV-mediated gene targeting methods for human cells. Nat. Protoc. 2011; 6: 482–501.
17. Miller DG, Petek LM, Russell DW. Adeno-associated virus vectors integrate at chromosome breakage sites. Nat. Genet. 2004; 36: 767–773.
18. Hunter DJ, Gurney EG. The genomic instability associated with integrated simian virus 40 DNA is dependent on the origin of replication and early control region. J. Virol.1994; 68: 787–796.
19. Hirata RK, Russell DW. Design and packaging of adeno-associated virus gene targeting vectors. J. Virol. 2000; 74: 4612–4620.
20. Hendrie PC, Hirata RK, Russell DW. Chromosomal integration and homologous gene targeting by replication-incompetent vectors based on the autonomous parvovirus minute virus of mice. J. Virol. 2003; 77: 13136–13145.
21. Liu X, Yan Z, Luo M, Zak R, Li Z, Driskell RR, Huang Y, Tran N, Engelhardt JF. Targeted correction of single-base-pair mutations with adeno-associated virus vectors under nonselective conditions. J. Virol. 2004; 78: 4165–4175.
22. Porteus MH, Cathomen T, Weitzman MD, Baltimore D. Efficient gene targeting mediated by adeno-associated virus and DNA double-strand breaks. Mol. Cell. Biol.2003; 23: 3558–3565.
23. Miller DG, Petek LM, Russell DW. Human gene targeting by adeno-associated virus vectors is enhanced by DNA double-strand breaks. Mol. Cell. Biol. 2003; 23: 3550–3557.
24. Vasileva A, Linden RM, Jessberger R. Homologous recombination is required for AAV-mediated gene targeting. Nucleic Acids Res. 2006; 34: 3345–3360.

CHAPTER 15

AN siRNA SCREEN IN PANCREATIC BETA CELLS REVEALS A ROLE FOR GPR27 IN INSULIN PRODUCTION

Gregory M. Ku,[1,2] Zachary Pappalardo,[1] Chun Chieh Luo,[1] Michael S. German,[1,2] and Michael T. McManus[1,3]

[1]Diabetes Center, University of California San Francisco, San Francisco, California, United States of America

[2]Division of Endocrinology, Metabolism, and Diabetes, Department of Medicine, University of California San Francisco, San Francisco, California, United States of America

[3]Department of Microbiology and Immunology, University of California San Francisco, San Francisco, California, United States of America

ABSTRACT

The prevalence of type 2 diabetes in the United States is projected to double or triple by 2050. We reasoned that the genes that modulate insulin production might be new targets for diabetes therapeutics. Therefore, we developed an siRNA screening system to identify genes important for the activity of the insulin promoter in beta cells. We created a subclone of the MIN6 mouse pancreatic beta cell line that expresses destabilized GFP under the control of a 362 base pair fragment of the human insulin promoter and the mCherry red fluorescent protein under the control of the constitutively active rous sarcoma virus promoter. The ratio of the GFP to mCherry fluorescence of a cell indicates its insulin promoter activity. As G protein coupled receptors (GPCRs) have emerged as novel targets for diabetes therapies, we used this cell line to screen an siRNA library targeting all known mouse GPCRs. We identified several known GPCR regulators of insulin secretion as regulators of the insulin promoter. One of the top positive regulators was *Gpr27*, an orphan GPCR with no known role in beta cell function. We show that knockdown of *Gpr27* reduces endogenous mouse insulin promoter activity and glucose stimulated insulin secretion. Furthermore, we show that *Pdx1* is important for *Gpr27*'s effect on the insulin promoter and insulin secretion. Finally, the over-expression of *Gpr27* in 293T cells increases inositol phosphate levels, while knockdown of *Gpr27* in MIN6 cells reduces inositol phosphate levels, suggesting this orphan GPCR might couple to Gq/11. In summary, we demonstrate a MIN6-based siRNA screening system that allows rapid identification of novel positive and negative regulators of the insulin promoter. Using this system, we identify *Gpr27* as a positive regulator of insulin production.

AUTHOR SUMMARY

Pancreatic beta cells are the only physiologic source of insulin. When these cells are destroyed in type 1 diabetics, there is uncontrolled hyperglycemia from complete insulin deficiency. In type 2 diabetes, these same cells fail to increase insulin secretion to compensate for peripheral insulin resistance leading to relative insulin deficiency. We constructed a novel screening system to find new regulators of insulin production in this critical cell type. Here, we describe a screen of the G protein coupled receptors (GPCRs) and show a role for orphan GPCR, *Gpr27*, in insulin promoter activity and insulin secretion. We propose that *Gpr27* is a novel target for diabetes therapeutics.

Citation: Gregory M. Ku, Zachary Pappalardo, Chun Chieh Luo, Michael S. German, and Michael T. McManus, An siRNA Screen in Pancreatic Beta Cells Reveals a Role for Gpr27 in Insulin Production. Copenhaver GP, ed. PLOS Genetics. 2012; 8(1):e1002449. DOI:10.1371/journal.pgen.1002449.

INTRODUCTION

Nearly 13% of American adults have diabetes and these numbers continue to rise, mostly from an increase in type 2 diabetes [1], [2]. Although insulin resistance is a cardinal feature of type 2 diabetes, most people with insulin resistance do not develop diabetes because their pancreatic beta cells are able to compensate by increasing insulin production. However, if insulin production cannot match the increased demand imposed by insulin resistance, hyperglycemia and frank diabetes ensues. Over time, beta cell function further declines in most people with type 2 diabetes, resulting in the eventual failure of oral medications and the necessity of insulin therapy [3].

Improving insulin production and beta cell function is therefore a universal goal of diabetes therapeutics. We reasoned that an unbiased search for regulators of insulin production might reveal new diabetes drug targets. Therefore, we constructed a novel screening system to screen for genes important for insulin promoter activity. By screening siRNAs targeting all GPCRs, we identify several GPCRs that regulate insulin promoter activity and specifically characterize *Gpr27* as a novel regulator of insulin production.

RESULTS

Generation of an Insulin Promoter Reporter Beta Cell Line

To allow rapid evaluation of insulin promoter activity, the MIN6 mouse beta cell line was infected with a lentivirus that stably expresses destabilized GFP under the control of the proximal 362 base pairs of the human insulin promoter (Figure 1A) [4]. This insulin promoter fragment maintains a substantial proportion of promoter activity and tissue specificity while being compact enough to allow lentiviral delivery [5].

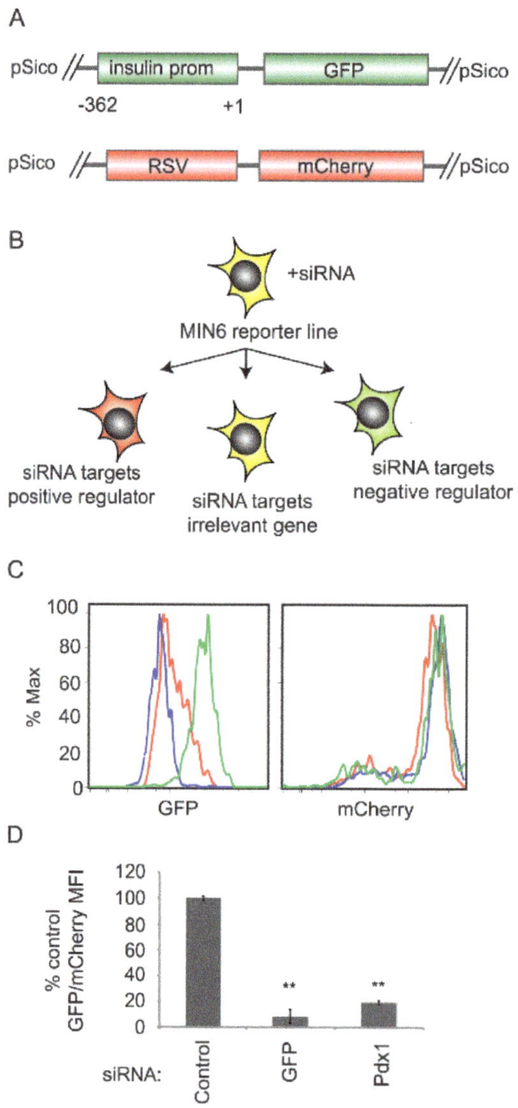

Figure 1. siRNA screening system to identify regulators of insulin promoter activity. A. Schema of reporter constructs integrated into the screening MIN6 cell line. These are pSico lentiviruses containing the proximal 362 bases of the human insulin promoter driving destabilized GFP and the rous sarcoma virus (RSV) promoter driving mCherry. B. After transfection with siRNAs, GFP and mCherry mean fluorescence intensity (MFI) are measured by flow cytometry. If the siRNA targets a positive regulator, GFP/mCherry MFI decreases. If the siRNA targets a negative regulator, the GFP/mCherry MFI increases. C. Control (green), Pdx1(red), or GFP(blue) siRNAs were transfected into the MIN6 reporter cell line. After 5 days, GFP (left panel) and mCherry (right panel) fluorescence were measured by flow cytometry. D. As in C. The ratio of GFP geometric mean fluorescence intensity (MFI) to mCherry MFI was calculated for each sample. The ratio was normalized to that of control siRNA transfected cells. Data shown are averages and standard error (n=3). ** $p<0.01$ versus control siRNA.

To favor single copy integration, the construct was delivered at a low multiplicity of infection (MOI) and a clonal line was selected. To generate an internal control reporter, the GFP positive subline was subsequently infected at a low MOI with a second lentivirus containing mCherry under the control of the constitutive rous sarcoma virus promoter (RSV) (Figure 1A). A stable clone expressing both constructs was isolated. In these cells, the ratio of GFP to mCherry fluorescence indicates human insulin promoter activity.

When transfected into this reporter line, siRNAs targeting activators of insulin gene transcription would be expected to reduce insulin promoter activity and reduce the GFP/mCherry ratio, while siRNAs targeting negative regulators of the insulin promoter should increase the GFP/mCherry ratio (Figure 1B). Indeed, transfection of an siRNA targeting the insulin gene transcription factor *Pdx1* reduced the GFP/mCherry ratio by 80% as compared to a non-targeting siRNAs (Figure 1C and 1D) [6].

siRNA Screen for Non-odorant GPCR Regulators of the Insulin Promoter

An RNAi library containing four independent siRNAs targeting the mouse GPCR-ome and selected GPCR related genes was transfected into the reporter cell line. The ratio of the GFP to mCherry fluorescence five days after transfection was calculated for each siRNA and the data were then analyzed using the redundant siRNA analysis (RSA) software [7]. To avoid off-target effects, each siRNA was transfected separately and only genes with more than one siRNA hit were selected for further analysis.

The top genes judged by RSA were then ranked by unsupervised clustering of each gene's RSA p value and its expression level in primary mouse islets, since only those genes expressed in primary cells are of biological interest (Figure 2A). Two publically available mouse islet mRNA-Seq data sets were used. One of these data sets has been previously published and consists of approximately four million mapped reads from islets isolated from female non-pregnant mice and approximately four million mapped reads from islets isolated from pregnant mice [8]. The second, submitted to the NCBI Short Read Archive by Merck, contains approximately 120 million reads from mouse islets (see methods). Because of these modest read numbers, some low abundance transcripts may be erroneously reported as being not expressed using this analysis [9].

siRNAs to the top six genes (*Ffar2, Gpr27, Grk5, p2ry6, Gpr109a, Bdkrb2*) that reduced insulin promoter activity and had detectable expression in primary mouse islets were transfected into the screening cell line for confirmation (Figure 2B). All six genes had at least 2 siRNAs confirm. For the siRNAs that increased GFP/mCherry, we retested the top three genes with high RSA scores and detectable expression in mouse primary islets –*Adra2a, Cckar, and Aplnr*. Of these three, only the known negative regulator of insulin secretion, *Adra2a*, confirmed with two independent siRNAs (Figure 2C).

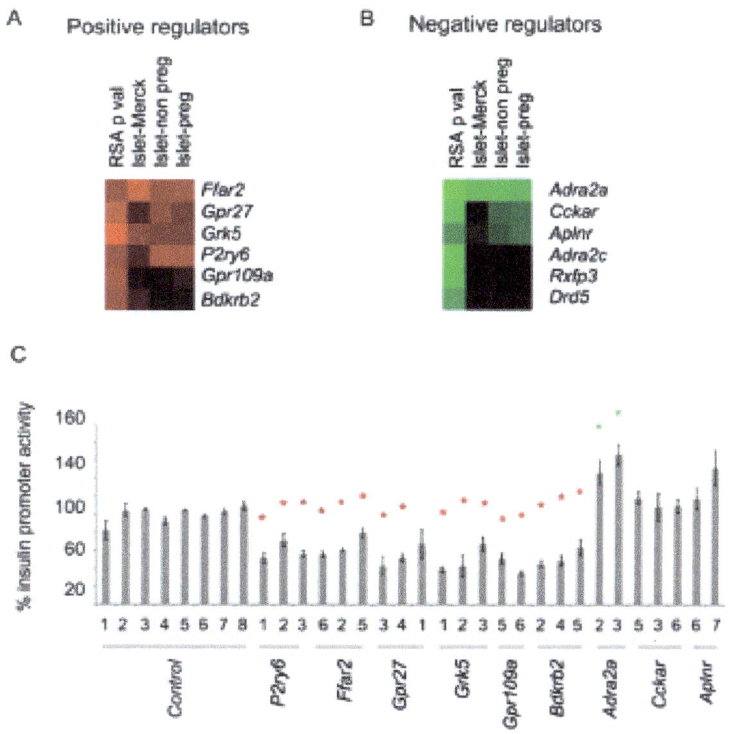

Figure 2. siRNA screen hit selection and initial confirmation. A. Heat map showing the top six putative positive regulators of the insulin promoter clustered based on RSA determined p value, and the base 10 log of the fragments per kilobase of exon model per million mapped reads (FPKM) value from three independent mRNA-seq data sets. B. As in B, but analyzed for negative regulators of insulin promoter activity. Islet-Merck refers to SRA008619 submitted by Merck, Islet-non preg refers to islets from non-pregnant mice, and Islet-preg refers to islets from pregnant mice [8]. C. Confirmation of hits from A and B. The indicated siRNAs were transfected into the screening cell line. GFP and mCherry was measured by flow cytometry. Error bars show standard error of three biological replicates performed (n=3). * indicates $p<0.05$ versus >6 of the negative control siRNAs.

Several of the positive regulators of the insulin promoter we identified were already known to stimulate insulin release in beta cells. Of particular interest was the orphan GPCR, *Gpr27*, which had no known role in insulin production but was previously found to be enriched in the mouse and human pancreatic islet [10], [11]. We subsequently tested all four siRNAs targeting *Gpr27* in the library set on an independently generated MIN6 reporter line expressing stable GFP under the control of the insulin promoter and mCherry under the control of the RSV promoter. All four siRNAs reduced GFP/mCherry fluorescence. Furthermore, all 4 siRNAs efficiently reduced expression of the *Gpr27* mRNA. We also confirmed that *Gpr27* is enriched in beta cell lines (beta TC and MIN6) compared to an alpha cell line (alpha TC) and is expressed in primary mouse beta cells.

Knockdown of *Gpr27* Reduces the Activity of the Endogenous Mouse Insulin Promoter in Cultured Cells and Primary Islets

Since the screen was based on a human insulin promoter fragment, we measured the effect of *Gpr27* knockdown on the endogenous mouse *Ins2* and *Ins1* genes. Because mature insulin mRNAs have a half-life of nearly 80 hours, we measured insulin pre-mRNAs as previously described [6], [12]. MIN6 cells infected with a *Gpr27* shRNA expressing adenovirus (Ad-shGpr27) had a 40–60% reduction in pre-ins2 and pre-ins1 levels compared to control adenovirus (Ad-control) (Figure 3A). To confirm these findings in primary beta cells, we infected intact primary mouse islets with these same adenoviruses. At a high MOI, we were only able to obtain 50% infection rates as measured by flow cytometry, presumably reflecting poor adenovirus penetration into the core of the mouse islet [13]. Therefore, intact islets were dissociated prior to adenovirus infection. Three days after infection, cells were isolated by flow cytometric sorting for GFP and RT-qPCR was performed. Knockdown of *Gpr27* produced a significant ~30% reduction of pre-ins2 (p=0.03). Concomitantly, there was a nearly significant 30% reduction in the less abundant pre-ins1 message (p=0.055) (Figure 3B).

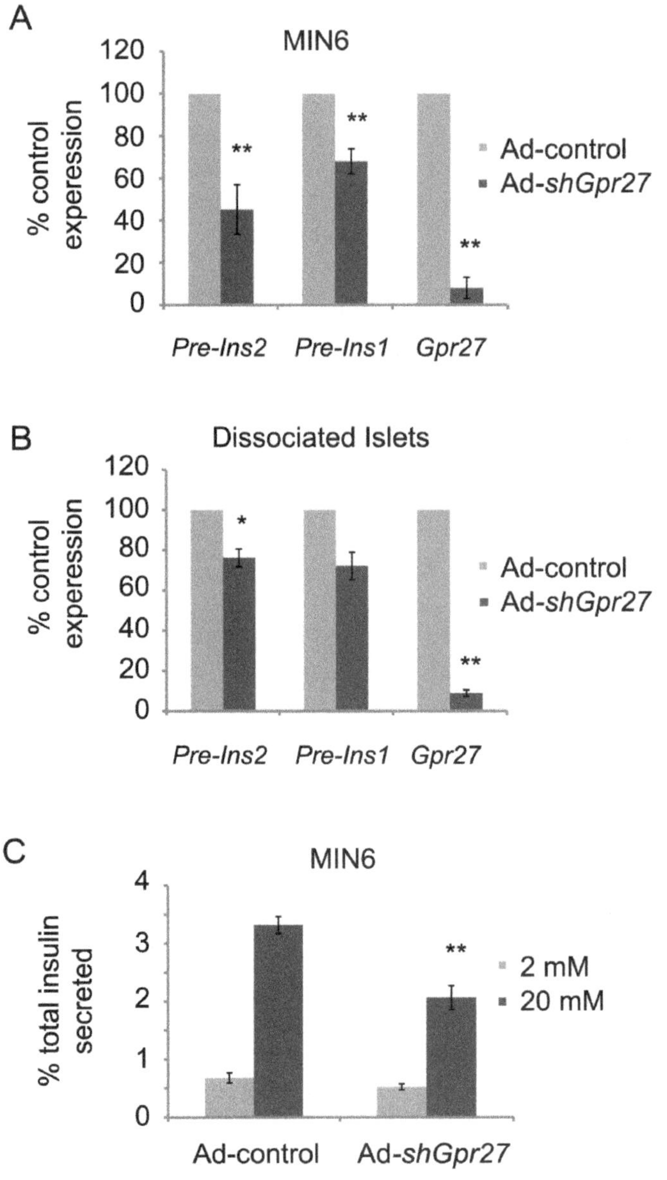

Figure 3. *Gpr27* is required for mouse insulin promoter activity and glucose stimulated insulin secretion. A. MIN6 cells were infected with either Ad-control or Ad-sh*Gpr27*. Three days after infection, RT-qPCR was performed for the indicated genes. The data are plotted as % expression compared to control adenovirus with standard error (n=3). B. As in A, but dispersed primary mouse islets were infected with Ad-control or Ad-sh*Gpr27*. After 3 days, infected cells were sorted by flow cytometry of GFP positive cells and RT-qPCR was performed for the indicated genes (n=3). C. MIN6 cells were infected with either Ad-control or Ad-sh*Gpr27*. Three days after infection, glucose stimulated insulin secretion was measured by ELISA after 1 hour of static incubation at either 2 mM or 20 mM glucose. Data are represented as the average of fractional insulin secretion with standard error (n=9). *p<0.05, **p<0.005 versus Ad-control.

Knockdown of *Gpr27* Impairs Glucose Stimulated Insulin Secretion

While insulin production requires insulin promoter activity, minute-to-minute changes in plasma insulin levels are controlled by insulin secretion. Therefore, we asked if *Gpr27*knockdown would affect glucose stimulated insulin secretion. Infection of MIN6 cells with Ad-control at an MOI necessary to get >90% infection inhibited glucose stimulated insulin secretion (data not shown). Therefore, we infected MIN6 cells at a lower MOI to achieve approximately 60% infection and measured glucose stimulated insulin secretion from this mixed population by batch incubation. Ad-shGpr27 infected MIN6 cells secreted ~40% less insulin at 20 mM glucose compared to Ad-control infected cells (Figure 3C). There was no statistically significant difference at 2 mM glucose. Notably, we did not detect a difference in total insulin as normalized to total protein concentration (Ad-control=27.9+/−1.1 mg insulin per g of total protein; Ad-shGpr27=29.4+/−0.94 mg insulin per g of total protein, p value=0.13). This was not unexpected since the half-life of insulin mRNA is ~80 hours and the knockdown of *Gpr27* was limited to 72 hours due to adenovirus toxicity after that time point. We conclude that *Gpr27* plays a measurable role in insulin secretion in addition to insulin promoter activity.

Gpr27's Effect on the Insulin Promoter and Insulin Secretion Requires *Pdx1*

To define the mechanism of *Gpr27* action, we measured transcript levels of selected regulators of the insulin promoter by RT-QPCR in MIN6 cells after Ad-shGpr27 infection.*Glis3*, *Pax6*, *Nkx6.1*, *HNF4a*, and *Pdx1* were reduced after *Gpr27* knockdown while others including *MafA*, *NeuroD1*, and *Pax4* were unchanged (Figure 4A). Concordant with this expression data, *Gpr27* knockdown reduced the transcriptional activity of mini-enhancers that bind to *Glis3* and *Pdx1* (Z, E1/A1, E2/A3) while *Gpr27* knockdown had no effect on mini-enhancers that bind to *MafA* and *NeuroD1* (C1/E1) (Figure 4B and 4C).

Since *Pdx1* is required for insulin promoter activity and insulin secretion [14], [15], we asked if *Pdx1* is required for the effect of *Gpr27's* on the insulin promoter. By luciferase assay, we found that the single knockdown of *Pdx1* reduced insulin promoter activity by 90% and*Gpr27* knockdown alone reduced insulin promoter activity by 40%. However, the knockdown of both *Gpr27* and *Pdx1* had no additional effect over the single knockdown of *Pdx1*, showing that *Pdx1* is important for the effect of *Gpr27* on the insulin promoter (Figure 4D). Importantly, double knockdown of both *Gpr27* and *Pdx1* was as efficient as single knockdown.

We then asked if *Pdx1* was required for the effect of *Gpr27* on insulin secretion. The knockdown of *Pdx1* reduced fractional insulin secretion at 20 mM glucose and total insulin content (Figure 4E). As with the adenoviral knockdown of *Gpr27*, an siRNA to*Gpr27* reduced glucose stimulated insulin secretion. However, the knockdown of *Gpr27* in addition to *Pdx1* did not further reduce insulin secretion at 20 mM glucose. We conclude that *Gpr27* plays a measurable role in insulin secretion and insulin promoter activity via a mechanism involving *Pdx1*.

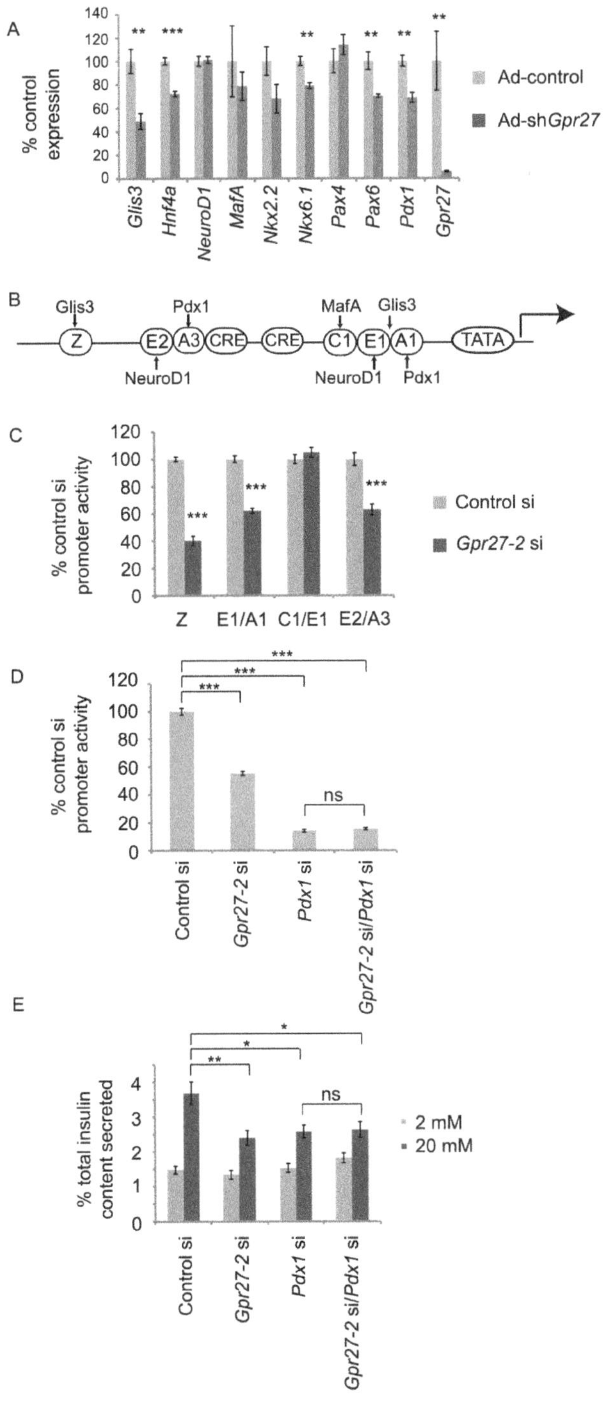

Figure 4. *Gpr27* knockdown affects multiple transcription factors and requires*Pdx1* for its effect on the insulin promoter. A. MIN6 cells were infected with Ad-control or Ad-sh*Gpr27*. Three days after infection RT-qPCR was performed for the indicated genes. Expression level normalized to that of Ad-control are plotted with standard error (n=6). B. Schematic of the human insulin promoter with selected regulatory sequences and transcription factors that bind to these elements. C. MIN6 cells were cotransfected with the indicated siRNA, insulin promoter firely luciferase construct and thymidine kinase renilla luciferase construct. Two days after transfection, firefly and renilla luciferase activity were measured. The ratio of firefly to renilla luciferase was normalized to the control siRNA. Average and standard error are plotted (n=6–9) D. As in C using human insulin promoter −362 firefly luciferase (n=6–9). E. MIN6 cells were transfected with the indicated siRNAs and glucose stimulated insulin secretion was measured after 5 days. Fractional insulin secretion is shown (n=12). * p<0.05 ** p<0.005 *** p<0.0005 versus control siRNA or control adenovirus (at high glucose part E).

Gpr27 Increases IP1 Levels

G protein coupling software analysis predicts that *Gpr27* could function via Gi or Gq/11 signaling pathways [16]. Since *Gpr27* is already expressed in MIN6 cells, we ectopically expressed mouse *Gpr27* in HEK293T cells. Robust expression of FLAG-tagged *Gpr27* was detected by 24 hours on the surface of the majority of cells (Figure 5B). We then measured cAMP and IP1 – higher cAMP would indicate Gs coupling, lower cAMP would indicate Gi coupling and higher IP1 would indicate Gq/11 coupling (Figure 5A). *Gpr27* expression resulted in a 2-fold elevation of IP1 levels while leaving cAMP levels unchanged (Figure 5C and 5D) showing that in this heterologous cell type, *Gpr27* may activate the Gq/11 pathway.

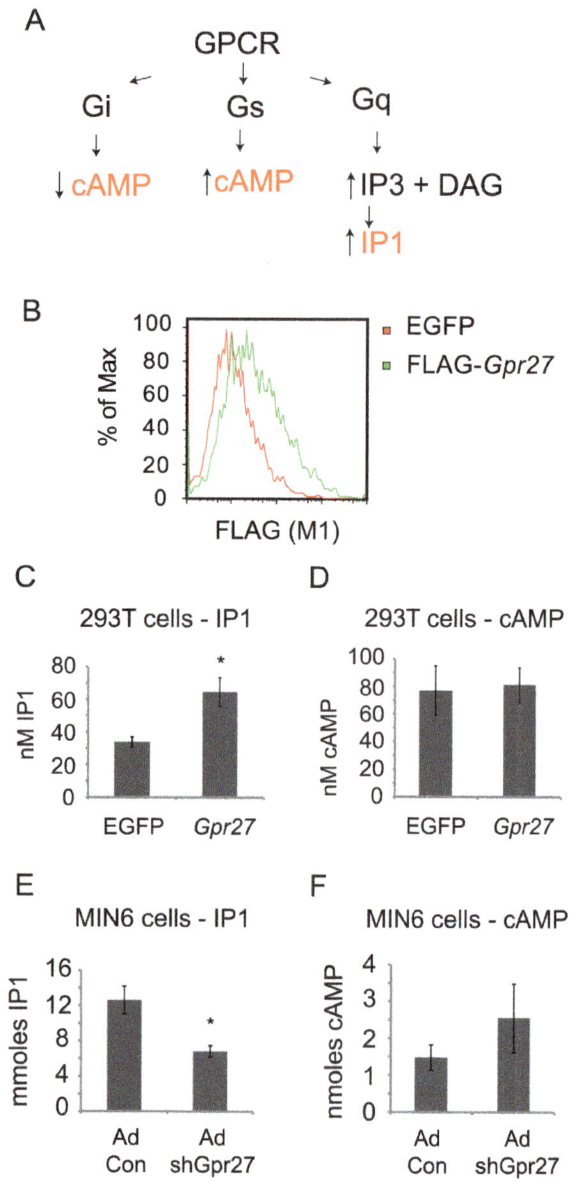

Figure 5. *Gpr27* positively regulates inositol phosphate levels. A. Schema of canonical GPCR signaling pathways and resulting expected changes in second messengers cAMP and IP3. B. HEK293T cells were transiently transfected with either GFP plasmid (control) or FLAG-*Gpr27* plasmid. 24 hours after transfection cells were analyzed by flow cytometry for extracellular FLAG. C. As in B, but cells were lysed and assayed for IP1. D. As in C, but lysates were analyzed for cAMP. E. MIN6 cells were infected with control or *Gpr27* knockdown adenovirus and 3 days later, cells were lysed and IP1 was measured. F. As in E, but cAMP was measured. For C–F, average and standard error is plotted (n=9); * p<0.005 versus control.

If *Gpr27* activates Gq/11 in beta cells, then IP1 levels should be reduced in MIN6 cells after knockdown of *Gpr27*. Therefore, we measured IP1 levels and cAMP levels in MIN6 cells after *Gpr27* knockdown. Indeed, knockdown of *Gpr27* resulted in reduced IP1 levels while cAMP levels were not significantly changed (Figure 5E and 5F). Taken together, these data show that *Gpr27* positively regulates inositol phosphate levels, supporting a role for *Gpr27*in activating the Gq/11 pathway.

DISCUSSION

To identify new regulators of the insulin promoter, we developed a novel siRNA screening system in MIN6 cells that allows rapid measurement of insulin promoter activity. As an initial test of the system, an siRNA screen of the GPCR-ome was performed. The RSA algorithm was used to select hits in order to capitalize on the four fold redundancy of the siRNA library[7]. To further increase the specificity of the screen, at least 2 siRNAs must have been identified for a gene to be a hit. The top RSA hits were then prioritized by expression level in mouse primary islets. Besides filtering out genes expressed in MIN6 but not in primary islets, this step also eliminates off-target hits. On the other hand, hit genes with low expression may have been erroneously eliminated because they were below the limit of detection of the mRNA-seq data available at this time [17]. Nonetheless, this filtering step allowed us to focus on genes with reasonable expression in primary cells.

While the confirmation rate for siRNAs to positive regulators was 100%, the confirmation rate for negative regulators was only 33%. This is likely due, in part, to the more modest effect of these siRNAs (~20–30% increase in GFP/mCherry ratio) as compared to the reconfirmed*Adra2a*(~50%), a known negative regulator of insulin secretion.

We identified several other known regulators of insulin secretion as regulators of the insulin promoter. The bradykinin receptor 2 mediates increases in insulin secretion in beta cells[18], [19]. Pyrimidinergic receptor 6 (*p2yr6*) agonists augment insulin release and this receptor participates in an autocrine feedback loop that potentiates insulin secretion [20],[21]. The free fatty acid receptor 2, which has been hypothesized to play a role in beta cells, was also identified as a positive regulator of the insulin promoter in our screen [22]. Several other receptors were identified in the screen that have no known role in beta cells and these may merit further investigation. Of note, *Glp1r* was not identified in this screen for a trivial reason; siRNAs targeting this gene were not included in the commercial screening set. Given the nature of our screen, hits would be predicted to either have basal activity or have ligand present in the culture conditions as has been described for *p2yr6* [21].

We were most intrigued by the orphan GPCR, *Gpr27* [23]. Previous studies have shown that it is enriched in the pancreatic islets of both human and mouse [10], [11]. Detailed mouse tissue profiling of *Gpr27* expression by RT-QPCR shows high expression in the mouse brain with lower expression in the islet and heart [24]. Furthermore, *Gpr27* mRNA is up-regulated in Neurogenin3 positive endocrine precursors in the developing mouse pancreas [11]. Conversely, *Gpr27* is 8-fold down regulated in the Neurogenin3 knockout pancreas [25],[26]. Taken together, these data suggest *Gpr27* is an endocrine pancreas specific gene.

We confirmed that knockdown of *Gpr27* reduces the activity of human insulin promoter reporters, levels of endogenous mouse *Ins2* pre-mRNA, and glucose stimulated insulin secretion. Importantly, *Gpr27* knockdown also reduces the levels of endogenous *Ins2* pre-mRNA in dissociated primary mouse islets. We also found that the mRNAs for multiple transcription factors that activate the insulin promoter (*Glis3, Pdx1, HNF4a*) were reduced by*Gpr27* knockdown. Other transcription factors critical for beta cell development were also reduced including *Nkx6.1* and *Pax6*. In agreement with the reduction in their expression, only *Pdx1* and *Glis3* binding mini-enhancers were affected by *Gpr27* knockdown (Figure 4C). Finally, there was no further reduction in insulin promoter activity when adding *Gpr27*knockdown to *Pdx1* knockdown. A limitation of this double knockdown experiment is that given the very strong effect of *Pdx1* knockdown alone on insulin promoter activity, a further reduction with *Gpr27/Pdx1* double knockdown may be either below our limit of detection or simply reflect no remaining insulin promoter activity.

How might *Gpr27* affect both insulin transcription and glucose stimulated insulin secretion? The Gq/11 pathway was an obvious candidate as the expression of *Gpr27* in HEK 293T cells increased IP1 levels while the knockdown of *Gpr27* reduced IP1 levels in MIN6 cells. Furthermore, triggering of an engineered Gq/11-coupled GPCR in beta cells increases steady state insulin mRNA levels and insulin secretion [27]. However, *Pdx1* levels did not change after triggering this Gq/11-coupled GPCR [27] and Gq/11 knockout beta cells have normal levels of *Ins1* and beta cell transcription factor mRNAs [21]. Therefore, even if *Gpr27*directly couples to Gq/11, *Gpr27* may affect insulin secretion and insulin promoter activity independent of Gq/11 as has recently been demonstrated for the M3 receptor [28].

Another candidate for mediating the effects of *Gpr27* on insulin promoter and insulin secretion was *Pdx1* since it is known to positively regulate both insulin transcription and insulin secretion [14], [15]. We found that the double siRNA knockdown of *Gpr27* and *Pdx1*produced no further reduction in insulin secretion over *Pdx1* knockdown alone, suggesting that *Pdx1* is important for *Gpr27's* effect on insulin secretion. In combination with the reduction in *Pdx1* mRNA by Gpr27 knockdown, these data suggest Gpr27 functions upstream of *Pdx1*. However, the double knockdown data do not exclude the possibility that a *Pdx1* lies in a parallel pathway to *Gpr27* and these two pathways intersect upstream of insulin secretion.

Taken together, these data suggest that a linear pathway connecting *Gpr27* to a single G protein and a single regulatory element in the insulin promoter is overly simplistic. Indeed, a single GPCR can trigger multiple G proteins (reviewed by [29]), can trigger a combination of G protein dependent and independent pathways [30], and can function as heterodimers[31]. Likewise, the insulin promoter contains multiple elements that are both redundant and cooperative [32]. The complexity of these systems highlights the advantage of using a broad, unbiased approach to finding new and unexpected regulators of the insulin promoter. Here, we used such a system to identify a novel GPCR regulator of both insulin secretion and insulin promoter activity – *Gpr27*. Based on its islet expression and its positive effects on the insulin promoter and insulin secretion, we suggest that *Gpr27* may be a novel target for diabetes therapies.

MATERIALS AND METHODS

Cell Culture

MIN6 cells were a gift from Dr. Miyazaki. Alpha TC and beta TC were a gift from Dr. Hanahan. Cells were maintained in high glucose DMEM with 10% fetal bovine serum, and 71.5 mM beta-mercaptoethanol. Sublines were isolated by limiting dilution. Original passage lines were used between passage 25–40. Sublines were used at passages 5–10.

Promoter Constructs

Human insulin promoter deletions have been previously described [5]. Promoters were subcloned from pFoxCAT into pFoxLuc [33]. For the lentiviral reporter, the human −362 promoter region was cloned upstream of destabilized GFP or GFP and this cassette was used to replace the U6/CMV-EGFP in pSicoR. pSicoR-RSV-mCherry was created by replacing the U6/CMV of pSicoR mCherry with the RSV promoter [5]. Mini-enhancer reporter constructs have also been previously described [5], [34]. They were subcloned upstream of a minimal thymidine kinase promoter-firefly luciferase reporter.

siRNA Transfection

Approximately 5,000 MIN6 cells were transfected in 96 well plates using HiPerfect (Qiagen) with a final siRNA concentration of 25 nM. Cells were analyzed by flow cytometry (LSRII, BD) 5 days after transfection and the geometric mean fluorescence intensity of GFP was normalized to that of mCherry. If the knockdown of GFP by an anti-GFP siRNA was not >80%, the transfection of that plate was considered to be a technical failure and the plate was discarded. This occurred on 1 out of 20 plates and for this reason some genes were only targeted by 3 siRNAs (including *Gpr27*). Each well was normalized to the negative control siRNA on that 96 well plate. For the confirmation assay for *Gpr27* siRNAs, a distinct MIN6 human insulin promoter-GFP/RSV-mCherry reporter line was transfected with the indicated siRNAs with Lipofectamine RNAiMax for 5 days and GFP and mCherry were measured.

GPCR Expression Analysis and Hierarchical Clustering

Mouse islet mRNA-seq data was downloaded from the NCBI Short Read Archive (SRP000752 and SRP002569) and FPKM values were calculated using the TopHat and Cufflinks software using the NCBI RefSeq as the reference. Log FPKM and negative log RSA p values were clustered using Cluster 3.0 and heat maps were plotted with JavaTreeView.

siRNA and qPCR Probes

siRNAs were obtained from Qiagen. All custom Taqman probes had a confirmed PCR efficiency of between 95–110%. Samples without reverse transcriptase did not amplify. Taqman probes to mouse *Glis3, MafA, Pdx1, NeuroD1, Pax4, Nkx6.1, HNF4a* were obtained from Applied Biosystems. Negative control siRNAs for the reconfirmation assay were All-Stars Negative Control (1027280), Negative Control (1022076), Unspecific-Luciferase-1 (1022070), Unspecific-Luciferase-2 (1022073), Hs_LMNA_11 (1022050), Mm_Lmna_5 (SI02655450), Hs_GAPD_5 (SI0253266), Hs_ACTB_1 (1022168).

RT–qPCR

Total RNA was isolated by Trizol (Invitrogen). The RNA was DNase I treated (Turbo DNase, Ambion) and reverse transcription was performed (Superscript III, Invitrogen) using a combination of random hexamers and oligo dT primers. For cell line experiments, each qPCR reaction used between 10–30 ng of total RNA equivalent. To convert to arbitrary linear units, the following formula was used: (2∧15)*(2∧(deltaCT to beta-glucuronidase).

Isolation of MIP-GFP Positive Cells

Islets from 12–30 week old MIP-GFP mice were isolated by the UCSF Islet Production Core. Islets were digested with trypsin until single cell suspensions were obtained. Cells were sorted by flow cytometry (Aria II, BD or MoFlo, DakoCytomation) into GFP positive and negative fractions and total RNA was isolated. 20 ng of total RNA equivalent was loaded per QRT-PCR reaction.

Luciferase Assays

140,000 MIN6 cells were transiently transfected in 24 well plates with the relevant siRNA (5 pmoles), the indicated insulin reporter firefly luciferase plasmid (100 ng), and pRL-TK(Promega) (25 ng) using Lipofectamine 2000 (Invitrogen). For double siRNA knockdowns, 5 pmoles of each siRNA or 10 pmoles of control (anti-GFP) were used. Two days after transient transfection, firely and renilla luciferase were measured using the Dual Luciferase Assay (Promega).

Gpr27 Expression in HEK 293T

Gpr27 was cloned by PCR from mouse genomic DNA downstream of a viral signal sequence and amino terminal FLAG epitope tag [35]. This cassette was used to replace the EGFP in pSicoR. HEK 293T cells were transiently transfected with either pSicoR-EGFP or pSicoR-FLAG-Gpr27 using LT1 (Mirus).

FLAG Flow Cytometry

293T cells were dissociated with PBS without Ca or Mg, stained with M1 anti-FLAG antibody and a Goat anti-mouse secondary antibody coupled to Alexa-594 (Invitrogen).

IP1 and cAMP Assays

For 293T, one day after transient transfection in 24 well plates, cells were placed in stimulation buffer (HTRF) for 30 minutes at 37 degrees. The stimulation buffer was then removed and the cells were lysed using the kit lysis buffer. IP1 and cAMP were then measured as directed by the protocol in 384 well plates (HTRF). IP1 and cAMP levels were normalized to live cells numbers counted from duplicate wells. Viable cells counts from*Gpr27* transfection were within 20% of control plasmid transfection. For MIN6 cells, 125,000 cells were infected with *Gpr27* shRNA or control adenovirus at an MOI of 200 and grown in 24 well dishes (resulting in nearly ~95% infection). Three days after infection, the cells were placed in stimulation buffer (HTRF) for 30 minutes at 37 degrees. The stimulation buffer was then removed and the cells were lysed in 1% Triton-X100, 50 mM HEPES pH 7.0, NaF 15 mM. The lysate was pre-cleared by centrifugation at 14,000 rpm for 10 minutes. A fraction of the lysate was taken for protein quantitation(micro-BCA, Pierce), IP1 or cAMP measurement (HTRF). Data were normalized to total protein content.

shRNA Adenovirus Construction

The *Gpr27* shRNA was cloned into a modified version of pSicoR with a BstXI site replacing the HpaI site. The mouse U6 promoter and *Gpr27* shRNA were then subcloned from pSicoR and placed upstream of the CMV-GFP marker in pAdTrack [36], [37]. Adenovirus was prepared and tittered as previously described [38].

Knockdown in Primary Mouse Islets

Islets were isolated by the UCSF Islet Production Core Facility from 8–12 week old C57Bl/6 male mice. After 24 hours of culture in RPMI and 10% FBS, islets were trypsinized until single cell suspensions were obtained. The dissociated islet cells were resuspended in RPMI+10% FBS and infected with adenovirus at multiplicity of infection (MOI) of 25. Three days after infection, the cells were sorted by flow cytometry (Aria II, BD) for GFP positive cells (50–75% of the live population) and RT-qPCR was performed. The knockdown of pre-ins2, pre-ins1 or *Gpr27* from *Gpr27* shRNA adenovirus infected cells was calculated by the delta-delta CT method compared to the control adenovirus infection.

Insulin Secretion Assays

For the adenovirus assays, approximately 500,000 MIN6 cells were infected with the indicated adenoviruses at an MOI of 100 in 6 cm dishes in complete media. Three days after infection, the infection rate was ~60% by FACS for GFP. Cells were washed 5 times in KRBH buffer (10 mM HEPES pH 7.4, 130 mM NaCl, 5 mM KCl, 1.25 mM KH2PO4, 1.25 mM MgSO4, 2.68 mM CaCl2, 5.26 mM NaHCO3) with 2 mM glucose and rested for 2 hours at 37 degrees. Cells were then washed an additional 3 times with 2 mM glucose KRBH and incubated in 3 mL of 2 mM glucose KRBH for 1 hour at 37 degrees. This supernatant was collected and replaced with 20 mM glucose KRBH for 1 hour at 37 degrees. Cells were washed with PBS before lysis in 50 mM Tris-HCl pH 8.0, 150 mM NaCl, 1% Triton X-100 with protease inhibitors. Lysates were spun at 14,000 rpm for 10 minutes and supernatants were spun at 5000 rpm for 5 minutes before analysis by an Ultrasensitive Insulin ELISA (Mercodia). Total protein was measured by Micro-BCA (Pierce). Total insulin was normalized to total protein in the lysate. For the siRNA transfections, 20,000 MIN6 cells were transfected per well of a Corning CellBIND 96 well plate with 25 nM of each siRNA (or 50 nM of control siRNA) using Lipofectamine RNAiMax. 5 days after transfection, cells were washed in KRBH with 2 mM glucose twice, then incubated for 2 hours at 37 degrees, then washed again with KRBH 2 mM glucose twice, then incubated for one hour with KRBH 2 mM, then KRBH with 20 mM glucose for 1 hour. Lysates were prepared in 75 uL of lysis buffer as above. Due to the lower cell numbers in the 96 well plate assay, total insulin was normalized to total genomic DNA measured by Qubit High Sensitivity DNA kit (Life Technologies).

Statistical Analysis

For siRNA primary confirmation assay, an independent, two sample, one tailed t-test was used. For the primary islet adenovirus knockdown of *Gpr27* an independent, one sample, two tailed t-test was used. All other p values were calculated with an independent, two sample, two tailed t-test.

Ethics Statement

Animal experiments were approved by the UCSF Institutional Animal Care and Use Committee (Protocol AN082433-02) with care taken to avoid any unnecessary suffering. Animals were maintained in accordance with the applicable portions of the Animal Welfare act and the DHHS Guide for the Care and Use of Laboratory Animals.

ACKNOWLEDGMENTS

We appreciate critical discussions with and reading of the manuscript by Dr. Nikki Shariat, Ha-il Kim, and Dr. Takeshi Miyatsuka of the McManus and German Labs at the UCSF Diabetes Center. We acknowledge the UCSF Viracore for production of lentiviruses and the UCSF Islet Production Core Facility for islet isolation. The glucuronidase QPCR probes were designed by the UCSF Cancer Center Genomics Core.

AUTHOR CONTRIBUTIONS

Conceived and designed the experiments: GMK MSG MTM. Performed the experiments: GMK ZP CCL. Analyzed the data: GMK ZP CCL MSG MTM. Contributed reagents/materials/analysis tools: GMK ZP CCL MSG MTM. Wrote the paper: GMK ZP CCL MSG MTM.

REFERENCES

1. Narayan KM, Boyle JP, Thompson TJ, Sorensen SW, Williamson DF (2003) Lifetime risk for diabetes mellitus in the United States. JAMA 290: 1884–1890.
2. Cowie CC, Rust KF, Ford ES, Eberhardt MS, Byrd-Holt DD, et al. (2009) Full accounting of diabetes and pre-diabetes in the U. S. population in 1988–1994 and 2005–2006. Diabetes Care 32: 287–294.
3. UKPDS (1998) Intensive blood-glucose control with sulphonylureas or insulin compared with conventional treatment and risk of complications in patients with type 2 diabetes (UKPDS 33). UK Prospective Diabetes Study (UKPDS) Group. Lancet 352: 837–853.
4. Miyazaki J, Araki K, Yamato E, Ikegami H, Asano T, et al. (1990) Establishment of a pancreatic beta cell line that retains glucose-inducible insulin secretion: special reference to expression of glucose transporter isoforms. Endocrinology 127: 126–132.
5. Odagiri H, Wang J, German MS (1996) Function of the human insulin promoter in primary cultured islet cells. J Biol Chem 271: 1909–1915.
6. Iype T, Francis J, Garmey JC, Schisler JC, Nesher R, et al. (2005) Mechanism of insulin gene regulation by the pancreatic transcription factor Pdx-1: application of pre-mRNA analysis and chromatin immunoprecipitation to assess formation of functional transcriptional complexes. J Biol Chem 280: 16798–16807.
7. Konig R, Chiang CY, Tu BP, Yan SF, DeJesus PD, et al. (2007) A probability-based approach for the analysis of large-scale RNAi screens. Nat Methods 4: 847–849.
8. Kim H, Toyofuku Y, Lynn FC, Chak E, Uchida T, et al. Serotonin regulates pancreatic beta cell mass during pregnancy. Nat Med 16: 804–808.
9. Griffith M, Griffith OL, Mwenifumbo J, Goya R, Morrissy AS, et al. Alternative expression analysis by RNA sequencing. Nat Methods 7: 843–847.
10. Gaulton KJ, Nammo T, Pasquali L, Simon JM, Giresi PG, et al. A map of open chromatin in human pancreatic islets. Nat Genet 42: 255–259.
11. Regard JB, Kataoka H, Cano DA, Camerer E, Yin L, et al. (2007) Probing cell type-specific functions of Gi in vivo identifies GPCR regulators of insulin secretion. J Clin Invest 117: 4034–4043.
12. Welsh M, Nielsen DA, MacKrell AJ, Steiner DF (1985) Control of insulin gene expression in pancreatic beta-cells and in an insulin-producing cell line, RIN-5F cells. II. Regulation of insulin mRNA stability. J Biol Chem 260: 13590–13594.
13. Takahashi R, Ishihara H, Takahashi K, Tamura A, Yamaguchi S, et al. (2007) Efficient and controlled gene expression in mouse pancreatic islets by arterial delivery of tetracycline-inducible adenoviral vectors. J Mol Endocrinol 38: 127–136.
14. Brissova M, Shiota M, Nicholson WE, Gannon M, Knobel SM, et al. (2002) Reduction in pancreatic transcription factor PDX-1 impairs glucose-stimulated insulin secretion. J Biol Chem 277: 11225–11232.
15. Gauthier BR, Wiederkehr A, Baquie M, Dai C, Powers AC, et al. (2009) PDX1 deficiency causes mitochondrial dysfunction and defective insulin secretion through TFAM suppression. Cell Metab 10: 110–118.
16. Sgourakis NG, Bagos PG, Hamodrakas SJ (2005) Prediction of the coupling specificity of GPCRs to four families of G-proteins using hidden Markov models and artificial neural networks. Bioinformatics 21: 4101–4106.
17. Griffith M, Tang MJ, Griffith OL, Morin RD, Chan SY, et al. (2008) ALEXA: a microarray design platform for alternative expression analysis. Nat Methods 5: 118.
18. Yang C, Lee B, Chen TH, Hsu WH (1997) Mechanisms of bradykinin-induced insulin secretion in clonal beta cell line RINm5F. J Pharmacol Exp Ther 282: 1247–1252.
19. Yang C, Hsu WH (1995) Stimulatory effect of bradykinin on insulin release from the perfused rat pancreas. Am J Physiol 268: E1027–1030.
20. Parandeh F, Abaraviciene SM, Amisten S, Erlinge D, Salehi A (2008) Uridine diphosphate (UDP) stimulates

insulin secretion by activation of P2Y6 receptors. Biochem Biophys Res Commun 370: 499–503.

21. Sassmann A, Gier B, Grone HJ, Drews G, Offermanns S, et al. The Gq/G11-mediated signaling pathway is critical for autocrine potentiation of insulin secretion in mice. J Clin Invest 120: 2184–2193.

22. Kebede MA, Alquier T, Latour MG, Poitout V (2009) Lipid receptors and islet function: therapeutic implications? Diabetes Obes Metab 11: Suppl 410–20.

23. Matsumoto M, Saito T, Takasaki J, Kamohara M, Sugimoto T, et al. (2000) An evolutionarily conserved G-protein coupled receptor family, SREB, expressed in the central nervous system. Biochem Biophys Res Commun 272: 576–582.

24. Regard JB, Sato IT, Coughlin SR (2008) Anatomical profiling of G protein-coupled receptor expression. Cell 135: 561–571.

25. Juhl K, Sarkar SA, Wong R, Jensen J, Hutton JC (2008) Mouse pancreatic endocrine cell transcriptome defined in the embryonic Ngn3-null mouse. Diabetes 57: 2755–2761.

26. Gu G, Wells JM, Dombkowski D, Preffer F, Aronow B, et al. (2004) Global expression analysis of gene regulatory pathways during endocrine pancreatic development. Development 131: 165–179.

27. Guettier JM, Gautam D, Scarselli M, Ruiz de Azua I, Li JH, et al. (2009) A chemical-genetic approach to study G protein regulation of beta cell function in vivo. Proc Natl Acad Sci U S A 106: 19197–19202.

28. Kong KC, Butcher AJ, McWilliams P, Jones D, Wess J, et al. M3-muscarinic receptor promotes insulin release via receptor phosphorylation/arrestin-dependent activation of protein kinase D1. Proc Natl Acad Sci U S A.

29. Millar RP, Newton CLThe year in G protein-coupled receptor research. Mol Endocrinol 24: 261–274.

30. Wei H, Ahn S, Shenoy SK, Karnik SS, Hunyady L, et al. (2003) Independent beta-arrestin 2 and G protein-mediated pathways for angiotensin II activation of extracellular signal-regulated kinases 1 and 2. Proc Natl Acad Sci U S A 100: 10782–10787.

31. Levoye A, Dam J, Ayoub MA, Guillaume JL, Jockers R (2006) Do orphan G-protein-coupled receptors have ligand-independent functions? New insights from receptor heterodimers. EMBO Rep 7: 1094–1098.

32. Ohneda K, Ee H, German M (2000) Regulation of insulin gene transcription. Semin Cell Dev Biol 11: 227–233.

33. Johnson JD, Zhang W, Rudnick A, Rutter WJ, German MS (1997) Transcriptional synergy between LIM-homeodomain proteins and basic helix-loop-helix proteins: the LIM2 domain determines specificity. Mol Cell Biol 17: 3488–3496.

34. Sander M, Griffen SC, Huang J, German MS (1998) A novel glucose-responsive element in the human insulin gene functions uniquely in primary cultured islets. Proc Natl Acad Sci U S A 95: 11572–11577.

35. Chang WC, Ng JK, Nguyen T, Pellissier L, Claeysen S, et al. (2007) Modifying ligand-induced and constitutive signaling of the human 5-HT4 receptor. PLoS ONE 2: e1317. doi:10. 1371/journal. pone. 0001317.

36. Ventura A, Meissner A, Dillon CP, McManus M, Sharp PA, et al. (2004) Cre-lox-regulated conditional RNA interference from transgenes. Proc Natl Acad Sci U S A 101: 10380–10385.

37. He TC, Zhou S, da Costa LT, Yu J, Kinzler KW, et al. (1998) A simplified system for generating recombinant adenoviruses. Proc Natl Acad Sci U S A 95: 2509–2514.

38. Luo J, Deng ZL, Luo X, Tang N, Song WX, et al. (2007) A protocol for rapid generation of recombinant adenoviruses using the AdEasy system. Nat Protoc 2: 1236–1247.

CHAPTER 16

COMPARATIVE ANALYSIS OF MICRORNA EXPRESSION IN MOUSE AND HUMAN BROWN ADIPOSE TISSUE

Isabelle Güller[1], Sarah McNaughton[1], Tamsyn Crowley[2,3], Vicente Gilsanz[4], Shingo Kajimura[5], Matthew Watt[6], and Aaron P. Russell[1]

[1]Centre for Physical Activity and Nutrition, School of Exercise and Nutrition Sciences, Deakin University, 221 Burwood highway, Burwood, VIC 3125, Australia.

[2]School of Medicine, Deakin University, Waurn Ponds, Australia.

[3]Australian Animal Health Laboratory, CSIRO Animal, Food and Health Sciences, Geelong, Victoria, Australia.

[4]Department of Radiology, Childrens Hospital Los Angeles, University of Southern California, Los Angeles, California, USA.

[5]UCSF Diabetes Center, Department of Cell and Tissue Biology, University of California, San Francisco, USA.

[6]Department of Physiology, Monash University, Clayton, Victoria 3800, Australia.

ABSTRACT

Background

In small mammals brown adipose tissue (BAT) plays a predominant role in regulating energy expenditure (EE) via adaptive thermogenesis. New-born babies require BAT to control their body temperature, however its relevance in adults has been questioned. Active BAT has recently been observed in adult humans, albeit in much lower relative quantities than small mammals. Comparing and contrasting the molecular mechanisms controlling BAT growth and development in mice and humans will increase our understanding or how human BAT is developed and may identify potential therapeutic targets to increase EE. MicroRNAs are molecular mechanisms involved in mouse BAT development however, little is known about the miRNA profile in human BAT. The aims of this study were to establish a mouse BAT-enriched miRNA profile and compare this with miRNAs measured in human BAT. To achieve this we firstly established a mouse BAT enriched-miRNA profile by comparing miRNAs expressed in mouse BAT, white adipose tissue and skeletal muscle. Following this the BAT-enriched miRNAs predicted to target genes potentially involved in growth and development were identified.

Methods

MiRNA levels were measured using PCR-based miRNA arrays. Results were analysed using ExpressionSuite software with the global mean expression value of all expressed miRNAs in a givensample used as the normalisation factor. Bio-informatic analyses was used to predict gene targets followed by Ingenuity Pathway Analysis.

Results

We identified 35 mouse BAT-enriched miRNAs that were predicted to target genes potentially involved in growth and development. We also identified 145 miRNAs expressed in both mouse and human BAT, of which 25 were enriched in mouse

Citation: Isabelle Güller, Sarah McNaughton, Tamsyn Crowley, Vicente Gilsanz, Shingo Kajimura, Matthew Watt, and Aaron P. Russell. Comparative analysis of microRNA expression in mouse and human brown adipose tissue, BMC Genomics 2015; 16:820. DOI: 10.1186/s12864-015-2045-8

BAT. Of these 25 miRNAs, miR-20a was predicted to target MYF5 and PPARγ, two important genes involved in brown adipogenesis, as well as BMP2 and BMPR2, genes involved in white adipogenesis. For the first time, 69 miRNAs were identified in human BAT but absent in mouse BAT, and 181 miRNAs were expressed in mouse but not in human BAT.

Conclusion

The present study has identified a small sub-set of miRNAs common to both mouse and human BAT. From this sub-set bioinformatics analysis suggested a potential role of miR-20a in the control of cell fate and this warrants further investigation. The large number of miRNAs found only in mouse BAT or only in human BAT highlights the differing molecular profile between species that is likely to influence the functional role of BAT across species. Nevertheless the BAT-enriched miRNA profiles established in the present study suggest targets to investigate in the control BAT development and EE.

BACKGROUND

Obesity is a world-wide epidemic that increases the risk of cardiovascular disorders, cancer and type 2 diabetes (T2D) [1]. Obesity results from an energy imbalance whereby energy intake (EI) exceeds energy expenditure (EE). Strategies to decrease obesity have been orientated towards a reduction in EI, unfortunately without success [2]. Consequently, new therapeutic strategies aimed at targeting EE are being investigated. Discovery of active human brown adipose tissue (BAT) has renewed interest in the possibility that it may regulate metabolism and EE in humans [3, 4, 5].

Skeletal muscle is another tissue that plays an important role in regulating EE and several lines of evidence suggest that skeletal muscle and BAT share a common lineage [6, 7]. microRNAs (miRNAs) are short non-coding RNAs that regulate gene expression and tissue development [8, 9]. Muscle-enriched microRNAs including miR-1, −133a and −206, are also expressed in murine brown pre and mature adipocytes, but not in white adipocytes [10]. The PRD1-BF-1-RIZ1 homologous domain protein containing protein-16 (PRDM16), a BAT-enriched protein controls the differentiation between BAT and skeletal muscle by forming a transcriptional complex with CCAAT/Enhancer binding protein β (C/EBPβ) [11, 12, 13]. Understanding the mechanisms controling BAT differentiation may identify potential therapeutic targets to enhance BAT-controlled metabolism and to increase EE.

MiRNA screenings and bioinformatics analyses have identified potential gene targets and biological functions controlled by certain miRNAs in tissues and diseases. Sun et al. observed 91 differentially expressed miRNAs when comparing mouse BAT, white adipose tissue (WAT) and skeletal muscle [14]. The oligonucleotide microchip technology used in this study allowed for the screening of only ~350 miRNAs however is not as sensitive as quantitative real time-polymerase chain reaction (RT-PCR) array technology [15]. Investigation of a broader number of BAT miRNA targets using highly sensitive qPCR-array methods will provide a more comprehensive overview of the potential miRNAs regulating BAT development and function. Several miRNAs controlling mouse brown adipocyte development and function have been identified in mice, including miR-27, −34a, −133, −155, −182, −193b-365, −196, −203 and miR-378 [14, 16, 17, 18, 19, 20, 21, 22, 23]. MiR-27 is decreased in mouse BAT following cold exposure and during brown adipogenesis [22]. As a consequence, several of the miR-27 target genes, including Prdm16, peroxisome proliferator-activated receptor alpha (Pparα), cAMP response element-binding protein (Creb) and peroxisome proliferator-activated receptor gamma coactivator 1-beta (Pgc1β) are upregulated, and enhance brown adipogenesis. MiR-34a inhibits beige and brown fat formation by targeting fibroblast growth factor-21 (Fgf21) and sirtuin 1 (Sirt1), two known activators of WAT browning [20]. miR-133 directly targets the 3'-UTR of *Prdm16* [19] and controls the differentiation of satellite cells within skeletal muscle towards an adipogenic or myogenic phenotype [17, 19]. While the cluster miR-193–365 is up-regulated by Prdm16, partially through Pparγ [14], they are not required for brown fat development and function [24]. MiR-182 and miR-203 are BAT-specific miRNAs, essential for the maintenance and differentiation of brown adipocytes *in vivo* [21]. Finally, miR-378 increases brown fat mass and as a consequence, suppresses development of beige adipocytes in subcutaneous WAT [23]. However it is unknown if these miRNAs are expressed in human BAT.

BAT plays a critical role in the regulation of energy balance and temperature in rodents and newborns [25, 26,27], however its role in human adult metabolism remains equivocal. While mice still remain the investigative of model of choice to understand the role and regulation BAT, studies establishing the similarities and differences in the molecular profile of mouse and human BAT are required.

Therefore the primary aims of this study were to (1) identify BAT-enriched miRNAs by comparing miRNA expression mouse BAT, skeletal muscle and WAT using PCR-based miRNA arrays (2); predict the BAT-enriched miRNA target genes potentially involved in growth, proliferation and development; (3) compare the miRNA profiles of mouse and human BAT.

RESULTS

Mouse BAT, WAT and Skeletal Muscle Tissues

miRNA Array Analysis on Mouse Tissues

To define miRNAs enriched in mouse BAT, comparisons were made between BAT, skeletal muscle (gastrocnemius) and WAT as described in the flowchart (Fig. 1). From the 750 miRNAs profiled 433 miRNAs were expressed at least in one tissue. The Venn diagram represents the tissue distribution of these 433 miRNAs. Six miRNAs were exclusively expressed in BAT. Nineteen miRNAs (circled) were expressed only in BAT and WAT, with four of these significantly higher in BAT tissue ($p<0.05$). Three miRNAs were expressed only in BAT and muscle and at similar levels. There were 298 miRNAs (circled) commonly expressed in the three tissues, of which 54 were significantly higher in BAT compared to WAT and muscle ($p<0.05$). When adding all the exclusively and highly expressed miRNAs found in BAT, 64 BAT-enriched miRNAs were obtained as shown in the heat map below (Fig. 2).

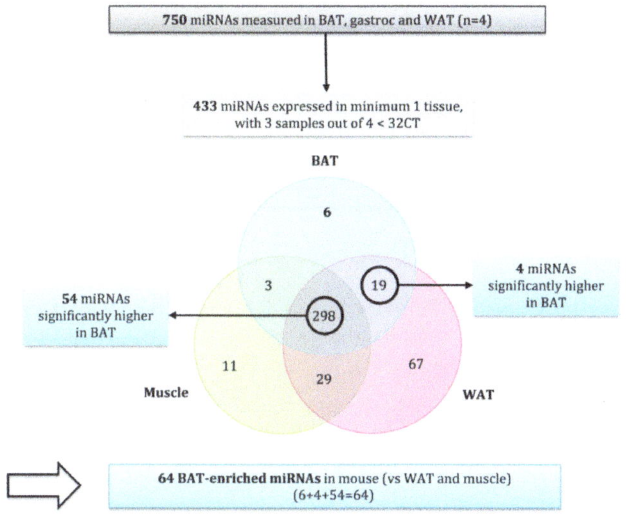

Figure 1. Flow chart of the miRNA array analysis from mouse tissues.

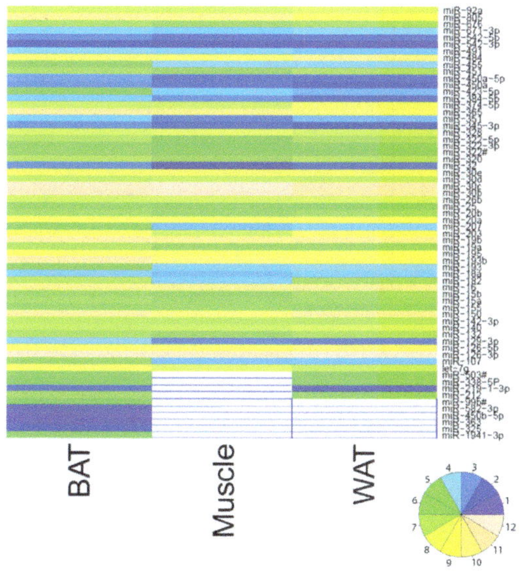

Figure 2. Heat map of the 64 miRNAs specifically and highly expressed in mouse BAT as compared with muscle and WAT. Values 1 and 12 represent the lowest and highest values, respectively. White values represent non-expressed miRNAs.

Eleven miRNAs were exclusively expressed in muscle, 67 miRNAs were exclusively expressed in WAT, while another 29 of miRNAs were expressed in both muscle and WAT only.

Bio-informatic Analyses on Mouse BAT-enriched miRNAs

To identify miRNAs potentially involved in BAT development a microRNA target filter was performed using IPA software (Ingenuity®Systems, www.ingenuity.com). Of the 64 BAT-enriched miRNAs, 35 miRNAs were predicted to target 859 genes potentially involved in cellular growth, proliferation and development. Table 1 presents these 35 miRNAs as well as the number of predicted genes targeted by each miRNA. miR-15a was predicted to target the most genes involved in cellular growth, proliferation and development pathways.

Table 1. Highly expressed miRNAs in mouse BAT. Number of targets predicted to regulate growth, proliferation and development. Relative expression values are arbitrary units (A.U.) and have been calculated relative to mouse miR-15a

miRNA	Number of targeted mRNAs	Relative expression (A.U)	miRNA	Number of targeted mRNAs	Relative expression (A.U.)
miR-15a	184	1.0	miR-140	48	2.7
miR-20a	149	8.1	miR-150	45	69.4
miR-19a	141	1.5	miR-193	39	0.4
miR-30b	139	239.2	miR-328	39	5.06
miR-182	135	1.8	miR-542-3p	39	0.004
let-7g	133	7.9	miR-491	38	0.07
miR-203	112	9.1	miR-18a	37	0.05
miR-351	108	0.0	miR-455	33	0.54
miR-26b	106	3.8	miR-484	11	17.6
miR-25	100	0.6	miR-451	10	1.10
miR-320	97	2.0	miR-129-3p	9	0.1
miR-107	92	0.6	miR-126-3p	6	337.2
miR-374-5p	89	3.8	miR-450a-5p	3	0.0

miR-155	82	0.6	miR-671-3p	3	0.1
miR-132	75	1.2	miR-345-3p	2	0.1
miR-423-5p	58	0.1	miR-582-3p	2	0.0
miR-142-3p	55	2.1	miR-338-5P	1	0.2
miR-365	49	67.1			

Comparison of miRNA Expressed in Human and Mouse BAT

To establish if human and mouse BAT express similar BAT-enriched miRNAs, miRNA array analysis was performed on 5 human and 4 mouse BAT samples. Data were analysed as described in the flowchart (Fig. 3). Of the 750 miRNAs measured 214 miRNAs and 326 miRNAs were expressed in a minimum of 3 human BAT samples and 3 mouse BAT samples, respectively. Of these miRNAs, 145 were expressed in both human and mouse BAT, including 23 miRNAs that had the same name but their sequences differed by few nucleotides between the species (i.e. miR-155, miR-193b, miR-455). Therefore, 69 miRNAs were expressed in human BAT but not mouse BAT, while 181 miRNAs were expressed in mouse BAT but not human BAT.

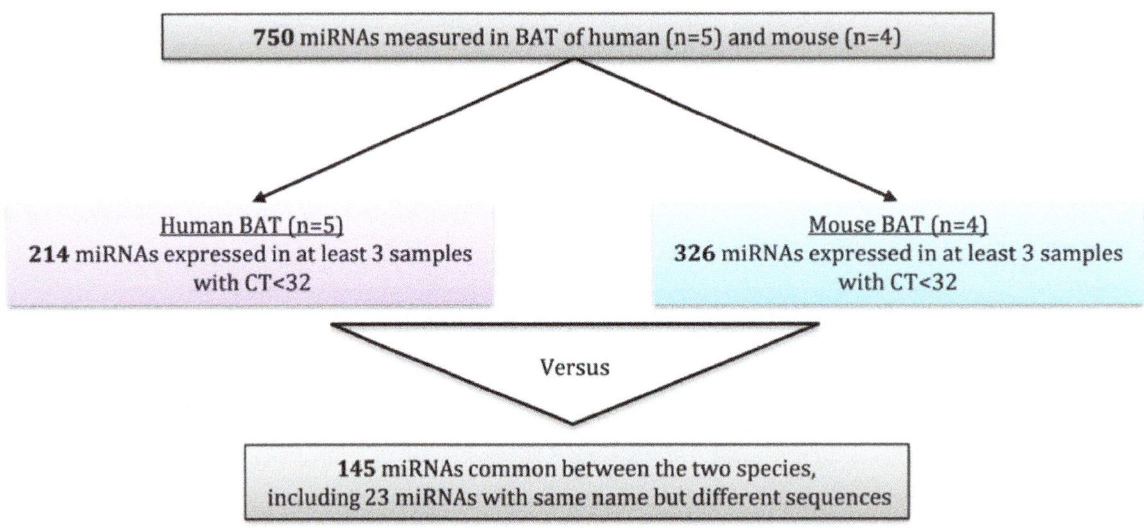

Figure 3. Flow chart of the miRNA array analysis for the comparisons of human and mouse BAT.

As shown in the flowchart in Fig. 4 a comparison was made between the 145 miRNAs common to both human and mouse BAT and with the 35 mouse BAT-enriched miRNAs identified from the bio-informatic analyses in Table 2. Twenty five of the 35 mouse BAT-enriched miRNAs were also expressed in human BAT and were predicted to regulate 788 genes involved growth, proliferation and development pathways. Ten miRNAs enriched in mouse BAT were not expressed in human BAT. Of the 25 BAT-enriched miRNAs commonly expressed in human and mouse BAT, 10 had previously been predicted to target gene involved in brown adipogenesis. These 10 miRNAs and their predicted regulatory network are presented in Fig. 5. *Bone morphogenetic protein 2* (*BMP2*) and *BMP7* are predicted to be targeted by miR-20a,-140 and miR-25, −30b, respectively. *Bone morphogenetic protein receptor 2* (*BMPR2*) is known to be targeted by miR-19a, −20a and miR-25 [28] and predicted to be targeted by miR-455. *Homeobox C9* (*HOXC9*) is predicted to be targeted by miR-193, 150 and -26b. Finally, *PPAR* is known to be targeted by miR-20a [29] and *myogenic factor 5 (MYF5)* is predicted to be targeted by miR-20a.

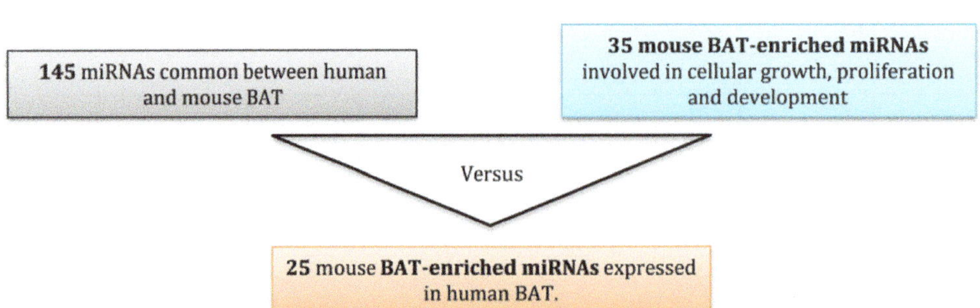

Figure 4. Flow chart of the miRNA analysis for the comparisons between human and mouse BAT-enriched miRNAs.

Table 2. miRNA expression in both human and mouse BAT. Number of targets predicted to regulate growth, proliferation and development. Relative expression values are arbitrary units (A.U.) and have been calculated relative to human miR-15a

miRNA	Number of targeted mRNAs	Relative expression (A.U)	miRNA	Number of targeted mRNAs	Relative expression (A.U)
miR-15a	184	1.0	miR-365	49	9.2
miR-20a	149	13.2	miR-140	48	7.2
miR-19a	141	12.7	miR-150	45	101.4
miR-30b	139	26.2	miR-193	39	6.8
let-7g	133	15.1	miR-328	39	0.4
miR-203	112	0.3	miR-491	38	0.5
miR-26b	106	15.5	miR-18a	37	0.2
miR-25	100	0.3	miR-455	33	0.4
miR-320	97	44.3	miR-484	11	65.6
miR-155	82	4.6	miR-451	10	1.9
miR-132	75	6.8	miR-126-3p	6	1438.0
miR-423-5p	58	0.4	miR-338-5P	1	1160.2
miR-142-3p	55	2.7			

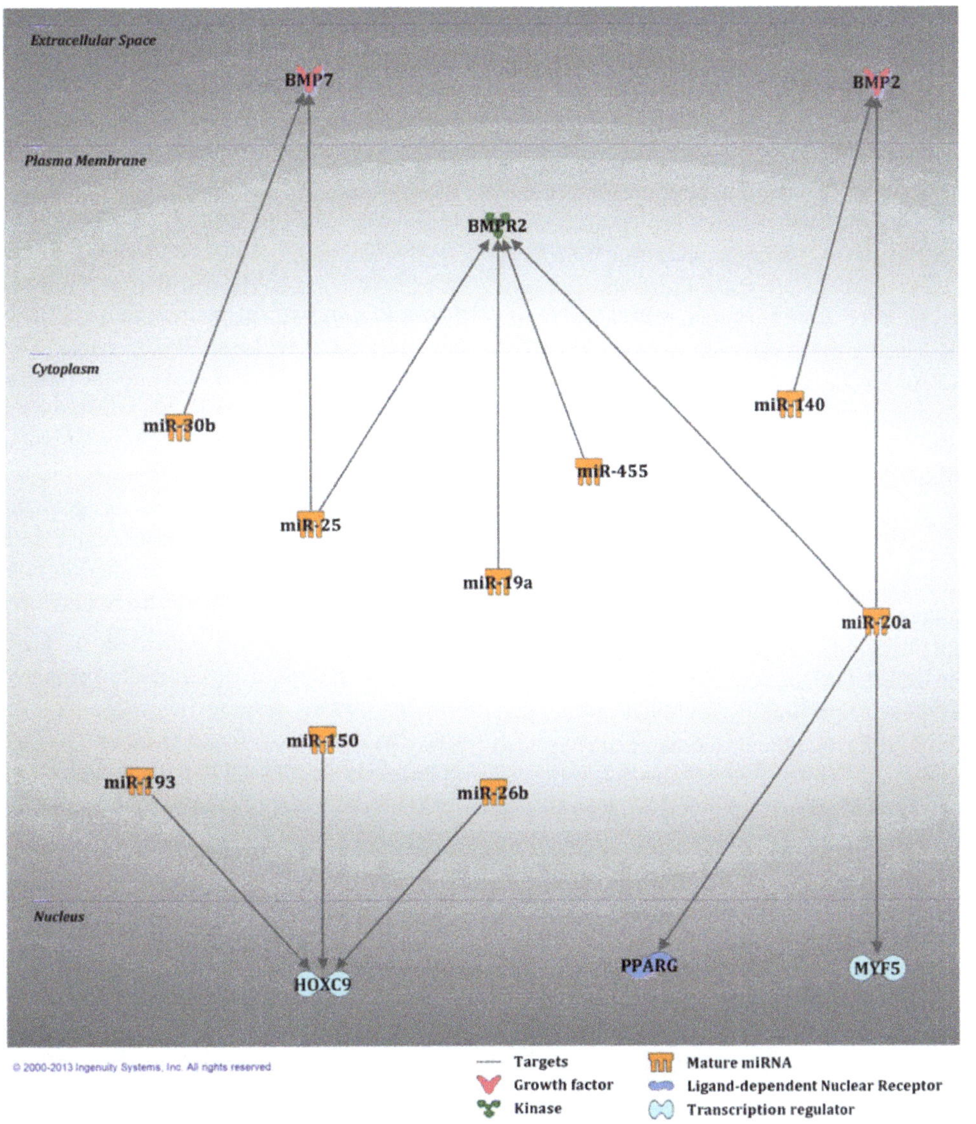

Figure 5. Selected miRNAs common to human and mouse BAT and their gene targets known to be involved in brown adipogenesis.

DISCUSSION

The recent discovery of active BAT in human adults [30, 31, 32, 33, 34, 35] has opened a new field of investigation for the treatment of obesity [3, 4, 5]. Therefore, understanding the molecular mechanisms, regulating human BAT development may identify novel therapeutic strategies to increase energy expenditure. miRNAs are important molecular switches that display high levels of tissue enrichment and can control cell differentiation and tissue growth and development. Studies have investigated miRNAs in mouse BAT [10, 14,16, 17, 18, 19, 20, 21, 22, 23, 24], however their expression levels have not been validated in human BAT. The present study measured and compared the expression of miRNAs highly enriched in mouse and human BAT, as well as predicting their gene targets potentially involved in BAT growth and development. Several novel observations were made. Firstly, 64 miRNAs were exclusively or highly expressed in mouse BAT in comparison to mouse skeletal muscle and WAT. Secondly, of these 64 BAT-enriched miRNAs, 35 were predicted to regulate genes involved in cellular growth, proliferation and development. Thirdly, 145 miRNAs were found to be commonly expressed in human and mouse BAT. Finally, 25 of these 145 miRNAs were also identified in the list of 35 mouse BAT-enriched miRNAs potentially involved in cellular growth, proliferation and development.

Identifying the molecular mechanisms (i.e. miRNAs) controlling BAT development has become a priority to develop therapeutic strategies in the fight against obesity. While BAT shares a common origin with skeletal muscle [6, 7], studies in mouse WAT recently discovered a phenomenon called "browning" that involves the formation of brown adipocytes in WAT after cold exposure or β-adrenergic stimulation [36]. Therefore, it was of interest to compare these three mouse tissues in order to identify BAT-enriched miRNAs. In the present study, qPCR-based miRNA screening of ~750 miRNAs identified 64 miRNAs exclusively or highly expressed in mouse BAT, in comparison to skeletal muscle and WAT. Consistent with a similar study [14], we identified that ~22% of the miRNAs commonly expressed between the three tissues were significantly higher in BAT. Of the 64 miRNAs observed to be enriched in mouse BAT in the present study, only 15 have previously in BAT [14].

miRNAs regulate various component of cellular networks and maintain cellular homeostasis [37]. Therefore it was of interest to select BAT-enriched miRNAs that were predicted to target genes specifically involved in cellular growth, proliferation and development and as such may also control BAT development. Thirty-five of the 64 miRNAs were predicted to target 859 genes involved in growth and development pathways. In cultured adipocytes, global deep sequencing identified 16 and 12 miRNAs that were differentially regulated in mouse primary pre-adipocytes when compared with differentiated mature brown adipocytes in vitro, [18]. Of the miRNAs upregulated in differentiated mature brown adipocytes, miR-142-3p and -19a were also up-regulated in the present list of 35 BAT-enriched miRNAs predicted to target genes involved in growth and differentiation. This supports a potential role of these miRNAs during BAT development in mice. While miR-155 and −328 were down-regulated in mature brown adipocytes when compared to pre-adipocytes, they were identified as being upregulated in our list of 35 BAT-enriched miRNAs. These discrepancies may be indicative of the differences between mature brown adipocytes *in vitro* and mature BAT.

Most studies have used mice as a model to study BAT development and function, due to the previously held view that adult humans did not possess BAT, as well as the limited access to human BAT from new born babies. As such, no study has looked at the miRNA profile of human BAT. It was therefore of interest to compare miRNA profiles of mouse and human BAT to establish the potential molecular similarities and differences between the two species. These comparisons also provide a snapshot of the potential to translate molecular observations from mice to humans. For the first time, a miRNA screening revealed the presence of 214 miRNAs expressed in human BAT, including 145 commonly expressed in mouse BAT. This screening also showed that 69 miRNAs expressed in human BAT were not expressed in mouse BAT and therefore were not taken into consideration when comparisons were made with the mouse model. Further investigations of these 69 miRNAs will be required to identify their potential unique role in human BAT development and function. In contrast, 181 miRNAs in mouse BAT were not present in human BAT, suggesting that translating results from mouse to human BAT needs careful consideration.

Of the 145 miRNAs common to both mouse and human BAT, 25 were identified in the list of 35 mouse BAT-enriched miRNAs potentially targeting genes involved in cellular growth, proliferation and development. 788 genes involved in these pathways were predicted to be targeted by these 25 miRNAs. In this present study, miR-20a was the only miRNA predicted to target *MYF5* and *PPAR* (a previously experimentally observed target [38]), two important factors directing cell fate towards a brown fat phenotype [11]. In contrast, miR-20a is also predicted to target *BMP2* and *BMPR2*, a growth factor and receptor increased in white fat differentiation [39]. These findings suggest that miR-20a may have the capacity to control cell fate toward a brown or white fat phenotype. Furthermore, miR-20a was the miRNA targeting the second most numbers of genes ($n = 149$) involved in growth and development, highlighting its potential importance in these pathways. To date, miR-20a has not been studied in relation to BAT growth and development and therefore requires further investigation.

miR-155, highly expressed in brown pre- adipocytes when compared to differentiated mature brown adipocytes, inhibits brown adipogenesis by targeting *C/EBP* [18], a transcription factor forming a complex with PRDM16 during brown adipogenesis [13]. In contrast to these observations, miR-155 was identified as a highly enriched miRNA in human and mouse BAT. This observation suggests that human BAT may be comprised of a mixed composition of cells with different stages of cell maturation. However, the presence of miR-20a, −25 and -30b, all predicted to inhibit expression of genes involved in brown fat development, may prevent the development of a "classical" brown fat phenotype in human BAT. Furthermore, miR-182 and −203 were also part of the top ten miRNAs targeting the most genes involved in growth and development in mouse BAT, suggesting their potentials roles in brown adipocyte development [21]. However, only miR-203 was conserved in human BAT. The absence of miR-

182 expression in human BAT might contribute to differences between mouse and human BAT development and function. A gene signature developed in human BAT identified a profile closely resembling a brite/beige phenotype [40, 41]. This phenotype is supported by the absence of miR-378, a key regulator controlling classical BAT-specific expansion and obesity resistance [23]. However the phenotypical signature may depend on the depth of the tissue taken with superficial tissue representing a more WAT-like signature while deeper tissue representing a more classical BAT-like signature [42].

The miR-193b-365 cluster and miR-455 are known to regulate mouse brown fat development and were identified in our selection of 25 BAT-enriched miRNAs common to mice and humans. Inhibition of miR-193a/b and/or miR-365 in mouse primary brown pre-adipocytes impairs brown adipogenesis *in vitro* [14]. However, miR-193b null mice, that are also deficient in miR-365, have normal BAT development, differentiation and function [24]. miR-193b null mice have elevated levels of miR-455 suggesting a compensatory effect in the absence of the miR-193b-365 cluster. miR-455 expression levels are also increased in mature murine brown adipocytes and positively correlate with uncoupling protein 1 (UCP1) levels suggesting a potential metabolic role in brown adipocytes [10]. miR-19a, −140, −150 and -26b are predicted to target white fat-related genes (*BMP2* or *BMPR2* or *HOXC9*) [39, 41] and therefore may have a role in selecting cell fate toward brown adipogenesis.

In the present study BAT from newborns was used to establish the human BAT miRNA profile. How precisely these findings translate to the human adult is presently unknown. Human newborns and adults have a similar distribution of BAT depots [43]. In newborns and infants BAT is mostly located in the interscapular, cervical and perirenal regions [43, 44]. In human adults the major depots of metabolically active BAT have been identified in the cervical, supraclavicular, axillary and paravertebral regions [30, 32, 33]. The location and quantity of human BAT alters with age, which may indicate an age-dependent change in its primary role; presently the precise functional role of adult BAT is yet to be determined. However, when the hands of both infants and adults are cooled there is a 0.3-0.7°C increase in subcutaneous cervical BAT, albeit to a lesser degree in older subjects [45]. This suggests that both infant and adult BAT may respond in a similar way to some external stress signals. Several studies have tried to develop a mRNA profile to establish how similarly infant and adult human BAT [40, 42] and brown adipocytes [44, 46] mimic the "classical" rodent BAT on a molecular level. Depending on the depth and location of the tissue analysed, infant and adult human BAT possess gene signatures with a mixture of "classical" BAT and "beige-like" adipocytes. The present study analysed newborn BAT taken from regions where active BAT is also found in the human adult. Therefore, it is most likely that the newborn human BAT miRNA profile established here would represent an adult profile with a mixture of "classic" BAT and/or "beige-like" adipocytes. Investigations are still required to clearly identify the mRNA and the miRNA species that are truly specific to brown and beige tissue; a task that that may complicated by the different levels of heterogeneity within BAT tissue depots.

CONCLUSION

This study has identified 64 miRNAs enriched in mouse BAT when compared to skeletal muscle and WAT, including 35 miRNAs specifically involved in cellular growth, proliferation and development. Furthermore, for the first time a comparison between human and mouse BAT revealed the presence of 145 commonly expressed miRNAs in BAT from both species. These included 25 miRNAs specifically expressed in mouse BAT and involved in cellular growth, proliferation and development. The identification of BAT-enriched miRNAs, conserved in both mouse and human BAT, such as miR-20a, may be a common factor controlling BAT development. Several of the newly identified miRNAs, common to mouse and human BAT, should be evaluated for their direct role in brown adipogenesis.

METHODS

Tissue Collection

The interscapular BAT, gastrocnemius and visceral WAT from 8 week old male C57BL/6J mice were donated by Professor Matthew Watt, Monash University. The tissue provided was approved by the School of Biomedical Sciences Animal Ethics Committee (SOBSA/2009/22) Monash University, Australia. RNA samples from human BAT, obtained from newborn humans aged between 10.6±4.3 days, were donated by Professor Vicente Gilsanz from the Children's Hospital Los Angeles (CHLA), USA. The RNA samples were originally obtained as a component of a previously published study [40]. This previously published human study was approved by the Institutional Review Board of the CHLA (CCI-10-00073). The Deakin University Human Research Ethics Committee Executive also reviewed the project (CCI-10-00073) and acknowledged the approval granted by the Institutional Review Board of the CHLA and allocated the project reference number 2015–189. Written informed consent was obtained by all parent/legal guardian(s) in the presence of a witness authorizing post-mortem examination and removal of tissues. Authorization was also given for the samples to be retained and preserved and to be used for diagnostic, therapeutic and other scientific purposes as deemed appropriate by the examining physician or surgeon. The research was carried out on in compliance with the Helsinki Declaration. The samples were surgically removed from the subcutaneous supraclavicular/supracervical area, intra-abdominal area or retroperitoneal area during post-mortem examination 24h after death and have been used previously of mRNA measurements [40]. miRNAs are exceptionally stable post-mortem [47] and therefore the miRNA profile generated would be representative of that observed in a living human. All tissue samples were snapped frozen in liquid nitrogen and kept at −80°C until required for RNA extraction.

RNA Extractions

Mouse Tissues

Approximately 20 mg of BAT and gastrocnemius and 100 mg of WAT were homogenised in 500 μl of Tri-Reagent® Solution (Life Technologies, CA, USA), using the MagNA Lyser (Roche, Basel, Switzerland) for 20s at a speed setting of 5500 revolutions per minute (rpm). After transferring the homogenate into a fresh tube, RNA extraction was completed according to the manufacturer's protocol. Quality of the RNA was assessed using a Nanodrop 1000 Spectrophotometer

(Thermo Fisher Scientific, MA, USA) and only samples with an O.D. 260/280 ratio >1.9 were used for the analyses.

Human Tissues

Approximately 20 mg of BAT were homogenised in 300 μl of the denaturation solution from the TōTALLY RNA kit™ (Life Technologies, CA, USA), using the MagNA Lyser (Roche, Basel, Switzerland) for 20s at a speed setting of 5500rpm. After transferring the homogenate into a fresh tube, RNA extraction was completed according to the manufacturer's protocol. Quality of the RNA was assessed using a Nanodrop 1000 Spectrophotometer (Thermo Fisher Scientific, MA, USA) and only samples with an O.D. 260/280 ratio >1.9 were used for the analyses.

Megaplex Reverse Transcription and miRNA Arrays

Mouse and Human Tissues

Rodent Megaplex™ Primer Pools and TaqMan® Array miRNA Cards A version 2.0 and B version 3.0 (Life Technologies, CA, USA) were used to reverse transcribe and amplify ~750 miRNAs according to the manufacturer's protocol. This was performed on 4 BAT, 4 gastrocnemius and 4 WAT samples. Human Megaplex™ Primer Pools and TaqMan® Array miRNA Cards A version 2.1 and B version 3.0 (Life Technologies, CA, USA) were used to reverse transcribe and amplify ~750 miRNAs according to the manufacturer's protocol. This was performed on human tissues from 5 different individuals. All cards were run on the 7900HT Fast Real-Time PCR System machine at Monash Institute of Medical Research (MIMR, Clayton, VIC, Australia). Results were analysed using ExpressionSuite software v1.0.1 (Life Technologies, CA, USA). The global mean expression value of all expressed miRNAs in a given samples was used as the normalisation factor in this study [48].

Bio-informatics

Bio-informatic analyses were performed on specific lists of miRNAs to determine their predicted gene targets. Data were analysed through the use of IPA (Ingenuity®Systems, www.ingenuity.com) and more specifically, the microRNA Target filter tool. Filters were used to refine the analysis based on experimentally observed and high predicted gene targets only in the column "confidence", as well as gene involved in cellular growth, proliferation and development pathways only.

Statistics

For the mouse samples, miRNAs were considered expressed in a given tissue when a minimum of 3 out of the 4 samples analysed had CT values below 32, as suggested in the manufacturer's protocol. miRNAs expressed in 2 tissues only were compared using an unpaired t-test following log transformation of their expression values. For the miRNAs expressed in the 3 tissues (BAT, gastrocnemius and WAT), statistical analysis was performed using STATA data analysis and statistical software using a Kruskal–Wallis test. When a significant difference was detected a two-sample Wilcoxon-rank sum test was used to locate the differences. A p-value of <0.05 was considered as significant for all analyses.

For the human samples, miRNAs were considered expressed in a given tissue when a minimum of 3 samples (out of 5) had CT values below 32, as mentioned in the manufacturer's protocol.

ACKNOWLEDGEMENTS

I.G. is supported by a Deakin University International Higher Degree by Research Scholarship.

AUTHORS' CONTRIBUTIONS

All authors were involved in the project design, interpretation of data, manuscript preparation and revisions and have agreed on the final submitted version. SK and GV provided human BAT samples. MW mouse BAT, WAT and muscle samples. IG completed the RNA extraction, sample preparation, miRNA arrays and analysed the miRNA array data. IG and SMcN performed the statistical analysis. TC and IG performed the bio-informatics and designed the figures. All authors read and approved the final manuscript.

REFERENCES

1. Haslam DW, James WP. Obesity. Lancet. 2005; 366: 1197–209.
2. Sullivan EL, Cameron JL. A rapidly occurring compensatory decrease in physical activity counteracts diet-induced weight loss in female monkeys. Am J Physiol Regul Integr Comp Physiol. 2010; 298: R1068–74.
3. Carey AL, Kingwell BA. Brown adipose tissue in humans: Therapeutic potential to combat obesity. Pharmacol Ther. 2013; 140((1)): 26–33.
4. Nedergaard J, Cannon B. The changed metabolic world with human brown adipose tissue: therapeutic visions. Cell Metab. 2010; 11: 268–72.
5. Vijgen GH, van Marken Lichtenbelt WD. Brown adipose tissue: clinical impact of a re-discovered thermogenic organ. Front Biosci. 2013; E5: 823–33.
6. Atit R, Sgaier SK, Mohamed OA, Taketo MM, Dufort D, Joyner AL, et al. Beta-catenin activation is necessary and sufficient to specify the dorsal dermal fate in the mouse. Dev Biol. 2006; 296: 164–76.
7. Timmons JA, Wennmalm K, Larsson O, Walden TB, Lassmann T, Petrovic N, et al. Myogenic gene expression signature establishes that brown and white adipocytes originate from distinct cell lineages. Proc Natl Acad Sci U S A. 2007; 104: 4401–6.
8. Reinhart BJ, Slack FJ, Basson M, Pasquinelli AE, Bettinger JC, Rougvie AE, et al. The 21-nucleotide let-7 RNA regulates developmental timing in Caenorhabditis elegans. Nature. 2000; 403: 901–6.
9. Hamilton AJ, Baulcombe DC. A species of small antisense RNA in posttranscriptional gene silencing in plants. Science. 1999; 286: 950–2.
10. Walden TB, Timmons JA, Keller P, Nedergaard J, Cannon B. Distinct expression of muscle-specific microRNAs (myomirs) in brown adipocytes. J Cell Physiol. 2009; 218: 444–9.

11. Seale P, Bjork B, Yang W, Kajimura S, Chin S, Kuang S, et al. PRDM16 controls a brown fat/skeletal muscle switch. Nature. 2008; 454: 961–7.

12. Seale P, Kajimura S, Yang W, Chin S, Rohas LM, Uldry M, et al. Transcriptional control of brown fat determination by PRDM16. Cell Metab. 2007; 6: 38–54.

13. Kajimura S, Seale P, Kubota K, Lunsford E, Frangioni JV, Gygi SP, et al. Initiation of myoblast to brown fat switch by a PRDM16-C/EBP-beta transcriptional complex. Nature. 2009; 460: 1154–8.

14. Sun L, Xie H, Mori MA, Alexander R, Yuan B, Hattangadi SM, et al. Mir193b-365 is essential for brown fat differentiation. Nat Cell Biol. 2011; 13: 958–65.

15. Chen Y, Gelfond JA, McManus LM, Shireman PK. Reproducibility of quantitative RT-PCR array in miRNA expression profiling and comparison with microarray analysis. BMC Genomics. 2009; 10: 407.

16. Mori M, Nakagami H, Rodriguez-Araujo G, Nimura K, Kaneda Y. Essential role for miR-196a in brown adipogenesis of white fat progenitor cells. PLoS Biol. 2012; 10: e1001314.

17. Yin H, Pasut A, Soleimani VD, Bentzinger CF, Antoun G, Thorn S, et al. MicroRNA-133 controls brown adipose determination in skeletal muscle satellite cells by targeting Prdm16. Cell Metab. 2013; 17: 210–24.

18. Chen Y, Siegel F, Kipschull S, Haas B, Frohlich H, Meister G, et al. miR-155 regulates differentiation of brown and beige adipocytes via a bistable circuit. Nat Commun. 2013; 4: 1769.

19. Trajkovski M, Ahmed K, Esau CC, Stoffel M. MyomiR-133 regulates brown fat differentiation through Prdm16. Nat Cell Biol. 2012; 14: 1330–5.

20. Fu T, Seok S, Choi S, Huang Z, Suino-Powell K, Xu HE, et al. MiR-34a inhibits beige and brown fat formation in obesity in part by suppressing adipocyte FGF21 signaling and SIRT1 function. Mol Cell Biol. 2014; 34(22): 4130–42.

21. Kim HJ, Cho H, Alexander R, Patterson HC, Gu M, Lo KA, et al. MicroRNAs Are required for the feature maintenance and differentiation of brown adipocytes. Diabetes. 2014; 63(12): 4045–56.

22. Sun L, Trajkovski M. MiR-27 orchestrates the transcriptional regulation of brown adipogenesis. Metabolism. 2014; 63: 272–82.

23. Pan D, Mao C, Quattrochi B, Friedline RH, Zhu LJ, Jung DY, et al. MicroRNA-378 controls classical brown fat expansion to counteract obesity. Nat Commun. 2014; 5: 4725.

24. Feuermann Y, Kang K, Gavrilova O, Haetscher N, Jang SJ, Yoo KH, et al. MiR-193b and miR-365-1 are not required for the development and function of brown fat in the mouse. RNA Biol. 2013; 10: 1807–14.

25. Lowell BB VSS, Hamann A, Lawitts JA, Himms-Hagen J, Boyer BB, et al. Development of obesity in transgenic mice after genetic ablation of brown adipose tissue. Nature. 1993; 366: 740–2.

26. Merklin RJ. Growth and distribution of human fetal brown fat. Anat Rec. 1974; 178: 637–45.

27. Silverman WA, Fertig JW, Berger AP. The influence of the thermal environment upon the survival of newly born premature infants. Pediatrics. 1958; 22: 876–86.

28. Brock M, Trenkmann M, Gay RE, Michel BA, Gay S, Fischler M, et al. Interleukin-6 modulates the expression of the bone morphogenic protein receptor type II through a novel STAT3-microRNA cluster 17/92 pathway. Circ Res. 2009; 104: 1184–91.

29. He J, Zhang JF, Yi C, Lv Q, Xie WD, Li JN, et al. miRNA-mediated functional changes through co-regulating function related genes. PLoS ONE. 2010; 5: e13558.

30. Cypess AM, Lehman S, Williams G, Tal I, Rodman D, Goldfine AB, et al. Identification and importance of brown adipose tissue in adult humans. N Engl J Med. 2009; 360: 1509–17.

31. Saito M, Okamatsu-Ogura Y, Matsushita M, Watanabe K, Yoneshiro T, Nio-Kobayashi J, et al. High incidence of metabolically active brown adipose tissue in healthy adult humans: effects of cold exposure and adiposity. Diabetes. 2009; 58: 1526–31.

32. van Marken Lichtenbelt WD, Vanhommerig JW, Smulders NM, Drossaerts JM, Kemerink GJ, Bouvy ND, et al. Cold-activated brown adipose tissue in healthy men. N Engl J Med. 2009; 360: 1500–8.

33. Virtanen KA, Lidell ME, Orava J, Heglind M, Westergren R, Niemi T, et al. Functional brown adipose tissue in healthy adults. N Engl J Med. 2009; 360: 1518–25.

34. Zingaretti MC, Crosta F, Vitali A, Guerrieri M, Frontini A, Cannon B, et al. The presence of UCP1 demonstrates that metabolically active adipose tissue in the neck of adult humans truly represents brown adipose tissue. FASEB J. 2009; 23: 3113–20.

35. Nedergaard J, Bengtsson T, Cannon B. Unexpected evidence for active brown adipose tissue in adult humans. Am J Physiol Endocrinol Metab. 2007; 293: E444–52.

36. Himms-Hagen J, Cui J, Danforth Jr E, Taatjes DJ, Lang SS, Waters BL, et al. Effect of CL-316,243, a thermogenic beta 3-agonist, on energy balance and brown and white adipose tissues in rats. Am J Physiol. 1994; 266: R1371–82.

37. Cui Q, Yu Z, Purisima EO, Wang E. Principles of microRNA regulation of a human cellular signaling network. Mol Syst Biol. 2006; 2: 46.

38. Zhang JF, Fu WM, He ML, Xie WD, Lv Q, Wan G, et al. MiRNA-20a promotes osteogenic differentiation of human mesenchymal stem cells by co-regulating BMP signaling. RNA Biol. 2011; 8: 829–38.

39. Stephens M, Ludgate M, Rees DA. Brown fat and obesity: the next big thing? Clin Endocrinol (Oxf). 2011; 74: 661–70.

40. Sharp LZ, Shinoda K, Ohno H, Scheel DW, Tomoda E, Ruiz L, et al. Human BAT possesses molecular signatures that resemble beige/brite cells. PLoS ONE. 2012; 7: e49452.

41. Wu J, Bostrom P, Sparks LM, Ye L, Choi JH, Giang AH, et al. Beige adipocytes are a distinct type of thermogenic fat cell in mouse and human. Cell. 2012; 150: 366–76.

42. Cypess AM, White AP, Vernochet C, Schulz TJ, Xue R, Sass CA, et al. Anatomical localization, gene expression profiling and functional characterization of adult human neck brown fat. Nat Med. 2013; 19: 635–9.

43. Heaton JM. The distribution of brown adipose tissue in the human. J Anat. 1972; 112: 35–9.

44. Lidell ME, Betz MJ, Dahlqvist Leinhard O, Heglind M, Elander L, Slawik M, et al. Evidence for two types of brown adipose tissue in humans. Nat Med. 2013; 19: 631–4.

45. Symonds ME, Henderson K, Elvidge L, Bosman C, Sharkey D, Perkins AC, et al. Thermal imaging to assess age-related changes of skin temperature within the supraclavicular region co-locating with brown adipose tissue in healthy children. J Pediatr. 2012; 161: 892–8.

46. Shinoda K, Luijten IH, Hasegawa Y, Hong H, Sonne SB, Kim M, et al. Genetic and functional characterization of clonally derived adult human brown adipocytes. Nat Med. 2015; 21: 389–94.

47. Fordyce SL, Kampmann ML, van Doorn NL, Gilbert MT. Long-term RNA persistence in postmortem contexts. Investig Genet. 2013; 4: 7.

48. Mestdagh P, Van Vlierberghe P, De Weer A, Muth D, Westermann F, Speleman F, et al. A novel and universal method for microRNA RT-qPCR data normalization. Genome Biol. 2009; 10: R64.

CHAPTER 17

HEIGHT, BODY MASS INDEX, AND SOCIOECONOMIC STATUS: MENDELIAN RANDOMISATION STUDY IN UK BIOBANK

Jessica Tyrrell,[1,2] Samuel E Jones,[1] Robin Beaumont,[1] Christina M Astley,[3,4] Rebecca Lovell,[2] Hanieh Yaghootkar,[1] Marcus Tuke,[1] Katherine S Ruth,[1] Rachel M Freathy,[1] Joel N Hirschhorn,[2,3,5] Andrew R Wood,[1] Anna Murray,[1] Michael N Weedon,[1] and Timothy M Frayling[1]

[1]Genetics of Complex Traits, Institute of Biomedical and Clinical Science, University of Exeter Medical School, Royal Devon and Exeter Hospital, Exeter EX2 5DW, UK

[2]European Centre for Environment and Human Health, University of Exeter Medical School, The Knowledge Spa, Truro TR1 3HD, UK

[3]Program in Medical and Population Genetics, Broad Institute of MIT and Harvard, Cambridge, MA 02142, USA

[4]Center for Basic and Translational Obesity Research and Division of Endocrinology, Boston Children's Hospital, Boston, MA 02115, USA

[5]Department of Genetics, Harvard Medical School, Boston, MA 02115, USA

ABSTRACT

Objective

To determine whether height and body mass index (BMI) have a causal role in five measures of socioeconomic status.

Design

Mendelian randomisation study to test for causal effects of differences in stature and BMI on five measures of socioeconomic status. Mendelian randomisation exploits the fact that genotypes are randomly assigned at conception and thus not confounded by non-genetic factors.

Setting

UK Biobank.

Participants

119 669 men and women of British ancestry, aged between 37 and 73 years.

Citation: Jessica Tyrrell, Samuel E Jones, Robin Beaumont, Christina M Astley, Rebecca Lovell, Hanieh Yaghootkar, Marcus Tuke, Katherine S Ruth, Rachel M Freathy, Joel N Hirschhorn, Andrew R Wood, Anna Murray, Michael N Weedon, and Timothy M Frayling. Height, body mass index, and socioeconomic status: mendelian randomisation study in UK Biobank, BMJ 2016; 352:i582. DOI.org/10.1136/bmj.i582.

Main Outcome Measures

Age completed full time education, degree level education, job class, annual household income, and Townsend deprivation index.

Results

In the UK Biobank study, shorter stature and higher BMI were observationally associated with several measures of lower socioeconomic status. The associations between shorter stature and lower socioeconomic status tended to be stronger in men, and the associations between higher BMI and lower socioeconomic status tended to be stronger in women. For example, a 1 standard deviation (SD) higher BMI was associated with a £210 (€276; $300; 95% confidence interval £84 to £420; $P=6\times10^{-3}$) lower annual household income in men and a £1890 (£1680 to £2100; $P=6\times10^{-15}$) lower annual household income in women. Genetic analysis provided evidence that these associations were partly causal. A genetically determined 1 SD (6.3 cm) taller stature caused a 0.06 (0.02 to 0.09) year older age of completing full time education (P=0.01), a 1.12 (1.07 to 1.18) times higher odds of working in a skilled profession ($P=6\times10^{-7}$), and a £1130 (£680 to £1580) higher annual household income ($P=4\times10^{-8}$). Associations were stronger in men. A genetically determined 1 SD higher BMI (4.6 kg/m^2) caused a £2940 (£1680 to £4200; $P=1\times10^{-5}$) lower annual household income and a 0.10 (0.04 to 0.16) SD (P=0.001) higher level of deprivation in women only.

Conclusions

These data support evidence that height and BMI play an important partial role in determining several aspects of a person's socioeconomic status, especially women's BMI for income and deprivation and men's height for education, income, and job class. These findings have important social and health implications, supporting evidence that overweight people, especially women, are at a disadvantage and that taller people, especially men, are at an advantage.

INTRODUCTION

Higher socioeconomic status is associated with better health and longer life.[1 2] For example, a recent article highlighted the strength of the association between wealth and health by pointing out the 18 and 20 year gaps in male life expectancy between the least and most wealthy parts of London, UK, and Baltimore, USA, respectively.[3]Two easily measured markers associated with socioeconomic status are adult height and body mass index (BMI).[4 56] In developed counties, taller stature and lower BMI are associated with higher socioeconomic status and better health.[4 5 6 7 8 9 10 11 12] Higher socioeconomic status is generally thought to cause taller stature and lower BMI owing to higher standards of nutrition in childhood, but there may also be effects in the opposite direction—taller stature and lower BMI may causally improve socioeconomic status through discrimination against shorter and fatter people or differences in self esteem that affect employability.[13 14] Evidence as to whether height and BMI have causal effects on socioeconomic status through these, or other, pathways is limited. For example, to our knowledge, no large studies have compared siblings or twins of different heights and BMIs, where childhood environment could be controlled for. If differences in BMI and height can lead to differences in socioeconomic status, this would have implications for policy makers. For example, evidence of a causal link would further highlight the need to adjust for unconscious biases in decision making in education and employment.

Gene based analyses, such as mendelian randomisation,[15] can be used to test for a causal relation between socioeconomic status and a genetically influenced phenotype such as BMI. Genetic variants can act as unconfounded proxies for the risk factors under investigation—here, BMI and height—because inherited genetic variation is randomly allocated at conception. The outcomes being tested—here, measures of socioeconomic status—cannot influence genetic variation, so reverse causality is avoided in genetic studies. Figure 1 illustrates the principle of mendelian randomisation. Previous studies have used genetic variants to test causal relation between health traits such as BMI and socioeconomic status related outcomes such as academic performance. However, these studies were limited by a lack of genetic variants robustly associated with BMI and by sample sizes of fewer than 2300 people.[16 17] Recent genome-wide association studies have identified many 10s and 100s of genetic variants associated with BMI and height, respectively,[18 19] and so provide the tools for mendelian randomisation tests.

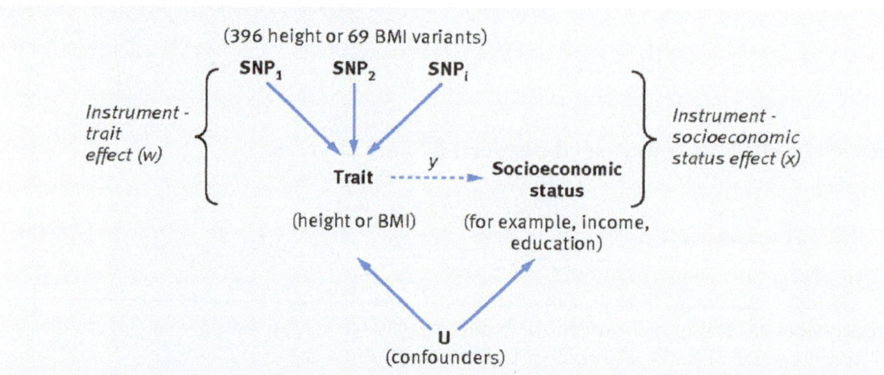

Figure 1. Principle of mendelian randomisation: if height or body mass index (BMI) causally influences socioeconomic status, genetic variants associated with that trait will also be associated with socioeconomic status. As genotype is assigned at conception, it should not be associated

with factors that normally confound the association between BMI and height and socioeconomic status (eg, environmental and behavioural factors). We can use our estimates of the genetic-height/BMI association (w) and the genetic-socioeconomic status association (x) to infer the causal effect of height or BMI on socioeconomic status (y=x/w), which is expected to be free from confounding. If the estimated causal effect (y) is different from the observational association between the height or BMI and socioeconomic status, this would suggest that the observational association is confounded (assuming that the assumptions of the mendelian randomisation analyses are valid). SNP=single nucleotide polymorphism.

Here, we used mendelian randomisation analysis to test the hypothesis that causal pathways link BMI and height to differences in five different measures of socioeconomic status. We used the first release of data from the UK Biobank. The UK Biobank has 119 669 participants of white British ancestry with genetic data, measures of socioeconomic status, and height and BMI measures. The UK Biobank thus represents a very powerful resource in which to investigate the causal relation between BMI, height, and socioeconomic status by using mendelian randomisation analysis.

METHODS

UK Biobank

The UK Biobank recruited more than 500 000 people aged 37-73 years (99.5% were between 40 and 69 years) from across the country in 2006-10. Participants provided a range of information via questionnaires and interviews (such as demographics, health status, and lifestyle); anthropometric measurements, blood pressure readings, and blood, urine and saliva samples were taken for future analysis. This has been described in more detail elsewhere.[20] We used 120 286 participants of white British descent from the initial UK Biobank dataset, of whom 119 669 had valid genetic data and both BMI and height measures available. We did not include other ethnic groups, because individually they were underpowered. Table 1 shows the basic characteristics of the sample. We defined people of white British descent as those who both self identified as white British and were confirmed as ancestrally "Caucasian" using principal components analyses of genome-wide genetic information. This dataset underwent extensive central quality control (http://biobank.ctsu.ox.ac.uk) (see supplementary methods).

Table 1. Summary of demographic characteristics of 119 669 participants of white British ancestry with valid genetic data and height and body mass index measures available, stratified by sex. Values are numbers (percentages) unless stated otherwise

Characteristic	All (n=119 669)	Men (n=56 652)	Women (n=63 017)	P value*
Mean (SD) age at recruitment, years	56.9 (7.9)	57.3 (8.0)	56.6 (7.8)	$<1\times10^{-15}$
Male sex	56 652 (47.3)	56 652 (100)	63 017 (100)	–
Mean (SD) body mass index, kg/m^2	27.5 (4.8)	27.9 (4.3)	27.2 (5.2)	$<1\times10^{-15}$
Mean (SD) height, cm	168.8 (9.2)	175.7 (6.7)	162.6 (6.2)	$<1\times10^{-15}$
Smoking status:				
Never	63 806 (53.3)	27 834 (49.1)	35 972 (57.1)	$<1\times10^{-15}$
Former	40 890 (34.2)	21 162 (37.4)	19 728 (31.3)	
Current	13 332 (11.1)	6767 (11.9)	6565 (10.4)	
Missing	1641 (1.4)	889 (1.6)	752 (1.2)	
Mean (SD) age completed full time education, years	16.6 (2.2)	16.6 (2.4)	16.5 (2.0)	2×10^{-9}
Degree level education	53 652 (44.8)	25 956 (45.8)	27 696 (44.0)	6×10^{-15}
Job class:				

Elementary occupations	3932 (3.3)	2054 (3.6)	1878 (3.0)	$<1\times10^{-15}$
Process plant and machine operatives	3740 (3.1)	3299 (5.8)	441 (0.7)	
Sales and customer service occupations	2658 (2.2)	588 (1.0)	2070 (3.3)	
Leisure and other personal service occupations	963 (0.8)	379 (0.7)	584 (0.9)	
Personal service occupations	3567 (3.0)	404 (0.7)	3163 (5.0)	
Skilled trades	6077 (5.1)	5404 (9.5)	673 (1.1)	
Administrative and secretarial roles	11 878 (9.9)	2329 (4.1)	9549 (15.2)	
Business and public sector associate professionals	4631 (3.9)	2548 (4.5)	2083 (3.3)	
Associate professionals	8388 (7.0)	3148 (5.6)	5240 (8.3)	
Professional occupations	17 044 (14.2)	8934 (15.8)	8110 (12.9)	
Senior officials	13 526 (11.3)	8521 (15.0)	5005 (7.9)	
Income:				
<£18 000	23 817 (19.9)	10 499 (18.5)	13 318 (21.1)	$<1\times10^{-15}$
£18 000 to £30 999	26 808 (22.4)	12 788 (22.6)	14 020 (22.3)	
£31 000 to £51 999	27 245 (22.8)	13 848 (24.4)	13 397 (21.3)	
£52 000 to £100 000	20 397 (17.0)	10 950 (19.3)	9447 (15.0)	
>£100 000	5060 (4.2)	2777 (4.9)	2283 (3.6)	
Mean (SD) Townsend deprivation index	−1.5 (3.0)	−1.51 (3.0)	−1.45 (2.9)	$<1\times10^{-15}$
Overall per allele height SNP association with height	0.021 (0.021 to 0.022); $P<1\times10^{-15}$	0.022 (0.022 to 0.023); $P<1\times10^{-15}$	0.020 (0.020 to 0.021); $P<1\times10^{-15}$	–
Overall per allele BMI SNP association with BMI	0.022 (0.021 to 0.023); $P<1\times10^{-15}$	0.022 (0.021 to 0.024); $P<1\times10^{-15}$	0.025 (0.023 to 0.026); $P<1\times10^{-15}$	–

BMI=body mass index; SNP=single nucleotide polymorphism.

Not all five socioeconomic status measures were available in all 119 669 individuals (see supplementary table A for further information).

*For comparison between men and women; models were adjusted for age at recruitment.

Patient Involvement

This study was conducted using the UK Biobank resource. Details of patient and public involvement in the UK Biobank are available online (www.ukbiobank.ac.uk/about-biobank-uk/ and https://www.ukbiobank.ac.uk/wp-content/uploads/2011/07/Summary-EGF-consultation.pdf?phpMyAdmin=trmKQlYdjjnQIgJ%2CfAzikMhEnx6). No patients were specifically involved in setting the research question or the outcome measures, nor were they involved in developing plans for recruitment, design, or implementation of this study. No patients were asked to advise on interpretation or writing up of results. There are no specific plans to disseminate the results of the research to study participants, but the UK Biobank disseminates key findings from projects on its website.

Exposure and Outcome Measures

Exposure and outcome measures were all collected at baseline when participants attended the assessment centre.

Height

Height (cm) was measured using a Seca 202 device in all participants in the UK Biobank (n=500 120). Sitting height was also measured (n=496 380). We excluded one person with a height more than 4.56 SD away from the mean and a sitting height to standing height ratio of greater than 0.75, which is not compatible with normal growth. In all, 119 669 people of white British ancestry with genetic data available also had a valid height and BMI measure.

Body Mass Index

The UK Biobank has two different measures of BMI—one calculated as weight/height^2 and one measured using electrical impedance to quantify mass. We excluded people with significant differences (>4.56 SD from the mean) between impedance and normal BMI measures (n=1172) where both variables were available. If only one measure of BMI was available, we used this (n=7290). Valid BMI was available for 119 669 people with genetic and height data available.

Socioeconomic Status

We used five different socioeconomic status variables.[1] Age in years at completion of full time education (questionnaire based; available only for people who did not go on to degree level education). Data were available for 82 543 people and missing in 37 126 people with valid height, BMI, and genetic data.[2] Education (coded as degree level or not, derived from the questionnaire); participants were asked, "Which of the following qualifications do you have? (You can select more than one)," with the options college or university degree, A levels or equivalent, O levels or GCSEs or equivalent, CSEs, NVQ/HND/HNC, professional qualifications (eg, nursing or teaching). We created a dichotomous variable comparing degree level education or professional qualifications (n=53 652) with other qualifications (n=64 913); 1104 people did not respond to this question.[3] Job class (coded as elementary occupations, process plant and machine operatives, sales and customer service occupations, leisure and other personal service occupations, personal service occupations, skilled trades, administrative and secretarial roles, business and public sector associate professionals, associate professionals, professional occupations, and managers and senior officials); this was coded from the UK Biobank job code variable. All participants were asked to select their current or most recent job. Data were available for 76 404 people, with missing data in 43 265. We dichotomised this variable into unskilled (n=21 036; elementary occupations to personal service occupations) and skilled (n=55 698; skilled trades to managers and senior officials).[4] A categorical income variable (questionnaire based), representing annual household income of <£18 000 (€23 600; $25 800), £18 000 to £30 999, £31 000 to £51 999, £52 000 to £100 000, and >£100 000. Data were available for 103 327 people and missing in 16 432 people with valid height, BMI, and genetic data.[5] Townsend deprivation index (a composite measure of deprivation based on unemployment, non-car ownership, non-home ownership, and household overcrowding; a negative value represents high socioeconomic status). This was calculated before participants joined the UK Biobank and was based on the preceding national census data, with each participant assigned a score corresponding to the postcode of their home dwelling. Data were available for 119 519 people and missing in 150 people with valid height, BMI, and genetic data.

For each of the five traits, we compared participants missing data with those reporting data; generally, those with missing data were older and shorter with higher BMIs (supplementary table A). We investigated the relation of these five socioeconomic status measures and four health outcomes: self reported coronary artery disease, hypertension (defined as a systolic blood pressure ≥140, a diastolic blood pressure ≥90, or the report of use of blood pressure lowering drugs), any self reported long term illness (based on the UK Biobank question "Do you have a long standing illness, disability or infirmity?"), and type 2 diabetes (based on self report, excluding people using insulin in the first year after diagnosis and those given a diagnosis before 35 years of age or within the previous year) (supplementary table B).

For three of the traits (the exceptions being education and job class, both binary traits), we converted the data to a normal distribution to limit the influence of any subtle population stratification and to provide standard deviation effect sizes. We took residuals of the exposure and outcome measures from standard linear regression by using nine covariates: age, sex, assessment centre location, five (within UK) ancestry principal components, and microarray used to measure genotypes. We then inverse normalised the residualised variables. To convert our results back to meaningful units after mendelian randomisation, we multiplied our SD βs by a 1 SD change in the socioeconomic status measure. For example, a 1 SD change in Townsend deprivation index was equivalent to 2.68 units. Therefore, a 0.05 SD change equated to a 0.134 unit change in deprivation.

Observational Associations

We regressed each socioeconomic status measure against height and BMI by using linear regression for continuous outcome variables and logistic regression for binary outcomes. We adjusted these associations for age and sex. We also investigated the association of each socioeconomic status measure with a range of health outcomes.

Genetic Variants

The genetic variants used were extracted genotypes from UK Biobank's imputation dataset (the supplementary methods provide more information on the UK Biobank's quality control). We excluded individual genotypes if the genotype probability was less than 0.9. We confirmed that the variants were imputed with high quality by comparing them with the directly genotyped data, where available. Details of imputation quality are given in supplementary table C.

Height

We selected 396 of 404 height genetic variants from independent loci that were associated with height at genome-wide significance in the GIANT studies of up to 253 288 people (supplementary table C).[19] We excluded eight variants that were unavailable (rs1420023, rs567401), were poorly imputed, had an imputation quality <0.9 (rs11683207, rs7534365), or were not in Hardy-Weinberg equilibrium ($P<1\times10^{-6}$; rs1401795, rs7692995, rs915506, rs3790086). The 396 variants explained 12.3% of the variance in adult height in the UK Biobank participants used.

Body Mass Index

We selected 69 of 76 common genetic variants that were associated with BMI at genome-wide significance in the GIANT consortium in studies of up to 339 224 people (supplementary table C).[18] We limited the BMI variants to those that were associated with BMI in the analysis of all people of European ancestry and did not include those that reached genome-wide levels of statistical confidence in only one sex or one stratum. We also excluded variants if they were known to be classified as a secondary signal within a locus. Three variants were excluded from the score owing to potential pleiotropy (rs11030104 (*BDNF* reward phenotypes), rs13107325 (*SLC39A8* lipids, blood pressure), rs3888190 (*SH2B1* multiple traits)), three were not in Hardy-Weinberg equilibrium ($P<1\times10^{-6}$; rs17001654, rs2075650, rs9925964), and one was unavailable (rs2033529). The 69 variants explained 1.5% of the variance in BMI in the UK Biobank participants.

We recoded individual variants as 0, 1, and 2 according to the number of height or BMI increasing alleles. We used the variants to create height and BMI genetic risk scores. Each variant was weighted by its relative effect size (β coefficient) obtained from the reported meta-analysis data.[18] We created a weighted score, in which β is the β coefficient of representing the association between each single nucleotide polymorphism (SNP) and height/BMI: weighted score=$\beta_1\times SNP_1+\beta_2\times SNP_2+\cdots\beta_n\times SNP_n$. We then rescaled the weighted score to reflect the number of trait increasing alleles: weighted genetic risk score=(weighted score×number of SNPs)/sum of β coefficients.

Mendelian Randomisation

The mendelian randomisation approach used in this study made the following assumptions[15]: the height and BMI genetic risk scores were robustly associated with measured height and BMI; the height and BMI genetic risk scores were not associated with confounding factors that bias conventional epidemiological associations between height/BMI and socioeconomic status; the height and BMI genetic risk scores were related to the outcome only via its association with the modifiable exposure; and the associations represented in figure 1 are linear and unaffected by statistical interactions.

Instrumental Variable Analysis

We used two methods that use genetic variants to assess causal relations between two traits. Firstly, to estimate the causal effect of height or BMI on individual socioeconomic status measures, we used instrumental variable analysis using the height or BMI genetic risk score.[15] We used the two stage, least squares estimator method that uses predicted levels of BMI or height per genotype and regresses each outcome against these predicted values.

For continuous socioeconomic status outcomes, we used the ivreg2 command in Stata to do the instrumental variable analysis. We compared results from observational and instrumental variable regressions by using the Durbin-Wu-Hausman test for endogeneity, which examines the difference between the estimates from linear regression (observational) and instrumental variable analysis.[21]

For binary outcomes, the instrumental variable analysis was done in two stages. Firstly, we assessed the association between the height or BMI genetic risk score and height or BMI, respectively. We saved the predicted values and residuals from this regression model. Secondly, we used the predicted values from stage 1 as the independent variable (reflecting an unconfounded estimate of variation in BMI or height) and degree status or job class as the dependent variable in a logistic or ordinal logistic regression model. We used robust standard errors to correct for uncertainty in the estimate. We examined the F statistics from first stage regressions to evaluate the strength of the instruments; weak instruments can bias results towards the (confounded) multivariable regression association or towards the null in a two stage design.[22,23]

Egger Method

We used a second method of mendelian randomisation, the Egger method,[24] as a sensitivity analysis if the instrumental variables test result was noteworthy. This method is more robust to potential violations of the standard instrumental variable assumptions. It uses a weighted regression with an unconstrained intercept to regress the effect sizes of variant-outcome associations (here, height or BMI variants versus socioeconomic status measures) against effect sizes of variant-risk factor associations (here, height or BMI variants versus height or BMI). The unconstrained intercept removes the assumption that all genetic variants are valid instrumental variables, so this method is less susceptible to confounding from potentially pleiotropic variants that will probably have stronger effects on outcomes compared with their effects on the primary trait. The approach is analogous to correcting for small study publication bias in meta-analyses.[24] Details of the Stata and R code used are provided in Bowden et al 2015.[24]

To ensure the robustness of our findings, we have highlighted results only where we see consistent results across the two different methods.

Differences between Men and Women

To test the hypothesis that the effects of height and BMI on socioeconomic status may differ in men and women, we repeated observational and genetic analyses separately in each sex. The selected height and BMI genetic variants have very similar effects in men and women, so we used the same genetic risk scores in all participants, men only, and women only. We compared the β values for men and women by using Fisher's z score method[25]: $z=(\beta_1-\beta_2)/\sqrt{(SE1^2+SE2^2)}$.

Sensitivity Analyses

In a sensitivity analysis to further confirm that our results were robust to any potential influence of population stratification, we used the linear mixed models approach as implemented in the software BOLT-LMM.[26] This approach corrects for all levels of inter-individual correlation of genotypes due to relatedness, from close relatives to cryptic relatedness caused by population stratification. We inverse normalised the socioeconomic status measures, then took the residuals using three covariates (age, sex, assessment centre location) and inverse normalised again.

RESULTS

Table 1 summarises the demographics of the 119 669 UK Biobank participants with valid genetic data and BMI and height measures. The height and BMI genetic risk scores were robustly associated with height and BMI (table 1). The association between the socioeconomic status measures and health outcomes and the associations between known height variants and height and known BMI variants and BMI in the UK Biobank are summarised in supplementary tables A-C.

Relation of Genetically Determined Taller Stature to Higher Socioeconomic Status Measures in UK Biobank

Education: Duration in Full Time Education

Using 82 543 participants, we found that taller stature was strongly correlated with participants spending longer in full time education (table 2). This association was similar in men and women. A 1 SD (6.3 cm) greater height was associated with a 0.11 (95% confidence interval 0.10 to 0.12) SD older age (approximately 0.2 years) at which full time education was completed. Genetic analyses provided evidence that this association was partly causal—a genetically determined 1 SD (6.3 cm) higher height was associated with a 0.03 (0.01 to 0.05) SD older age (approximately 0.06 (0.02 to 0.10) years) at which full time education was completed (table 2, supplementary figure A-I).

Table 2. Associations between taller stature and five measures of socioeconomic, using linear or logistic regression and instrumental variable analysis

Socioeconomic status measures and subcategories	No	Observational*: Change in socioeconomic status (95%CI) per SD taller stature	Observational*: P value	Genetic†: Change in socioeconomic status (95%CI) per SD taller stature	Genetic†: P value	Genetic: Egger‡: Change in socioeconomic status (95%CI) per SD taller stature	Genetic: Egger‡: P value
Age completed full time education:							
All	82 543	0.11 (0.10 to 0.12)	$<1\times10^{-15}$	0.03 (0.01 to 0.05)	0.01	0.07 (0.03 to 0.11)	0.0004
Men only	38 342	0.11 (0.10 to 0.12)	$<1\times10^{-15}$	0.04 (0.01 to 0.07)	0.009	0.08 (0.02 to 0.14)	0.004
Women only	44 201	0.11 (0.10 to 0.12)	$<1\times10^{-15}$	0.01 (−0.02 to 0.04)	0.40	–	
Degree level education:							
All	118 565	OR: 1.25 (1.24 to 1.27)	$<1\times10^{-15}$	1.02 (0.99 to 1.05)	0.22	–	
Men only	56 111	OR: 1.25 (1.23 to 1.27)	$<1\times10^{-15}$	1.04 (1.00 to 1.09)	0.08	–	
Women only	62 454	OR: 1.26 (1.24 to 1.28)	$<1\times10^{-15}$	1.00 (0.95 to 1.05)	0.97	–	
Job class (skilled/unskilled):							
All	76 404	OR: 1.29 (1.27 to 1.32)	$<1\times10^{-15}$	1.12 (1.07 to 1.18)	6E−7	1.18 (1.08 to 1.29)	0.0002
Men only	37 608	OR: 1.31 (1.28 to 1.34)	$<1\times10^{-15}$	1.13 (1.07 to 1.21)	2E−5	1.23 (1.10 to 1.37)	0.0004

Women only	38 796	OR: 1.27 (1.24 to 1.31)	$<1\times10^{-15}$	1.14 (1.05 to 1.24)	0.003	1.21 (1.08 to 1.36)	0.002
Annual household income:							
All	103 327	0.13 (0.12 to 0.14)	$<1\times10^{-15}$	0.05 (0.03 to 0.07)	4E−8	0.05 (0.02 to 0.08)	0.0009
Men only	50 862	0.15 (0.14 to 0.16)	$<1\times10^{-15}$	0.07 (0.05 to 0.10)	1E−9	0.08 (0.04 to 0.12)	0.0002
Women only	52 465	0.11 (0.10 to 0.12)	$<1\times10^{-15}$	0.02 (0.00 to 0.05)	0.09	–	
Townsend deprivation index:							
All	119 519	−0.08 (−0.09 to −0.07)	$<1\times10^{-15}$	0.00 (−0.02 to 0.01)	0.71	–	
Men only	56 582	−0.10 (−0.10 to −0.09)	$<1\times10^{-15}$	−0.02 (−0.05 to 0.00)	0.05	−0.08 (−0.12 to −0.04)	0.0004
Women only	62 937	−0.07 (−0.07 to −0.06)	$<1\times10^{-15}$	0.02 (−0.01 to 0.04)	0.19	–	

OR=odds ratio.

For age completed full time education, annual household income, and Townsend deprivation index, changes reported are standard deviation. For degree and job class, odds ratios are shown, representing odds of higher socioeconomic status per SD greater height.

*Age, assessment centre, and sex adjusted associations.

†Uses instrumental variable analysis via ivreg2 command in Stata for continuous variables and two step procedure for binary outcomes using height genetic risk score. F statistic when considering all participants is ≥10 898 for each socioeconomic status measure; in men only, F statistic is ≥5308 for each socioeconomic status measure; in women only, F statistic is ≥5615 for each socioeconomic status measure.

‡Alternative genetic approach detailed in Bowden et al 2015,[24] used as sensitivity analysis when instrumental variable was P<0.05.

Education: Degree Level (or Equivalent) or Not

Using 118 565 participants, we found that taller stature was strongly correlated with participants' chances of having obtained a degree. A 1 SD (6.3 cm) greater height was associated with 1.25 (1.24 to 1.27) increased odds of reporting degree level education. This association was similar in men and women (P for comparison>0.05). Genetic analyses provided no consistent evidence for a causal role of height in obtaining degree level education (table 2, supplementary figure A-II).

Job Class

Using 76 404 participants, we found that taller stature was strongly correlated with job class. A 1 SD (6.3 cm) greater height was associated with increased odds of working in skilled job roles (odds ratio 1.29, 1.27 to 1.32). Genetic analyses provided evidence that this association was partly causal—a 1 SD (6.3 cm) genetically determined higher height was associated with increased odds of working in more professional roles (odds ratio 1.12, 1.07 to 1.18) (table 2, supplementary figure A-III). This association was consistent when we analysed the data as 11 ordered job classes (supplementary table D). We found no genetic evidence that the effect was stronger in men or women.

Annual Household Income

Using 103 327 participants, we found that taller stature was strongly correlated with higher household income. The correlation was approximately 50% stronger in men (table 2). A 1 SD (6.3 cm) greater height was associated with a 0.13 (0.12 to 0.14) SD increase in income (table 2). This difference is approximately equivalent to a £2940 (£2730 to £3185) higher annual household income. Genetic analyses provided evidence that this association was partly causal—a genetically determined 1 SD (6.3 cm) greater height was associated with a 0.05 (0.03 to 0.07) SD increase in annual household income, equivalent to £1130 (£680 to £1580) (table 2 and supplementary figure AIV). The genetic analyses showed that the effect was approximately twice as strong in men as in women (P for comparison=5×10^{-4}); a 1 SD greater height in men caused a £1580 (£1130 to £2260) increase in household income (supplementary figure A-IV). This association was consistent when we analysed the data as five ordered income classes (supplementary table D).

Townsend Deprivation Index

Using 119 519 participants, we found that taller stature was strongly correlated with lower levels of social deprivation, as measured by the Townsend deprivation index. This association was stronger in men than women. A 1 SD (6.3 cm) greater height was associated with a 0.08 (0.07 to 0.09) SD lower deprivation, which is equivalent to a 0.21 (0.18 to 0.24) unit reduction in Townsend deprivation index (table 2). Genetic analyses provided evidence that this association was partly causal in men but not when all participants or women were considered. In all participants, genetically determined height was not associated with deprivation (table 2, supplementary figure A-V). In men, a genetically determined 1 SD (6.3 cm) greater height was associated with a 0.02 (0.00 to 0.05) SD reduction in deprivation (supplementary figure AV). This difference is equivalent to a 0.05 (0.00 to 0.13) unit lower Townsend index.

Relation of Genetically Determined Higher BMI to Reduced Income and Deprivation Measures of Socioeconomic Status in UK Biobank

Education: Duration in Full Time Education

Using 82 543 participants, we found that higher BMI was strongly correlated with participants finishing full time education at a younger age. The association was similar in men and women (P for comparison>0.05) (table 3). A 1 SD (4.6 kg/m^2) higher BMI was associated with a 0.08 (0.07 to 0.08) SD younger age (approximately 0.15 years) at which full time education was completed. We found no genetic evidence that this association was causal when considering all participants, men only, or women only (supplementary figure B-I).

Table 3. Associations between higher BMI and five measures of socioeconomic, using linear or logistic regression and instrumental variable analysis

Socioeconomic status measures and subcategories	No	Observational*		Genetic†		Genetic: Egger‡	
		Change in socio-economic status (95%CI) per SD higher BMI	P value	Change in socioeconomic status (95%CI) per SD higher BMI	P value	Change in socioeconomic status (95%CI) per SD higher BMI	P value
Age completed full time education:							
All	82 543	−0.08 (−0.08 to −0.07)	$<1\times10^{-15}$	−0.01 (−0.07 to 0.04)	0.63	–	
Men only	38 342	−0.07 (−0.08 to −0.06)	$<1\times10^{-15}$	0.00 (−0.09 to 0.09)	0.98	–	
Women only	44 201	−0.08 (−0.09 to −0.07)	$<1\times10^{-15}$	−0.02 (−0.09 to 0.05)	0.56	–	
Degree level education:							
All	118 565	OR: 0.83 (0.82 to 0.84)	$<1\times10^{-15}$	0.94 (0.85 to 1.03)	0.18	–	
Men only	56 111	OR: 0.82 (0.81 to 0.84)	$<1\times10^{-15}$	0.94 (0.81 to 1.09)	0.43	–	
Women only	62 454	OR: 0.83 (0.82 to 0.84)	$<1\times10^{-15}$	0.93 (0.82 to 1.06)	0.28	–	
Job class (skilled/unskilled):							
All	76 404	OR: 0.91 (0.89 to 0.92)	$<1\times10^{-15}$	0.90 (0.79 to 1.02)	0.10	–	
Men only	37 608	OR: 0.93 (0.91 to 0.95)	8×10^{-9}	0.88 (0.73 to 1.08)	0.22	–	
Women only	38 796	OR: 0.89 (0.87 to 0.91)	$<1\times10^{-15}$	0.91 (0.76 to 1.08)	0.29	–	
Annual household income:							
All	103 327	−0.06 (−0.06 to −0.05)	$<1\times10^{-15}$	−0.05 (−0.10 to −0.00)	0.041	−0.03 (−0.11 to 0.05)	0.58
Men only	50 862	−0.01 (−0.02 to −0.00)	$<1\times10^{-15}$	0.06 (−0.02 to 0.14)	0.15	–	
Women only	52 465	−0.09 (−0.10 to −0.08)	$<1\times10^{-15}$	−0.14 (−0.20 to −0.08)	1×10^{-5}	−0.17 (−0.25 to −0.05)	0.004

Townsend deprivation index:							
All	119 519	0.08 (0.07 to 0.08)	<1×10⁻¹⁵	0.05 (0.01 to 0.10)	0.024	−0.00 (−0.08 to 0.08)	0.96
Men only	56 582	0.05 (0.04 to 0.05)	<1×10⁻¹⁵	−0.01 (−0.08 to 0.06)	0.78	–	
Women only	62 937	0.10 (0.09 to 0.11)	<1×10⁻¹⁵	0.10 (0.04 to 0.16)	0.001	0.10 (−0.01 to 0.21)	0.08

BMI=body mass index; OR=odds ratio.

For age completed full time education, annual household income, and Townsend deprivation index, changes reported are standard deviation. For degree and job class, odds ratios are shown, representing odds of higher socioeconomic status per SD higher BMI.

*Age, assessment centre, and sex adjusted associations.

†Uses instrumental variable analysis, via ivreg2 command in Stata for continuous variables and two step approach for binary outcomes, using BMI genetic risk score. F statistic for all participants is ≥1257 for each socioeconomic status measure; in men only, F statistic is ≥591 for each socioeconomic status measure; in women only, F statistic is ≥666 for each socioeconomic status measure.

‡Alternative genetic approach detailed in Bowden et al 2015,[24] used as sensitivity analysis when instrumental variable was P<0.05.

Education: Degree Level (or Equivalent or Not)

Using 118 565 participants, we found that higher BMI was associated with lower odds of having obtained a degree. A 1 SD higher BMI was associated with lower odds of obtaining degree level education (odds ratio 0.83, 0.82 to 0.84). We found no consistent genetic evidence that this association was causal when considering all participants, men only, or women only (supplementary figure B-II).

Job Class

Using 76 404 participants, we found that higher BMI was associated with employment in less skilled professions. A 1 SD (4.6 kg/m^2) higher BMI was associated with lower odds of working in skilled job roles (0.91, 0.89 to 0.92), and the association was stronger in women. We found no consistent genetic evidence that this association was causal when considering all participants, men only, or women only (supplementary figure B-III). However, we found some evidence of causality when we analysed the data as 11 ordered job classes (supplementary table D).

Annual Household Income

Using 103 327 participants, we found that higher BMI was associated with a lower annual household income, but this effect was very strongly driven by the association in women. A 1 SD higher BMI was associated with a 0.09 (0.08 to 0.10) SD lower household income for women. This effect equates to £1890 (£1680 to £2100) less income per annum for women. In men, a 1 SD higher BMI approximated to a £210 (£84 to £420) lower annual household income. Genetic analyses were consistent with these observations being causal in women but not in men (P for comparison with men=9×10^{-5})—a genetically determined 1 SD higher BMI was associated with an annual household income of 0.14 (0.08 to 0.20) SD less in women. This effect is equivalent to £2940 (£1680 to £4200) less for women (table 3, supplementary figure B-IV). This association was consistent when we analysed the data as five ordered income classes (supplementary table D). The association between higher BMI and lower income was consistent in women who worked, with or without a husband/partner at home, and women who did not work with a husband/partner at home (supplementary table E). It was also consistent when we considered only women without health conditions (supplementary table E).

Townsend Deprivation Index

Higher BMI was associated with higher levels of deprivation as assessed by the Townsend deprivation index. A 1 SD higher BMI was associated with a 0.08 (0.07 to 0.08) SD higher deprivation value, which is equivalent to a 0.21 (0.19 to 0.21) unit increase in Townsend index (table 3). This association was twice as strong in women. We found limited genetic evidence of a causal relation between BMI and deprivation in men, but some evidence in women. A 1 SD genetically higher BMI was associated with a 0.10 (0.04 to 0.16) SD higher level of deprivation in women (table 3, Supplementary figure B-V).

Sensitivity Analyses

The Egger method provided consistent results for causal relations between height and duration in full time education, job class, income, and Townsend deprivation index in men (table 2, supplementary table F). The Egger method also provided evidence of consistent associations between higher BMI and income in women (table 3, supplementary table G). Use of genome-wide methods to account for genetic and socioeconomic status correlations between close and distant relatives did not alter our findings (supplementary table H).

DISCUSSION

Using genetic variants as unconfounded proxies for height and BMI, our study provides evidence that shorter stature and higher BMI lead to lower measures of several aspects of socioeconomic status. It is important to note that our data are consistent with the height and BMI to socioeconomic status associations being only partly causal—we have not excluded a causal effect in the other direction. The study adds causal evidence to a large number of observational studies. This work may have important implications for public health, as low socioeconomic status increases mortality and morbidity.[2,3] The association between socioeconomic status measures and health was strong in the UK Biobank data, where, for example, people possessing a university degree had a 38% lower odds of coronary artery disease compared with those without degree level education. Our study also showed sex differences in the causal relations between height or BMI and socioeconomic status that are consistent with observational data. Height effects were stronger in men, but the BMI effects tended to be stronger in women.

Evidence for Taller Stature Leading to Higher Socioeconomic Status

The causal effect, as estimated using genetics, of taller stature on higher socioeconomic status was present in four of the five measures of socioeconomic status. For income, where the statistical evidence was strongest, the estimated causal effects were approximately two to three times stronger in men than in women. The causal evidence for greater height leading to higher levels of socioeconomic status is consistent with observational studies, in which taller stature was associated with higher job class, earnings, and educational attainment.[4,5] One US based study showed a reduction in earnings of $789 per annum per inch of height. This equates to £1250 per SD (6.3 cm) of height in our study, which is very similar to our genetic estimate of £1130.[13] Despite the strong evidence that taller stature directly influences measures of socioeconomic status, the genetic estimates were consistently smaller than the observational estimates. These differences indicate that the observed association between taller stature and higher socioeconomic status is a mixture of direct causal effects and other factors that could include a causal effect in the opposite direction.

A range of factors could link taller stature to higher socioeconomic status, although this study did not consider which of these factors were involved. Some of the possibilities include complex interactions between self esteem, stigma, positive discrimination,[13] and increased intelligence.[4,27,28] Evidence shows that self esteem, leadership perception, and height discrimination tend to be greater in men than in women, which fits with our findings.[29,30,31]

Evidence for Higher BMI Leading to Lower Socioeconomic Status

Higher BMI, as estimated using genetics, was causally associated with having a lower annual household income and higher levels of deprivation. These associations were stronger in women, with no consistent evidence of a causal relation between higher BMI and lower socioeconomic status measures in men. These findings were consistent with previous literature, in which most associations of BMI with socioeconomic status were observed in women only.[7,32] We found no evidence that the associations between higher BMI and educational outcomes were causal, a result consistent with a review of the effect of BMI on social outcomes.[32] Our findings add to evidence from observational studies, in which BMI is associated with lower levels of employment, less skilled work, and lower income.[32,33] A range of factors could link higher BMI to lower income and higher deprivation in women, although this study did not consider which of these factors were involved. One of the possibilities is discrimination in the workplace, with overweight job applicants and employees being evaluated more negatively.[32] The disparity between the sexes may be partially explained by discrimination, which may occur at lower weight levels for women than for men.[34,35] Additionally, cultivation theory in social science indicates that very thin women are idealised and more socially valued, compared with their normal weight and overweight peers.[33] In contrast, a very different set of social standards exists regarding men's weight, so discrimination based on body size could well be different in men and women.[33] Two of the strongest measures in women were household income and Townsend deprivation, which are not just specific to the individual but also indicative of partner's income. However, additional analyses showed that genetically determined higher BMI was associated with lower income both in non-working women with partners and in working women without a partner, suggesting that the associations were not just driven by partner's income.

Limitations of Study

Although our results are consistent with a direct causal effect of shorter stature and higher BMI on lower socioeconomic status, some qualifications should be considered. Firstly, the UK Biobank participants were born between 1938 and 1971, and the causal associations may not remain in today's society or be generalisable to societies outside of the United Kingdom. The causal associations may have been influenced by parental genotype-socioeconomic status associations. For example, the causal pathway could reflect parental genetic predisposition to higher BMI resulting in families moving to a more obese and lower socioeconomic status neighbourhood, which in turn could lower children's socioeconomic status. Because parents' and children's genotypes are correlated, this pathway could lead to a genetic association between UK Biobank participants' socioeconomic status and BMI that reflects parental factors during the 20th century. However, such a pathway would be unlikely to result in genetic associations between BMI and socioeconomic status that were stronger in women than in men.

Secondly, higher BMI leads to poorer health; this could affect productivity, which in turn could affect socioeconomic status. However, we saw evidence of genetic associations between higher BMI and lower socioeconomic status in women reporting no adverse health outcomes as well as in those reporting health problems (supplementary table E). We also need to take care in interpreting negative results; although the large sample size of the UK Biobank provided greater than 95% power for investigating the causal relations between height and socioeconomic status, power was limited for some of the causal associations for BMI. The SNPs selected for height and BMI may have effects on socioeconomic status not mediated by their effects on height or BMI (pleiotropy), which were not measured but could potentially affect socioeconomic status. However, to minimise this possibility, we selected SNPs carefully and used the Egger method, which can detect and adjust for pleiotropy bias in many scenarios[24] (hence the broader confidence intervals observed).

The educational, job status, and income data used in this study were self reported, which may result in measurement bias. However, Townsend deprivation index was derived by the UK Biobank and we observed consistent trends across the different socioeconomic status constructs, suggesting limited bias due to self report. Socioeconomic status is a very complex multidimensional construct. We looked at a range of individual components and observed similar trends for each, but the selected variables may not cover the entirety of social status. This study used a homogenous population, so the results may not be generalisable to other ethnic groups. We note that the genetic variants associated with BMI together explain only 1.5% of the variation in BMI, but collectively the variants provide a robust test, as reflected by the strong F statistic of the genetic risk score. The 69 genetic variants associated with BMI provided a stronger "instrument" than those used in previous mendelian randomisation studies that inferred a causal effect of higher BMI leading to lower vitamin D concentrations and higher risk of heart failure and markers of poor metabolic health.[36 37]

Finally, height, BMI, and socioeconomic status are subtly stratified across the United Kingdom, with people living and working in the north of the country having lower socioeconomic status, higher BMI, and shorter stature, on average, than those in the south. If genetic variants are also subtly different between north and south, this could have confounded our results. However, several factors mean that this population stratification should not have caused false positive results. Firstly, we would not have expected to have seen differences between men and women (because gene allele frequencies do not differ between the sexes). Secondly, we used both within UK genetic ancestry principal components and a second method that corrects for all levels of relatedness, and our results did not change.

CONCLUSION

In summary, using up to 119 000 participants from the UK Biobank, we provide evidence that high BMI and short stature, as estimated by genetics, are causally related to lower socioeconomic status. Further work is needed to understand the factors that lead to and from anthropometric traits and socioeconomic status.

REFERENCES

1. Kennedy BP, Kawachi I, Glass R, Prothrow-Stith D. Income distribution, socioeconomic status, and self rated health in the United States: multilevel analysis. *BMJ* 1998;317:917-21. doi:10.1136/bmj.317.7163.917. 9756809.
2. Adler NE, Ostrove JM. Socioeconomic status and health: what we know and what we don't. *Ann N Y Acad Sci* 1999;896:3-15. doi:10.1111/j.1749-6632.1999.tb08101.x. 10681884.
3. Marmot M. The health gap: the challenge of an unequal world. *Lancet* 2015;386:2442-4. doi:10.1016/S0140-6736(15)00150-6. 26364261.
4. Case A, Paxson C. Stature and status: Height, ability, and labor market outcomes. *J Polit Econ* 2008;116:499-532. doi:10.1086/589524. 19603086.
5. Magnusson PK, Rasmussen F, Gyllensten UB. Height at age 18 years is a strong predictor of attained education later in life: cohort study of over 950,000 Swedish men. *Int J Epidemiol* 2006;35:658-63. doi:10.1093/ije/dyl011. 16446353.
6. Grabner M. BMI trends, socioeconomic status, and the choice of dataset. *Obes Facts* 2012;5:112-26. doi:10.1159/000337018. 22433623.
7. Gortmaker SL, Must A, Perrin JM, Sobol AM, Dietz WH. Social and economic consequences of overweight in adolescence and young adulthood. *N Engl J Med* 1993;329:1008-12. doi:10.1056/NEJM199309303291406. 8366901.
8. Lundborg P, Nystedt P, Rooth DO. Body size, skills, and income: evidence from 150,000 teenage siblings. *Demography* 2014;51:1573-96. doi:10.1007/s13524-014-0325-6. 25199549.
9. Reading R, Raybould S, Jarvis S. Deprivation, low birth weight, and children's height: a comparison between rural and urban areas. *BMJ* 1993;307:1458-62. doi:10.1136/bmj.307.6917.1458. 8281087.
10. Sobal J, Stunkard AJ. Socioeconomic status and obesity: a review of the literature. *Psychol Bull* 1989;105:260-75. doi:10.1037/0033-2909.105.2.260. 2648443.
11. Asao K, Kao WH, Baptiste-Roberts K, Bandeen-Roche K, Erlinger TP, Brancati FL. Short stature and the risk of adiposity, insulin resistance, and type 2 diabetes in middle age: the Third National Health and Nutrition Examination Survey (NHANES III), 1988-1994. *Diabetes Care* 2006;29:1632-7. doi:10.2337/dc05-1997. 16801590.
12. Whitley E, Martin RM, Davey Smith G, Holly JM, Gunnell D. The association of childhood height, leg length and other measures of skeletal growth with adult cardiovascular disease: the Boyd-Orr cohort. *J Epidemiol Community Health* 2012;66:18-23. doi:10.1136/jech.2009.104216. 20736489.
13. Rosenburg IB. Height discrimination in employment. *Utah Law Rev* 2009;3:907.
14. Andreyeva T, Puhl RM, Brownell KD. Changes in perceived weight discrimination among Americans, 1995-1996 through 2004-2006. *Obesity (Silver Spring)* 2008;16:1129-34. doi:10.1038/oby.2008.35. 18356847.
15. Lawlor DA, Harbord RM, Sterne JA, Timpson N, Davey Smith G. Mendelian randomization: using genes as instruments for making causal inferences in epidemiology. *Stat Med* 2008;27:1133-63. doi:10.1002/sim.3034. 17886233.
16. Fletcher JM, Lehrer SF. Genetic lotteries within families. *J Health Econ* 2011;30:647-59.doi:10.1016/j.jhealeco.2011.04.005. 21664708.
17. Ding W, Lehrer SF, Rosenquist JN, Audrain-McGovern J. The impact of poor health on academic performance: New evidence using genetic markers. *J Health Econ* 2009;28:578-97. doi:10.1016/j.jhealeco.2008.11.006. 19217678.
18. Locke AE, Kahali B, Berndt SI, et al. LifeLines Cohort Study ADIPOGen Consortium AGEN-BMI Working Group CARDIOGRAMplusC4D Consortium CKDGen Consortium GLGC ICBP MAGIC Investigators MuTHER Consortium MIGen Consortium PAGE

Consortium ReproGen Consortium GENIE Consortium International Endogene Consortium. Genetic studies of body mass index yield new insights for obesity biology. *Nature* 2015;518:197-206.25673413.

19. Wood AR, Esko T, Yang J, et al. Electronic Medical Records and Genomics (eMEMERGEGE) Consortium MIGen Consortium PAGEGE Consortium LifeLines Cohort Study. Defining the role of common variation in the genomic and biological architecture of adult human height. *Nat Genet* 2014;46:1173-86. doi:10.1038/ng.3097. 25282103.
20. Collins R. What makes UK Biobank special?*Lancet* 2012;379:1173-4. doi:10.1016/S0140-6736(12)60404-8. 22463865.
21. Baum C, Schaffer M, Stillman S. IVENDOG: Stata module to calculate Durbin-Wu-Hausman endogeneity test after ivreg. Statistical Software Components S494401. Boston College Department of Economics, 2007.
22. Staiger D, Stock JH. Instrumental variables regression with weak instruments. *Econometrica* 1997;65:557-86doi:10.2307/2171753. .
23. Stock J, Yogo M, Wright J. A Survey of weak instruments and weak identification in generalized method of moments. *J Bus Econ Stat* 2002;20:518-29doi:10.1198/073500102288618658. .
24. Bowden J, Davey Smith G, Burgess S. Mendelian randomization with invalid instruments: effect estimation and bias detection through Egger regression. *Int J Epidemiol* 2015;44:512-25. doi:10.1093/ije/dyv080. 26050253.
25. Paternoster R, Brame R, Mazerolle P, et al. Using the correct statistical test for the equality of regression coefficients.*Criminology* 1998;36:859-66doi:10.1111/j.1745-9125.1998.tb01268.x. .
26. Loh PR, Tucker G, Bulik-Sullivan BK, et al. Efficient Bayesian mixed-model analysis increases association power in large cohorts. *Nat Genet* 2015;47:284-90. doi:10.1038/ng.3190. 25642633.
27. Marioni RE, Batty GD, Hayward C, et al. Generation Scotland. Common genetic variants explain the majority of the correlation between height and intelligence: the generation Scotland study. *Behav Genet* 2014;44:91-6. doi:10.1007/s10519-014-9644-z. 24554214.
28. Sundet JM, Tambs K, Harris JR, Magnus P, Torjussen TM. Resolving the genetic and environmental sources of the correlation between height and intelligence: a study of nearly 2600 Norwegian male twin pairs. *Twin Res Hum Genet* 2005;8:307-11. doi:10.1375/twin.8.4.307. 16176713.
29. Orth U, Robins RW, Widaman KF. Life-span development of self-esteem and its effects on important life outcomes. *J Pers Soc Psychol* 2012;102:1271-88. doi:10.1037/a0025558. 21942279.
30. Judge TA, Cable DM. The effect of physical height on workplace success and income: preliminary test of a theoretical model. *J Appl Psychol* 2004;89:428-41. doi:10.1037/0021-9010.89.3.428. 15161403.
31. Blaker NM, Rompa I, Dessing IH, et al. The height leadership advantage in men and women: Testing evolutionary psychology predictions about the perceptions of tall leaders. *Group Process Intergroup Relat* 2013;16:17-27doi:10.1177/1368430212437211. .
32. Puhl RM, Heuer CA. The stigma of obesity: a review and update. *Obesity (Silver Spring)* 2009;17:941-64. doi:10.1038/oby.2008.636. 19165161.
33. Judge TA, Cable DM. When it comes to pay, do the thin win? The effect of weight on pay for men and women. *J Appl Psychol* 2011;96:95-112. doi:10.1037/a0020860. 20853946.
34. Puhl RM, Andreyeva T, Brownell KD. Perceptions of weight discrimination: prevalence and comparison to race and gender discrimination in America. *Int J Obes (Lond)* 2008;32:992-1000. doi:10.1038/ijo.2008.22. 18317471.
35. Morris S. Body mass index and occupational attainment. *J Health Econ* 2006;25:347-64.doi:10.1016/j.jhealeco.2005.09.005. 16290234.
36. Vimaleswaran KS, Berry DJ, Lu C, et al. Genetic Investigation of Anthropometric Traits-GIANT Consortium. Causal relationship between obesity and vitamin D status: bi-directional Mendelian randomization analysis of multiple cohorts. *PLoS Med* 2013;10:e1001383. doi:10.1371/journal.pmed.1001383. 23393431.
37. Fall T, Hägg S, Mägi R, et al. European Network for Genetic and Genomic Epidemiology (ENGAGE) consortium. The role of adiposity in cardiometabolic traits: a Mendelian randomization analysis. *PLoS Med* 2013;10:e1001474.doi:10.1371/journal.pmed.1001474. 23824655.

CHAPTER 18

GENETIC SUSCEPTIBILITY TO VITILIGO: GWAS APPROACHES FOR IDENTIFYING VITILIGO SUSCEPTIBILITY GENES AND LOCI

Changbing Shen,[1] Jing Gao,[2] Yujun Sheng,[1] Jinfa Dou,[1] Fusheng Zhou,[1] Xiaodong Zheng,[1] Randy Ko,[3] Xianfa Tang,[1] Caihong Zhu,[1] Xianyong Yin,[4] Liangdan Sun,[1] Yong Cui,[5] and Xuejun Zhang[1,2]

[1]Institute and Department of Dermatology, The First Affiliated Hospital, Anhui Medical University, Hefei, China

[2]Department of Dermatology, The Second Affiliated Hospital, Anhui Medical University, Hefei, China

[3]Department of Biochemistry, University of New Mexico, Albuquerque, NM, USA

[4]Department of Genetics and Renaissance Computing Institute, University of North Carolina at Chapel Hill, Chapel Hill, NC, USA

[5]Department of Dermatology, China-Japan Friendship Hospital, Beijing, China

ABSTRACT

Vitiligo is an autoimmune disease with a strong genetic component, characterized by areas of depigmented skin resulting from loss of epidermal melanocytes. Genetic factors are known to play key roles in vitiligo through discoveries in association studies and family studies. Previously, vitiligo susceptibility genes were mainly revealed through linkage analysis and candidate gene studies. Recently, our understanding of the genetic basis of vitiligo has been rapidly advancing through genome-wide association study (GWAS). More than 40 robust susceptible loci have been identified and confirmed to be associated with vitiligo by using GWAS. Most of these associated genes participate in important pathways involved in the pathogenesis of vitiligo. Many susceptible loci with unknown functions in the pathogenesis of vitiligo have also been identified, indicating that additional molecular mechanisms may contribute to the risk of developing vitiligo. In this review, we summarize the key loci that are of genome-wide significance, which have been shown to influence vitiligo risk. These genetic loci may help build the foundation for genetic diagnosis and personalize treatment for patients with vitiligo in the future. However, substantial additional studies, including gene-targeted and functional studies, are required to confirm the causality of the genetic variants and their biological relevance in the development of vitiligo.

INTRODUCTION

Vitiligo is a relatively common skin disease, and is an acquired pigmentary disorder characterized by areas of depigmented skin resulting from loss of epidermal melanocytes. The prevalence of this disease varies from 0.2% to 1% in various global populations without sex predilection (Spritz, 2008). The pathogenesis of vitiligo remains elusive, although many theories such as autoimmune hypothesis, genetics theory, reactive oxygen species model, zinc-α2-glycoprotein deficiency hypothesis, viral theory, intrinsic theory and biochemical, molecular and cellular alterations accounting for loss of functioning melanocytes in vitiligo were elaborated to clarify vitiligo pathogenesis and showed that it was a multifactorial disease involving many different interactions (Mohammed et al., 2015). In this review, we mainly summarize the recent studies in the genetics of vitiligo through genome-

Citation: Changbing Shen, Jing Gao, Yujun Sheng, Jinfa Dou, Fusheng Zhou, Xiaodong Zheng, Randy Ko, Xianfa Tang, Caihong Zhu, Xianyong Yin, Liangdan Sun, Yong Cui, and Xuejun Zhang. Genetic Susceptibility to Vitiligo: GWAS Approaches for Identifying Vitiligo Susceptibility Genes and Loci. Frontiers in Genetics. 2016; 7:3. DOI:10.3389/fgene.2016.00003.

wide association studies (GWASs), with a focus on the susceptibility genes or loci (**Table 1**) that have been identified to date, which implicate important pathways in the pathogenesis of vitiligo.

Table 1. A summary of loci associated with vitiligo through GWAS, GWAS-MA studies up to 2015-10.

Chr	Reported Gene(s)	SNP-risk allele	Context	*P*-value	OR[95% CI]	Population	Reference
1p13.2	*PTPN22*	rs2476601-A	Intron	1.31E-07	1.39[1.23–1.57]	European	Jin et al., 2010a
1p36.23	*RERE*	rs4908760-G	Intron	7.07E-15	1.36[1.26–1.48]	European	Jin et al., 2010a
2q24.2	*IFIH1*	rs2111485-G	Intergenic	4.91E-15	1.30[NR]	European	Jin et al., 2012a
3q13.33	*CD80*	rs59374417-C	Intergenic	3.78E-10	1.34[NR]	European	Jin et al., 2012a
3q28	*LPP*	rs9851967-?	intron	8.57E-08	1.14[1.09–1.19]	Han Chinese	Tang et al., 2013
		rs1464510-T	intron	1.01E-11	1.31[1.21–1.41]	European	Jin et al., 2010a
4p16.1	*CLNK*	rs16872571-C	Intergenic	1.96E-08	1.21[NR]	European	Jin et al., 2012a
6p21.32	*C6orf10, BTNL2*	rs7758128-A	Intergenic	3.29E-16	2.19[1.80–2.65]	European	Jin et al., 2010a
		rs7758128-A	Intergenic	1.36E-09	1.5[NR]	European	Jin et al., 2011
	BTNL2, HLA-DRA	rs3806156-T	Intron	7.22E-19	1.42[1.32–1.54]	European	Jin et al., 2010a
6p21.33	*HLA-C, HLA-B*	rs11966200-A	Intron	1.48E-48	1.90[1.74–2.07]	East Asian	Quan et al., 2010
	HLA	rs9468925-?	Intergenic	2.21E-33	1.35[1.28–1.41]	East Asian	Quan et al., 2010
6p22.1	*HLA-A, HCG9*	rs3823355-T	Near-Gene-5	9.05E-23	1.50[1.39–1.63]	European	Jin et al., 2010a
6q15	*BACH2*	rs3757247-A	Intron	2.53E-08	1.20[NR]	European	Jin et al., 2012a
6q27	*RNASET2, FGFR1OP, CCR6*	rs2236313-T	Intron	9.72E-17	1.20[1.15–1.25]	East Asian	Quan et al., 2010
		rs6902119-C	Intergenic	9.75E-14	1.17[1.13–1.23]	East Asian	Quan et al., 2010
	SMOC2	rs13208776-?	Intron	8.51E-08	NR	Romanian	Birlea et al., 2010
8q24.22	*SLA*	rs853308-G	Intron	1.58E-08	1.20[NR]	European	Jin et al., 2012a
10p15.1	*IL2RA*	rs706779-A	Intron	2.78E-09	1.27[1.17–1.37]	European	Jin et al., 2010a
10q22.1	*SLC29A3, CDH23*	rs1417210-C	Intergenic	1.83E-08	1.14[1.09–1.19]	Han Chinese	Tang et al., 2013

10q22.3	*ZMIZ1*	rs11593576-?	Intron	8.31E-07	1.14[1.09–1.20]	East Asian	Quan et al., 2010
10q25.3	*CASP7*	rs3814231-G	Intron	3.56E-08	1.23[NR]	European	Jin et al., 2012a
11p13	*CD44*	rs10768122-G	UTR-3	1.78E-09	1.22[NR]	European	Jin et al., 2012a
11q14.3	*TYR*	rs1393350-G	Intron	1.60E-18	1.53[1.39–1.68]	European	Jin et al., 2010a
11q21	*TYR*	rs4409785-C	Intergenic	1.57E-13	1.34[NR]	European	Jin et al., 2012a
11q23.3	*CXCR5, DDX6*	rs638893-C	Intergenic	2.47E-09	1.22[1.14–1.30]	Han Chinese	Tang et al., 2013
12q13.2	*PMEL, IKZF4*	rs10876864-G	Near-Gene-5	8.07E-12	1.18[1.13–1.24]	Han Chinese	Tang et al., 2013
	IKZF4	rs2456973-C	Intron	2.75E-14	1.29[NR]	European	Jin et al., 2012a
12q24.12	*SH2B3*	rs4766578-T	Intron	3.54E-18	1.32[NR]	European	Jin et al., 2012a
14q12	*GZMB*	rs8192917-G	Missense	3.44E-08	1.28[1.17–1.39]	European	Jin et al., 2010a
		rs2273844-A	Near-Gene-5	6.78E-08	1.27[1.17–1.39]	European	Jin et al., 2010a
15q13.1	*OCA2, HERC2*	rs1129038-C	UTR-3	3.91E-08	1.22[NR]	European	Jin et al., 2012a
16q12.2	*KIAA1005*	rs3213758-A	Missense	6.20E-11	2.77[2.04–3.76]	Korean	Cheong et al., 2013
16q24.3	*MC1R*	rs9926296-A	Intron	1.82E-13	1.27[NR]	European	Jin et al., 2012a
19p13.3	*TICAM1*	rs6510827-T	Intron	8.80E-08	1.19[NR]	European	Jin et al., 2012a
21q22.3	*UBASH3A*	rs11203203-A	Intron	1.26E-09	1.27[1.18–1.38]	European	Jin et al., 2010a
22q13.1	*C1QTNF6*	rs229527-T	Missense	2.21E-16	1.38[1.28–1.50]	European	Jin et al., 2010a
22q13.2	*TOB2*	rs4822024-G	Intergenic	6.81E-10	1.28[NR]	European	Jin et al., 2012a

THE CONCEPT OF VITILIGO GENETICS

The earliest evidence relating to the genetic basis of vitiligo was a description provided by Addison in the year of 1855, Addison presented a patient with idiopathic adrenal insufficiency, generalized vitiligo (GV), and pernicious anemia (PA; Addison, 1855). In the 1950s, perhaps the first time to address the possible inheritance of GV, Stüttgen and Teindel described multigenerational families with multiple cases of GV and other autoimmune diseases, concluding the possibility that GV is likely an autosomal dominant inheritance disease (Teindel, 1950; Stüttgen, 1950). Many subsequent studies around the world confirmed frequent clustering of vitiligo cases within families (Alkhateeb et al., 2003; Zhang et al., 2004b; Sun et al., 2006; Zhang et al., 2009). The genetics of vitiligo cannot be explained by simple Mendelian genetics, and it is characterized by incomplete penetrance, multiple susceptibility loci, and genetic heterogeneity factors. Interestingly, many of the risk loci that have been identified are shared between vitiligo and other autoimmune diseases by GWASs (**Table 2**), implying that common molecular pathways exist among various autoimmune disorder processes.

Table 2. Vitiligo risk loci involved in pathways and shared with other autoimmune diseases.

Function	Chromosome	Genes	Other autoimmune disease#
HLA regulation for vitiligo	6p21	HLA region	PS, SLE, RA, T1D, IBD, CD
Immunoregulatory genes for vitiligo	1p13.2 2q24	*PTPN22* *IFIH1*	RA, SLE T1D, GD, MS, Lupus
	4p16.1	*CLNK*	Gout
	6q15	*BACH2*	Asthma, CD, MS, T1D
	6p21.3	*BTNL2*	T1D, RA, SLE, PS, GD
	6q27	*CCR6*	IBD
	8q24	*SLA*	ATD, ALL
	10p15	*IL2RA*	T1D, RA, SLE
	11q23.3	*CXCR5*	CC, SLE, MS
	12q24	*SH2B3*	T1D, RA, Lupus
	12q13	*IKZF4*	T1D, AA
	21q22.3	*UBASH3A*	SLE
Melanocyte related genes for vitiligo	6q27 10q22.3	*FGFR1OP* *ZMIZ1*	CD, GD T2D, CD, IBD
	15q13.1	*OCA2*	AS, GDD, ASD
Apoptotic and cytotoxic genes	14q11.2 10q22.1	*GZMB* *SLC29A3*	JIA, BD HS
	10q25	*CASP7*	T1D, RA
Susceptibility loci with unknown functions for vitiligo	3q2822q13.1 6q27	*LPP C1QTNF6* *SMOC2*	RAGD, T1D AITD

PS, Psoriasis; SLE, Systemic lupus erythematosus; RA, Rheumatoid arthritis; T1D, Type I diabetes; T2D, Type II diabetes; IBD, Inflammatory bowel disease; CD, Crohn's disease; GD, Graves' disease; CC, Colorectal cancer; MS, Multiple sclerosis; AS, Angelman syndrome; GDD, Global developmental delay; ASD, Autism spectrum disorder; ALL, Acute lymphoblastic leukemia; JIA, Juvenile idiopathic arthritis; HS, H syndrome; BD, Behcet's disease; AITD, Autoimmune thyroid disease; AA, Alopecia areata.

THE GENETIC APPROACHES FOR VITILIGO

Over the past several decades, a large number of genes and genomic regions involved in vitiligo susceptibility have been revealed through linkage analysis and candidate gene studies (Spritz, 2012). Candidate gene association studies are best suited to detect genetic signals that represent relatively common causal variants with modest effect sizes. Moreover, candidate gene association studies are relatively easy to carry out, usually involving a simple comparison of allele frequencies in cases and controls. At least 33 different candidate genes for vitiligo have been reported on the basis of such studies (Birlea et al., 2011). However, such studies are usually subject to false positive results because of the ethnic differences in case-control analyses, inadequate statistical power and statistical fluctuation, and inadequate correction for multiple testing both within and across studies.

Genome-wide linkage study is a method, when executed correctly, identifies genetic loci of vitiligo in multiplex families. These studies help to determine the position of the genetic marker inherited together with a specific disease. Genome-wide linkage studies in the Caucasian population multiplex vitiligo families identified additional linkage signals on chromosomes 7, 8, 9, 11, 13, 17, 19, and 22 (Fain et al., 2003; Spritz et al., 2004). In addition, this parallels with the genetic linkage studies of vitiligo in Chinese Han population which detected linkage signals on chromosomes 1, 4, 6, 14, and 22 (Chen et al., 2005; Liang et al., 2007). Normally, genetic loci discovered through genome-wide linkage studies encompass several megabases. The diversity of proposed regions has created a challenge in fine mapping.

Genome-wide association studies is a high through put technology, capable of "pin-pointing" disease-causing genes. Since 2005, GWAS has been proved to be the most powerful and efficient study design thus far in identifying genetic variants that are associated with complex diseases. More than 1000 types of complex diseases and traits have been investigated by this approach. Since 2010, several GWASs have been performed in patients with vitiligo in various ethnic populations (Birlea et al., 2010;Quan et al., 2010; Jin et al., 2011, 2010a, 2012a; Cheong et al., 2013), these studies have confirmed genetic associations of almost 40 genes and loci with vitiligo during the past 5 years (**Table 1**; **Figure 1**). Recently, Next-Generation DNA Re-Sequencing and exome sequencing analysis have also been used for identifying variants of genes for vitiligo (Jin et al., 2012b).

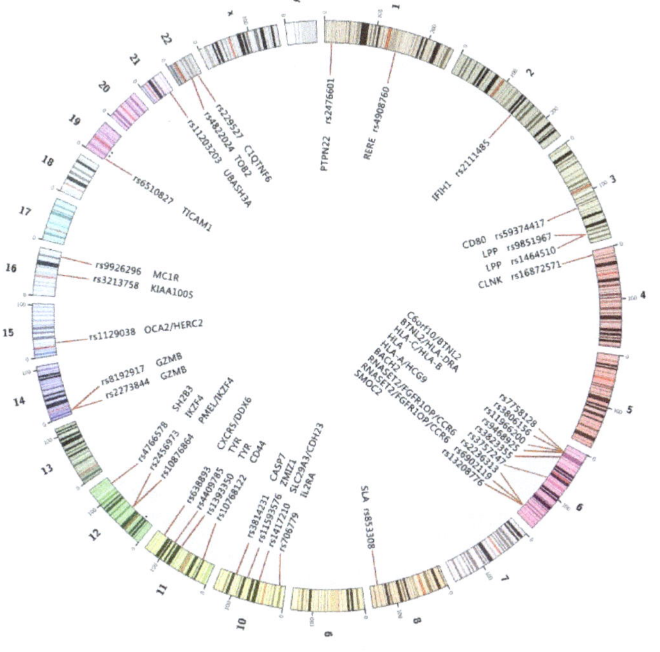

Figure 1. The reported genes and loci on the chromosomes associated with vitiligo through GWAS and GWAS meta-analysis up to 2015-10. All of the above marked regions and genes on the chromosomes achieved genome-wide significance ($p < 5E\text{-}8$) in at least one study.

THE SUSCEPTIBLE GENES FOR VITILIGO FROM GWAS

Human Leukocyte Antigen Genes

The human leukocyte antigen (HLA) is the most gene-dense region of the genome, encoding more than 120 functional genes in humans which are distributed over a 3.6 Mbp region. Many previous studies have found genes in the *HLA* region associated with vitiligo, such as *HLA-A*02*, *HLA-A*30*, *HLA-B*13*, *HLA-C*0602*, *HLA-DRB1*04*, *HLA-DRB1*07* and *HLA-DQB1*03* (Tastan et al., 2004; Zhang et al., 2004a; Liu et al., 2007). GWASs for vitiligo have detected major association signals in the MHC on chromosome 6p21.3 in Caucasian and Chinese populations, and there are some specific associations differed between the two populations. In Caucasians, two major association peaks in MHC region were detected, the peak in class I gene region, principally between *HLA-A*and *HCG9*, and in the class II region, mainly between HLA-DRB1 and HLA-DQA1, in linkage disequilibrium (LD) with HLA-DRB1*04 (Jin et al., 2010a). In the Chinese population study, the major MHC association signal was in the class III gene region, though there was also some evidence for independent association in the class II region (Quan et al., 2010). In Chinese Han population, GWAS for vitiligo identified two independent association signals (rs11966200 and rs9468925) within the *HLA* region. Further analyses have suggested that the strong association at rs11966200 might reflect the reported association of the *HLA-A*3001*, *HLA-B*1302*, *HLA-C*0602* and *HLA-DRB1*0701* alleles and the association at rs9468925 might represent a previously unknown *HLA* susceptibility allele (*HLA-C/HLA-B*) (Quan et al., 2010). Additional studies found that rs9468925 is associated with clinical features of GV (Liu et al., 2012), and more importantly, rs9468925 in *HLA-C/HLA-B* is associated with both psoriasis (PS) and vitiligo (Zhu et al., 2011), providing first important evidence that two major skin diseases share a common genetic locus in the MHC, and revealing that this genetic locus may share the same molecular mechanism for PS and vitiligo. Genome-wide analysis has identified rs532098 in the vicinity of *HLA-DRB1-DQA1*, showing suggestive evidence of the age of onset for GV (Jin et al., 2011). All evidences shown here clearly suggest that HLA genes represent attractive therapeutic targets for vitiligo pathogenesis.

Immunoregulatory Genes for Vitiligo

Vitiligo is a common autoimmune disease, around 20% of vitiligo patients manifest concomitant occurrence of other autoimmune diseases, particularly autoimmune thyroid disease (AMD), rheumatoid arthritis (RA), late-onset type I diabetes (T1D), PS, PA, systemic lupus erythematosus (SLE), and Addison's disease (AD; Birlea et al., 2010). Outside MHC region, some susceptibility genes encode immunoregulatory proteins which involve in biological pathways that are most likely influencing the development of vitiligo.

CD44 and *CD80*

CD44 encodes a cell surface glycoprotein with various functions, including a role in T cell development (Baaten et al., 2010), and is associated with lupus (Ramos et al., 2011). *CD80* encodes a surface protein on activated B-cells, monocytes, and dendritic cells that co-stimulates T cell priming (Peach et al., 1995). Vitiligo is a CD8 T cell-mediated autoimmune disease and promotes the longevity of memory T cell responses to melanoma. Studies have found that HA-specific CD8 T cells are adoptively transferred into mice expressing HA as a self-antigen in the pancreas, and HA-specific T cells proliferate in draining lymph nodes and upregulated *CD44* (Hernandez et al., 2001). GWAS meta analysis (GWAS-MA) of vitiligo has showed suggestive association of SNP rs4330287 and imputed SNP rs59374417 in *CD80*, which was confirmed by replication study and overall meta-analysis (Jin et al., 2012a). Flow cytometric analysis has found that the percentage of CD80+ monocytes are significantly increased in the vitiligo group compared with the controls (Basak et al., 2008), which may indicate alterations of monocyte function in the pathogenesis of vitiligo.

SLA and *BACH2*

GWAS-MA has identified vitiligo-associated loci *TG/SLA* and *BACH2*in European-derived white (CEU) population (Jin et al., 2012a). *TG* encodes thyroglobulin, *SLA* encodes Src-like adaptor protein, an inhibitor of T- and B-cell receptor signaling. It is not apparent what role thyroglobulin might play in vitiligo pathogenesis, suggesting association of vitiligo with the *TG/SLA* locus may derive from*SLA*, rather than *TG* (Jin et al., 2012a). Studies have suggested that *SLA* might likewise account for reported association with autoimmune thyroid disease (ATD; Tomer and Greenberg, 2004) and acute lymphoblastic leukemia (ALL; Mansha et al., 2010). The GWAS-MA for vitiligo showed suggestive association of SNPs (nt 90941239-91915693) spanning *BACH2*, particularly rs3757247, confirmed by the replication study and overall meta-analysis (Jin et al., 2012a). *BACH2* encodes a transcriptional repressor of B cells (Sasaki et al., 2000), which is a key regulator of CD4(+) T-cell differentiation that prevents inflammatory disease by controlling the balance between tolerance and immunity. Genetic polymorphisms analysis shows that *BACH2* is associated with asthma, Crohn's disease (CD), multiple sclerosis (MS) and T1D (Roychoudhuri et al., 2013).

IFIH1 and *TICAM1*

IFIH1 and *TICAM1* are involved in the innate immune response system (Kato et al., 2006). GWAS-MA for vitiligo has shown genome-wide level association with SNP rs2111485 in *IFIH1* (Jin et al., 2012a). *IFIH1* encodes an interferon-induced RNA helicase involved in antiviral innate immune responses, associated with T1D (Smyth et al., 2006), Graves' disease (GD) (Sutherland et al., 2007), MS (Martinez et al., 2008), PS (Li et al., 2010), and possibly lupus (Gateva et al., 2009). *TICAM1*, also known as TIR domain-containing adaptor-inducing IFN-b (TRIF), eventually activates transcription factors (TF), interferon regulatory factor-3 (IRF-3), NF-κB and AP-1, leading to the induction of type I interferons and inflammatory cytokines (Kumeta et al., 2014). *TICAM1* encodes a toll-like receptor adaptor molecule 1, which mediates innate immune responses to viral pathogens (Seya et al., 2009). Viral factor has been implicated in the etiopathogenesis of vitiligo, we speculate that *TICAM1*might act as a viral factor in the pathogenesis of vitiligo.

PTPN22, UBASH3A, and *CLNK*

PTPN22, *UBASH3A*, and *CLNK* are the T-cell–receptor signaling pathway genes. *PTPN22* 620W allele plays a role in autoimmune disorders, and underscores the importance of a genetically mediated autoimmune mechanism in the pathogenesis of vitiligo. Evidence shown that the *PTPN22* 1858C/T variants contribute to risk of GV in European Caucasian and Mexican populations (Laberge et al., 2008;Garcia-Melendez et al., 2014), but it does not appear to play a similar role in the Jordanian population and Turkish generalized-vitiligo patients (Alkhateeb et al., 2013;Akbas et al., 2014). A meta-analysis was conducted of the association between *PTPN22*1858 C/T polymorphisms and vitiligo and found *PTPN22* C1858T polymorphism is associated with vitiligo susceptibility in European population (Song et al., 2013). Variants of *PTPN22* also are associated with a number of different autoimmune diseases, including RA (Begovich et al., 2004) and SLE (Kyogoku et al., 2004). There was an association between GV and nine SNPs in the region spanning *UBASH3A* on chromosome 21q22.3; of these SNPs, rs2839511 showed genome wide significance (Jin et al., 2010a). Functional prediction of the variants in non-MHC vitiligo loci identified predicted deleterious variants at *UBASH3A* confer protection from vitiligo (Jin et al., 2012a). *UBASH3A* encodes a regulator of T-cell–receptor signaling and is associated with T1D (Concannon et al., 2008), the genotype distribution of rs2277798 is significantly associated with hematuria in SLE patients (Liu et al., 2015). At 4p16.1, the GWAS-MA for vitiligo showed suggestive association of SNPs (nt 10702156–10729386) upstream of *CLNK*, including rs16872571 and several imputed SNPs, particularly rs11940117, confirmed by the replication study and overall meta-analysis.*CLNK* encodes a mast cell immunoreceptor signal transducer, a positive regulator in immunoreceptor signaling (Wu and Koretzky, 2004). Haplotype analysis has shown that the TCATTCTGA haplotype of *CLNK* is more frequent among patients with gout (Jin et al., 2015).

IKZF4, IL2RA, and *BTNL2*

IKZF4, *IL2RA*, and *BTNL2* are involved in T-cell activation. Two GWASs have identified *IKZF4* as one of the susceptible genes for GV (Jin et al., 2012a; Tang et al., 2013). *IKZF4*, is a critical mediator of Foxp3-dependent gene silencing in T cell reguration (Treg), interacts directly with Foxp3 and is necessary for gene silencing without affecting the expression of Foxp3 activated genes (Pan et al., 2009). *IKZF4* maybe another biological candidate gene for vitiligo and influences the development of vitiligo. Further fine mapping and function analysis required to determine the causal variants within this locus for vitiligo. Besides, studies have also found *IKZF4* is associated with T1D (Hakonarson et al., 2008) and alopecia areata (AA;Petukhova et al., 2010). There are 25 SNPs in the region of *IL2RA* (encoding the interleukin-2-receptor alpha chain) on chromosome 10p15.1, 8 of which showed genome wide significance, SNPs rs706779 and rs7090530 had the strongest association with GV (Jin et al., 2010a). Elevated serum interleukin-2-receptor levels indicate T-cell activation in GV (Honda et al., 1997; Wang et al., 2009). The variants of *IL2RA* have been shown to be associated with T1D (Vella et al., 2005), RA (Hinks et al., 2009), and SLE (Carr et al., 2009). Study identified a quantitative trait locus for vitiligo age of onset, located near*c6orf10-BTNL2* (rs7758128), a region that is also associated with GV susceptibility (Jin et al., 2011). *BTNL2* encodes an immunoglobulin superfamily membrane protein implicated in T-cell activation. Variants in *BTNL2* may play a role which involved in vitiligo susceptibility versus vitiligo age of onset. The *BTNL2* gene region has been associated with susceptibility to many other autoimmune diseases, such as

T1D (He et al., 2009), RA (Cui et al., 2009), SLE (Orozco et al., 2005), PS (Feng et al., 2009), and GD (Simmonds et al., 2006).

CXCR5, *CCR6*, and *SH2B3*

CXCR5, *CCR6*, and *SH2B3* encode chemokine or cytokine receptors. Association analyses identified that rs638893 at 11q23.3 is associated with vitiligo in the Chinese Han population, and rs638893 is located in an intergenic region between *CXCR5* and *DDX6* (Tang et al., 2013). *CXCR5* encodes a multi-pass membrane protein that belongs to the CXC chemokine receptor family. This cytokine receptor binds to B-lymphocyte chemoattractant (BLC), and is involved in B-cell migration into B-cell follicles of spleen and Peyer patches. *CXCR5* has also been shown to have an important role in the pathogenesis of colorectal cancer (CC) (Qi et al., 2014), SLE (Zhang et al., 2014), and MS (Lill et al., 2013). At 6q27, GWAS of GV in the Chinese populations identified rs6902119 in CCR6 with genome wide significance (Quan et al., 2010). Another study showed that the most significant association SNPs rs6902119 and rs2301436 in *CCR6* were observed, SNPs rs6902119 and rs2301436 are in moderate LD, and logistic regression analysis indicated that association of GV with rs2301436 might be secondary to LD with rs6902119 (Jin et al., 2010b). *CCR6* encodes chemokine receptor 6 and is favorably expressed by immature dendritic cells and memory T cells (Schutyser et al., 2003). When binding to its ligand (CCL20), *CCR6* may result in the chemoattraction of immune cells, which might have a role in skin and mucosal surfaces under homeostatic and inflammatory conditions (Le Borgne et al., 2006). *CCR6* is also associated with inflammatory bowel disease (IBD) (Barrett et al., 2008). The GWAS-MA showed association with SNPs (nt 111708458-112906415) within and near *SH2B3*, particularly rs3184504 and imputed SNP rs4766578, located downstream, within *ATXN2*(Jin et al., 2012a). *ATXN* encodes Ataxin-2, and is causal for spinocerebellar ataxia type 2. *SH2B3* encodes a member of the *SH2B* adaptor family of proteins, which are involved in a range of signaling activities by growth factor and cytokine receptors. *SH2B3* thus seems more likely relevant to vitiligo susceptibility than *ATXN2*. *SH2B3* is also involved in the development regulation of both B and T cells, and associated with some immune diseases, including T1D (Devalliere and Charreau, 2011), RA (Coenen et al., 2009), and lupus (Li et al., 2010).

Melanocyte Related Genes for Vitiligo

Vitiligo is one of the most common pigment disorders of the skin and hair and results from a selective destruction of melanocytes. Vitiligo patients have a progressive loss of melanocytes, predominantly in areas of skin subject to physical abrasion or at pressure points, leading to white patches appearing on the body. In normal physiological circumstances, melanin pigment is generated by the melanocytes and transferred to the surrounding keratinocytes to produce skin complexion and hair coloration (Smith and Sturm, 2010). GWASs for vitiligo have also identified some susceptible genes which showed genome-wide significant association level with the related function that influence the activity of melanocyte.

ZMIZ1

ZMIZ1 locus at 10q22, encodes a protein related to protein inhibitor of activated STAT (PIAS). PIAS3, a related member of the PIAS protein family, can inhibit the transcriptional activity of microphthalmia transcription factor (MITF), which has been demonstrated to be a key regulator of melanocyte development, function and survival (Garraway et al., 2005). Another study confirmed *ZMIZ1* as a novel susceptibility locus for vitiligo and further suggested rs1408944 to be the putative causal variant that potentially interrupts TF binding and thus the transcriptional regulation of *ZMIZ1* (Sun et al., 2014). In addition, *ZMIZ1* might be associated with type II diabetes (T2D) (Matsuba et al., 2015), CD (Yang et al., 2015), and IBD (Jakobsen et al., 2014).

PMEL

PMEL encodes a melanocyte-specific type I transmembrane glycoprotein. The encoded protein is enriched in melanosomes, which are the melanin-producing organelles in melanocytes, and plays an essential role in the structural organization of premelanosomes (McGlinchey et al., 2009). Skin biopsy transcriptome analysis found that *PMEL* has a decreased expression in vitiligo lesional skin compared to vitiligo perilesional normal skin (Tang et al., 2013). In addition, the antigen-specific CD8+ T cells exhibit reactivity to modified *PMEL* peptide epitopes in HLA-A2-positive vitiligo patients (Mandelcorn-Monson et al., 2003), which also supports the notion that there is a cell-mediated immunopathologic mechanism in vitiligo.

TYR

GWAS identifies two SNPs (rs1847134 and rs1393350) in association with the *TYR*gene region which showed genome-wide significance with vitiligo, and haplotype analysis reveals a strong association with a block of six SNPs (rs1018528, rs10765198, rs1847134, rs1393350, rs1126809, and rs1806319) in tight LD (Jin et al., 2010a). *TYR*encodes tyrosinase, an enzyme of the melanocyte that catalyzes the rate-limiting steps of melanin biosynthesis and constitutes a major autoantigen in GV (Rezaei et al., 2007). Next-generation DNA re-sequencing identifies common variants of *TYR* and *HLA-A* that modulate the risk of GV via antigen presentation (Jin et al., 2012b). The biological interaction between *HLA-A* and *TYR* shows an apparent inverse relationship between susceptibility to GV versus malignant melanoma (Spritz, 2010), suggesting that GV may result from dysregulation of normal processes of immune surveillance against melanoma.

MC1R

MC1R, encoding the receptor protein for melanocyte-stimulating hormone (MSH), is a regulator of melanogenesis and minor vitiligo autoantigen, associating with malignant melanoma and with skin and hair color (Dessinioti et al., 2011). Expression shows *MC1R* is marked significantly different between lesional and non-lesional vitiligo skin (Kingo et al., 2007). The frequency of five *MC1R* coding region SNPs: Arg67Gln (G200A), Val92Met (G274A), Ile120Thr (T359C), Arg160Arg (C478A), and Gln163Arg (A488G) in Korean vitiligo patients and normal controls did not reach statistical significance (Na et al., 2003). However, another study shows that the Arg160Trp allele of *MC1R* gene may be able to protect against vitiligo (Szell et al., 2008). Further study need to be conducted to confirm this conclusion between vitiligo and *MC1R* coding region SNPs.

RNASET2 and *FGFR1OP*

In the Chinese Han population, GWAS for vitiligo identifies a risk locus at 6q27, which contains three genes: *RNASET2*, *FGFR1OP* and *CCR6.RNASET2* is potentially involved in tumorigenesis and associated with human malignancies and chromosomal rearrangement. Overexpression of *RNASET2*hypersensitizes cells to oxidative stress, thus promoting cell death during peroxide exposure and stationary-phase onset (Thompson and Parker, 2009). Therefore, *RNASET2*regulates the oxidative stress and intervenes the initial pathogenic event in melanocyte destruction in vitiligo. *FGFR1OP* encodes a largely hydrophilic centrosomal protein that is required for anchoring microtubules to subcellular structures. Loss of *FGFR1OP*causes apoptosis in the G1 phase of the cell cycle, with accumulation of a 32-kDa p53 tumor suppressor isoform and NOXA and FAS transcripts, suggesting that *FGFR1OP* is necessary for cell-cycle progression and survival (Acquaviva et al., 2009). Mutations in this gene not only associate with vitiligo (Quan et al., 2010), but also are associated with CD (Yang et al., 2014), GD (Chu et al., 2011).

OCA2 and *HERC2*

At 15q12–q13.1, GWAS-MA shows suggestive association of SNPs spanning *OCA2* upstream to within *HERC2*, especially SNP rs12913832 and imputed SNP rs1129038 (Jin et al., 2012a). *OCA2* encode melanocyte antigens presented by HLA-A*0215, for vitiligo protection is associated with reduced functional protein, and for susceptibility to vitiligo and melanoma constitute genetic opposites, perhaps modulating immune surveillance for melanoma (Jin et al., 2012a). Mutations in *OCA2*gene associate with oculocutaneous albinism type II (Rimoldi et al., 2014) and melanoma (Hawkes et al., 2013). *HERC2* belongs to the *HERC* gene family that encodes a group of unusually large proteins, which contain multiple structural domains. Variants within *HERC2* down-regulate transcription of the *OCA2* allele in *cis* (Jin et al., 2012a). Mutations in *HERC2* are associated with the development of angelman syndrome (AS;Harlalka et al., 2013), global developmental delay (GDD), and autismspectrum disorder (AD) (Puffenberger et al., 2012).

Apoptotic and Cytotoxic Genes for Vitiligo

Vitiligo is an acquired and progressive hypomelanotic disease that manifests as circumscribed depigmented patches on the skin. The interactions between melanocytes and other typical skin cells, particularly keratinocytes, may be an interpretation for the cause of vitiligo. Some genes have the role in gene repression, apoptosis and cell survival, inflammation, and cytotoxic cells, its possible involvement in the progression vitiligo.

RERE

RERE is the locus on chromosome 1p36.23, which encodes the arginine–glutamic acid dipeptide repeats protein. The protein is a transcriptional corepressor that is highly expressed in lymphoid cells and is thought to regulate apoptosis (Wang and Tsai, 2008). There are 40 SNPs in *RERE* were identified as having an association with GV especially rs301819 showed genome-wide significance, the most strongly associated SNPs in the*RERE* region were in tight LD; no haplotype showed a significantly stronger association than any component SNP (Jin et al., 2012a). Another SNP rs4908760 in *RERE* is the strongest associated SNP with genome-wide significance in the GWAS study (Jin et al., 2010a).

CASP7

At 10q25.3, GWAS-MA for vitiligo shows suggestive association of SNPs spanning *CASP7*, particularly rs3814231, which is confirmed by the replication and meta-analysis study (Jin et al., 2012a). *CASP7* encodes a member of the cysteine-aspartic acid protease (caspase) family, and sequential activation of caspases plays a central role in the execution-phase of cell apoptosis and inflammation (Lamkanfi and Kanneganti, 2010). *CASP7* also is associated with RA (Garcia-Lozano et al., 2007), and may be a candidate gene for T1D (Babu et al., 2003).

GZMB

Genome-wide association studies in European-derived whites have demonstrated genetic association between vitiligo and *GZMB* (Jin et al., 2012a; Jin et al., 2010a).*GZMB* is a caspase much alike serine protease that mediates two processes: immune-induced target-cell apoptosis mediated by cytotoxic T cells (CTLs), and natural killer cells and activation-induced cell death or type 2 helper T cells, which terminates the immune response (Trapani and Sutton, 2003; Devadas et al., 2006). Next-generation DNA re-sequencing has identified a direct causal role for the *GZMB* rs8192917-C-rs11539752-C haplotype (55R-94A) in the pathogenesis of GV (Ferrara et al., 2013).*GZMB* is only genetically associated with juvenile idiopathic arthritis (JIA; Donn et al., 2008) and Behcet's disease (BD; Kucuksezer et al., 2009), which suggests the possibility that *GZMB* may be relatively specific for melanocyte-directed autoimmune susceptibility.

Susceptibility Loci with Unknown Functions for Vitiligo

There are some vitiligo susceptible genes identified by GWAS and reach at the genome-wide significance level, but the function of these genes in the pathogenesis and development of vitiligo is still unclear.

TOB2

At 22q13.2, the GWAS-MA showed association with SNPs in a broad region (nt 41707054-42062822), particularly rs79008, upstream of *TOB2*, and several imputed SNPs, including rs4822024, between *ZC3H7B* and *TEF* (Jin et al., 2012a). *TOB2* locus on 22q13.2, encodes a regulator of cell cycle progression involved in T cell tolerance (Jia and Meng, 2007). However, the assignment of *TOB2* as causal remains uncertain in the pathogenesis of vitiligo.

SMOC2

Genome-wide association studies in an isolated European population identified*SMOC2* as a risk locus for GV (Birlea et al., 2010). However, another study shows that the variant rs13208776 in *SMOC2* gene does not play a major role in increasing the risk of vitiligo in Jordanian Arab patients (Alkhateeb et al., 2010), maybe due to the different genetic background in these two populations. In the skin, *SMOC2* is mainly present in the basal levels of the epidermis, and *SMOC2*-stimulated attachment of primary keratinocytes in culture (Maier et al., 2008). The study has also found that *SMOC2* may play a role in autoimmune thyroid disease (AITD) susceptibility as a dominant polymorphism (Alkhateeb et al., 2013).

KIAA1005

KIAA1005, also known as retinitis pigmentosa GTPase regulator-interacting protein 1-like (RPGRIP1L) gene, encodes a protein that can localize to the basal body-centrosome complex or to primary cilia and centrosomes in ciliated cells. In *KIAA1005*, the genotype and allele frequencies of 3854 G > A (1264 Asp > Asn) in vitiligo patients are significantly different compared to healthy controls. The GG frequency is lower and AA frequency is higher in vitiligo, suggesting the A allele at the *KIAA1005* G3854A may increase susceptibility to vitiligo (Cheong et al., 2013). Multiple variants of the*KIAA1005* gene have also been associated with certain clinical manifestations, particularly ciliopathies as in *DNAH5* with neurological, renal and ocular manifestations (Delous et al., 2007).

SLC29A3 and *CDH23*

GWAS has revealed rs1417210 at 10q22.1 to have a strong association with vitiligo (Tang et al., 2013). This SNP is located in an LD block that contains *SLC29A3* and *CDH23*. *SLC29A3* encodes a nucleoside transporter. The encoded protein plays a role in cellular uptake of nucleosides, nucleobases, and their related correspondents. Mutations in this gene have been associated with H syndrome (Molho-Pessach et al., 2008). *CDH23* is a member of the cadherin superfamily, whose genes encode calcium dependent cell-cell adhesion glycoproteins. The encoded protein is thought to be involved in stereocilia organization and hair bundle formation. Whole-exome sequencing (WES) has identified that *CDH23* mutations cause hearing loss in Koreans families (Woo et al., 2014).

LPP, *DDX6*, and *C1QTNF6*

The *LPP* gene locus on 3q28, encodes a member of a subfamily of LIM domain proteins that are characterized by a N-terminal proline-rich region and three C-terminal LIM domains. *LPP* has also been associated with celiac disease and RA (Coenen et al., 2009). Rs638893 located in an intergenic region between*DDX6* and *CXCR5* is associated with vitiligo (Tang et al., 2013). *DDX6* encodes a member of the DEAD box protein family, which is a RNA helicase found in P-bodies and stress granules, and functions in translation suppression and mRNA degradation (Weston and Sommerville, 2006). Rs229527 in *C1QTNF6* has shown to have an association with GV (Jin et al., 2010a). *C1QTNF6* denotes the C1q and tumor necrosis factor–related protein 6 gene. *C1QTNF6-RAC2* at 22q12.3-13.1 has reached a genome-wide significant association is a novel susceptibility loci for GD (Zhao et al., 2013), and GWAS-MA also identifies *C1QTNF6* as one of risk loci for T1D (Cooper et al., 2008).

GENES AND LOCUS INTERACTIONS IN SUSCEPTIBILITY TO VITILIGO

In general, genes and locus only explain partial variation of heritability (Manolio et al., 2009), gene–gene (or genetic variants) interactions are strongly believed to contribute to the genetic risk of common diseases (Cordell, 2009). Transmission disequilibrium and family based association statistical tests found the SNP markers in regions 7p13, 7q11, and 9q22 were significantly associated with GV, tagging SNPs for these regions represented by rs6960920, rs734930, and rs4744411, respectively (Spritz et al., 2004). The investigators examined the potential genetic interactions for these independently identified loci using two-way tests (and three-way tests in the context of the previously identified *NLRP1* gene tagged by the rs6502867). Notably, all three SNPs showed significant interaction with the *NLRP1* gene in predicting the GV phenotype (Jin et al., 2010c). The pairwise interaction analysis between 6q27, 10q22 and the two MHC SNPs (rs11966200 and rs9468925) were performed in the Chinese Han population, but no significant genetic interaction ($P > 0.05$ after correction for multiple testing) was identified (Quan et al., 2010).

It is observed that the autoimmunity feature of vitiligo is supported through the significant linkage to the MHC region on 6p21-p22 and evidence provided for the association of *HLA-DR* with vitiligo. The epistatic interaction between rs2269577 (*XBP1*) and *HLADRB1*07* was tested by using logistic regression analysis and found that the full model with both the main and interactive effects was better than the model with only the main effect to fit the data (Ren et al., 2009). Stratified association analysis of rs2269577 by *HLA-DRB1*07* allele has shown that the association at rs2269577 is significant in both the patients carrying and not carrying the -DRB1*07 allele. However, the association seems to be stronger in patients carrying the *HLA-DRB1*07* allele (Ren et al., 2009). *XBP1* as one of biological candidate genes for vitiligo may be due to its plausible role in the development of the disease through its interaction with *HLA-DR*.

GWAS identifies *RNASET2* as a susceptible gene to vitiligo in the Chinese Han population (Quan et al., 2010), but the function of *RNASET2* in vitiligo pathogenesis or in melanocyte apoptosis is unknown. *In vitro* analyses have indicated that overexpression of *RNASET2* is inducible in cultured primary human melanocytes and keratinocytes through stressful conditions, exposure to ultraviolet irradiation, hydrogen peroxide, and inflammatory factors, respectively, and lead to increased cell apoptosis via the tumor necrosis factor receptor-associated factor 2 (TRAF2)-caspases pathway through the physical interaction of *RNASET2* with *TRAF2* (Wang et al., 2014). Hence, *RNASET2*may contribute to vitiligo pathogenesis by inhibiting *TRAF2* expression.

CONCLUSION AND PROSPECTION

In the past five years, GWASs have contributed tremendously to the identification of key loci that were associated with the risk of developing vitiligo. These genes may provide novel therapeutic and prophylactic targets for new interventional approaches to treat and prevent vitiligo. Developments in this area will be exciting and influence the therapeutic approaches for the suppression of vitiligo in the future. We summarized and evaluated the importance of these loci in their respective molecular signaling pathways, and suggested new etiologic clues to vitiligo development. Considering that most of these genetic associations are restricted to moderate effects, large sample size studies are required in future investigations in order for these subtle variations to be detected. With the increasing number of GWASs being conducted, it is desirable to combine these findings across these studies to improve the statistical power. Meta-analysis of multiple GWASs improves the power to detect more associations, and to investigate the heterogeneity or consistency of these associations across different datasets and study populations. Beyond gene-association approaches, functional and gene-targeted assays, whole exome sequencing are required to identify the causal variants and understand their biological function.

AUTHOR CONTRIBUTIONS

XZ is the principal investigator for this review article and has contributed to the concept and planning of the article, collection of data, and reporting of the work described. CS, JG, YS, FZ, XZ, RK, XT, CZ, XY, and JD contributed to the planning of the article, collection of data, and reporting of the work described. XY, LS, and YC are the other principal investigators for this review article and have contributed to the concept of the manuscript, planning of the article, collection of data, and reporting of the work described. All authors contributed to drafting the article or revising it critically for important intellectual content.

CONFLICT OF INTEREST STATEMENT

The authors declare that the research was conducted in the absence of any commercial or financial relationships that could be construed as a potential conflict of interest.

ACKNOWLEDGMENTS

This review article was supported by grants from the General Program of National Natural Science Foundation of China (81402591, 1408085QH177, 81171506, 81273302, 81000691, 81072460, 8127330).

REFERENCES

1. Acquaviva C., Chevrier V., Chauvin J. P., Fournier G., Birnbaum D., Rosnet O. (2009). The centrosomal FOP protein is required for cell cycle progression and survival. *Cell Cycle* 8 1217–1227. 10.4161/cc.8.8.8248

2. Addison T. (1855). On the constitutional and local effects of disease of the suprarenal capsules. *Med. Classics* 1937 244–293.

3. Akbas H., Dertlioglu S. B., Dilmec F., Atay A. E. (2014). Lack of association between PTPN22 gene +1858 C>T polymorphism and susceptibility to generalized vitiligo in a Turkish population. *Ann. Dermatol.* 26 88–91. 10.5021/ad.2014.26.1.88

4. Alkhateeb A., Al-Dain Marzouka N., Qarqaz F. (2010). SMOC2 gene variant and the risk of vitiligo in Jordanian Arabs. *Eur. J. Dermatol.* 20 701–704.

5. Alkhateeb A., Fain P. R., Thody A., Bennett D. C., Spritz R. A. (2003).Epidemiology of vitiligo and associated autoimmune diseases in Caucasian probands and their families. *Pigment Cell Res.* 16 208–214. 10.1034/j.1600-0749.2003.00032.x

6. Alkhateeb A., Marzouka N. A., Tashtoush R. (2013). Variants in PTPN22 and SMOC2 genes and the risk of thyroid disease in the Jordanian Arab population. *Endocrine* 44 702–709. 10.1007/s12020-013-9908-z

7. Baaten B. J., Li C. R., Bradley L. M. (2010). Multifaceted regulation of T cells by CD44. *Commun. Integr. Biol.* 3 508–512. 10.4161/cib.3.6.13495

8. Babu S. R., Bao F., Roberts C. M., Martin A. K., Gowan K., Eisenbarth G. S., et al. (2003). Caspase 7 is a positional candidate gene for IDDM 17 in a Bedouin Arab family. *Ann. N. Y. Acad. Sci.* 1005 340–343. 10.1196/annals.1288.054

9. Barrett J. C., Hansoul S., Nicolae D. L., Cho J. H., Duerr R. H., Rioux J. D., et al. (2008). Genome-wide association defines more than 30 distinct susceptibility loci for Crohn's disease. *Nat. Genet.* 40 955–962. 10.1038/ng.175

10. Basak P. Y., Adiloglu A. K., Koc I. G., Tas T., Akkaya V. B. (2008). Evaluation of activatory and inhibitory natural killer cell receptors in non-segmental vitiligo: a flow cytometric study. *J. Eur. Acad. Dermatol. Venereol.* 22 970–976. 10.1111/j.1468-3083.2008.02681.x

11. Begovich A. B., Carlton V. E., Honigberg L. A., Schrodi S. J., Chokkalingam A. P., Alexander H. C., et al. (2004). A missense single-nucleotide polymorphism in a gene encoding a protein tyrosine phosphatase (PTPN22) is associated with rheumatoid arthritis. *Am. J. Hum. Genet.* 75 330–337. 10.1086/422827

12. Birlea S. A., Gowan K., Fain P. R., Spritz R. A. (2010). Genome-wide association study of generalized vitiligo in an isolated European founder population identifies SMOC2, in close proximity to IDDM8. *J. Invest. Dermatol.* 130 798–803. 10.1038/jid.2009.347

13. Birlea S. A., Jin Y., Bennett D. C., Herbstman D. M., Wallace M. R., McCormack W. T., et al. (2011). Comprehensive association analysis of candidate genes for generalized vitiligo supports XBP1, FOXP3, and TSLP. *J. Invest. Dermatol.* 131371–381. 10.1038/jid.2010.337

14. Carr E. J., Clatworthy M. R., Lowe C. E., Todd J. A., Wong A., Vyse T. J., et al. (2009). Contrasting genetic association of IL2RA with SLE and ANCA-associated vasculitis. *BMC Med. Genet.* 10:22 10.1186/1471-2350-10-22

15. Chen J. J., Huang W., Gui J. P., Yang S., Zhou F. S., Xiong Q. G., et al. (2005). A novel linkage to generalized vitiligo on 4q13-q21 identified in a genomewide linkage analysis of Chinese families. *Am. J. Hum. Genet.* 76 1057–1065. 10.1086/430279

16. Cheong K. A., Kim N. H., Noh M., Lee A. Y. (2013). Three new single nucleotide polymorphisms identified

by a genome-wide association study in Korean patients with vitiligo. *J. Korean Med. Sci.* 28 775–779. 10.3346/jkms.2013.28.5.775

17. Chu X., Pan C. M., Zhao S. X., Liang J., Gao G. Q., Zhang X. M., et al. (2011). A genome-wide association study identifies two new risk loci for Graves' disease. *Nat. Genet.* 43 897–901. 10.1038/ng.898

18. Coenen M. J., Trynka G., Heskamp S., Franke B., van Diemen C. C., Smolonska J., et al. (2009). Common and different genetic background for rheumatoid arthritis and coeliac disease. *Hum. Mol. Genet.* 18 4195–4203. 10.1093/hmg/ddp365

19. Concannon P., Onengut-Gumuscu S., Todd J. A., Smyth D. J., Pociot F., Bergholdt R., et al. (2008). A human type 1 diabetes susceptibility locus maps to chromosome 21q22.3. *Diabetes* 57 2858–2861. 10.2337/db08-0753

20. Cooper J. D., Smyth D. J., Smiles A. M., Plagnol V., Walker N. M., Allen J. E., et al. (2008). Meta-analysis of genome-wide association study data identifies additional type 1 diabetes risk loci. *Nat. Genet.* 40 1399–1401. 10.1038/ng.249

21. Cordell H. J. (2009). Detecting gene-gene interactions that underlie human diseases. *Nat. Rev. Genet.* 10 392–404. 10.1038/nrg2579

22. Cui J., Taylor K. E., Destefano A. L., Criswell L. A., Izmailova E. S., Parker A., et al. (2009). Genome-wide association study of determinants of anti-cyclic citrullinated peptide antibody titer in adults with rheumatoid arthritis. *Mol. Med.*15 136–143. 10.2119/molmed.2009.00008

23. Delous M., Baala L., Salomon R., Laclef C., Vierkotten J., Tory K., et al. (2007).The ciliary gene RPGRIP1L is mutated in cerebello-oculo-renal syndrome (Joubert syndrome type B) and Meckel syndrome. *Nat. Genet.* 39 875–881. 10.1038/ng2039

24. Dessinioti C., Antoniou C., Katsambas A., Stratigos A. J. (2011). Melanocortin 1 receptor variants: functional role and pigmentary associations. *Photochem. Photobiol.* 87 978–987. 10.1111/j.1751-1097.2011.00970.x

25. Devadas S., Das J., Liu C., Zhang L., Roberts A. I., Pan Z., et al. (2006).Granzyme B is critical for T cell receptor-induced cell death of type 2 helper T cells. *Immunity* 25 237–247. 10.1016/j.immuni.2006.06.011

26. Devalliere J., Charreau B. (2011). The adaptor Lnk (SH2B3): an emerging regulator in vascular cells and a link between immune and inflammatory signaling. *Biochem. Pharmacol.* 82 1391–1402. 10.1016/j.bcp.2011.06.023

27. Donn R., Ellison S., Lamb R., Day T., Baildam E., Ramanan A. V. (2008). Genetic loci contributing to hemophagocytic lymphohistiocytosis do not confer susceptibility to systemic-onset juvenile idiopathic arthritis. *Arthritis Rheum.* 58869–874. 10.1002/art.23270

28. Fain P. R., Gowan K., LaBerge G. S., Alkhateeb A., Stetler G. L., Talbert J., et al. (2003). A genomewide screen for generalized vitiligo: confirmation of AIS1 on chromosome 1p31 and evidence for additional susceptibility loci. *Am. J. Hum. Genet.* 72 1560–1564. 10.1086/375451

29. Feng B. J., Sun L. D., Soltani-Arabshahi R., Bowcock A. M., Nair R. P., Stuart P., et al. (2009). Multiple Loci within the major histocompatibility complex confer risk of psoriasis. *PLoS Genet* 5:14 10.1371/journal.pgen.1000606

30. Ferrara T. M., Jin Y., Gowan K., Fain P. R., Spritz R. A. (2013). Risk of generalized vitiligo is associated with the common 55R-94A-247H variant haplotype of GZMB (encoding granzyme B). *J. Invest. Dermatol.* 2013 Jun;133(6):1677-9 10.1038/jid.2013.5.

31. Garcia-Lozano J. R., Torres B., Fernandez O., Orozco G., Alvarez-Marquez A., Garcia A., et al. (2007). Caspase 7 influences susceptibility to rheumatoid arthritis. *Rheumatology* 46 1243–1247. 10.1093/rheumatology/kem096

32. Garcia-Melendez M. E., Salinas-Santander M., Sanchez-Dominguez C., Gonzalez-Cardenas H., Cerda-Flores R. M., Ocampo-Candiani J., et al. (2014). Protein tyrosine phosphatase PTPN22 +1858C/T polymorphism is associated with active vitiligo. *Exp. Ther. Med.* 8 1433–1437.

33. Garraway L. A., Widlund H. R., Rubin M. A., Getz G., Berger A. J., Ramaswamy S., et al. (2005). Integrative genomic analyses identify MITF as a lineage survival oncogene amplified in malignant melanoma. *Nature* 436 117–122. 10.1038/nature03664

34. Gateva V., Sandling J. K., Hom G., Taylor K. E., Chung S. A., Sun X., et al. (2009). A large-scale replication study identifies TNIP1, PRDM1, JAZF1, UHRF1BP1 and IL10 as risk loci for systemic lupus erythematosus. *Nat. Genet.*41 1228–1233. 10.1038/ng.468

35. Hakonarson H., Qu H. Q., Bradfield J. P., Marchand L., Kim C. E., Glessner J. T., et al. (2008). A novel susceptibility locus for type 1 diabetes on Chr12q13 identified by a genome-wide association study. *Diabetes Metab. Res. Rev.* 571143–1146. 10.2337/db07-1305

36. Harlalka G. V., Baple E. L., Cross H., Kuhnle S., Cubillos-Rojas M., Matentzoglu K., et al. (2013). Mutation of HERC2 causes developmental delay with Angelman-like features. *J. Med. Genet.* 50 65–73. 10.1136/jmedgenet-2012-101367

37. Hawkes J. E., Cassidy P. B., Manga P., Boissy R. E., Goldgar D., Cannon-Albright L., et al. (2013). Report of a novel OCA2 gene mutation and an investigation of OCA2 variants on melanoma risk in a familial melanoma pedigree. *J. Dermatol. Sci.* 69 30–37. 10.1016/j.jdermsci.2012.09.016

38. He C., Hamon S., Li D., Barral-Rodriguez S., Ott J. (2009). MHC fine mapping of human type 1 diabetes using the T1DGC data. *Diabetes. Obes. Metab.* 1 53–59. 10.1111/j.1463-1326.2008.01003.x

39. Hernandez J., Aung S., Redmond W. L., Sherman L. A. (2001). Phenotypic and functional analysis of CD8(+) T cells undergoing peripheral deletion in response to cross-presentation of self-antigen. *J. Exp. Med.* 194 707–717. 10.1084/jem.194.6.707

40. Hinks A., Ke X., Barton A., Eyre S., Bowes J., Worthington J., et al. (2009).Association of the IL2RA/

CD25 gene with juvenile idiopathic arthritis. *Arthritis Rheum.* 60 251–257. 10.1002/art.24187

41. Honda Y., Okubo Y., Koga M. (1997). Relationship between levels of soluble interleukin-2 receptors and the types and activity of vitiligo. *J. Dermatol.* 24 561–563. 10.1111/j.1346-8138.1997.tb02292.x

42. Jakobsen C., Cleynen I., Andersen P. S., Vermeire S., Munkholm P., Paerregaard A., et al. (2014). Genetic susceptibility and genotype-phenotype association in 588 Danish children with inflammatory bowel disease. *J. Crohns Colitis* 8 678–685. 10.1016/j.crohns.2013.12.010

43. Jia S., Meng A. (2007). Tob genes in development and homeostasis. *Dev. Dyn.* 236913–921. 10.1002/dvdy.21092

44. Jin T. B., Ren Y., Shi X., Jiri M., He N., Feng T., et al. (2015). Genetic variations in the CLNK gene and ZNF518B gene are associated with gout in case-control sample sets. *Rheumatol. Int.* 35 1141–1147. 10.1007/s00296-015-3215-3

45. Jin Y., Birlea S. A., Fain P. R., Ferrara T. M., Ben S., Riccardi S. L., et al. (2012a).Genome-wide association analyses identify 13 new susceptibility loci for generalized vitiligo. *Nat. Genet.* 44 676–680. 10.1038/ng.2272

46. Jin Y., Ferrara T., Gowan K., Holcomb C., Rastrou M., Erlich H. A., et al. (2012b).Next-generation DNA re-sequencing identifies common variants of TYR and HLA-A that modulate the risk of generalized vitiligo via antigen presentation. *J. Invest. Dermatol.* 132 1730–1733. 10.1038/jid.2012.37

47. Jin Y., Birlea S. A., Fain P. R., Gowan K., Riccardi S. L., Holland P. J., et al. (2010a). Variant of TYR and autoimmunity susceptibility loci in generalized vitiligo. *N. Engl. J. Med.* 362 1686–1697. 10.1056/NEJMoa0908547

48. Jin Y., Birlea S. A., Fain P. R., Mailloux C. M., Riccardi S. L., Gowan K., et al. (2010b). Common variants in FOXP1 are associated with generalized vitiligo. *Nat. Genet.* 42 576–578. 10.1038/ng.602

49. Jin Y., Riccardi S. L., Gowan K., Fain P. R., Spritz R. A. (2010c). Fine-mapping of vitiligo susceptibility loci on chromosomes 7 and 9 and interactions with NLRP1 (NALP1). *J. Invest. Dermatol.* 130 774–783. 10.1038/jid.2009.273

50. Jin Y., Birlea S. A., Fain P. R., Gowan K., Riccardi S. L., Holland P. J., et al. (2011). Genome-wide analysis identifies a quantitative trait locus in the MHC class II region associated with generalized vitiligo age of onset. *J. Invest. Dermatol.* 131 1308–1312. 10.1038/jid.2011.12

51. Kato H., Takeuchi O., Sato S., Yoneyama M., Yamamoto M., Matsui K., et al. (2006). Differential roles of MDA5 and RIG-I helicases in the recognition of RNA viruses. *Nature* 441 101–105. 10.1038/nature04734

52. Kingo K., Aunin E., Karelson M., Philips M. A., Ratsep R., Silm H., et al. (2007).Gene expression analysis of melanocortin system in vitiligo. *J. Dermatol. Sci.* 48113–122. 10.1016/j.jdermsci.2007.06.004

53. Kucuksezer U. C., Duymaz-Tozkir J., Gul A., Saruhan-Direskeneli G. (2009). No association of granzyme B gene polymorphism with Behcet's disease. *Clin. Exp. Rheumatol.* 27 (2 Suppl 53), S102.

54. Kumeta H., Sakakibara H., Enokizono Y., Ogura K., Horiuchi M., Matsumoto M., et al. (2014). The N-terminal domain of TIR domain-containing adaptor molecule-1, TICAM-1. *J. Biomol. NMR* 58 227–230. 10.1007/s10858-014-9819-1

55. Kyogoku C., Langefeld C. D., Ortmann W. A., Lee A., Selby S., Carlton V. E., et al. (2004). Genetic association of the R620W polymorphism of protein tyrosine phosphatase PTPN22 with human SLE. *Am. J. Hum. Genet.* 75 504–507. 10.1086/423790

56. Laberge G. S., Birlea S. A., Fain P. R., Spritz R. A. (2008). The PTPN22-1858C>T (R620W) functional polymorphism is associated with generalized vitiligo in the Romanian population. *Pigment Cell Melanoma Res.* 21 206–208. 10.1111/j.1755-148X.2008.00443.x

57. Lamkanfi M., Kanneganti T. D. (2010). Caspase-7: a protease involved in apoptosis and inflammation. *Int. J. Biochem. Cell Biol.* 42 21–24. 10.1016/j.biocel.2009.09.013

58. Le Borgne M., Etchart N., Goubier A., Lira S. A., Sirard J. C., van Rooijen N., et al. (2006). Dendritic cells rapidly recruited into epithelial tissues via CCR6/CCL20 are responsible for CD8+ T cell crosspriming in vivo. *Immunity* 24 191–201. 10.1016/j.immuni.2006.01.005

59. Li Y., Liao W., Cargill M., Chang M., Matsunami N., Feng B. J., et al. (2010).Carriers of rare missense variants in IFIH1 are protected from psoriasis. *J. Invest. Dermatol.* 130 2768–2772. 10.1038/jid.2010.214

60. Liang Y., Yang S., Zhou Y., Gui J., Ren Y., Chen J., et al. (2007). Evidence for two susceptibility loci on chromosomes 22q12 and 6p21-p22 in Chinese generalized vitiligo families. *J. Invest. Dermatol.* 127 2552–2557. 10.1038/sj.jid.5700904

61. Lill C. M., Schjeide B. M., Graetz C., Ban M., Alcina A., Ortiz M. A., et al. (2013). MANBA, CXCR5, SOX8, RPS6KB1 and ZBTB46 are genetic risk loci for multiple sclerosis. *Brain* 136 1778–1782. 10.1093/brain/awt101

62. Liu J., Ni J., Leng R. X., Pan H. F., Ye D. Q. (2015). Association of UBASH3A gene polymorphisms and systemic lupus erythematosus in a Chinese population. *Gene* 565 116–121. 10.1016/j.gene.2015.04.005

63. Liu J., Tang H., Zuo X., Liang B., Wang P., Sun L., et al. (2012). A single nucleotide polymorphism rs9468925 of MHC region is associated with clinical features of generalized vitiligo in Chinese Han population. *J. Eur. Acad. Dermatol. Venereol.* 26 1137–1141. 10.1111/j.1468-3083.2011.04259.x

64. Liu J. B., Li M., Chen H., Zhong S. Q., Yang S., Du W. D., et al. (2007).Association of vitiligo with HLA-A2: a meta-analysis. *J. Eur. Acad. Dermatol. Venereol.* 21 205–213. 10.1111/j.1468-3083.2006.01899.x

65. Maier S., Paulsson M., Hartmann U. (2008). The widely expressed extracellular matrix protein SMOC-2 promotes keratinocyte attachment and migration. *Exp. Cell Res.* 314 2477–2487. 10.1016/j.yexcr.2008.05.020

66. Mandelcorn-Monson R. L., Shear N. H., Yau E., Sambhara S., Barber B. H., Spaner D., et al. (2003). Cytotoxic T lymphocyte reactivity to gp100, MelanA/MART-1, and tyrosinase, in HLA-A2-positive vitiligo patients. *J. Invest. Dermatol.* 121 550–556. 10.1046/j.1523-1747.2003.12413.x

67. Manolio T. A., Collins F. S., Cox N. J., Goldstein D. B., Hindorff L. A., Hunter D. J., et al. (2009). Finding the missing heritability of complex diseases. *Nature* 461747–753. 10.1038/nature08494

68. Mansha M., Carlet M., Ploner C., Gruber G., Wasim M., Wiegers G. J., et al. (2010). Functional analyses of Src-like adaptor (SLA), a glucocorticoid-regulated gene in acute lymphoblastic leukemia. *Leuk. Res.* 34 529–534. 10.1016/j.leukres.2009.06.029

69. Martinez A., Santiago J. L., Cenit M. C., de Las Heras V., de la Calle H., Fernandez-Arquero M., et al. (2008). IFIH1-GCA-KCNH7 locus: influence on multiple sclerosis risk. *Eur. J. Hum. Genet.* 16 861–864. 10.1038/ejhg.2008.16

70. Matsuba R., Sakai K., Imamura M., Tanaka Y., Iwata M., Hirose H., et al. (2015).Replication study in a japanese population to evaluate the association between 10 SNP loci, identified in european genome-wide association studies, and type 2 diabetes. *PLoS ONE* 10:e0126363 10.1371/journal.pone.0126363

71. McGlinchey R. P., Shewmaker F., McPhie P., Monterroso B., Thurber K., Wickner R. B. (2009). The repeat domain of the melanosome fibril protein Pmel17 forms the amyloid core promoting melanin synthesis. *Proc. Natl. Acad. Sci. U.S.A.* 10613731–13736. 10.1073/pnas.0906509106

72. Mohammed G. F., Gomaa A. H., Al-Dhubaibi M. S. (2015). Highlights in pathogenesis of vitiligo. *World J. Clin. Cases* 3 221–230. 10.12998/wjcc.v3.i3.221

73. Molho-Pessach V., Lerer I., Abeliovich D., Agha Z., Abu Libdeh A., Broshtilova V., et al. (2008). The H syndrome is caused by mutations in the nucleoside transporter hENT3. *Am. J. Hum. Genet.* 83 529–534. 10.1016/j.ajhg.2008.09.013

74. Na G. Y., Lee K. H., Kim M. K., Lee S. J., Kim D. W., Kim J. C. (2003).Polymorphisms in the melanocortin-1 receptor (MC1R) and agouti signaling protein (ASIP) genes in Korean vitiligo patients. *Pigment. Cell Res.* 16 383–387. 10.1034/j.1600-0749.2003.00062.x

75. Orozco G., Eerligh P., Sanchez E., Zhernakova S., Roep B. O., Gonzalez-Gay M. A., et al. (2005). Analysis of a functional BTNL2 polymorphism in type 1 diabetes, rheumatoid arthritis, and systemic lupus erythematosus. *Hum. Immunol.*66 1235–1241. 10.1016/j.humimm.2006.02.003

76. Pan F., Yu H., Dang E. V., Barbi J., Pan X., Grosso J. F., et al. (2009). Eos mediates Foxp3-dependent gene silencing in CD4+ regulatory T cells. *Science* 3251142–1146. 10.1126/science.1176077

77. Peach R. J., Bajorath J., Naemura J., Leytze G., Greene J., Aruffo A., et al. (1995).Both extracellular immunoglobin-like domains of CD80 contain residues critical for binding T cell surface receptors CTLA-4 and CD28. *J. Biol. Chem.* 270 21181–21187. 10.1074/jbc.270.36.21181

78. Petukhova L., Duvic M., Hordinsky M., Norris D., Price V., Shimomura Y., et al. (2010). Genome-wide association study in alopecia areata implicates both innate and adaptive immunity. *Nature* 466 113–117. 10.1038/nature09114

79. Puffenberger E. G., Jinks R. N., Wang H., Xin B., Fiorentini C., Sherman E. A., et al. (2012). A homozygous missense mutation in HERC2 associated with global developmental delay and autism spectrum disorder. *Hum. Mutat.* 33 1639–1646. 10.1002/humu.22237

80. Qi X. W., Xia S. H., Yin Y., Jin L. F., Pu Y., Hua D., et al. (2014). Expression features of CXCR5 and its ligand, CXCL13 associated with poor prognosis of advanced colorectal cancer. *Eur. Rev. Med. Pharmacol. Sci.* 18 1916–1924.

81. Quan C., Ren Y. -Q., Xiang L. -H., Sun L. -D., Xu A. -E., Gao X. -H., et al. (2010). Genome-wide association study for vitiligo identifies susceptibility loci at 6q27 and the MHC. *Nat. Genet.* 42 614–618. 10.1038/ng.603

82. Ramos P. S., Williams A. H., Ziegler J. T., Comeau M. E., Guy R. T., Lessard C. J., et al. (2011). Genetic analyses of interferon pathway-related genes reveal multiple new loci associated with systemic lupus erythematosus. *Arthritis Rheum.*63 2049–2057. 10.1002/art.30356

83. Ren Y., Yang S., Xu S., Gao M., Huang W., Gao T., et al. (2009). Genetic variation of promoter sequence modulates XBP1 expression and genetic risk for vitiligo.*PLoS Genet* 5:e1000523 10.1371/journal.pgen.100052319

84. Rezaei N., Gavalas N. G., Weetman A. P., Kemp E. H. (2007). Autoimmunity as an aetiological factor in vitiligo. *J. Eur. Acad. Dermatol. Venereol.* 21 865–876. 10.1111/j.1468-3083.2007.02228.x

85. Rimoldi V., Straniero L., Asselta R., Mauri L., Manfredini E., Penco S., et al. (2014). Functional characterization of two novel splicing mutations in the OCA2 gene associated with oculocutaneous albinism type II. *Gene* 537 79–84. 10.1016/j.gene.2013.11.102

86. Roychoudhuri R., Hirahara K., Mousavi K., Clever D., Klebanoff C. A., Bonelli M., et al. (2013). BACH2 represses effector programs to stabilize T(reg)-mediated immune homeostasis. *Nature* 498 506–510. 10.1038/nature12199

87. Sasaki S., Ito E., Toki T., Maekawa T., Kanezaki R., Umenai T., et al. (2000).Cloning and expression of human B cell-specific transcription factor BACH2 mapped to chromosome 6q15. *Oncogene* 19 3739–3749. 10.1038/sj.onc.1203716

88. Schutyser E., Struyf S., Van Damme J. (2003). The CC chemokine CCL20 and its receptor CCR6. *Cytokine Growth Factor. Rev.* 14 409–426. 10.1016/S1359-6101(03)00049-2

89. Seya T., Matsumoto M., Ebihara T., Oshiumi H. (2009). Functional evolution of the TICAM-1 pathway for extrinsic RNA sensing. *Immunol. Rev.* 227 44–53. 10.1111/j.1600-065X.2008.00723.x

90. Simmonds M. J., Heward J. M., Barrett J. C., Franklyn J. A., Gough S. C. (2006).Association of the BTNL2 rs2076530 single nucleotide polymorphism with Graves' disease appears to be secondary to DRB1 exon 2 position beta74. *Clin. Endocrinol.* 65 429–432. 10.1111/j.1365-2265.2006.02586.x

91. Smith A. G., Sturm R. A. (2010). Multiple genes and locus interactions in susceptibility to vitiligo. *J. Invest. Dermatol.* 130 643–645. 10.1038/jid.2009.403

92. Smyth D. J., Cooper J. D., Bailey R., Field S., Burren O., Smink L. J., et al. (2006). A genome-wide association study of nonsynonymous SNPs identifies a type 1 diabetes locus in the interferon-induced helicase (IFIH1) region. *Nat. Genet.*38 617–619. 10.1038/ng1800

93. Song G. G., Kim J. H., Lee Y. H. (2013). The CTLA-4 +49 A/G, CT60 A/G and PTPN22 1858 C/T polymorphisms and susceptibility to vitiligo: a meta-analysis.*Mol. Biol. Rep.* 40 2985–2993. 10.1007/s11033-012-2370-9

94. Spritz R. A. (2008). The genetics of generalized vitiligo. *Curr. Dir. Autoimmun.* 10244–257. 10.1159/000131501

95. Spritz R. A. (2010). The genetics of generalized vitiligo: autoimmune pathways and an inverse relationship with malignant melanoma. *Genome Med* 2 7810.1186/gm199

96. Spritz R. A. (2012). Six decades of vitiligo genetics: genome-wide studies provide insights into autoimmune pathogenesis. *J. Invest. Dermatol.* 132 268–273. 10.1038/jid.2011.321

97. Spritz R. A., Gowan K., Bennett D. C., Fain P. R. (2004). Novel vitiligo susceptibility loci on chromosomes 7 (AIS2) and 8 (AIS3), confirmation of SLEV1 on chromosome 17, and their roles in an autoimmune diathesis.*Am. J. Hum. Genet.* 4 188–191. 10.1086/381134

98. Stüttgen G. (1950). Die Vitilgio in erbbiologischer Betrachtung. *Z. Haut. Geschlenkskr.* 9 451–456.

99. Sun X., Xu A., Wei X., Ouyang J., Lu L., Chen M., et al. (2006). Genetic epidemiology of vitiligo: a study of 815 probands and their families from south China. *Int. J. Dermatol.* 45 1176–1181. 10.1111/j.1365-4632.2006.02907.x

100. Sun Y., Zuo X., Zheng X., Zhou F., Liang B., Liu H., et al. (2014). A comprehensive association analysis confirms ZMIZ1 to be a susceptibility gene for vitiligo in Chinese population. *J. Med. Genet.* 51 345–353. 10.1136/jmedgenet-2013-102233

101. Sutherland A., Davies J., Owen C. J., Vaikkakara S., Walker C., Cheetham T. D., et al. (2007). Genomic polymorphism at the interferon-induced helicase (IFIH1) locus contributes to Graves' disease susceptibility. *J. Clin. Endocrinol. Metab.* 923338–3341. 10.1210/jc.2007-0173

102. Szell M., Baltas E., Bodai L., Bata-Csorgo Z., Nagy N., Dallos A., et al. (2008).The Arg160Trp allele of melanocortin-1 receptor gene might protect against vitiligo. *Photochem. Photobiol.* 84 565–571. 10.1111/j.1751-1097.2008.00296.x

103. Tang X. F., Zhang Z., Hu D. Y., Xu A. E., Zhou H. S., Sun L. D., et al. (2013).Association analyses identify three susceptibility Loci for vitiligo in the Chinese Han population. *J. Invest. Dermatol.* 133 403–410. 10.1038/jid.2012.320

104. Tastan H. B., Akar A., Orkunoglu F. E., Arca E., Inal A. (2004). Association of HLA class I antigens and HLA class II alleles with vitiligo in a Turkish population. *Pigment. Cell Res.* 17 181–184. 10.1111/j.1600-0749.2004.00141.x

105. Teindel H. (1950). Familiäre. *Z. Haut. Geschlenkskr.* 9 456–462.

106. Thompson D. M., Parker R. (2009). The RNase Rny1p cleaves tRNAs and promotes cell death during oxidative stress in Saccharomyces cerevisiae. *J. Cell Biol.* 185 43–50. 10.1083/jcb.200811119

107. Tomer Y., Greenberg D. (2004). The thyroglobulin gene as the first thyroid-specific susceptibility gene for autoimmune thyroid disease. *Trends Mol. Med.* 10306–308. 10.1016/j.molmed.2004.05.002

108. Trapani J. A., Sutton V. R. (2003). Granzyme B: pro-apoptotic, antiviral and antitumor functions. *Curr. Opin. Immunol.* 15 533–543. 10.1016/S0952-7915(03)00107-9

109. Vella A., Cooper J. D., Lowe C. E., Walker N., Nutland S., Widmer B., et al. (2005). Localization of a type 1 diabetes locus in the IL2RA/CD25 region by use of tag single-nucleotide polymorphisms. *Am. J. Hum. Genet.* 76 773–779. 10.1086/429843

110. Wang J., Wicker L. S., Santamaria P. (2009). IL-2 and its high-affinity receptor: genetic control of immunoregulation and autoimmunity. *Semin. Immunol.* 21 363–371. 10.1016/j.smim.2009.04.004

111. Wang L., Tsai C. C. (2008). Atrophin proteins: an overview of a new class of nuclear receptor corepressors. *Nucl. Recept. Signal.* 6 31.

112. Wang Q., Jiang M., Wu J., Ma Y., Li T., Chen Q., et al. (2014). Stress-induced RNASET2 overexpression mediates melanocyte apoptosis via the TRAF2 pathway in vitro. *Cell Death Dis.* 23 539 10.1038/cddis.2013.539

113. Weston A., Sommerville J. (2006). Xp54 and related (DDX6-like) RNA helicases: roles in messenger RNP assembly, translation regulation and RNA degradation. *Nucleic Acids Res.* 34 3082–3094. 10.1093/nar/gkl409

114. Woo H. M., Park H. J., Park M. H., Kim B. Y., Shin J. W., Yoo W. G., et al. (2014). Identification of CDH23 mutations in Korean families with hearing loss by whole-exome sequencing. *BMC Med. Genet.* 15:46 10.1186/1471-2350-15-46

115. Wu J. N., Koretzky G. A. (2004). The SLP-76 family of adapter proteins. *Semin. Immunol.* 16 379–393. 10.1016/j.smim.2004.08.018

116. Yang S. K., Hong M., Choi H., Zhao W., Jung Y., Haritunians T., et al. (2015).Immunochip analysis identification of 6 additional susceptibility loci for Crohn's disease in Koreans. *Inflamm. Bowel Dis.* 21 1–7. 10.1097/MIB.0000000000000268

117. Yang S. K., Hong M., Zhao W., Jung Y., Baek J., Tayebi N., et al. (2014).Genome-wide association study of Crohn's disease in Koreans revealed three new susceptibility loci and common attributes of genetic

susceptibility across ethnic populations. *Gut* 63 80–87. 10.1136/gutjnl-2013-305193

118. Zhang J., Zhang Y., Yang J., Zhang L., Sun L., Pan H. F., et al. (2014). Three SNPs in chromosome 11q23.3 are independently associated with systemic lupus erythematosus in Asians. *Hum. Mol. Genet.* 23 524–533. 10.1093/hmg/ddt424

119. Zhang X. J., Liu H. S., Liang Y. H., Sun L. D., Wang J. Y., Yang S., et al. (2004a).Association of HLA class I alleles with vitiligo in Chinese Hans. *J. Dermatol. Sci.*35 165–168. 10.1016/j.jdermsci.2004.05.003

120. Zhang X. J., Liu J. B., Gui J. P., Li M., Xiong Q. G., Wu H. B., et al. (2004b).Characteristics of genetic epidemiology and genetic models for vitiligo. *J. Am. Acad. Dermatol.* 51 383–390. 10.1016/j.jaad.2003.12.044

121. Zhang Z., Xu S. X., Zhang F. Y., Yin X. Y., Yang S., Xiao F. L., et al. (2009). The analysis of genetics and associated autoimmune diseases in Chinese vitiligo patients. *Arch. Dermatol. Res.* 301 167–173. 10.1007/s00403-008-0900-z

122. Zhao S. X., Xue L. Q., Liu W., Gu Z. H., Pan C. M., Yang S. Y., et al. (2013).Robust evidence for five new Graves' disease risk loci from a staged genome-wide association analysis. *Hum. Mol. Genet.* 22 3347–3362. 10.1093/hmg/ddt183

123. Zhu K. J., Lv Y. M., Yin X. Y., Wang Z. X., Sun L. D., He S. M., et al. (2011).Psoriasis regression analysis of MHC loci identifies shared genetic variants with vitiligo. *PLoS ONE* 6:18 10.1371/journal.pone.0023089

CHAPTER 19

CHROMOSOMAL AND GENETIC ANALYSIS OF A HUMAN LUNG ADENOCARCINOMA CELL LINE OM

Yong-Wu Li,[1] Lin Bai,[2] Lyu-Xia Dai,[2] Xu He,[2] and Xian-Ping Zhou[2]

[1]Department of Emergency Surgery, First Affiliated Hospital of Jiamusi University, Jiamusi, Heilongjiang 154002, China

[2]Experiment Teaching Center of Clinical Medicine, The First Affiliated Hospital of Chengdu Medical College, Chengdu, Sichuan 610500, China

ABSTRACT

Background

Lung cancer has become the leading cause of death in many regions. Carcinogenesis is caused by the stepwise accumulation of genetic and chromosomal changes. The aim of this study was to investigate the chromosome and gene alterations in the human lung adenocarcinoma cell line OM.

Methods

We used Giemsa banding and multiplex fluorescence in situ hybridization focusing on the human lung adenocarcinoma cell line OM to analyze its chromosome alterations. In addition, the gains and losses in the specific chromosome regions were identified by comparative genomic hybridization (CGH) and the amplifications of cancer-related genes were also detected by polymerase chain reaction (PCR).

Results

We identified a large number of chromosomal numerical alterations on all chromosomes except chromosome X and 19. Chromosome 10 is the most frequently involved in translocations with six different interchromosomal translocations. CGH revealed the gains on chromosome regions of 3q25.3-28, 5p13, 12q22-23.24, and the losses on 3p25-26, 6p25, 6q26-27, 7q34-36, 8p22-23, 9p21-24, 10q25-26.3, 12p13.31-13.33 and 17p13.1-13.3. And PCR showed the amplification of genes: Membrane metalloendopeptidase (MME), sucrase-isomaltase (SI), butyrylcholinesterase (BCHE), and kininogen (KNG).

Citation: Li Y-W, Bai L, Dai L-X, He X, and Zhou X-P. Chromosomal and Genetic Analysis of a Human Lung Adenocarcinoma Cell Line OM. Chinese Medical Journal. 2016; 129(4):405-409. DOI:10.4103/0366-6999.176066.

Conclusions

The lung adenocarcinoma cell line OM exhibited multiple complex karyotypes, and chromosome 10 was frequently involved in chromosomal translocation, which may play key roles in tumorigenesis. We speculated that the oncogenes may be located at 3q25.3-28, 5p13, 12q22-23.24, while tumor suppressor genes may exist in 3p25-26, 6p25, 6q26-27, 7q34-36, 8p22-23, 9p21-24, 10q25-26.3, 12p13.31-13.33, and 17p13.1-13.3. Moreover, at least four genes (MME, SI, BCHE, and KNG) may be involved in the human lung adenocarcinoma cell line OM.

INTRODUCTION

Approximately, 1.6 million new cases of lung cancer are diagnosed each year throughout the world.[1] Both genetic and epigenetic changes contribute to the development of human cancer.[2] Since carcinogenesis is caused by the stepwise accumulation of genetic changes,[3] it is very important to understand the alterations of chromosome and gene in cancer cells. The traditional Giemsa banding (G-banding) technique is applied to detect cell sample with a higher mitotic index, but lung cancer cells have a lower mitotic index,[4] which makes karyotypic analysis difficult. In our study, combining with the latest molecular biology techniques, we investigated the chromosomal and genetic alterations in the human lung adenocarcinoma cell line OM. Here, we summarized the chromosomal and genetic abnormalities detected in lung cancer cells and discussed the possible implication of those alterations in the processes of tumorigenesis.

METHODS

Cell Line

The lung adenocarcinoma cell line OM was established in the second Department of Biology Kochi Medical School, Kochi University, Japan. The cells in cell line OM were from human lung adenocarcinoma tissues. After cultured and purified, these cells developed into a special cell line for cancer research.

Chromosome Preparation and Giemsa Banding Analysis

The cultures were harvested and G-banding of the chromosomes was performed by hypo-osmolality, fixation, trypsinization, and Giemsa (Invitrogen Corp., Carlsbad, CA, USA) staining. The subsequent karyotype description followed the recommendations of the ISCN (1995).

Multiplex Fluorescence *in situ* Hybridization

The slides after denaturation were dehydrated in graded series of ethanol and air dried. The 24 XCyte mFISH kit (Vysis Corp., USA) with five fluorophores were used for hybridization. Samples were evaluated, and images were captured using an Olympus BX60 fluorescence microscope (Olympus, Tokyo, Japan) and analyzed with the aid of image analysis software (CytoVision, Applied Imaging Corp., Santa Clara, CA, USA).

Comparative Genomic Hybridization

DNA extraction was performed with 2-ProPanol (Sigma-Aldrich, Japan). After washing with ethanol, dried, concentrated, and purified, the sample and reference genomic DNAs were digested and random primed labeled with Cyanine-5/Cyanine 3-dUTP (Vysis Corp., USA). Cohybridization of these DNAs to a comparative genomic hybridization (CGH) arrays was performed for 72 h at 37°C. After washing, the chromosomes were counterstained with DAPI. Array CGH was scanned on Olympus BX60 (OLYMPUS Corp., Japan) and data were extracted and analyzed using Multiscan medical image analysis system (SONY Corp., Tokyo, Japan).

Genomic DNA Polymerase Chain Reaction

Extracted DNA from a normal human was used as the internal control. All the primers were provided by the Invitrogen Corp., (Invitrogen Corp., Carlsbad, CA, USA). The amplified samples were analyzed by electrophoresis at 100 V on 1.5% agarose gels stained with ethidium bromide. The visual of the bands of amplified DNA were captured by a camera in the device.

Statistical Analysis

SAS 9.3 (SAS Institute Inc., Cary, NC, USA) is useful statistical software to analyze this kind of data. An exact 95% confidence interval (*CI*) using the Clopper–Pearson method was given for the positive rate of abnormal chromosomes and genes. Qualitative data were expressed as a frequency. The *CI* value of reference standard was 0.3%.

RESULTS

Karyotype Analysis

Characterizations of complex chromosomal abnormalities in the human lung adenocarcinoma cell line OM were showed by G-banding analysis, and the wrong number of chromosomes was present in almost all cells. Either aneuploid or polyploid karyotype was observed which displayed chromosome instability in the human lung adenocarcinoma cell line OM. G-banding showed chromosomal aberrations on all chromosomes except chromosome X (95% *CI*: 0–64%) and 19 (95% *CI*: 0–64%) [Figure 1].

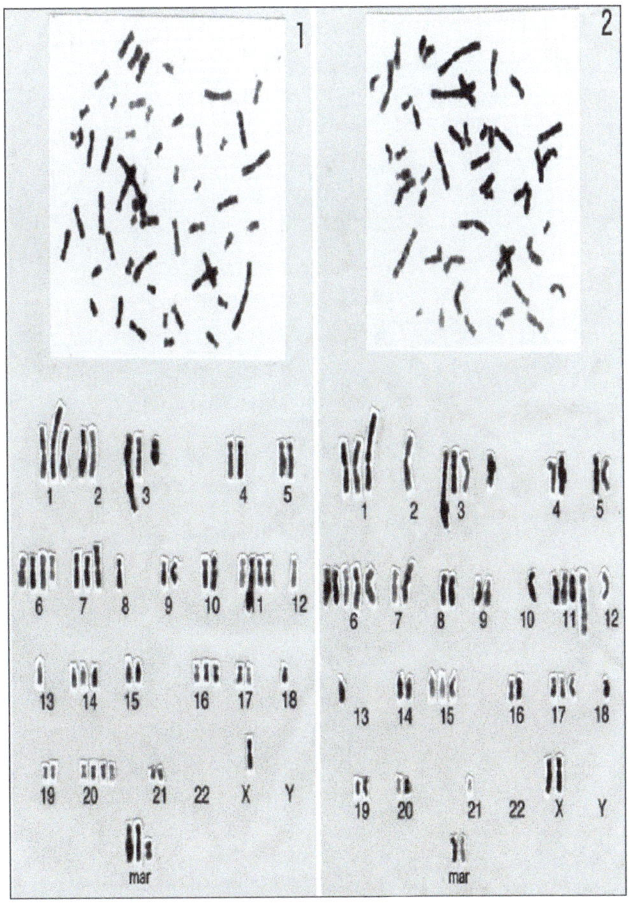

Figure 1. Giemsa-banding analysis revealed chromosomal aberrations on all chromosomes except, 4, 8, 14, 15, 17, and 19. The wrong number of chromosomes was present in almost all cells. Either aneuploid or polyploid karyotype was observed.

Multiplex Fluorescence *in situ* Hybridization Analysis

The structurally abnormal chromosomes were analyzed by multiplex fluorescence *in situ*hybridization (M-FISH). As was illustrated in Figure 2, chromosome 10 is the most frequently (95% *CI*: 29–100%) involved in translocations with six different interchromosomal translocations, and its frequency was 4–6 times in each sample. The other chromosomes involved in translocation were chromosomes 1, 2, 3, 5, 6, 7, 8, 9, 11, 16, and 18, and the 95% *CI* minimum value of each statistical datum was more than 9%.

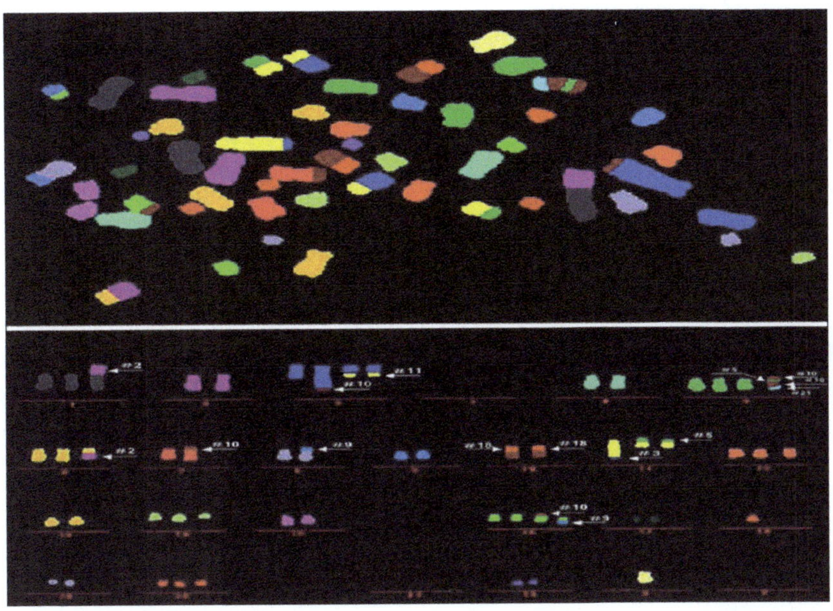

Figure 2. Multiple numerical and structural aberrations were observed in the majority of cells from multiplex fluorescence *in situ* hybridization image. Chromosome 10 is the most frequently involved in translocations with four different interchromosomal translocations. The other chromosomes involved in translocation were chromosomes 2, 5, 9, 11, 18 and 3, 21.

Comparative Genomic Hybridization Analysis

CGH revealed that the gains in the region of chromosome bands 3q25.3-28, 5p13, 12q22-23.24, and the losses in that of 3p25-26, 6p25, 6q26-27, 7q34-36, 8p22-23, 9p21-24, 10q25-26.3, 12p13.31-13.33, and 17p13.1-13.3 [Figure 3]. The 95% *CI* minimum value of each statistical datum was more than 15%.

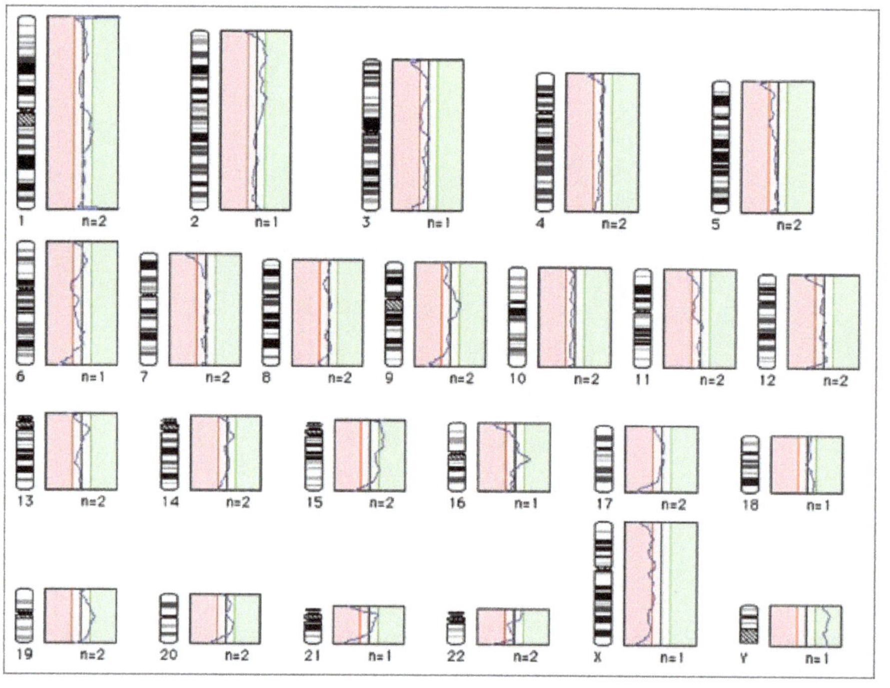

Figure 3. Array comparative genomic hybridization analysis showed partial deletion of in 3p24.1-26, 6p25, 6q25.2-27, 7q32-36, 8p22-23, 9p21-24, 10q25-26.3, 12p13.31-13.33, and 17p131-13.3 together with duplication at 3q24-28, 5p13, and 12q22-23.24.

Polymerase Chain Reaction Analysis of Genomic DNA

According to the results obtained by CGH, gene analysis was performed in the 3q24-28 region. The results of the polymerase chain reaction (PCR) experiments showed that the genes membrane metalloendopeptidase (*MME*) (95% *CI*: 1–88%), sucrase-isomaltase (*SI*) (95% *CI*: 54–100%), butyrylcholinesterase (*BCHE*) (95% *CI*: 16–100%), and kininogen (*KNG*) (95% *CI*: 1–72%) were amplified in this region. However, the amplification of genes Histidine-rich glycoprotein (*HRG*, 95% *CI*: 0–60%) and Mucin (95% *CI*: 0–52%) were not observed [Figure 4].

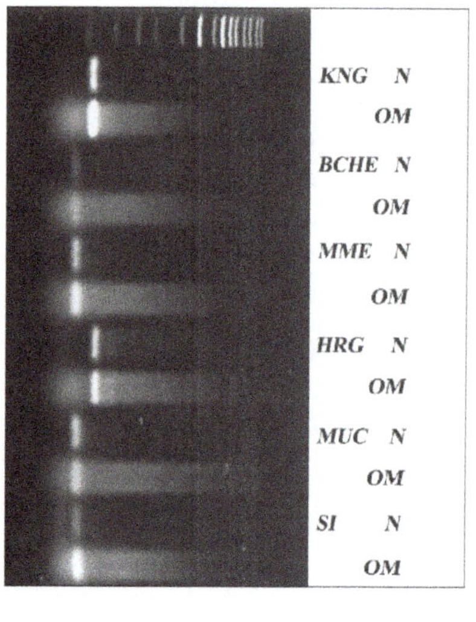

Figure 4. Agarose gel electrophoresis of polymerase chain reaction amplification products from the human lung adenocarcinoma cell line OM. N: Negative control (normal human lung cell line); OM: Human lung adenocarcinoma cell line OM. *MME*: Membrane metalloendopeptidase; *SI*: Sucrase-isomaltase; *BCHE*: Butyrylcholinesterase; *KNG*: Kininogen; *HRG*: Histidine-rich glycoprotein; *MUC*: Mucin.

DISCUSSION

Lung cancer remains the leading cause of cancer-related morbidity and mortality worldwide.[5] Unlike in leukemia research, lung cancer cell has a lower cell mitotic index,[6] which makes karyotypic analysis difficult. Using modern genetic technologies, we investigated the chromosome and gene alterations in human lung adenocarcinoma of cell line OM. The goal of our study is to identify the alterations of chromosome and gene in human lung adenocarcinoma cell line OM, which may provide new insights into the mechanisms of tumorigenesis.

With high-resolution G-banding techniques, both the numerical and structural chromosomal abnormalities were observed. G-banding analysis revealed that the changes in chromosome number were present in most cells, and either aneuploidy or polyploidy was identified within them. In OM cell line, certain chromosomes frequently participated the structural rearrangement, prominent among them being chromosome 3, 6, 7, 9, 10, 11, 18, and Y.

Although G-banding is a fundamental technique for cytogenetic study, it has limited utility in instances of cryptic or very complex rearrangements.[7] M-FISH is a powerful technique that can be used for identifying uniquely all 24 chromosome types of the human genome.[8] The M-FISH image showed that the highest frequency of translocation was in chromosome 10. In genetics, a chromosome translocation is caused by rearrangement of parts between nonhomologous chromosomes, and the distinct classes of chromosomal rearrangements create proto-oncogenes.[9] Loci on chromosome 10 include microseminoprotein beta, which encodes beta-microseminoprotein, a primary constituent of semen and a proposed prostate cancer biomarker.[10] Moreover, multiple tumor-suppressive genes are located in chromosome 10,[11] and the translocations of chromosome 10 may lead to damage of these tumor-suppressive genes. Therefore, frequent translocations of chromosome 10 in cell line OM may be catastrophic genomic events and play key roles in tumorigenesis.

In general, regional gains on chromosome indicate that oncogenes may exist in this region, while the chromosome losses have been interpreted as an evidence that the region has affected a certain tumor suppressor genes.[12] CGH analysis revealed that gain regions were observed on chromosomes 3q25.3-28, 5p13, and 12q22-23.24, which indicated that there may exist oncogenes that were activated in the gain regions. While loss regions were detected in chromosome 3p25-26, 6p25, 6q26-27, 7q34-36, 8p22-23, 9p21-24, 10q25-26.3, 12p13.31-13.33, and 17p13.1-13.3, which indicated that some tumor suppressor genes may be located in these regions. It is well-known that tumor suppressor genes p16 and p53 were located in chromosome 9p21 and 17p13, respectively.[13]

Based on the above results, we detected amplification of genes at chromosome band 3q24-28 by PCR in the cell lines. The genes *BCHE*, *MME*, *KNG*, and *SI* were amplified in the chromosomal region. The *BCHE* gene provides instructions for making the pseudocholinesterase enzyme,[14] which has a variety of physiological effects such as the metabolism of the cocaine and heroin[15] and the breakdown of organophosphorus esters.[14] Recent studies indicated that *BCHE* gene was amplified in the cells of lung carcinomas,[16] which were consistent with our results. Brass *et al.*[17] observed the amplification of genes *BCHE* and *SLC2A2* at 3q26 in squamous cell lung carcinomas, whereas no amplification has been found in genes *MME* and *KNG*. *MME* gene is expressed in both normal and neoplastic cells as a common acute leukemia antigen.[18] In squamous cell lung carcinoma, the high *MME* gene expression was significantly associated with poor overall survival.[19]

There were many studies on the expression of *MME* gene in lung cancer cells; however, these results were quite conflicting. Many factors may be responsible for the difference among the results, such as experimental conditions, experimental methods,[19] and pathological types of lung cancer. The *KNG* gene, which is one of the members of a cystatin-like superfamily, has potential roles in angiogenesis and (or) tumor development.[20] Kinins that derived from *KNG* involved in cell proliferation, leukocyte activation, cell migration, endothelial cell activation, and nociception.[21] Kashuba *et al.*[22] thought that *KNG* may be a novel therapeutic target in chronic lymphocytic leukemia and the possible association with prognosis. There are few reports on the expression of *KNG* gene in lung cancer cell. We observed that *KNG* gene was amplified in lung adenocarcinoma cell line OM, which suggested that *KNG* gene was likely to be involved in tumorigenesis of lung cancer cells. *SI* gene encodes an *SI* enzyme that is expressed in the intestinal brush border.[23] Although there are numerous studies elucidating the role of *SI* gene in the digestion of dietary carbohydrates,[24] there are limited data indicating a specific involvement of *SI* gene in the genesis of human cancers. Our experiments demonstrated that *SI* gene was amplified on chromosome band 3q24-28. Since cancer is attributed to multi-genetic alterations accumulated in the cells, we speculated *SI* gene may be involved in tumorigenesis.In summary, our findings suggested that most of those examined cells exhibited multiple complex karyotypes in human lung adenocarcinoma cell line OM. Chromosome 10 was frequently involved in chromosomal translocation, which may be catastrophic genomic events and play key roles in tumorigenesis. According to the results of CGH, we speculated that the oncogenes may be located at 3q25.3-28, 5p13, and 12q22-23.24, while tumor suppressor genes may exist in 3p25-26, 6p25, 6q26-27, 7q34-36, 8p22-23, 9p21-24, 10q25-26.3, 12p13.31-13.33, and 17p13.1-13.3. In human lung adenocarcinoma cell line OM, at least four genes (*MME*, *SI*, *BCHE*, and *KNG*) were involved in carcinogenesis. It may be helpful to deeply understand the tumorigenesis of lung cancer.

ACKNOWLEDGMENT

The authors thank Professor Shiro Sasaguli and Lecturer Hiroshi Taguchi Yet (the Second Department of Biology Kochi Medical School, Kochi University, Japan) for kindly providing help and guidance.

REFERENCES

1. Ramalingam SS, Owonikoko TK, Khuri FR. Lung cancer: New biological insights and recent therapeutic advances. CA Cancer J Clin. 2011;61:91–112. doi: 10.3322/caac.20102.
2. Sadikovic B, Al-Romaih K, Squire JA, Zielenska M. Cause and consequences of genetic and epigenetic alterations in human cancer. Curr Genomics. 2008;9:394–408.doi: 10.2174/138920208785699580.
3. Hahn WC, Weinberg RA. Rules for making human tumor cells. N Engl J Med.2002;347:1593–603. doi: 10.1056/NEJMra021902.
4. Mitsuuchi Y, Testa JR. Cytogenetics and molecular genetics of lung cancer. Am J Med Genet. 2002;115:183–8. doi: 10.1002/ajmg.10692.
5. Jemal A, Siegel R, Ward E, Hao Y, Xu J, Thun MJ. Cancer statistics, 2009. CA Cancer J Clin. 2009;59:38–42. doi: 10.3322/caac.20006.
6. Testa JR, Liu Z, Feder M, Bell DW, Balsara B, Cheng JQ, et al. Advances in the analysis of chromosome alterations in human lung carcinomas. Cancer Genet Cytogenet.1997;95:20–32. doi: 10.1038/sj.ejhg.5200659.
7. Varella-Garcia M. Chromosomal and genomic changes in lung cancer. Cell Adh Migr.2010;4:100–6. doi: 10.4161/cam.4.1.10884.
8. Li J, Lin D, Cao H, Wang YP. An improved sparse representation model with structural information for multicolour fluorescence *in-situ* hybridization (M-FISH) image classification. BMC Syst Biol. 2013;7(Suppl 4):S5. doi: 10.1186/1752-0509-7-S4-S5.

9. Bishop JM. The molecular genetics of cancer. Science. 1987;235:305–11. doi: 10.1126/science.3541204.

10. Thomas G, Jacobs KB, Yeager M, Kraft P, Wacholder S, Orr N, et al. Multiple loci identified in a genome-wide association study of prostate cancer. Nat Genet.2008;40:310–5. doi: 10.1038/ng.91.

11. Kwong LN, Chin L. Chromosome 10, frequently lost in human melanoma, encodes multiple tumor-suppressive functions. Cancer Res. 2014;74:1814–21. doi: 10.1158/0008-5472.CAN-13-1446.

12. Marsit CJ, Wiencke JK, Nelson HH, Kim DH, Hinds PW, Aldape K, et al. Alterations of 9p in squamous cell carcinoma and adenocarcinoma of the lung: Association with smoking, TP53, and survival. Cancer Genet Cytogenet. 2000;162:115–21. doi: 10.1016/j.cancergencyto.2005.04.001.

13. Yoo WJ, Cho SH, Lee YS, Park GS, Kim MS, Kim BK, et al. Loss of heterozygosity on chromosomes 3p, 8p, 9p and 17p in the progression of squamous cell carcinoma of the larynx. J Korean Med Sci. 2004;19:345–51. doi: 10.3346/jkms.2004.19.3.345.

14. Lockridge O. Review of human butyrylcholinesterase structure, function, genetic variants, history of use in the clinic, and potential therapeutic uses. Pharmacol Ther.2015;148:34–46. doi: 10.1016/j.pharmthera.2014.11.011.

15. Munshaw S, Hwang HS, Torbenson M, Quinn J, Hansen KD, Astemborski J, et al. Laser captured hepatocytes show association of butyrylcholinesterase gene loss and fibrosis progression in hepatitis C-infected drug users. Hepatology. 2012;56:544–54. doi: 10.1002/hep.25655.

16. Rácz A, Brass N, Heckel D, Pahl S, Remberger K, Meese E. Expression analysis of genes at 3q26-q27 involved in frequent amplification in squamous cell lung carcinoma. Eur J Cancer. 1999;35:641–6. doi: 10.1016/S0959-8049(98)00419-5.

17. Brass N, Rácz A, Heckel D, Remberger K, Sybrecht GW, Meese EU. Amplification of the genes BCHE and SLC2A2 in 40% of squamous cell carcinoma of the lung. Cancer Res. 1997;57:2290–4.

18. Gürel D, Kargi A, Karaman I, Onen A, Unlü M. CD10 expression in epithelial and stromal cells of non-small cell lung carcinoma (NSCLC): A clinic and pathologic correlation. Pathol Oncol Res. 2012;18:153–60. doi: 10.1007/s12253-011-9421-8.

19. Leithner K, Wohlkoenig C, Stacher E, Lindenmann J, Hofmann NA, Gallé B, et al. Hypoxia increases membrane metallo-endopeptidase expression in a novel lung cancer*ex vivo* model – Role of tumor stroma cells. BMC Cancer. 2014;14:40. doi: 10.1186/1471-2407-14-40.

20. Hsu SJ, Nagase H, Balmain A. Identification of fetuin-B as a member of a cystatin-like gene family on mouse chromosome 16 with tumor suppressor activity. Genome.2004;47:931–46. doi: 10.1139/g04-043.

21. da Costa PL, Sirois P, Tannock IF, Chammas R. The role of kinin receptors in cancer and therapeutic opportunities. Cancer Lett. 2014;345:27–38. doi: 10.1016/j.canlet.2013.12.009.

22. Kashuba E, Eagle GL, Bailey J, Evans P, Welham KJ, Allsup D, et al. Proteomic analysis of B-cell receptor signaling in chronic lymphocytic leukaemia reveals a possible role for kininogen. J Proteomics. 2013;91:478–85. doi: 10.1016/j.jprot.2013.08.002.

23. Nichols BL, Avery S, Sen P, Swallow DM, Hahn D, Sterchi E. The maltase-glucoamylase gene: Common ancestry to sucrase-isomaltase with complementary starch digestion activities. Proc Natl Acad Sci U S A. 2003;100:1432–7. doi: 10.1073/pnas.0237170100.

24. Diaz-Sotomayor M, Quezada-Calvillo R, Avery SE, Chacko SK, Yan LK, Lin AH, et al. Maltase-glucoamylase modulates gluconeogenesis and sucrase-isomaltase dominates starch digestion glucogenesis. J Pediatr Gastroenterol Nutr. 2013;57:704–12. doi: 10.1097/MPG.0b013e3182a27438.

CHAPTER 20

A SUMMARY OF MOLECULAR GENETIC FINDINGS IN FRUCTOSE-1, 6-BISPHOSPHATASE DEFICIENCY WITH A FOCUS ON A COMMON LONG-RANGE DELETION AND THE ROLE OF MLPA ANALYSIS

René Santer[1], Marcel du Moulin[1], Tatevik Shahinyan[2], Inga Vater[3], Esther Maier[4], Ania C. Muntau[1], and Beat Steinmann[5]

[1]Department of Pediatrics, University Medical Center Hamburg-Eppendorf, Martinistraße 52, D - 20246 Hamburg, Germany

[2]Arabkir Institute of Child and Adolescent Health, Yerevan, Armenia

[3]Institute of Human Genetics, University of Kiel, Kiel, Germany

[4]Department of Pediatrics, University Children's Hospital, Munich, Germany

[5]Department of Pediatrics, University of Zurich, Zurich, Switzerland

ABSTRACT

Background

Fructose-1,6-bisphosphatase deficiency is a rare inborn error of metabolism affecting gluconeogenesis with only sporadic reports on its molecular genetic basis.

Results

We report our experience with mutation analysis in 14 patients (13 families) with fructose-1,6-bisphosphatase deficiency using conventional Sanger sequencing and multiplex ligation-dependent probe amplification analysis, and we provide a mutation update for the fructose bisphosphatase-1 gene (*FBP1*). Mutations were found on both chromosomes in all of our 14 patients including 5 novel mutations. Among the novel mutations is a 5412-bp deletion (c.-24-26_170+5192del) including the entire coding sequence of exon 2 of *FBP1* that was repeatedly found in patients from Turkey and Armenia which may explain earlier poorly defined findings in patients from this area. This deletion can be detected with specific primers by generation of a junction fragment and by MLPA and SNP array assays. MLPA analysis was able to detect copy number variations in two further patients, one heterozygous for a deletion within exon 8, another heterozygous for a novel deletion of the entire *FBP1* gene.

Citation: René Santer, Marcel du Moulin, Tatevik Shahinyan, Inga Vater, Esther Maier, Ania C. Muntau, and Beat Steinmann. A summary of molecular genetic findings in fructose-1,6-bisphosphatase deficiency with a focus on a common long-range deletion and the role of MLPA analysis. Orphanet Journal of Rare Diseases. 2016; 11:44. DOI:10.1186/s13023-016-0415-1.

Conclusions

Based on our update for the *FBP1* gene, currently listing 35 mutations worldwide, and knowledge of PCR conditions that allow simple detection of a common *FBP1* deletion in the Armenian and Turkish population, molecular genetic diagnosis has become easier in FBP1 deficiency. Furthermore, MLPA analysis may plays a useful role in patients with this disorder.

BACKGROUND

Fructose-1,6-bisphosphatase (FBP1) deficiency [OMIM: 229700], first described in 1970 [1], is an inborn error of gluconeogenesis. Patients present with ketotic hypoglycemia and lactic acidosis triggered by catabolic episodes such as prolonged fasting and/or febrile infections [2]. Laboratory findings may include hyperalaninemia, hyperketonemia, and an increased lactate/pyruvate ratio, an elevated plasma concentration of uric acid, glyceroluria [2], and pseudo-hypertriglyceridemia [3]. FBP1 deficiency is generally believed to be very rare with an estimated incidence between 1: 350,000 and <1 : 900,000 in the Dutch and French population, respectively [4, 5]; but it may be more frequent in populations with a higher rate of consanguinity.

FBP1 deficiency is inherited in an autosomal recessive way. It is caused by mutations within the *FBP1* gene (OMIM: 611570) which spans approx. 31 kb on chromosome 9q22.2-q22.3 and consists of 8 exons [6]. To date, only a small number of mutations has been published and among them, c.959dupG has been reported to be the most common one in Caucasians but also in patients from Japan and China [5, 7–9].

We report results of mutation analysis of our laboratory, describe how we have characterized a common exon 2 deletion detected in patients with Turkish or Armenian ethnic background, and provide PCR conditions for verification of this deletion which is otherwise not detectable by standard sequencing techniques. Finally, we show for the first time that MLPA analysis may play a useful role in the diagnosis of FBP1 deficiency.

METHODS

Fourteen patients with FBP1 deficiency from 13 families with typical clinical and laboratory results were diagnosed in our laboratory between 2006 and 2014 (Table 1). Not all of them had enzymatic studies performed but all parents gave their informed consent to search for the molecular basis of the disease of their children and to be investigated for their own carrier status. In all of them, all 8 exons and adjacent intronic segments of the *FBP1* gene were amplified by PCR and sequenced according to standard Sanger techniques (primer sequences and PCR conditions available upon request). In those patients in whom we assumed a deletion of exon 2 (the first coding exon), we were able to generate a junction fragment with primers 5′-taaaggtttccgcgattcac-3′ (sn) and 5′-gaccatcctggccaacac-3′ (asn). Results of sequencing studies were compared to our *FBP1*reference sequence NM_001127628.1. Nomenclature for the description of sequence variants follows the recommendations of the Human Genome Variation Society [10]. The bioinformatic tools Polyphen-2 [11] and Mutation Taster [12] were used to predict effects of sequence aberrations.

Table 1. Ethnic origin and molecular genetic findings of the 14 patients of this study

Patient number		SANGER sequencing results			MLPA results	
	Ethnic origin	**Mutation**	**Mutation effect**		**Deletion of …**	**Reference for first report**
1	Armenia[a]	**c.-24–26_170+5192del** [b]	**p.0?**	homo	exon 2 **(homo)**	**This study**
2	Turkey[a]	**c.-24–26_170+5192del** [b]	**p.0?**	homo	exon 2 **(homo)**	**This study**
3	Turkey[a]	**c.-24–26_170+5192del** [b]	**p.0?**	homo	exon 2 **(homo)**	**This study**
4	Pakistan	c.841G>A	p.(E281K)[c]	homo	*n.a.*	[3]
5	Pakistan	c.841G>A	p.(E281K)[c]	homo	*n.a.*	[3]
6	Pakistan	c.881G>A	p.(G294E)	homo	*n.a.*	[18]
7	Pakistan	c.841G>A	p.(E281K)[c]	homo	*n.a.*	[3]
8-1	Germany	c.490G>A	p.G164S	homo	*n.a.*	[8]
8-2	Germany	c.490G>A	p.G164S	homo	*n.a.*	[8]
9	Germany	c.704dupC	p.(D236Rfs*2)	homo	*n.a.*	[7]
10	Turkey /	**c.359C>T**	**p.(P120L)**	hetero	*n.a.*	**This study**

	Turkey	c.881G>A	p.(G294E)	hetero	*n.a.*	[4]
11	Turkey	c.841G>A	p.(E281K)[d]	homo	*n.a.*	[4]
12	Germany /	**c.619G>C**	**p.(G207R)**	hetero	-	**This study**
	Germany	*n.d.*	?	?	exon 8 **(hetero)**	**This study**
13	Germany /	c.959dupG	p.(S321Ifs*13)	hetero	-	[14]
	Germany	deletion[e]	?	hetero	exons 1–8 **(hetero)**	**This study**

[a]see Additional file 1: Fig. S1

[b]only detectable when sequencing a junction fragment with specific primers

[c,d]represents different haplotypes

[e]long range deletion (larger than exon 08) suggested by haplotype analysis

n.a., not applied

n.d., not detected

Novel mutations are shown in bold

In those patients in whom the diagnosis of FBP1 deficiency was not confirmed by Sanger sequencing and the detection of 2 biallelic mutations within *FBP1*, MLPA analysis was performed. We used the reaction mixtures SALSA MLPA probemix P255-B1 ALDOB-FBP1 (MRC Holland, Amsterdam, The Netherlands) according to the manufacturer's recommendations. Acquired data were normalized with 3–5 control DNA samples isolated in our laboratory. Calculations were performed with the SeqPilot software for genetic analyses version 4.1.2 (JSI Medical Systems, Ettenheim, Germany). SNP array analysis was performed using the Genome-Wide Human SNP Array 6.0 (Affymetrix, Santa Clara, CA, USA) evaluated by the Genotyping Console software version 4.1.

RESULTS AND DISCUSSION

Conventional Sanger sequencing analysis of all coding exons allowed the diagnosis of FBP1 deficiency in 9 out of the 14 patients (patients 4–11 in Table 1). These patients were found to be homozygous or compound heterozygous for mutations within *FBP1*.Among them, we found two novel missense mutations, p.(Pro120Leu) and p.(Gly207Arg) in exons 4 and 6, respectively, each in single families. Each of these two amino acid positions are part of highly conserved stretches of amino acids. Polyphen-2 predicts both of these 2 missense mutations to be 'probably damaging' (score 1.00). Mutation Taster classifies them as 'disease-causing' (with probability scores of 0.99999999999648 and 0.999999999878082, resp.). To our knowledge, p.(Pro120Leu) has never been reported to databases before; according to the ExAC database, the p.(Gly207Arg) variant has been observed in 10 European (non-Finnish) individuals in the heterozygous state with an allele frequency of 0.0001498 [13].

To date, only a limited number of *FBP1* mutations has been detected worldwide; our study brings up the total number to 35 (Table 2). Only few mutations have been found that do not have the characteristics of a private mutation. Among them is c.959dupG, originally found in the Japanese population [14] that has also been detected in patients from Europe [5] and North America [7], and recently also in patients from China [9]. Another example is c.841G>A which has been detected in several unrelated patients from Pakistan [3] but also, with a different haplotype, in patients from Turkey [this study]. Furthermore, c.685C>T has repeatedly been found in seemingly unrelated families from Morocco [5, 15].

Table 2. Summary of the 35 *FBP1* mutations reported in fructose-1,6-bisphosphatase deficiency

	Nucleotide change	Amino acid change	Ethnic origin	Reference[a]
Single nucleotide changes				
Exon 2	c.88G>T	p.(E30*)	Japan	[8]
Exon 4	c.359C>T	p.(P120L)	Turkey	**This study**
Exon 5	c.472C>T	p.(R158W)	France	[5]
	c.490G>A	p.G164S	Japan/South Korea/?	[5, 8, 19]
	c.530C>A	p.A177D	Japan	[8]

Exon 6	c.581 T>C	p.(F194S)	Japan	[20]
	c.619G>C	p.(G207R)	Germany	**This study**
	c.639C>G	p.(N213K)	?	[5, 7]
	c.648C>G	p.(Y216*)	Sweden	[18]
	c.685C>T	p.(Q229*)	Morocco	[5, 15]
Exon 7	c.778G>A	p.G260R	Pakistan/Sweden	[18, 21]
Exon 8	c.841G>A	p.(E281K)	Pakistan[b]/Turkey[b]	[3]
	c.841G>T	p.(E281*)	Saudi Arabia	[22]
	c.851C>G	p.(P284R)	Japan	[20]
	c.881G>T	p.(G294V)		[7]
	c.881G>A	p.(G294E)	Sweden/Pakistan	[18/3]
Deletions				
Complete Deletion of the *FBP1* gene[c]		p.0?	Sweden	[18]
Complete Deletion of the *FBP1* gene[d]		p.0?	Germany	**This study**
Exon 2	**c.-24–26_170+5192del**	p.?	**Turkey/Armenia**	**This study**
	c.35delA	p.N12Tfs*2	Turkey/Germany (?)	[21]
	c.48delC	p.(F17Sfs*15)	France	[5]
Exon 3–7	complete deletion	p.?	?	[5]
Exon 6	c.616_619delAAAG	p.(K206V*70)	Turkey	[23]
	c.660delT	p.(F220Lfs*57)	Turkey	[24]
Exon 7	c.807delG	p.(K270Rfs*7)	?	[7]
Exon 8	deletion[e]	p.?	Germany	**This study**
	c.838delT	p.Y280Tfs*25	South Korea	[19]
	c.966delC	p.D323Tfs*7	Iran	[21]
Insertions/Duplication				
Exon 2	c.114_119dupCTGCAC	p.(C39_T40dup)	Saudi Arabia	[22]
Exon 6	c.704dupC	p.(D236Rfs*2)	?	[7]
Exon 8	c.865dupA	p.(M289Nfs*45)	Greece	[5]
	c.959dupG[f]	p.S321Ifs*13	Japan/Europe/China	[5, 7–9]
Indel				
Exon 7	c.731_738delins20	p.(R244_Y245delins6)	Turkey	[5]
Splicing				
Intron 4	c.427–1del	p.(K143_P189del)	?	[5]
Intron 7	c.825+1G>A	p.?	?	[5]

[a]slash (/) refers to slash in column 'ethnic origin'

[b]with different haplotypes

[c]together with deletion of *FBP2* and parts of *ONPEP* (hg19 chr9:g.(97295486_97300076)_(97571249_97571455), approx. 0.28 Mb)

[d]together with deletion of *FBP2* (hg19 chr9:g.(97281072_97289359)_(97419146_97420857), approx. 0.13 Mb)

[e]exon 8 only according to additional SNP array analysis (hg19 chr9:g.(97364379_97365560)_(97365642_97365985))

[f]originally named c.960_1insG

Novel mutations are shown in bold

In two of our patients, #12 and #13, only one mutation was detected by conventional Sanger sequencing analysis, however, haplotype analysis in the parents of patient #13 already suggested a long range deletion of the paternal allele (*detailed results not shown*). Of note, in 3 consecutive unrelated patients, one from Armenia and two from Turkey, no PCR product could be generated for exon 2 of the *FBP1* gene. This observation prompted us to further investigate these patients. This was of particular interest since earlier reports on mutations in *FBP1* had speculated that deletions within exon 2 (which at that time was termed exon 1) are common in the Turkish population, although the authors were not able to further characterize them [7]. Since we assumed the presence of a long-range deletion in these 3 patients, extensive modification of primer pairs was performed with the aim to generate a PCR product of acceptable size to be visible on polyacrylamide gel electrophoresis and eventually allowed the successful generation of a junction fragment (Fig. 1). All 3 patients in whom exon 2 could not be amplified with standard primers were thus found to be homozygous for a large deletion spanning 5412 base pairs and including the entire coding sequence of exon 2 (c.-24-26_170+5192del). All these patients were seemingly homozygous for the following polymorphisms that are all known from databases and have also been detected in our lab both in healthy and diseased controls: c.426+7T [rs8192689], c.567+31G [rs3739747], c.651T [p.(=), rs1042144], c.653A [p.(Arg218Lys), rs1769259], c.705+14C [rs2297084], c.960G [p.(=), rs1769257], c.*213T [rs9695]. Segregation analysis showed that all the patients' parents carried the deletion in the heterozygous state and indicated that a single haplotype was associated with this deletion (Additional file 2: Fig. S3). These results are compatible with our assumption that this mutation represents a founder mutation in the Armenian and Turkish population. We believe that this mutation plays quite an important role in that geographical area since, in addition to Herzog et al. [7] (*see above*) who supposed deletions in exon 2 in patients originating from Turkey, also Lebigot et al. [5], in a most recent study, reported exon 2 deletions by gene dose assays in Turkish patients; again, no further details regarding its length and location were provided. Furthermore, a preliminary communication from Turkey reported a relatively high number of FBP1-deficient cases from this region and, again, mentioned poorly defined exon 2 deletions [16]. It may therefore be speculated that the deletion characterized in detail in this paper is the same deletion as originally mentioned by several authors [5, 7, 16] and it may be concluded that this deletion of exon 2 is a relatively common cause of FBP1 deficiency in patients of Turkish and Armenian origin. Patients with this ethnic background should primarily be screened for this deletion and Sanger sequencing is now possible when using specific primers that allow sequencing of a junction fragment.

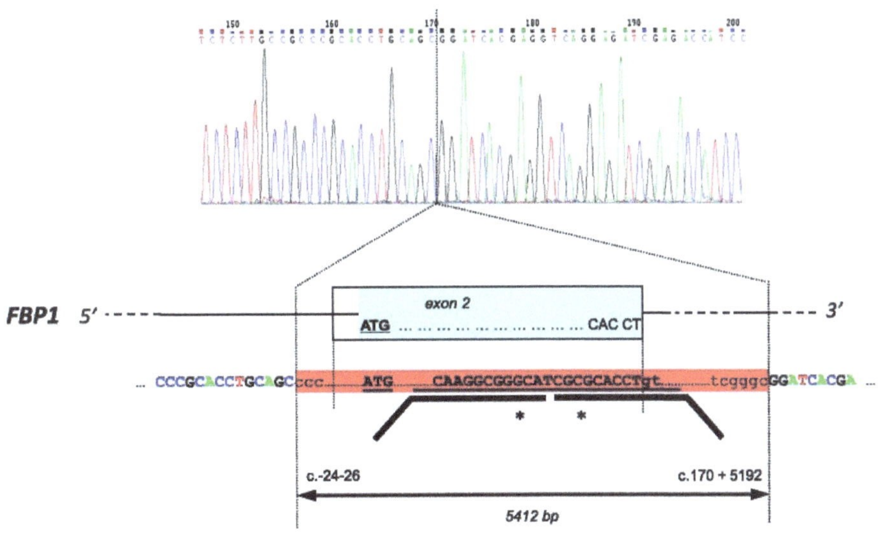

Figure 1. Characterization of a common long-range deletion of the *FBP1* gene. A junction fragment including a deletion in the range of exon 2 was generated from DNA of patient 1. The result of the sequencing reaction is shown. The novel deletion (*indicated in red*) comprises 26 bp of intron 1, another 24 bp of the untranslated region (5′-UTR) before the ATG initiation codon of exon 2, the entire coding region of 170 bp of exon 2 (*blue*), and another 5192 bp of intron 2. The bold black lines (*indicated by the asterisks*) describe the position of the MLPA probes for exon 2 used in this study

Such long-range deletions and other variations in copy number, particularly when present in the heterozygous state, may escape conventional sequencing techniques. Multiplex ligation-dependent probe amplification (MLPA), originally described in 2002 [17], is increasingly used for the targeted screening for copy number variations and has recently become commercially available for the *FBP1* gene. Therefore, we applied this method to the five patients in whom we had not arrived at a diagnosis with standard sequencing techniques. Patients #1 to #3 all showed the typical pattern of homozygosity for an exon 2 deletion (Fig. 2b), thus, MLPA analysis was in accordance with our sequencing results. In patient #12, we found that MLPA for exon 8 was diminished to approximately 50 % of normal controls (Fig. 2c). Therefore, heterozygosity for a long-range deletion was supposed, which was subsequently confirmed by SNP array analysis (Table 2). In patient #13, heterozygosity for a deletion on the paternal allele was confirmed and we could show that the deletion affects all 8 exons (Fig. 2d). Furthermore, we were able to demonstrate by SNP array analysis that the mutation in pt #13 affecting the entire *FBP1* gene is not identical to the one reported by Asberg [18] who described a patient with a deletion of the entire *FBP1* gene together with the neighboring *FBP2* and*ONPEP* genes (Table 2).

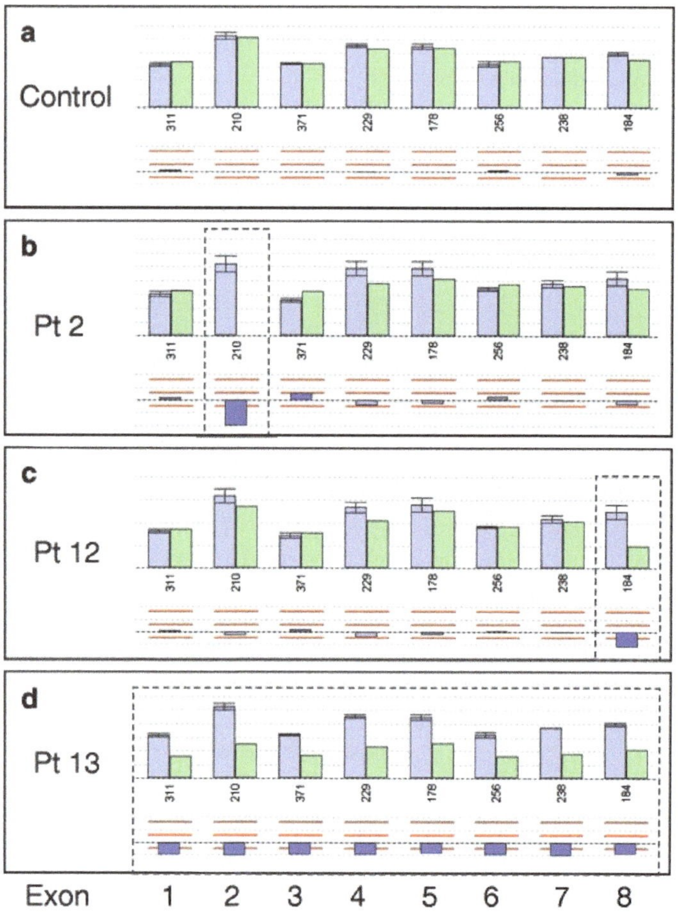

Figure 2. Results of MLPA analysis in FBP1 deficiency. Results are shown for a control sample (**a**), patient 2 (**b**), patient 12 (**c**), and patient 13 (**d**). For patient details *see* Table 1. Each panel shows the results for the intensity of probe amplification for the eight exons of *FBP1*. Patients' results are depicted in green bars, while means (± SD) of concomitantly measured controls are shown in blue. The presentation below these bars shows the deviation of patients' results as a percentage of control with the dotted line representing 0, and the horizontal red bars −25 %, +25 %, and +50 %, respectively. *Note* the missing probe amplification for exon 2 in patient 2 which is in line with homozygosity for the novel exon 2 deletion. Patient 12 shows a signal intensity for exon 8 of approximately 50 % suggesting heterozygosity for a deletion. In patient 13, signal intensities for all 8 exons are reduced to approximately 50 % of controls suggesting heterozygosity for a deletion of the entire *FBP1* gene

CONCLUSIONS

In summary, we provide an update of the 35 *FBP1* mutations reported to date, present PCR conditions that allow detection of a common *FBP1* mutation in the Armenian and Turkish population, and more generally, demonstrate for the first time the useful role of MLPA analysis in the diagnosis of FBP1 deficiency.

ACKNOWLEDGEMENTS

We thank Juliane Bergmann and Barbara Schröder for excellent technical support. This study was supported by the Partnership Program of the Arabkir Institute of Child and Adolescent Health, Yerevan (Armenia) with the University Children's Hospital, Zurich. Further support was received from Nutricia Metabolics, Friedrichsdorf, Germany.

Partly presented in abstract form at the Annual Symposium 2014 of the Society for the Study of Inborn Errors of Metabolism (SSIEM), Innsbruck, Austria.

AUTHORS' CONTRIBUTIONS

RS is responsible for the design of the study, coordinated all investigations, performed molecular genetic analyses and drafted the manuscript. MdM helped to compile the literature data. TS, EM, and ACM had important roles in acquistion of clinical data and DNA samples. IV performed SNP array analyses. BS was involved in data acquisition, critical revision and finalisation of the manuscript. All authors read and approved the final manuscript.

REFERENCES

1. Baker L, Winegrad AI. Fasting hypoglycemia and metabolic acidosis associated with deficiency of hepatic fructose-1,6-bisphosphatase activity. Lancet. 1970; 2: 13–6. doi: 10.1016/S0140-6736(70)92474-8.
2. Steinmann B, Santer R. Disorders of fructose metabolism. In: Saudubray JM, van den Berghe G, Walter H, editors. Inborn metabolic diseases. Heidelberg: Springer; 2012. pp. 162–5.
3. Afroze B, Yunus Z, Steinmann B, Santer R. Transient pseudo-hypertriglyceridemia: a useful biochemical marker of fructose-1,6-bisphosphatase deficiency. Eur J Pediatr.2013; 172: 1249–53. doi: 10.1007/s00431-013-2084-6.
4. Visser G, Bakker HD, de Klerk JBC, Smeitink JAM, Smit GPA, Wijburg FA. Natural history and treatment of fructose-1,6,-bisphosphatase deficiency in the Netherlands (abstract) J Inherit Metab Dis. 2004; 27(Suppl 1): 207.
5. Lebigot E, Brassier A, Zater M, Imanci D, Feillet F, Thérond P, et al. Fructose-1,6-bisphosphatase deficiency: clinical, biochemical and genetic features in French patients.J Inherit Metab Dis. 2015
6. El-Maghrabi MR, Lange AJ, Jiang W, Yamagata K, Stoffel M, Takeda J, et al. Human fructose-1,6-bisphosphatase gene (FBP1): exon-intron organization, localization to chromosome bands 9q22.2-q22.3, and mutation screening in subjects with fructose-1,6-bisphosphatase deficiency. Genomics. 1995; 27: 520–5. doi: 10.1006/geno.1995.1085.
7. Herzog B, Morris AAM, Saunders C, Eschrich K. Mutation spectrum in patients with fructose-1,6-bisphosphatase deficiency. J Inherit Metab Dis. 2001; 24: 87–8. doi: 10.1023/A: 1005691730058.
8. Kikawa Y, Inuzuka M, Jin BY, Kaji S, Koga J, Yamamoto Y, et al. Identification of genetic mutations in Japanese patients with fructose 1,6-bisphosphatase deficiency. Am J Hum Genet. 1997; 61: 852–61. doi: 10.1086/514875.
9. Xu K, Liu XQ, Zhang CY, Wang Y, Li X, Wu Y, et al. Genetic diagnosis of fructose-1,6-bisphosphatase deficiency: a case report. Beijing Da Xue Xue Bao. 2014; 46: 681–5.
10. den Dunnen JT, Antonarakis SE. Mutation nomenclature extensions and suggestions to describe complex

mutations: a discussion. Hum Mutat. 2000; 15: 7–12. doi: 10.1002/(SICI)1098-1004(200001)15: 1<7: AID-HUMU4>3.0.CO; 2-N.

11. Adzhubei IA, Schmidt S, Peshkin L, Ramensky VE, Gerasimova A, Bork P, et al. A method and server for predicting damaging missense mutations. Nat Methods.2010; 7: 248–9. doi: 10.1038/nmeth0410-248.

12. Schwarz JM, Cooper DN, Schuelke M, Seelow D. MutationTaster2: mutation prediction for the deep-sequencing age. Nat Methods. 2014; 11: 361–2. doi: 10.1038/nmeth.2890.

13. Exome Aggregation Consortium (ExAC), Cambridge, MA,http: //exac.broadinstitute.org/variant/9-97369183-C-G. Accessed 4 Jan 2016.

14. Kikawa Y, Inuzuka M, Jin BY, Kaji S, Yamamoto Y, Shigematsu Y, et al. Identification of a genetic mutation in a family with fructose-1,6-bisphosphatase deficiency. Biochem Biophys Res Comm. 1995; 210: 797–804. doi: 10.1006/bbrc.1995.1729.

15. Prahl P, Christensen E, Hansen L, Mortensen HB. Fructose-1,6-bisphosphatase deficiency as cause of recessive serious hypoglycemia. Ugeskr Laeger. 2006; 46: 4014–5.

16. Gokçay G, Shin YS, Podskarbi T, Balci MC, Karaca M, Demirkol M. Fructose-1,6-bisphosphatase deficiency: natural course of the disease with relevance to diagnosis and treatment in 23 patients (abstract) J Inherit Metab Dis. 2015; 38(Suppl 1): S179.

17. Schouten JP, McElgunn CJ, Waaijer R, Zwijnenburg D, Diepvens F, Pals G. Relative quantification of 40 nucleic acid sequences by multiplex ligation-dependent probe amplification. Nucleic Acids Res. 2002; 30: e57. doi: 10.1093/nar/gnf056.

18. Asberg C, Hjalmarson O, Alm J, Martinsson T, Waldenström J, Hellerud C. Fructose-1,6-bisphosphatase deficiency: enzyme and mutation analysis performed on calcitriol-stimulated monocytes with a note on prognosis. J Inherit Metab Dis. 2010; 33: S113–21. doi: 10.1007/s10545-009-9034-5.

19. Moon S, Kim JH, Han JH, Ko SH, Ahn YB, Kim JH, et al. Novel compound heterozygous mutations in the fructose-1,6-bisphosphatase gene cause hypoglycemia and lactic acidosis. Metab Clin Exp. 2011; 60: 107–13. doi: 10.1016/j.metabol.2009.12.021.

20. Matsuura T, Chinen Y, Arashiro R, Katsuren K, Tamura T, Hyakuna N, et al. Two newly identified genomic mutations in a Japanese female patient with fructose-1,6 bisphosphatase (FBPase) deficency. Mol Genet Metab. 2002; 76: 207–10. doi: 10.1016/S1096-7192(02)00038-0.

21. Herzog B, Wendel U, Morris AAM, Eschrich K. Novel mutations in patients with fructose-1,6-bisphosphatase deficiency. J Inherit Metab Dis. 1999; 22: 132–8. doi: 10.1023/A: 1005489617843.

22. Faiyaz-Ul-Haque M, Al-Owain M, Al-Dayel F, Al-Hassnan Z, Al-Zaidan H, Rahbeeni Z, et al. Novel FBP1 gene mutations in Arab patients with fructose-1,6-bisphosphatase deficiency. Eur J Pediatr. 2009; 168: 1467–71. doi: 10.1007/s00431-009-0953-9.

23. Ali BR, Hertecant JL, Al-Jasmi FA, Hamdan MA, Khuri SF, Akawi NA, et al. New and known mutations associated with inborn errors of metabolism in a heterogeneous Middle Eastern population. Saudi Med J. 2011; 32: 353–9.

24. Eren E, Edgunlu T, Abuhandan M, Yetkin I. Novel fructose-1,6-bisphosphatase gene mutation in two siblings. DNA Cell Biol. 2013; 32: 635–9. doi: 10.1089/dna.2013.2119.

INDEX

H

I

K

L

M

N

O

P

Q

R

S

T

U

V

W

X